Communications
in Computer and Information Science   **564**

Commenced Publication in 2007
Founding and Former Series Editors:
Alfredo Cuzzocrea, Dominik Ślęzak, and Xiaokang Yang

More information about this series at http://www.springer.com/series/7899

Alexander Dudin · Anatoly Nazarov
Rafael Yakupov (Eds.)

Information Technologies and Mathematical Modelling

Queueing Theory and Applications

14th International Scientific Conference, ITMM 2015
named after A.F. Terpugov
Anzhero-Sudzhensk, Russia, November 18–22, 2015
Proceedings

 Springer

Editors

Alexander Dudin
Applied Mathematics and Computer Science
Belarusian State University
Minsk
Russia

Rafael Yakupov
Kemerovo State University
Anzhero-Sudzhensk
Russia

Anatoly Nazarov
Tomsk State University
Tomsk
Russia

ISSN 1865-0929   ISSN 1865-0937 (electronic)
Communications in Computer and Information Science
ISBN 978-3-319-25860-7   ISBN 978-3-319-25861-4   (eBook)
DOI 10.1007/978-3-319-25861-4

Library of Congress Control Number: 2015952063

Springer Cham Heidelberg New York Dordrecht London

Springer International Publishing AG Switzerland is part of Springer Science+Business Media
(www.springer.com)

Preface

The series of scientific conferences on "Information Technologies and Mathematical Modelling" (ITMM) was started in 2002. In the beginning, it was a national conference, but in 2012, it attained international status. The conference series is named after Alexander Terpugov, one of the first organizers of the conference, an outstanding scientist of Tomsk State University and a leader of the famous Siberian school on applied probability, queueing theory, and applications.

Traditionally, the conference has ten tracks in various fields of mathematical modelling and information technologies. Throughout the years, the tracks on probabilistic methods and models, queueing theory, telecommunication systems, and software engineering have proven to be the most popular ones at the conference. There is an international presence at the events with participants from many countries including: Austria, Azerbaijan, Belarus, Bulgaria, China, Hungary, India, Italy, Kazakhstan, Korea, Poland, and the USA. Many of these foreign participants come to this Siberian conference every year because of its warm welcome and serious scientific discussions.

This volume presents selected papers devoted to new results in queueing theory and its applications. It is aimed at specialists in probabilistic theory, random processes, mathematical modelling, as well as engineers engaged in logical and technical design and operational management of telecommunication and computer networks.

November 2015   Anatoly Nazarov

Organization

The ITMM conference is organized by the Anzhero-Sudzhensk Branch of Kemerovo State University together with Kemerovo State University, National Research Tomsk State University, Institute of Computational Technologies of the Siberian Branch of the Russian Academy of Sciences, Trapeznikov Institute of Control Sciences of the Russian Academy of Sciences.

International Program Committee

A. Dudin, Belarus, Chair
A. Nazarov, Russia, Co-chair
R. Yakupov, Russia, Co-chair
I. Atencia, Spain
S. Chakravarthy, USA
B.D. Choi, Korea
T. Czachórski, Poland
D. Efrosinin, Austria
A. Gortsev, Russia
K. Al-Begain, UK
Yu. Kharin, Belarus
C.S. Kim, Korea
A. Krishnamurthy, India
U. Krieger, Germany

G. Medvedev, Belarus
Q.-L. Li, China
A. Melikov, Azerbaijan
R. Nobel, The Netherlands
D. Novikov, Russia
E. Orsingher, Italy
M. Pagano, Italy
K. Samuylov, Russia
Y. Shokin, Russia
J. Sztrik, Hungary
H. Tijms, The Netherlands
O. Tikhonenko, Poland
V. Vishnevsky, Russia

Local Organizing Committee

A. Dudin, Chair
A. Nazarov, Co-chair
R. Yakupov, Co-chair
I. Garayshina
B. Gladkikh
E. Glukhova
A. Gudov
R. Ivanovskiy
M. Farkhadov
V. Ivnitskii
Yu. Kostyuk

T. Lyubina
A. Moiseev
S. Moiseeva
V. Poddubnyi
S. Rozhkova
V. Rykov
S. Senashov
A. Shkurkin
S. Suschenko
K. Voytikov
O. Zmeev

Contents

A Multi-server Queueing Model with Markovian Arrivals and Phase Type Cooperative Services - Simulation Approach

Srinivas R. Chakravarthy[(✉)]

Department of Industrial and Manufacturing Engineering,
Kettering University, Flint, MI 48504, USA
schakrav@kettering.edu

Abstract. In Chakravarthy [5] a new class of queueing models in which one type of customers opt for cooperative services with fellow customers was introduced in the context of a single server. Under the assumption of versatile Markovian point process for the arrivals, exponential services, and with a limit of no more than two groups of cooperative customers be present in the system, the model was analyzed in steady-state and some interesting numerical examples were illustrated in [5]. In this paper we generalize that cooperative services model by relaxing the assumptions of single server, exponential services, and only two groups be present at any given time. Thus, we consider a multi-server queueing model in which the customers arrive according to a versatile Markovian point process. One type of customers require individual services whereas the second type of customers opt for a cooperative service (to be offered along with other similar customers). We assume that at any given time there can be at most $K, 2 < K \leq \infty$, groups of customers needing cooperative services and that the services are of phase type with representation depending on the type of service offered. While this model can be analyzed using matrix-analytic method with a very large state space, in this paper we will study the model using simulation to bring out a few salient features of this new class of queueing models.

Keywords: Markovian arrival process · Phase type distribution · Cooperative services · Phase type distribution · Simulation

1 Introduction

In Chakravarthy [5] a new class of queueing models in which customers can opt for cooperative services with fellow customers was introduced in the context of a single server. This type of queueing models has a number of applications in real word situations. For example, in online shopping which has become so prevalent that even small and upcoming companies are becoming integral part of this e-commerce business. According to the U.S. Census Bureau (see www.census. gov) the annual retail e-commerce (which includes online shopping) has grown

© Springer International Publishing Switzerland 2015
A. Dudin et al. (Eds.): ITMM 2015, CCIS 564, pp. 1–12, 2015.
DOI: 10.1007/978-3-319-25861-4_1

from 4.984 billion dollars in 1998 to 260.669 billion dollars in 2013. Thus, more and more retailers want a piece of action in this business. Consider a small to medium retailer handling customers' orders over the phone. Suppose that two types of customers, say, Type 1 and Type 2, place orders over the phone. The orders are processed by a receiving attendant (server) and the processing times (not including the shipping and receiving times) are the actual service times. Type 1 customers prefer to get their orders shipped directly to their addresses whereas Type 2 customers, in order to save money in shipping and processing, prefer to get their orders shipped to a particular location to be picked up by them. The orders from Type 2 customers will be processed only when the number of orders hits, say, L, a pre-determined threshold, on a non-preemptive priority (over Type 1) basis provided the server is free. Otherwise, the Type 2 customers need to wait until the server is available. The requirement of needing exactly L orders is to make sure they are put in the same package for shipping to a specific location for all Type 2 customers to individually pick up their order. Thus, the orders are processed on a FIFO basis within Type 1 and on a non-preemptive priority basis for Type 2. Similar applications can be found in other areas notably in service systems requiring parts for offering services and these parts are ordered as and when the inventory level reaches a certain point.

While cooperative services are widely used in inventory modeling, they are very rarely considered in queueing literature. The rare situations where cooperative services are considered in queueing literature arise from the point of view of the service providers as opposed to the customers' points of view (see e.g., [1,6]). Basically, in these queueing situations the resources (i.e., service providers) of many queueing systems are pooled in one form or the other to efficiently provide services to the pooled customers. This motivated the study of the cooperative service queueing model in [5], wherein it was pointed out that the model studied here does not belong to any of the various types of queueing models with batch (or group) services studied extensively in the literature. Under the assumption of versatile Markovian point process for the arrivals, exponential services, and with a limit of no more than two groups of cooperative customers be present in the system, the model was analyzed in steady-state and some interesting numerical examples were illustrated. We refer the reader to [5] for details on this.

In this paper we generalize the cooperative services model of [5] by relaxing the assumptions of single server, exponential services, and only two groups be present at any given time. The rest of the paper is organized as follows. In Sect. 2 we give a description of the model under study. Some key system performance measures used in this study are listed in Sect. 3. The roles of some key parameters for the model under study are discussed in Sect. 4. Some concluding remarks are given in Sect. 5.

2   Model Description

We consider a c-server queueing system in which customers arrive according to a Markovian arrival process (MAP) with representation (D_0, D_1, D_2) of order m.

The generator D, defined by $D = D_0 + D_1 + D_2$, governs the underlying Markov chain of the MAP such that D_0 accounts for the transitions corresponding to no arrival; D_1 governs those corresponding to an arrival of a customer who requires individual services, and D_2 governs those corresponding to an arrival of a customer who requires cooperative services. By assuming D_0 to be a nonsingular matrix, the interarrival times will be finite with probability one and the arrival process does not terminate. Hence, we see that D_0 is a stable matrix. Henceforth, we will refer to customers requiring individual services as Type 1 customers and those requiring cooperative services will be referred to as Type 2 customers.

A MAP is a tractable class of Markov renewal processes. It should be noted that by appropriately choosing the parameters of the MAP the underlying arrival process can be made as a renewal process. The MAP is a rich class of point processes that includes many well-known processes such as Poisson, PH-renewal processes, and Markov-modulated Poisson process. One of the most significant features of the MAP is the underlying Markovian structure and fits ideally in the context of matrix-analytic solutions to stochastic models. Matrix-analytic methods were first introduced and studied by Neuts [10]. The idea of the MAP is to significantly generalize the Poisson processes and still keep the tractability for modelling purposes. Furthermore, MAP is a convenient tool to model both renewal and non-renewal arrivals. It can be shown that MAP is equivalent to Neuts' versatile Markovian point process. The point process described by the MAP is a special class of semi-Markov processes. For further details on MAP and their usefulness in stochastic modelling, we refer to [8,9,11,12] and for a review and recent work on MAP we refer the reader to [2–4].

Type 1 customers require services individually whereas Type 2 customers opt for cooperative services to be offered along with a group of other similar customers. The group size is fixed to be, say, $L, 1 \leq L < \infty$. Strictly speaking L should be a finite number greater than one as it represents a group size. However, the case $L = 1$ is included so that this special case will lead to some interesting applications in non-preemptive priority model [5]. We assume that Type 1 customers have an infinite waiting space while Type 2 customers have a restriction that at any given time there can be at most $K, 2 \leq K \leq \infty$, groups of Type 2 customers in the system. Note that we also allow for the possibility of K to be infinite.

An arriving Type 1 customer finding an server idle will get into service immediately; otherwise will enter into Type 1 buffer and will wait for a server to be free to offer Type 1 services. An arriving Type 2 customer will either (a) get into service immediately along with other similar customers provided a server is available and there are $L - 1$ such customers already waiting in the system to receive cooperative services; (b) get into Type 2 buffer provided there is enough space available irrespective of whether a server is available or not; (c) with probability $\gamma, 0 \leq \gamma \leq 1$, will become a Type 1 customer and enter into Type 1 buffer since Type 2 buffer is full; or (d) be lost with probability $1 - \gamma$ since Type 2 buffer is full. On becoming a Type 1, the customer remains as Type 1 until leaving the system with a service.

We assume that the service times for both types of customers are of phase $(PH-)$ type with representation $(\alpha(1), T(1))$ of dimension n_1 for Type 1 and with representation $(\alpha(2), T(2))$ of dimension n_2 for Type 2. Recall that a PH-distribution is obtained as the time until absorption in a finite state Markov chain with one absorption state. It is characterized by an initial probability vector α and a square matrix T governing the transitions to various transient states. PH-distributions are defined for both discrete and continuous time. For details on PH-distributions and their properties, we refer the reader to [10,11,13]. We also assume that Type 2 customers have a non-preemptive priority over Type 1 customers. Thus, upon completion of a service the free server will offer a service to a group of Type 2 customers if there are L Type 2 customers waiting; otherwise the server will offer a service to a Type 1 customer, if any, or becomes idle.

Let η be the stationary probability vector of the Markov process with irreducible generator D. That is, η is the unique (positive) probability vector satisfying $\eta D = 0$, ηe=1, where e is a column vector of 1's of appropriate dimension. We denote the average arrival rate and the average service rates, respectively, by λ, μ_1, and μ_2 and these are given by $\lambda = \eta(D_1 + D_2)\ e$, $\mu_1 = [\alpha(1)(-T(1))^{-1}\ e]^{-1}$, $\mu_2 = [\alpha(2)(-T(2))^{-1}\ e]^{-1}$.

The model outlined in this section can be studied as a Markov process by keeping track of quantities such as the number of Type 1 and Type 2 customers waiting for service, the phases of the services, the number of servers busy with Type 1 and Type 2 customers, and the phase of the arrival process. The generator of this Markov process can be set up with the help of Kronecker products and sums of matrices. However, it is clear that the steady-state analysis requires some form of approximation or truncation due to many (sub)states that grow without bound (in the case when $K = \infty$). The accuracy of the approximation or truncation depends on the degree to which these are carried out. Our focus in this paper is not in providing an approximation or truncation or a combination of both in performing the steady-state analysis. These are currently work-in-process and the results will be reported elsewhere. Instead, our goal is to see how the type of distributional assumption affects some selected system performance measures through simulation. Further, this simulated results can be used to compare any approximation/truncation methods possibly proposed in the future. Thus, the rest of the paper is based on simulating the cooperative services model described in Sect. 1 with the help of ARENA [7].

3   Selected System Performance Measures

In this section we will list three key system performance measures among many for our illustration.

1. The probability, $PCHA$, that an arriving Type 2 customer becomes a Type 1 customer due to not having enough space in Type 2 buffer.
2. The mean, $MWTS$, waiting time in the system of a customer.

3. The fraction, $F_{below} = P(Y_{WTS} \leq \mu_{WTS}^{CQ})$, where μ_{WTS}^{CQ} is the mean waiting time in the system of a customer at an arrival epoch in the classical $MAP/PH/c$ queue and Y_{WTS} to be the waiting time in the system of a customer at an arrival epoch for the current model under study.

The reasons for using the fraction mentioned above are [5]: (a) from a customer's point of view, $MWTS$ should be less than μ_{WTS}^{CQ} in order to request cooperative services; and (b) one would like to see more customers leave the system by spending less time (as compared to μ_{WTS}^{CQ}) in the system.

To our knowledge there is no (analytical or numerical) result reported in the literature for the fraction of customers whose waiting time in the system is less than the average waiting time for the classical $MAP/PH/c$ queue. Only for the most simplest case involving $M/M/1$ queue, an analytical expression is available and can easily be verified to be $1 - e^{-1} = 0.6321$. For other cases, one has to depend on numerical methods and or simulation. In Table 1 (see Sect. 5 below) we display the simulated values of this fraction along with the mean waiting time in the system for the classical $MAP/PH/c$ queue under various scenarios.

4   Validation of the Simulated Model

Before we proceed to discuss the simulated results, it is important to validate our simulated model by comparing our results with the published results in the literature. The only case for which analytical results are available for the model under study is for the single server model with exponential services as presented in [5]. In that paper to obtain the fraction F_{below} simulation was employed and a comparison of simulated and analytical results was presented (see Table 5 in [5]); however, the simulation was done only for the single server with exponential services case. Since this cooperative services model is introduced recently [5] we do not have other models with analytical results to validate our simulation model.

5   Illustrative Simulated Examples

In this section we will illustrate the model under study with simulated results using the system performance measures of Sect. 3 under different scenarios for $MAP/PH/c$ type cooperative services.

For the arrival process, we consider the following five sets of values for D_0 and D as follows. Note that here we take $D_1 = pD$ and $D_2 = (1 - p)D$ and $p, 0 \leq p \leq 1$, is used as another parameter of the model. It should be pointed that one can take D_1 and D_2 differently so as to incorporate any correlation between the arrivals of Type 1 and Type 2 customers.

1. Erlang (*ERLA*):

$$
D_0 = \begin{pmatrix} -5 & 5 & 0 & 0 & 0 \\ 0 & -5 & 5 & 0 & 0 \\ 0 & 0 & -5 & 5 & 0 \\ 0 & 0 & 0 & -5 & 5 \\ 0 & 0 & 0 & 0 & -5 \end{pmatrix}, D = \begin{pmatrix} 0 & 0 & 0 & 0 & 0 \\ 0 & 0 & 0 & 0 & 0 \\ 0 & 0 & 0 & 0 & 0 \\ 0 & 0 & 0 & 0 & 0 \\ 5 & 0 & 0 & 0 & 0 \end{pmatrix}
$$

2. Exponential (*EXPA*):

$$D_0 = (-1), D = (1)$$

3. Hyperexponential (*HEXA*):

$$D_0 = \begin{pmatrix} -1.90 & 0 \\ 0 & -0.19 \end{pmatrix}, D = \begin{pmatrix} 1.710 & 0.190 \\ 0.171 & 0.019 \end{pmatrix}$$

4. *MAP* with negative correlation (*MNCA*):

$$D_0 = \begin{pmatrix} -1.00222 & 1.00222 & 0 \\ 0 & -1.00222 & 0 \\ 0 & 0 & -225.75 \end{pmatrix}, D = \begin{pmatrix} 0 & 0 & 0 \\ 0.01002 & 0 & 0.9922 \\ 223.4925 & 0 & 2.2575 \end{pmatrix}$$

5. *MAP* with positive correlation (*MPCA*):

$$D_0 = \begin{pmatrix} -1.00222 & 1.00222 & 0 \\ 0 & -1.00222 & 0 \\ 0 & 0 & -225.75 \end{pmatrix}, D = \begin{pmatrix} 0 & 0 & 0 \\ 0.9922 & 0 & 0.01002 \\ 2.2575 & 0 & 223.4925 \end{pmatrix}.$$

The above *MAP* processes will be normalized so as to have a specific arrival rate. However, these are qualitatively different in thaty they have different variance and correlation structure. The first three arrival processes, namely, *ERLA*, *EXPA*, and *HEXA*, have zero correlation for two successive inter-arrival times. The arrival processes labeled *MNCA* and *MPCA*, respectively, have negative and positive correlation for two successive inter-arrival times with values -0.4889 and 0.4889. The ratio of the standard deviation of the inter-arrival times of these five arrival processes with respect to *ERLA* are, respectively, 1, 2.23607, 5.01935, 3.15178, and 3.15178.

For both the service times $((\alpha(1), T(1)))$ and $((\alpha(2), T(2)))$, we consider the following three *PH*−distributions.

A. Erlang (ERLS): $\alpha(1) = \alpha(2) = (1, 0, 0, 0, 0)$,

$$T(1) = T(2) = \begin{pmatrix} -5 & 5 & 0 & 0 & 0 \\ 0 & -5 & 5 & 0 & 0 \\ 0 & 0 & -5 & 5 & 0 \\ 0 & 0 & 0 & -5 & 5 \\ 0 & 0 & 0 & 0 & -5 \end{pmatrix}.$$

B. Exponential (EXPS): $\alpha(1) = \alpha(2) = 1$,

$$T(1) = T(2) = (-1).$$

C. Hyperexponential (HEXS): $\alpha(1) = \alpha(2) = (0.9, 0.1)$,

$$T(1) = T(2) = \begin{pmatrix} -10 & 0 \\ 0 & -d \end{pmatrix}.$$

For all cases considered we fix $\lambda = 1, \mu_1 = \mu_2 = 1.2, p = 0.5$, and $\gamma = 1.0$, and vary other parameters as follows: $L = 2, 4, 5, 10$, $K = 1, 2, 5, 100$, and $c = 1, 2, 5, 10$. Note that the PH-representations will be normalized except in the case of hyperexponential in which case d will be chosen to have the specific mean. In all our examples we simulated the model using 5 replications and for 100,000 units (which in our case is minutes) for each replicate.

Before we discuss a few illustrative examples for the model under study we display the simulated mean waiting time in the system as well as the fraction, F_{below}, for the classical $MAP/PH/c$ queue under various scenarios using the same applicable parameters listed above. Note that the values of the fractions are displayed within parentheses in the table. It should be pointed out that as the number of servers increases the mean waiting time in the system, $MWTS$, decreases to the mean service time for all except $MPCA$ arrivals. For $MPCA$ arrivals more than 10 servers are needed to see the $MWTS$ approach the mean service time. This is as expected and reported in the literature since positively correlated arrivals tend to behave differently than the others.

Table 1. $MWTS$ and F_below for the classical $MAP/PH/c$ queue

c	PHS	Arrival process				
		$ERLA$	$EXPA$	$HEXA$	$MNCA$	$MPCA$
1	ERLS	2.21(0.630)	3.36(0.634)	11.72(0.632)	3.62(0.629)	238.63(0.634)
	EXPS	3.81(0.630)	4.88(0.632)	13.18(0.636)	5.22(0.634)	249.03(0.612)
	HEXS	34.31(0.640)	34.10(0.635)	42.66(0.622)	34.21(0.630)	247.82(0.617)
2	ERLS	0.87(0.560)	0.95(0.573)	1.23(0.605)	0.96(0.576)	35.56(0.666)
	EXPS	0.92(0.625)	1.01(0.622)	1.38(0.618)	1.04(0.619)	35.09(0.667)
	HEXS	2.12(0.786)	2.21(0.779)	2.88(0.749)	2.22(0.780)	36.57(0.682)
5	ERLS	0.83(0.557)	0.83(0.557)	0.84(0.557)	0.83(0.557)	10.17(0.690)
	EXPS	0.83(0.633)	0.83(0.633)	0.83(0.632)	0.83(0.632)	10.68(0.686)
	HEXS	0.84(0.910)	0.84(0.910)	0.84(0.908)	0.84(0.910)	8.37(0.718)
10	ERLS	0.83(0.557)	0.83(0.557)	0.83(0.557)	0.83(0.557)	4.40(0.700)
	EXPS	0.83(0.633)	0.83(0.633)	0.83(0.632)	0.83(0.632)	4.35(0.700)
	HEXS	0.83(0.911)	0.83(0.911)	0.83(0.911)	0.83(0.910)	3.10(0.764)

Now we will discuss our simulated results for the model under study with regard to the three measures listed in Sect. 3. First, we look at the measure, $PCHA$, whose matrix plot for various scenarios is displayed in Fig. 1.

A quick at this figure reveals the following observations.

– The impact of K appears to decrease as c is increased in all but $MPCA$ arrivals. This is as expected since the saturation level of the system is significantly reduced when c is increased. We didn't adjust the service rate to

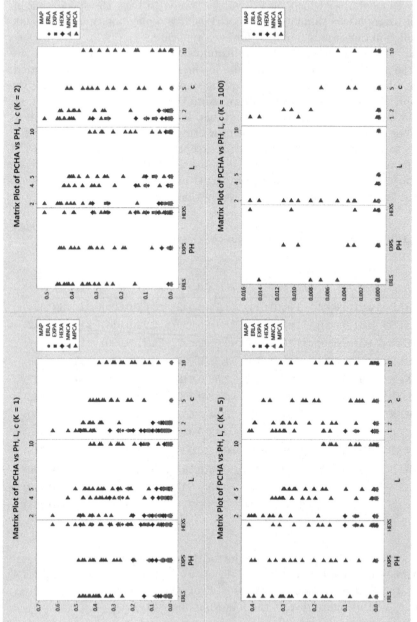

Fig. 1. P(a Type 2 customer becomes a Type 1 customer due to lack of space)

maintain the same saturation level when going from single to multiple servers in this particular case.

- In the case of positively correlated arrivals we notice that even for a very low saturation queue there is a significant number of customers changing from Type 2 to Type 1.

We now look at the mean waiting time in the system under various scenarios. Since this measure is very large for some scenarios, the matrix plot of $ln(MWTS)$, the natural logarithm of the mean waiting time in the system, is displayed in Fig. 2.

From Fig. 2, it can be seen that the $MWTS$ is sensitive to L for all combinations and this is intuitively explained as follows. When L is increased there is more chance for Type 2 customers to wait longer in the queue to achieve the requisite group size resulting in a higher mean waiting time. However, the sensitivity of this measure on K can be seen only in the cases of (a) $HEXA$ arrivals with $HEXS$ services and $c = 1$; and (b) $MPCA$ arrivals with all three types of services and for $c = 1, 2, 5$. However, when c is increased to 2 to more the sensitivity goes down significantly for $HEXA$ case, and for $MPCA$ one needs more than 5 servers. This illustrates the significant role played by (positive) correlation in the arrivals.

Finally, we look at the measure F_{below} by comparing this to the corresponding classical $MAP/PH/c$ queue. Towards this end, we look at the ratio, $\frac{F_{below}^{Coop}}{F_{below}^{CQ}}$, where F_{below}^{Coop} is the probability that a customer's waiting time in the system with cooperative services is less than the mean waiting time in the system of the corresponding classical $MAP/PH/c$ queue, and F_{below}^{CQ} is the probability that a customer's waiting in the system is less than the mean waiting time in the classical $MAP/PH/c$ queue. In Fig. 3 we display the bubble charts for this ratio for the cases when $\mu_1 = \mu_2 = 1.2$ and $\mu_1 = \mu_2 = \frac{1.2}{c}$. That is, the left side of Fig. 3 is for a low saturated system and the right side of this figure is for a reasonably saturated system. The sizes of the bubbles correspond to the values of the ratios.

In order for the cooperative service model to be better in the sense of more customers' waiting time in the system is less than the average waiting time in the corresponding classical queue, we want this ratio to be at least 1.0. Note that in these figures we indicate those ratios that are less than 1 by enclosing them in rectangular boxes.

A quick examination of this figure reveals that this ratio is greater than 1.0 in the case of single server for all scenarios. In the case of 2-server system we see that highly variable process ($HEXS$) and positively correlated arrival process ($MPCA$) yield this ratio to be greater than one. In the case of other multiple-server system, the ratio is greater than one under many scenarios only when the system is moderate to highly saturated.

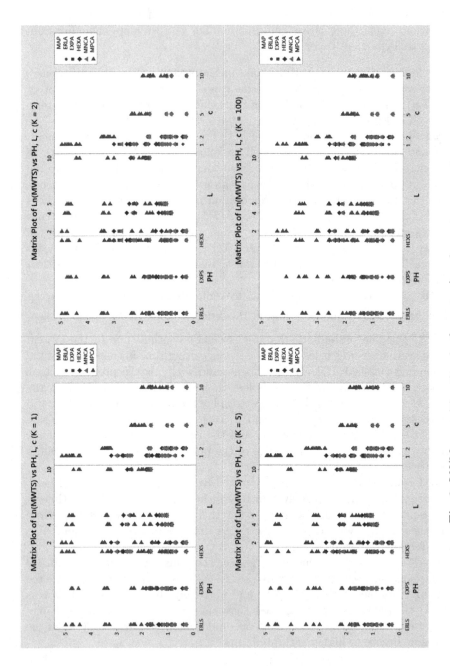

Fig. 2. LN(Mean waiting time in the system) under various scenarios

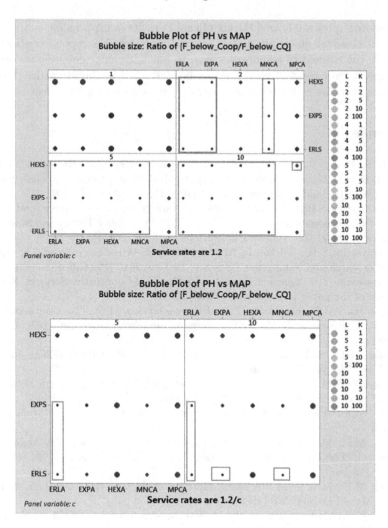

Fig. 3. Ratio of the fractions under various scenarios

6 Concluding Remarks

In this paper we considered a multi-server queueing system in which customers, who arrive according to a versatile Markovian point process, may request for cooperative services which are offered in groups of fixed size. The purpose of such requests is to minimize the waiting time in the system or to maximize the chances of staying in the system shorter than the average time it takes in the corresponding classical queueing model with first-come-first-served basis or to share the costs associated with services with fellow customers or a combination of two or more of these. We also employed simulation approach in order to quantify certain measures, which otherwise, are very difficult to obtain. Through simulated examples we showed the significant benefits in introducing cooperative services in classical queueing models. Even though the concept of cooperative services in the context of inventory exists, to our knowledge cooperative service from a customer's point of view has not been studied in the literature. Due to space restriction we restricted our illustrative examples to a few select cases.

References

1. Anily, S., Haviv, M.: Cooperation in service systems. Oper. Res. **58**, 660–673 (2010)
2. Artalejo, J.R., Gomez-Correl, A., He, Q.M.: Markovian arrivals in stochastic modelling: a survey and some new results. SORT **34**(2), 101–144 (2010)
3. Chakravarthy, S.R.: The batch Markovian arrival process: a review and future work. In: Krishnamoorthy, A. (ed.) Advances in Probability Theory and Stochastic Processes, pp. 21–39. Notable Publications Inc., NJ (2001)
4. Chakravarthy, S.R.: Markovian Arrival Processes. Wiley Encyclopedia of Operations Research and Management Science. 15 June 2010 (Published Online)
5. Chakravarthy, S.R.: Queueing Models with Optional Cooperative Services. Revised version of the paper is under review with EJOR
6. Garcia-Sanz, M.D., Fernandez, F.R., Fiestras-Janeiro, M.G., Garca-Jurado, I., Pu erto, J.: Cooperation in Markovian queueing models. J. Oper. Res. **188**, 485–495 (2008)
7. Kelton, W.D., Sadowski, R.P., Swets, N.B.: Simulation with ARENA, 5th edn. McGraw-Hill, New York (2010)
8. Lucantoni, D.M.: New results on the single server queue with a batch Markovian arrival process. Stochast. Models **7**, 1–46 (1991)
9. Neuts, M.F.: A versatile Markovian point process. J. Appl. Prob. **16**, 764–779 (1979)
10. Neuts, M.F.: Matrix-Geometric Solutions in Stochastic Models: An Algorithmic Approach. The Johns Hopkins University Press, Baltimore (1994). [1994 version is Dover Edition]
11. Neuts, M.F.: Structured Stochastic Matrices of $M/G/1$ type and their Applications. Marcel Dekker, NY (1989)
12. Neuts, M.F.: Models based on the Markovian arrival process. IEICE Trans. Commun. **E75B**, 1255–1265 (1992)
13. Neuts, M.F.: Algorithmic Probability: A collection of problems. Chapman and Hall, NY (1995)

Joint Probability Density of the Intervals Length of Modulated Semi-synchronous Integrated Flow of Events in Conditions of a Constant Dead Time and the Flow Recurrence Conditions

Maria Bakholdina$^{(\boxtimes)}$ and Alexander Gortsev

Department of Operations Research, Faculty of Applied Mathematics
and Cybernetics, National Research Tomsk State University,
36 Lenina Avenue, Tomsk 634050, Russia
maria.bakholdina@gmail.com

Abstract. This paper is focused on studying the modulated semi-synchronous integrated flow of events which is one of the mathematical models for incoming streams of events (claims) in computer communication networks and is related to the class of doubly stochastic Poisson processes (DSPPs). The flow is considered in conditions of its incomplete observability, when the dead time period of a constant duration T is generated after every registered event. In this paper we propose a technique for obtaining the formulas for calculation the probability density of the interval length between two neighboring flow events and the joint probability density of the length of two successive intervals. Also we find the conditions of the flow recurrence.

Keywords: Modulated semi-synchronous integrated flow of events · Doubly stochastic poisson process (DSPP) · Markovian arrival process (MAP) · Dead time · Flow parameters estimation · Probability density · Joint probability density · Flow recurrence conditions

1 Introduction

Mathematical models of the queueing theory have found wide application in describing real physical, technical and other objects and systems. It is worthwhile to note that the conditions of the real objects and systems operation are such that we can assert that the servers parameters are known and stable as time goes, but we can not tell this about the intensity processes of the input flows of events that come to the servers. Moreover, the intensities of the input flows usually vary within time and frequently their changes are accidental. As a result, it is necessary to consider the mathematical models of doubly stochastic Poisson processes (DSPPs), which are characterized by having the number of events in any given time interval as being Poisson distributed, conditionally to another positive stochastic process $\lambda(t)$ called intensity [1–5].

© Springer International Publishing Switzerland 2015
A. Dudin et al. (Eds.): ITMM 2015, CCIS 564, pp. 13–27, 2015.
DOI: 10.1007/978-3-319-25861-4_2

There are two known classes of doubly stochastic flows of events. The first class contains the flows of events, which intensity process is a continuous random process. The second class contains flows, which intensity is a piecewise constant stationary random process with a finite number of states. The second-class flows are most typical for telecommunication networks. They were considered for the first time and independently presented in works [6,7]. Since the early 1990s to date, these flows of events are called as doubly stochastic flows of events or MAP-flows, or MC-flows [8–13].

In turn, MC-flows may be divided into three groups depending on how the intensity process changes its state from one to another: (1) synchronous flows – flows, which intensity process changes its state from one state to another at random times, which are the time moments of the flow events arrival [14–16]; (2) asynchronous flows – flows, which intensity process changes its state from one state to another at random times, which do not depend on the time moments of the flow events arrival [17–19]; (3) semi-synchronous flows – flows, for which for the one set of states the first definition is valid and for another set of states the second definition is valid [20–22]. We shall emphasize that synchronous, asynchronous and semi-synchronous flows can be presented as the mathematical models of MAP-flows of events with the constraints on the flow parameters [23].

In the recent literature, the problem of estimating the intensity process from observations of doubly stochastic Poisson processes (DSPPs) has been of a great interest, since DSPPs have found applications in many fields such as network theory, peer-to-peer streaming networks and adaptive data streaming, optical communication systems, statistical modeling, quantitative finance, spatial epidemiology, etc. [24–29]. As has been mentioned above, in the real situations the input flow parameters can be unknown or partially known or, worse, may vary in time in a random way. That is why, the central problems faced when modeling these processes are: (1) flow states estimation on monitoring the time moments of the events occurrence (the filtering of the underlying and unobservable intensity process) [30–33]; (2) flow parameters estimation on monitoring the time moments of the events occurrence [34–37].

It is worth noting, that in most of the cases researchers consider the mathematical models of the flows, where time moments of the flow events occurrence are observable. In practice, however, any recording device (server in this context) spends some finite time for event measurement and registration, during which server can not handle the next event correctly. In other words, every event registered by a server causes the period which is called the period of a dead time [38], during which no other events are observed (they are lost). We may suppose that this period has a fixed duration (constant dead time). Particularly, we may find examples of this mathematical model in the real computer networks using CSMA/CD (Carrier Sense Multiple Access with Collision Detection) protocol. At the moment of a conflict recording at the in-port of a network node, a jam signal is transmitted across the network. During the signal transmission, calls coming to a node of the network are declined and sent to a source of repeated calls. Here time, during which the network node is closed for calls serving after a conflict recording, can be interpreted as a dead time of a server, which registers the conflict in the network nodes.

In this paper we continue to study the modulated semi-synchronous integrated flow of events [31–33], which is a generalization of the semi-synchronous flow of events [20] and semi-synchronous integrated flow of events [39] and belonging to the class of Markovian arrival processes (MAPs). The rest of the paper is organized as follows. In Sect. 2 we present the modulated semi-synchronous integrated flow of events, which provides our modeling framework. In Sects. 3 and 4 we obtain the expressions for probability density of the interval length between two neighboring flow events $p_T(\tau)$, $\tau \geq 0$, and the joint probability density of the length of two successive intervals $p_T(\tau_1, \tau_2)$, $\tau_1 \geq 0$, $\tau_2 \geq 0$, explicitly. And finally, in Sect. 5 we obtain the recurrence conditions of the observable flow of events.

2   Problem Statement

In this paper we consider the modulated semi-synchronous integrated flow of events (further flow of events), which intensity process is a piecewise constant stationary random process $\lambda(t)$ with two states 1, 2 (first, second correspondingly). In the state 1 $\lambda(t) = \lambda_1$ and in the state 2 $\lambda(t) = \lambda_2$ ($\lambda_1 > \lambda_2$). The duration of the process $\lambda(t)$ staying in the first (second) state is distributed according to the exponential law with parameter β (α). If at the time moment t the process $\lambda(t)$ is found in the first (second) state, then at the interval $[t, t + \Delta t)$, where Δt (hereinafter) is sufficiently small, with probability $\beta \Delta t + o(\Delta t)$ ($\alpha \Delta t + o(\Delta t)$) the sojourn time of the process $\lambda(t)$ in the first (second) state comes to the end and process $\lambda(t)$ transits to the second (first) state. During the time interval when $\lambda(t) = \lambda_i$, a Poisson flow of events with intensity λ_i, $i = 1, 2$, arrives. Also at any moment of an event occurrence in state 1 of the process $\lambda(t)$, the process can change its state to state 2 with probability p ($0 \leq p \leq 1$) or continue to stay in state 1 with complementary probability $1 - p$. I.e., after an event occurrence the process $\lambda(t)$ can change or not change its state from state 1 to state 2. The transition of the process $\lambda(t)$ from state 2 to state 1 at the moment of an event occurrence in the second state is impossible. At the moment when the state changes from the second to the first state, an additional event is assumed to be initiated with probability δ ($0 \leq \delta \leq 1$). Such flows with additional events initiation are called integrated flows. Under the made assumptions we can assert that $\lambda(t)$ is a Markovian process. So the flow can be characterized by $\{D_0, D_1\}$, in terms of the rate matrices,

$$D_0 = \left\| \begin{matrix} -(\lambda_1 + \beta) & \beta \\ (1 - \delta)\alpha & -(\lambda_2 + \alpha) \end{matrix} \right\|, \ D_1 = \left\| \begin{matrix} (1 - p)\lambda_1 & p\lambda_1 \\ \delta\alpha & \lambda_2 \end{matrix} \right\|.$$

Intensities of the process $\lambda(t)$ transitions from state to state with the event occurrence fill in the matrix D_1. Nondiagonal elements of the matrix D_0 are intensities of the process $\lambda(t)$ transitions from state to state without the event occurrence. Diagonal elements of the matrix D_0 are intensities of the process $\lambda(t)$ output from its states taken with the opposite signs. Note also that if $\beta = 0$, then the integrated semi-synchronous flow of events will take place [39].

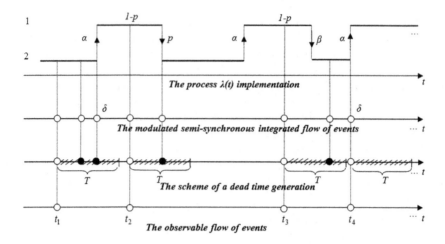

Fig. 1. The formation of an observable flow of events

The registration of the flow events is considered in condition of a constant dead time (of its incomplete observability). The dead time period of a constant duration T begins after every registered at the moment t_k, $k \geq 1$, event. During this period no other events are observed. When the dead time period is over, the first coming event causes the next interval of a dead time of duration T and so on. Figure 1 shows the possible variant of the flow operation and observation. Here 1, 2 are the states of the process $\lambda(t)$; additional events, that may occur at the moment of the process $\lambda(t)$ transition from state 2 to state 1, are marked with letter δ; dead time periods of duration T are marked with hatching; unobserved events are displayed as black circles, observed events t_1, t_2, ..., are shown as white circles.

It should be mentioned that it is not specified exactly, in which state an additional event is assumed to be initiated with probability δ, when the process $\lambda(t)$ changes its state from the second to the first one. This fact is inessential for further formulas derivation as the event occurrence and the process $\lambda(t)$ transition to the first state happens instantly. In practical situations, two variants are possible: (1) first an additional event is initiated with probability δ in state 2 and thereafter the process $\lambda(t)$ transition from state 2 to state 1 is made; (2) first the process $\lambda(t)$ transition from state 2 to state 1 is made and thereafter an additional event is initiated with probability δ in state 1. But to obtain numerical results during simulation procedure, we should take the mentioned details into account and fix, what occurs first, event or transition.

We should note that the process $\lambda(t)$ is basically unobservable. We register only time moments t_1, t_2, ... of the events occurrence in observable flow. The process $\lambda(t)$ is considered in a steady-state conditions. So under the made assumptions we can assert that the sequence of the time moments t_1, t_2, ... corresponds to an embedded Markov chain $\{\lambda(t_k)\}$, i.e. the flow has the Markov property if the evolution of the flow is considered from the time moment t_k,

$k = 1, 2, ...$, of the event occurrence. Denote by $\tau_k = t_{k+1} - t_k$, $k = 1, 2, ...$, the value of the k interval length between two neighboring flow events. In a steady-state conditions we may take that the probability density of the k interval length is $p_T(\tau_k) = p_T(\tau)$, $\tau \geq 0$, for any k (the index T stresses that the probability density depends on the dead time period duration). Thereby we may also take that the time moment t_k is equal to zero, i.e. the moment of the event occurrence is $\tau = 0$. Now let (t_k, t_{k+1}), (t_{k+1}, t_{k+2}) be the successive intervals with the corresponding values of interval length $\tau_k = t_{k+1} - t_k$, $\tau_{k+1} = t_{k+2} - t_{k+1}$. Due to the stationary of the flow, the arrangement of the intervals on a time axis is arbitrarily. That is why we may consider two successive intervals (t_1, t_2), (t_2, t_3) with the corresponding values of the interval length $\tau_1 = t_2 - t_1$, $\tau_2 = t_3 - t_2$, $\tau_1 \geq 0$, $\tau_2 \geq 0$, wherein $\tau_1 = 0$ corresponds to the time moment t_1 and $\tau_2 = 0$ corresponds to the time moment t_2 of the flow events arrival. The respective joint probability density is defined as $p_T(\tau_1, \tau_2)$, $\tau_1 \geq 0$, $\tau_2 \geq 0$.

In that way, the main problem is to obtain the expressions for probability density $p_T(\tau)$, $\tau \geq 0$, and the joint probability density $p_T(\tau_1, \tau_2)$, $\tau_1 \geq 0$, $\tau_2 \geq 0$, explicitly, and also to find the recurrence conditions of the observable flow of events.

3   The Expressions for Probability Density $p_T(\tau)$

Let us consider the interval $(0, \tau)$ between two neighboring events of the observable flow, which length can be written as $\tau = T + t$, where t is a duration of the interval between the end of the dead time period and the next observable event $(t \geq 0)$. Let $p_{jk}(t)$ be the conditional probability that there is no observable events at the interval $(0, t)$ and $\lambda(t) = \lambda_k$ in condition that at the time moment $t = 0$ the value of the process $\lambda(t)$ is $\lambda(0) = \lambda_j$, $j, k = 1, 2$. Denote the corresponding probability density by $\widetilde{p}_{jk}(t)$, $j, k = 1, 2$. Next introduce into consideration probability $q_{ij}(T)$ – the transitional probability that the process $\lambda(\tau)$ changes its state from the state i (at the time moment $\tau = 0$) to the state j (at the time moment $\tau = T$), $i, j = 1, 2$, during the dead time period of the duration T, and probability $\pi_i(0|T)$ – the conditional probability that the process $\lambda(\tau)$ sojourns in the state i $(i = 1, 2)$ at the time moment $\tau = 0$ in condition that at this time moment the event of the observable flow arrived and the dead time period of a constant duration T was generated. With the above-stated notations the desired probability density $p_T(\tau)$ can be written as

$$p_T(\tau) = \begin{cases} 0, & 0 \leq \tau < T, \\ \sum_{i=1}^{2} \pi_i(0|T) \sum_{j=1}^{2} q_{ij}(T) \sum_{k=1}^{2} \widetilde{p}_{jk}(\tau - T), & \tau \geq T. \end{cases} \tag{1}$$

Let us obtain the explicit expressions for $\widetilde{p}_{jk}(\tau - T)$, $q_{ij}(T)$, $\pi_i(0|T)$, $i, j, k = 1, 2$.

The probabilities $p_{jk}(t)$ satisfy the following systems of differential equations:

$$p'_{11}(t) = -(\lambda_1 + \beta)p_{11}(t) + \alpha(1 - \delta)p_{12}(t), \quad p'_{12}(t) = \beta p_{11}(t) - (\lambda_2 + \alpha)p_{12}(t);$$
$$p'_{21}(t) = -(\lambda_1 + \beta)p_{21}(t) + \alpha(1 - \delta)p_{22}(t), \quad p'_{22}(t) = \beta p_{21}(t) - (\lambda_2 + \alpha)p_{22}(t);$$

with the boundary conditions: $p_{11}(0) = 1$, $p_{12}(0) = 0$; $p_{21}(0) = 0$, $p_{22}(0) = 1$.
Solving these systems, we find

$$
\begin{aligned}
&p_{11}(t) = \tfrac{1}{z_2 - z_1}[(\lambda_2 + \alpha - z_1)e^{-z_1 t} - (\lambda_2 + \alpha - z_2)e^{-z_2 t}], \\
&p_{12}(t) = \tfrac{\beta}{z_2 - z_1}(e^{-z_1 t} - e^{-z_2 t}),\ p_{21}(t) = \tfrac{\alpha(1-\delta)}{z_2 - z_1}(e^{-z_1 t} - e^{-z_2 t}), \\
&p_{22}(t) = \tfrac{1}{z_2 - z_1}[(\lambda_1 + \beta - z_1)e^{-z_1 t} - (\lambda_1 + \beta - z_2)e^{-z_2 t}], \\
&z_1 = \tfrac{1}{2}[\lambda_1 + \lambda_2 + \alpha + \beta - \sqrt{(\lambda_1 - \lambda_2 - \alpha + \beta)^2 + 4\alpha\beta(1 - \delta)}], \\
&z_2 = \tfrac{1}{2}[\lambda_1 + \lambda_2 + \alpha + \beta + \sqrt{(\lambda_1 - \lambda_2 - \alpha + \beta)^2 + 4\alpha\beta(1 - \delta)}], \\
&0 < z_1 < z_2.
\end{aligned}
\tag{2}
$$

According to the definition of the modulated semi-synchronous integrated flow of events we introduce the probability $p_{11}(t)\,e^{-\beta \Delta t}(1 - e^{-\lambda_1 \Delta t})(1 - p) = p_{11}(t)\lambda_1(1 - p)\Delta t + o(\Delta t)$ – the joint probability that the process $\lambda(t)$ changes its state from the first state to the first one at the interval $(0, t)$ without the event occurring ($\lambda(0) = \lambda_1$, $\lambda(t) = \lambda_1$), and at the half-interval $[t, t + \Delta t)$ the duration of the first state of the process $\lambda(t)$ does not come to the end, the event of the Poisson flow with intensity λ_1 arrives and the process $\lambda(t)$ remains in the first state. The joint probabilities take the following form for different j and k $(j, k = 1, 2)$

$$
\begin{array}{ll}
p_{11}(t)\lambda_1(1 - p)\Delta t + o(\Delta t), & p_{12}(t)\alpha\delta\,\Delta t + o(\Delta t), \\
p_{11}(t)\lambda_1 p\Delta t + o(\Delta t), & p_{12}(t)\lambda_2\Delta t + o(\Delta t), \\
p_{21}(t)\lambda_1(1 - p)\Delta t + o(\Delta t), & p_{22}(t)\alpha\delta\,\Delta t + o(\Delta t), \\
p_{21}(t)\lambda_1 p\Delta t + o(\Delta t), & p_{22}(t)\lambda_2\Delta t + o(\Delta t).
\end{array}
$$

The corresponding probability densities take the form

$$
\begin{array}{ll}
\widetilde{p}_{11}^{(1)}(t) = p_{11}(t)\lambda_1(1 - p), & \widetilde{p}_{11}^{(2)}(t) = p_{12}(t)\alpha\delta, \\
\widetilde{p}_{12}^{(1)}(t) = p_{11}(t)\lambda_1 p, & \widetilde{p}_{12}^{(2)}(t) = p_{12}(t)\lambda_2, \\
\widetilde{p}_{21}^{(1)}(t) = p_{21}(t)\lambda_1(1 - p), & \widetilde{p}_{21}^{(2)}(t) = p_{22}(t)\alpha\delta, \\
\widetilde{p}_{22}^{(1)}(t) = p_{21}(t)\lambda_1 p, & \widetilde{p}_{22}^{(2)}(t) = p_{22}(t)\lambda_2.
\end{array}
$$

Then the probability densities $\widetilde{p}_{jk}(t)$ that the process $\lambda(t)$ changes its state from the state j to the state k without the event occurrence at the interval $(0, t)$ and with the event occurrence at the time moment t, can be written for different j and k $(j, k = 1, 2)$ as

$$
\begin{aligned}
&\widetilde{p}_{11}(t) = p_{11}(t)\lambda_1(1 - p) + p_{12}(t)\alpha\delta, \quad \widetilde{p}_{12}(t) = p_{11}(t)\lambda_1 p + p_{12}(t)\lambda_2, \\
&\widetilde{p}_{21}(t) = p_{21}(t)\lambda_1(1 - p) + p_{22}(t)\alpha\delta, \quad \widetilde{p}_{22}(t) = p_{21}(t)\lambda_1 p + p_{22}(t)\lambda_2.
\end{aligned}
\tag{3}
$$

Substituting (2) into (3), we obtain the explicit formulas for probability densities $\widetilde{p}_{jk}(t)$, $j, k = 1, 2$.

The probabilities $q_{ij}(\tau)$, $0 \leq \tau \leq T$, satisfy the following systems of differential equations:

$$
\begin{aligned}
&q'_{11}(\tau) = -(p\lambda_1 + \beta)q_{11}(\tau) + \alpha q_{12}(\tau), \quad q'_{12}(\tau) = (p\lambda_1 + \beta)q_{11}(\tau) - \alpha q_{12}(\tau); \\
&q'_{21}(\tau) = -(p\lambda_1 + \beta)q_{21}(\tau) + \alpha q_{22}(\tau), \quad q'_{22}(\tau) = (p\lambda_1 + \beta)q_{21}(\tau) - \alpha q_{22}(\tau);
\end{aligned}
$$

with the boundary conditions: $q_{11}(0) = 1$, $q_{12}(0) = 0$; $q_{21}(0) = 0$, $q_{22}(0) = 1$. Solving these systems, we obtain for $\tau = T$

$$q_{11}(T) = \pi_1 + \pi_2 e^{-(p\lambda_1+\beta+\alpha)T}, \qquad q_{12}(T) = \pi_2 - \pi_2 e^{-(p\lambda_1+\beta+\alpha)T},$$
$$q_{21}(T) = \pi_1 - \pi_1 e^{-(p\lambda_1+\beta+\alpha)T}, \qquad q_{22}(T) = \pi_2 + \pi_1 e^{-(p\lambda_1+\beta+\alpha)T}, \qquad (4)$$
$$\pi_1 = \frac{\alpha}{p\lambda_1+\beta+\alpha}, \qquad \pi_2 = \frac{p\lambda_1+\beta}{p\lambda_1+\beta+\alpha}.$$

Turn now to obtaining the probabilities $\pi_i(0|T)$, $i = 1, 2$. Denote by π_{ij} the transitional probability that the process $\lambda(\tau)$ changes its state from state i to state j ($i, j = 1, 2$) during the time from the moment $\tau = 0$ till the moment of the next event arrival in observable flow. Since the sequence of the time moments of the events occurrence in observable flow corresponds to an embedded Markov chain, the following system of differential equations for $\pi_i(0|T)$ takes place:

$$\pi_1(0|T) = \pi_1(0|T)\pi_{11} + \pi_2(0|T)\pi_{21},$$
$$\pi_2(0|T) = \pi_1(0|T)\pi_{12} + \pi_2(0|T)\pi_{22}; \ \pi_1(0|T) + \pi_2(0|T) = 1. \qquad (5)$$

Let us introduce into consideration probability p_{ij} – a transitional probability that the process $\lambda(t)$ changes its state from state i to state j ($i, j = 1, 2$) during the time from the time moment $t = 0$ (the end of the dead time period) till the moment of the next observable flow event arrival. Here the probabilities p_{ij} are determined as

$$p_{ij} = \int_0^\infty \tilde{p}_{ij}(t)\, dt, \qquad (6)$$

where $\tilde{p}_{ij}(t)$ are defined by (3), $p_{ij}(t)$ are defined by (2) ($i, j = 1, 2$). Calculating the corresponding integrals (6) for different i and j ($i, j = 1, 2$)

$$p_{11} = \int_0^\infty \tilde{p}_{11}(t)\, dt = \lambda_1(1-p)\int_0^\infty p_{11}(t)\, dt + \alpha\delta\int_0^\infty p_{12}(t)\, dt,$$
$$p_{12} = \int_0^\infty \tilde{p}_{12}(t)\, dt = \lambda_1 p\int_0^\infty p_{11}(t)\, dt + \lambda_2\int_0^\infty p_{12}(t)\, dt,$$
$$p_{21} = \int_0^\infty \tilde{p}_{21}(t)\, dt = \lambda_1(1-p)\int_0^\infty p_{21}(t)\, dt + \alpha\delta\int_0^\infty p_{22}(t)\, dt,$$
$$p_{22} = \int_0^\infty \tilde{p}_{22}(t)\, dt = \lambda_1 p\int_0^\infty p_{21}(t)\, dt + \lambda_2\int_0^\infty p_{22}(t)\, dt,$$

we obtain

$$p_{11} = \frac{1}{z_1 z_2}[\lambda_1(1-p)(\lambda_2+\alpha) + \alpha\delta\beta],$$
$$p_{12} = \frac{1}{z_1 z_2}[\lambda_1 p(\lambda_2+\alpha) + \lambda_2\beta],$$
$$p_{21} = \frac{1}{z_1 z_2}[\lambda_1\alpha(1-p+p\delta) + \alpha\delta\beta], \qquad (7)$$
$$p_{22} = \frac{1}{z_1 z_2}[\lambda_2(\lambda_1+\beta) + p\lambda_1\alpha(1-\delta)],$$

where $z_1 z_2 = \lambda_1\lambda_2 + \lambda_1\alpha + \lambda_2\beta + \alpha\delta\beta$.

Since the process $\lambda(t)$ is a Markovian process, the obtained earlier transitional probabilities $q_{ij}(T)$ and p_{ij}, $i, j = 1, 2$, allow us to write the expressions for transitional probabilities π_{ij}, $i, j = 1, 2$, in the following form

$$\pi_{11} = q_{11}(T)p_{11} + q_{12}(T)p_{21}, \qquad \pi_{12} = q_{11}(T)p_{12} + q_{12}(T)p_{22},$$
$$\pi_{21} = q_{21}(T)p_{11} + q_{22}(T)p_{21}, \qquad \pi_{22} = q_{21}(T)p_{12} + q_{22}(T)p_{22}. \qquad (8)$$

Substituting first (4) into (8) and next (7) into (8), we obtain

$$
\begin{aligned}
\pi_{11} &= \tfrac{1}{z_1 z_2}\left\{\lambda_1(1-p)(\lambda_2+\alpha)+\alpha\delta\beta-\lambda_1\pi_2[\lambda_2-p(\lambda_2+\alpha\delta)]\left[1-e^{-(p\lambda_1+\beta+\alpha)T}\right]\right\}, \\
\pi_{12} &= \tfrac{1}{z_1 z_2}\left\{p\lambda_1(\lambda_2+\alpha)+\lambda_2\beta+\lambda_1\pi_2[\lambda_2-p(\lambda_2+\alpha\delta)]\left[1-e^{-(p\lambda_1+\beta+\alpha)T}\right]\right\}, \\
\pi_{21} &= \tfrac{1}{z_1 z_2}\left\{\alpha[\lambda_1(1-p+p\delta)+\delta\beta]+\lambda_1\pi_1[\lambda_2-p(\lambda_2+\alpha\delta)]\left[1-e^{-(p\lambda_1+\beta+\alpha)T}\right]\right\}, \\
\pi_{22} &= \tfrac{1}{z_1 z_2}\left\{\lambda_2(\lambda_1+\beta)+p\,\lambda_1\alpha(1-\delta)-\lambda_1\pi_1[\lambda_2-p(\lambda_2+\alpha\delta)]\left[1-e^{-(p\lambda_1+\beta+\alpha)T}\right]\right\}.
\end{aligned}
\tag{9}
$$

Then, substituting (9) into (5), we obtain the expressions for $\pi_i(0|T)$, $i, j = 1, 2$:

$$
\begin{aligned}
\pi_1(0|T) &= \frac{\alpha[\lambda_1(1-p+p\delta)+\delta\beta]+\lambda_1\pi_1[\lambda_2-p(\lambda_2+\alpha\delta)]\left[1-e^{-(p\lambda_1+\beta+\alpha)T}\right]}{\lambda_1\alpha+(p\lambda_1+\beta)(\lambda_2+\alpha\delta)+\lambda_1[\lambda_2-p(\lambda_2+\alpha\delta)]\left[1-e^{-(p\lambda_1+\beta+\alpha)T}\right]}, \\
\pi_2(0|T) &= \frac{p\lambda_1(\lambda_2+\alpha)+\lambda_2\beta+\lambda_1\pi_2[\lambda_2-p(\lambda_2+\alpha\delta)]\left[1-e^{-(p\lambda_1+\beta+\alpha)T}\right]}{\lambda_1\alpha+(p\lambda_1+\beta)(\lambda_2+\alpha\delta)+\lambda_1[\lambda_2-p(\lambda_2+\alpha\delta)]\left[1-e^{-(p\lambda_1+\beta+\alpha)T}\right]},
\end{aligned}
\tag{10}
$$

where π_1, π_2 are defined in (4).

Substituting first (3) into (1) and next (2), (4) and (10) into (1), carrying out laborious transformations and considering that $t = \tau - T$, we obtain

$$
\begin{aligned}
p_T(\tau) &= \begin{cases}
0, & 0 \le \tau < T, \\
\gamma(T)z_1 e^{-z_1(\tau-T)} + (1-\gamma(T))z_2 e^{-z_2(\tau-T)}, & \tau \ge T, \\
\end{cases} \\
\gamma(T) &= \tfrac{1}{z_2-z_1}\left[z_2-\lambda_1+(\lambda_1-\lambda_2-\alpha\delta)\pi_2(T)\right],
\end{aligned}
\tag{11}
$$

$$
\begin{aligned}
\pi_1(T) &= \pi_1 + [\pi_2 - \pi_2(0|T)]\, e^{-(p\lambda_1+\beta+\alpha)T}, \\
\pi_2(T) &= \pi_2 - [\pi_2 - \pi_2(0|T)]\, e^{-(p\lambda_1+\beta+\alpha)T},
\end{aligned}
\tag{12}
$$

where z_i are defined in (2); π_i – in (4); $\pi_i(0|T)$ – in (10), $i = 1, 2$.

In particular, by setting $T=0$ in (11), (12), we obtain the formulas for $p(\tau)$ that were presented in [40].

4   The Expressions for Joint Probability Density $p_T(\tau_1, \tau_2)$

Let $\tau_1 = T + t^{(1)}$, $\tau_2 = T + t^{(2)}$ be the values of the intervals length for two successive intervals between the time moments of the events arrival in observable flow of events, where $\tau_1 = 0$ is the arrival time for the first flow event, $\tau_2 = 0$ is the arrival time for the second flow event. Since the sequence of the time moments of the events arrival in observable flow corresponds to an embedded Markov chain, then with the above notation (see Sect. 3) the joint probability density $p_T(\tau_1, \tau_2)$ takes the following form

$$
p_T(\tau_1, \tau_2) = \begin{cases}
0, & 0 \le \tau_1 < T, \quad 0 \le \tau_2 < T, \\
\sum_{i=1}^{2}\pi_i(0|T)\sum_{j=1}^{2}q_{ij}(T)\sum_{k=1}^{2}\widetilde{p}_{jk}(\tau_1-T) \\
\times\sum_{s=1}^{2}q_{ks}(T)\sum_{n=1}^{2}\widetilde{p}_{sn}(\tau_2-T), & \tau_1 \ge T, \ \tau_2 \ge T,
\end{cases}
\tag{13}
$$

where $\widetilde{p}_{jk}(\tau_1 - T) = \widetilde{p}_{jk}(t^{(1)})$, $\widetilde{p}_{sn}(\tau_2 - T) = \widetilde{p}_{sn}(t^{(2)})$ are defined by (3) and t should be replaced by $t^{(1)}$ and $t^{(2)}$ in expressions for $\widetilde{p}_{ij}(t)$, $i, j = 1, 2$. Then substituting first $\widetilde{p}_{jk}(t^{(1)})$, $\widetilde{p}_{sn}(t^{(2)})$, that are defined by (3), next $p_{jk}(t^{(1)})$, $p_{sn}(t^{(2)})$, that are defined by (2) for $t = t^{(1)}$ and $t = t^{(2)}$, next $q_{ij}(T)$, $q_{ks}(T)$, that are defined by (4), and finally $\pi_i(0|T)$, $i = 1, 2$, that are defined by (10), into (13) and carrying out laborious transformations, we obtain

$$p_T(\tau_1, \tau_2) = 0, \quad 0 \leq \tau_1 < T, \quad 0 \leq \tau_2 < T,$$

$$
\begin{aligned}
p_T(\tau_1, \tau_2) &= p_T(\tau_1)p_T(\tau_2) + e^{-(p\lambda_1+\beta+\alpha)T}\gamma(T)\left[1 - \gamma(T)\right]\frac{\lambda_1[\lambda_2 - p(\lambda_2+\alpha\delta)]}{z_1 z_2} \\
&\times \left[z_1 e^{-z_1(\tau_1-T)} - z_2 e^{-z_2(\tau_1-T)}\right]\left[z_1 e^{-z_1(\tau_2-T)} - z_2 e^{-z_2(\tau_2-T)}\right], \quad \tau_1 \geq T, \quad \tau_2 \geq T,
\end{aligned}
\tag{14}
$$

where $z_1 z_2 = \lambda_1\lambda_2 + \lambda_1\alpha + \lambda_2\beta + \alpha\delta\beta$ and $\gamma(T)$, $p_T(\tau_k)$ are defined by (11) for $\tau = \tau_k$, $k = 1, 2$.

It follows from (14) that in general case the modulated semi-synchronous integrated flow of events is a correlated flow. By taking in (14) $T = 0$, we get the formula for the joint probability density $p(\tau_1, \tau_2)$ presented in [40].

There is no difficulty in obtaining the probabilistic characteristics of the observable flow of events, such as mathematical expectation of the interval length between the neighboring flow events, variance and covariance:

$$M\tau = T + \frac{\gamma(T)}{z_1} + \frac{1-\gamma(T)}{z_2}, \quad D\tau = 2\left[\frac{\gamma(T)}{z_1^2} + \frac{1-\gamma(T)}{z_2^2}\right] - \left[\frac{\gamma(T)}{z_1} + \frac{1-\gamma(T)}{z_2}\right]^2,$$

$$cov(\tau_1, \tau_2) = e^{-(p\lambda_1+\beta+\alpha)T}\lambda_1\gamma(T)\left[1 - \gamma(T)\right]\left[\lambda_2 - p(\lambda_2+\alpha\delta)\right]\frac{(z_2-z_1)^2}{(z_1 z_2)^3}.$$

It is worthwhile to note that there are three types of events in the modulated semi-synchronous integrated flow of events: (1) events of a Poisson flow with intensity λ_1; (2) events of a Poisson flow with intensity λ_2; (3) additional events, which are indistinguishable. Introduce into consideration probabilities $q_1^{(i)}(T)$ – stationary probability that the event appeared is the event of a Poisson flow with intensity λ_1 (first type event) and the process $\lambda(t)$ changes its state from the state 1 to the state i ($i = 1, 2$); $q_2(T)$ – stationary probability that the event appeared is the event of a Poisson flow with intensity λ_2 (second type event); $q_3(T)$ – stationary probability that the event appeared is an additional event (third type event). Now it is not difficult to obtain the explicit expressions for the introduced probabilities on the basis of the above results:

$$q_1^{(1)}(T) = \lambda_1(1 - p)\frac{\alpha + [(\lambda_2 + \alpha\delta)\pi_1 - \alpha\delta]\left[1 - e^{-(p\lambda_1+\beta+\alpha)T}\right]}{z_1 z_2 - \lambda_1\left[\lambda_2 - p(\lambda_2 + \alpha\delta)\right]e^{-(p\lambda_1+\beta+\alpha)T}},$$

$$q_1^{(2)}(T) = \lambda_1 p\frac{\alpha + [(\lambda_2 + \alpha\delta)\pi_1 - \alpha\delta]\left[1 - e^{-(p\lambda_1+\beta+\alpha)T}\right]}{z_1 z_2 - \lambda_1\left[\lambda_2 - p(\lambda_2 + \alpha\delta)\right]e^{-(p\lambda_1+\beta+\alpha)T}},$$

$$q_2(T) = \lambda_2\frac{p\lambda_1 + \beta + \lambda_1(1 - p - \pi_1)\left[1 - e^{-(p\lambda_1+\beta+\alpha)T}\right]}{z_1 z_2 - \lambda_1\left[\lambda_2 - p(\lambda_2 + \alpha\delta)\right]e^{-(p\lambda_1+\beta+\alpha)T}},$$

$$q_3(T) = \alpha\delta\frac{p\lambda_1 + \beta + \lambda_1(1 - p - \pi_1)\left[1 - e^{-(p\lambda_1+\beta+\alpha)T}\right]}{z_1 z_2 - \lambda_1\left[\lambda_2 - p(\lambda_2 + \alpha\delta)\right]e^{-(p\lambda_1+\beta+\alpha)T}}.$$

Then the stationary probability $q_1(T)$ that the event appeared is the event of a Poisson flow with intensity λ_1 (first type event) can be written as

$$q_1(T) = q_1^{(1)}(T) + q_1^{(2)}(T) = \lambda_1 \frac{\alpha + [(\lambda_2 + \alpha\delta)\pi_1 - \alpha\delta]\left[1 - e^{-(p\lambda_1 + \beta + \alpha)T}\right]}{z_1 z_2 - \lambda_1\left[\lambda_2 - p(\lambda_2 + \alpha\delta)\right]e^{-(p\lambda_1 + \beta + \alpha)T}}.$$

Finally, note that $\pi_1(0|T) = q_1^{(1)}(T) + q_3(T)$, $\pi_2(0|T) = q_1^{(2)}(T) + q_2(T)$.

5   The Conditions of the Observable Flow Recurrence

Let us consider the specific cases, when the modulated semi-synchronous integrated flow of events becomes the recurrent flow. It can be shown by using the expressions (11), (12) for $\gamma(T)$, $\pi_1(T)$, $\pi_2(T)$ and (10) for $\pi_1(0|T)$, $\pi_2(0|T)$ that

$$\gamma(T)\left[1 - \gamma(T)\right] = \frac{(\lambda_1 - \lambda_2 - \alpha\delta)[\lambda_1\alpha + (p\lambda_1 + \beta)(\lambda_2 + \alpha\delta)][(p\lambda_1 + \beta)\pi_1(0) - \alpha\pi_2(0)]z_1 z_2}{(z_2 - z_1)^2 (p\lambda_1 + \beta + \alpha)^2\left[z_1 z_2 - \lambda_1[\lambda_2 - p(\lambda_2 + \alpha\delta)]e^{-(p\lambda_1 + \beta + \alpha)T}\right]^2}$$

$$\times\left\{z_1 z_2 - [2z_1 z_2 - (p\lambda_1 + \beta + \alpha)(z_1 + z_2)]\,e^{-(p\lambda_1 + \beta + \alpha)T}\right. \tag{15}$$

$$\left.+[z_1 z_2 - (p\lambda_1 + \beta + \alpha)(\lambda_1(1 - p) + \lambda_2)]e^{-2(p\lambda_1 + \beta + \alpha)T}\right\},$$

where $\pi_i(0)$ is the conditional stationary probability that the process $\lambda(\tau)$ sojourns in the state i ($i = 1, 2$) at the time moment $\tau = 0$ in condition that at this time moment the flow event has arrived ($\pi_1(0) + \pi_2(0) = 1$). And $\pi_i(0)$, $i = 1, 2$, are defined as follows

$$\pi_1(0) = \alpha\frac{\lambda_1(1 - p + p\delta) + \delta\beta}{\lambda_1\alpha + (p\lambda_1 + \beta)(\lambda_2 + \alpha\delta)}, \quad \pi_2(0) = \frac{p\lambda_1(\lambda_2 + \alpha) + \lambda_2\beta}{\lambda_1\alpha + (p\lambda_1 + \beta)(\lambda_2 + \alpha\delta)}.$$

Note, that the expression enclosed in braces in formula (15), which we denote by $f(T)$, after the transformation can be written in form

$$f(T) = z_1 z_2\left[1 - e^{-(p\lambda_1 + \beta + \alpha)T}\right]^2 + (p\lambda_1 + \beta + \alpha)e^{-(p\lambda_1 + \beta + \alpha)T}[z_1 + z_2$$

$$-(\lambda_1(1 - p) + \lambda_2)e^{-(p\lambda_1 + \beta + \alpha)T}] = f_1(T) + f_2(T) = f_1(T) + \varphi_1(T)\varphi_2(T).$$

It is easy to show, that for any $T \geq 0$ we have $f_1(T) \geq 0$, $\varphi_1(T) > 0$ and $\varphi_2(T) > 0$ and thus $f_2(T) > 0$. Hence, for any $T \geq 0$ we have $f(T) > 0$. It follows from (15) that:

(1) if $\lambda_1 - \lambda_2 - \alpha\delta = 0$, then the joint probability density (14) becomes factorable: $p_T(\tau_1, \tau_2) = p_T(\tau_1)p_T(\tau_2)$; and it follows from (2) that $z_1 = \lambda_1$, $z_2 = \lambda_2 + \alpha + \beta$; (11) implies $\gamma(T) = 1$, and then $p_T(\tau_k) = \lambda_1 e^{-\lambda_1(\tau_k - T)}$, $\tau_k \geq T$, $k = 1, 2$, i.e. $p_T(\tau) = \lambda_1 e^{-\lambda_1(\tau - T)}$, $\tau \geq T$.

(2) if $(p\lambda_1 + \beta)\pi_1(0) - \alpha\pi_2(0) = 0$, then the joint probability density (14) becomes factorable: $p_T(\tau_1, \tau_2) = p_T(\tau_1)p_T(\tau_2)$; and it follows from (2) that $z_1 = \lambda_1(1 - p + p\delta) + \delta\beta$; (11) implies $\gamma(T) = 1$, and then $p_T(\tau_k) = z_1 e^{-z_1(\tau_k - T)}$, $\tau_k \geq T$, $k = 1, 2$, i.e. $p_T(\tau) = z_1 e^{-z_1(\tau - T)}$, $\tau \geq T$.

The third condition of the joint probability density $p_T(\tau_1, \tau_2)$ factorization follows from (14): $\lambda_2 - p(\lambda_2 + \alpha\delta) = 0$. In this case $p_T(\tau)$ is defined by the formula (11), where

$$\pi_2(0|T) = p; \ \pi_2(T) = \frac{p\lambda_1 + \beta}{p\lambda_1 + \beta + \alpha} + \left[p - \frac{p\lambda_1 + \beta}{p\lambda_1 + \beta + \alpha} \right] e^{-(p\lambda_1 + \beta + \alpha)T}; \ p \neq 1.$$

In particular, if we put $p = 1$ in the third condition, we have $\delta = 0$. Then $p_T(\tau)$ is defined by the formula (11), where

$$\pi_2(0|T) = 1; \qquad \pi_2(T) = \frac{1}{\lambda_1 + \beta + \alpha} \left[\lambda_1 + \beta + \alpha e^{-(p\lambda_1 + \beta + \alpha)T} \right].$$

Since the sequence of the time moments $t_1, t_2, ..., t_k, ...$ corresponds to an embedded Markov chain, then upon meeting one of the above-mentioned conditions or their combination we may show that the joint probability density $p_T(\tau_1, ..., \tau_k)$ becomes factorable for any k. This suggests that in this case the observable flow of events is a recurrent flow. For, let $p_T(\tau_1, ..., \tau_k, \tau_{k+1})$ be the joint probability density of $\tau_1, ..., \tau_k, \tau_{k+1}$, where $\tau_k = t_{k+1} - t_k$, $k = 1, 2, ...$. For $k = 2$ we have $p_T(\tau_1, \tau_2) = p_T(\tau_1)p_T(\tau_2)$. Now we proceed by mathematical induction. Assume that $p_T(\tau_1, ..., \tau_k) = p_T(\tau_1)...p_T(\tau_k)$. Since the sequence of the time moments $t_1, t_2, ..., t_k, t_{k+1}$ of the flow events occuring is an embedded Markov chain, then the flow has the Markov property at the moments of the flow events arrival. Then $p_T(\tau_1, ..., \tau_k, \tau_{k+1}) = p_T(\tau_1, ..., \tau_k)p_T(\tau_{k+1}|\tau_1, ..., \tau_k) = p_T(\tau_1, ..., \tau_k)p_T(\tau_{k+1}|\tau_k)$, where $p_T(\tau_{k+1}|\tau_k) = p_T(\tau_k, \tau_{k+1})/p_T(\tau_k)$. Since for the neighboring intervals (t_k, t_{k+1}) and (t_{k+1}, t_{k+2}), $k = 1, 2, ...$, which location on the time axis is arbitraraly, we have $p_T(\tau_k, \tau_{k+1}) = p_T(\tau_k)p_T(\tau_{k+1})$, then $p_T(\tau_{k+1}|\tau_k) = p_T(\tau_{k+1})$. This proves the factorization of the joint probability density $p_T(\tau_1, ..., \tau_k, \tau_{k+1})$.

Note that the factorization conditions are identical for $T = 0$ [40] and $T \neq 0$.

In further discussion of the flow recurrence conditions we should consider results obtained in [31–33].

For the first recurrence condition a posteriori probability $w(\lambda_1|t)$ behavior at the intervals (t_k, t_{k+1}), $k = 1, 2, ...$, is determined with the explicit formulas:

$$w(\lambda_1|t) = \pi_1 - [\pi_1 - w(\lambda_1|t_k + 0)] \, e^{-(p\lambda_1 + \beta + \alpha)(t - t_k)}, \ t_k < t \leq t_k + T,$$

$$w(\lambda_1|t) = \frac{w_1[w_2 - w(\lambda_1|t_k + T)] - w_2[w_1 - w(\lambda_1|t_k + T)] \, e^{-b(t - t_k - T)}}{w_2 - w(\lambda_1|t_k + T) - [w_1 - w(\lambda_1|t_k + T)] \, e^{-b(t - t_k - T)}}, \ t_k + T < t \leq t_{k+1}, \tag{16}$$

where

$$w(\lambda_1|t_k + 0) = \frac{\alpha\delta + [\lambda_1(1 - p) - \alpha\delta] \, w(\lambda_1|t_k - 0)}{\lambda_2 + \alpha\delta},$$

$$w_1 = \frac{\lambda_1 - \lambda_2 + \alpha + \beta - 2\alpha\delta - b}{2(\lambda_1 - \lambda_2 - \alpha\delta)}, \qquad w_2 = \frac{\lambda_1 - \lambda_2 + \alpha + \beta - 2\alpha\delta + b}{2(\lambda_1 - \lambda_2 - \alpha\delta)},$$

$$b = \sqrt{(\lambda_1 - \lambda_2 - \alpha + \beta)^2 + 4\alpha\beta(1 - \delta)}, \tag{17}$$

and π_1 is defined by (4). In spite of the fact that the flow becomes recurrent and probability density $p_T(\tau)$ is exponential, a posteriori probability $w(\lambda_1|t)$ depends on prehistory, i.e. it depends on the time moments $t_1, t_2, ..., t_k$ of the events occurrence in observable flow. In fact, $w(\lambda_1|t)$ depends on the initial condition at the time moment t_k – the value of $w(\lambda_1|t_k + 0)$, $k = 1, 2, ...$. In turn $w(\lambda_1|t_k + 0)$ depends on the value of $w(\lambda_1|t_k - 0)$, of probability $w(\lambda_1|t)$ at the moment t_k, when $w(\lambda_1|t)$ changes at the half-interval $[t_{k-1}, t_k)$ preceding the half-interval $[t_k, t_{k+1})$, $k = 1, 2, \dots$. Thereby, all prehistory of the flow observation from the time moment $t_0 = 0$ to t_k is concentrated in the value of $w(\lambda_1|t_k + 0)$. And it may be stated that the flow is close to a simple stream. If to add an additional condition $\lambda_1(1-p) - \alpha\delta = 0$, then a posteriori probability $w(\lambda_1|t)$ will not depend on prehistory, it will depend on the value of $w(\lambda_1|t)$ at the moment of the event occurrence t_k, i.e. on $w(\lambda_1|t_k + 0) = \alpha\delta/(\lambda_2 + \alpha\delta)$, $k = 1, 2, \dots$. In this case we may state that the flow is more close to a simple stream.

For the second recurrence condition a posteriori probability $w(\lambda_1|t)$ behavior at the intervals (t_k, t_{k+1}), $k = 1, 2, \dots$, is determined with the explicit formulas (16), where

$$w(\lambda_1|t_k + 0) = \frac{\alpha\delta + [\lambda_1(1 - p) - \alpha\delta]\, w(\lambda_1|t_k - 0)}{\lambda_2 + \alpha\delta + (\lambda_1 - \lambda_2 - \alpha\delta)\, w(\lambda_1|t_k - 0)}, \quad k = 1, 2, \dots.$$

In spite of the fact that the flow becomes recurrent and probability density $p_T(\tau)$ is exponential, a posteriori probability $w(\lambda_1|t)$ also depends on prehistory, i.e. it depends on the time moments t_1, t_2, ..., t_k of the events occurrence in observable flow. In this case we may state that the flow is close to a simple stream.

For the third recurrence condition probability density $p_T(\tau)$ is defined by the formula (11) and it is not exponential, so there is no closeness with a simple stream of events.

6   Conclusion

The obtained results provide the possibility to solve the problem of parameters estimation of the modulated semi-synchronous integrated flow of events in condition of a constant dead time. One of the most interesting and important problems of the flow parameters estimation is estimating the dead time period duration. This is necessary to estimate the quantity of the lost flow events (events carrying useful information). To solve this problem we can apply the following methods: (1) maximum-likelihood technique; (2) method of moments.

To estimate duration of the dead time period with maximum-likelihood technique, first of all, the likelihood function is constructed

$$L(T\,|\tau_1, \dots, \tau_n) = \prod_{k=1}^{n} p_T(\tau_k),$$

where τ_k, $k = \overline{1, n}$, are the measured values of the intervals length duration $\tau_k = t_{k+1} - t_k$, $k = \overline{1, n}$. Then the following task of optimization is resolved

$$L(T\,|\tau_1, \dots, \tau_n) \Longrightarrow \max_{T}, \quad 0 \leq T \leq \tau_{min},$$

where $\tau_{min} = min\, \tau_k$, $k = \overline{1, n}$. The point of global maximum T^* of the likelihood function $L(T\,|\tau_1, \dots, \tau_n)$ will be the desired estimation $\hat{T}$ of the dead time period duration.

To solve the estimation problem with the method of moments $\hat{cov}(\tau_1, \tau_2)$ statistic is constructed. $\hat{cov}(\tau_1, \tau_2)$ is the estimation of theoretical covariance

$$cov_T(\tau_1, \tau_2) = \int_T^\infty \int_T^\infty [\tau_1 - M\tau_1][\tau_2 - M\tau_2]\, p_T(\tau_1, \tau_2)d\tau_1 d\tau_2,$$

where $M\tau_k$, $k = 1, 2$, are mathematical expectations of the intervals length $\tau_1 = t_2 - t_1$ and $\tau_2 = t_3 - t_2$. Then the equation of moments $cov_T(\tau_1, \tau_2) = \hat{cov}(\tau_1, \tau_2)$ is solved for the unknown T and a solution of this equation is chosen as $\hat{T}$.

Acknowledgments. The work is supported by Tomsk State University Competitiveness Improvement Program.

References

1. Cox, D.R.: Some statistical methods connected with series of events. J. R. Stat. Soc. B **17**, 129–164 (1955)
2. Kingman, Y.F.C.: On doubly stochastic poisson process. Proc. Camb. Phylosophical Soc. **60**(4), 923–930 (1964)
3. Cox, D.R., Isham, V.: Point Processes. Chapman & Hall, London (1980)
4. Bremaud, P.: Point Processes and Queues: Martingale Dynamics. Springer-Verlag, New York (1981)
5. Last, G., Brandt, A.: Marked Point Process on the Real Line: The Dynamic Approach. Springer-Verlag, New York (1995)
6. Basharin, G.P., Kokotushkin, V.A., Naumov, V.A.: Method of equivalent substitutions for calculating fragments of communication networks for digital computer. Eng. Cybern. **17**(6), 66–73 (1979)
7. Neuts, M.F.: A versatile Markov point process. J. Appl. Probab. **16**, 764–779 (1979)
8. Lucantoni, D.M.: New results on the single server queue with a batch markovian arrival process. Commun. Stat. Stoch. Models **7**, 1–46 (1991)
9. Lucantoni, D.M., Neuts, M.F.: Some steady-state distributions for the MAP/SM/1 queue. Commun. Stat. Stoch. Models **10**, 575–598 (1994)
10. Breuer, L.: An EM algorithm for batch Markovian arrival processes and its comparison to a simpler estimation procedure. Ann. Oper. Res. **112**, 123–138 (2002)
11. Telek, M., Horvath, G.: A minimal representation of Markov arrival processes and a moments matching method. Perform. Eval. **64**, 1153–1168 (2007)
12. Okamura, H., Dohi, T., Trivedi, K.S.: Markovian arrival process parameter estimation with group data. IEEE/ACM Trans. Networking **17**(4), 1326–1339 (2009)
13. Horvath, A., Horvath, G., Telek, M.: A joint moments based analysis of networks of MAP/MAP/1 queues. Perform. Eval. **67**(9), 759–778 (2010)
14. Bushlanov, I.V., Gortsev, A.M., Nezhelskaya, L.A.: Estimating parameters of the synchronous twofold-stochastic flow of events. Autom. Remote Control **69**(9), 1517–1533 (2008)
15. Gortsev, A.M., Nezhelskaya, L.A.: Estimation of the dead-time period and parameters of a synchronous alternating flow of events. Bull. Tomsk State Univ. **6**, 232–239 (2003). (in Russian)
16. Gortsev, A.M., Nezhelskaya, L.A.: Parameters estimation of a synchronous doubly stochastic flow of events using method of moments. Bull. Tomsk State Univ. **1**, 24–29 (2002). (in Russian)

17. Gortsev, A.M., Nezhelskaya, L.A.: An asynchronous double stochastic flow with initiation of superfluous events. Discrete Math. Appl. **21**(3), 283–290 (2011)
18. Gortsev, A.M., Nissenbaum, O.V.: Estimation of the dead time period and parameters of an asynchronous alternative flow of events with unextendable dead time period. Russ. Phys. J. **48**(10), 1039–1054 (2005). (in Russian)
19. Gortsev, A.M., Nezhelskaya, L.A., Shevchenko, T.I.: States estimation of the MC flow of events in the presence of measurement errors. Russ. Phys. J. **12**, 67–85 (1993). (in Russian)
20. Gortsev, A.M., Nezhelskaya, L.A.: Estimation of the dead-time period and parameters of a semi-synchronous double-stochastic stream of events. Meas. Tech. **46**(6), 536–545 (2003)
21. Gortsev, A.M., Nezhelskaya, L.A.: Semi-synchronous doubly stochastic flow of events in condition of prolonged dead time. Comput. Technol. **13**(1), 31–41 (2008) (in Russian)
22. Gortsev, A.M., Nezhelskaya, L.A.: Parameters estimation of a semi-synchronous doubly stochastic flow of events using method of moments. Bull. Tomsk State Univ. **1**, 18–23 (2002). (in Russian)
23. Gortsev, A.M., Nezhelskaya, L.A.: On connection of MC flows and MAP flows of events. Bull. Tomsk State Univ. Control, Comput. Eng. Inform. **1**(14), 13–21 (2011). (in Russian)
24. Adamu, A., Gaidamaka, Y., Samuylov, A.: Discrete Markov chain model for analyzing probability measures of P2P streaming network. In: Balandin, S., Koucheryavy, Y., Hu, H. (eds.) NEW2AN 2011 and ruSMART 2011. LNCS, vol. 6869, pp. 428–439. Springer, Heidelberg (2011)
25. Bouzas, P.R., Valderrama, M.J., Aguilera, A.M., Ruiz-Fuentes, N.: Modelling the mean of a doubly stochastic poisson process by functional data analysis. Comput. Stat. Data Anal. **50**(10), 2655–2667 (2006)
26. Centanni, S., Minozzo, M.: A Monte Carlo approach to filtering for a class of marked doubly stochastic poisson processes. J. Am. Stat. Assoc. **101**, 1582–1597 (2006)
27. Dubois, J.-P.: Traffic estimation in wireless networks using filtered doubly stochastic point processes (conference paper). In: Proceedings - 2004 International Conference on Electrical, Electronic and Computer Engineering, ICEEC 2004, pp. 116–119 (2004)
28. Hossain, M.M., Lawson, A.B.: Approximate methods in Bayesian point process spatial models. Comput. Stat. Data Anal. **53**(8), 2831–2842 (2009)
29. Snyder, D.L., Miller, M.I.: Random Point Processes in Time and Space. Springer-Verlag, Heidelberg (1991)
30. Gortsev, A.M., Nezhelskaya, L.A., Solovev, A.A.: Optimal state estimation in MAP event flows with unextendable dead time. Autom. Remote Control **73**(8), 1316–1326 (2012)
31. Bakholdina, M.A.: Optimal estimation of the states of modulated semi-synchronous integrated flow of events. Bull. Tomsk State Univ. Control Comput. Eng. Inf. **2**(23), 10–21 (2013). (in Russian)
32. Bakholdina, M.A., Gortsev, A.M.: Optimal estimation of the states of modulated semi-synchronous integrated flow of events in condition of a constant dead time. Bull. Tomsk State Univ. Control Comput. Eng. Inf. **1**(26), 13–24 (2014) (in Russian)
33. Bakholdina, M.A., Gortsev, A.M.: Optimal estimation of the states of modulated semi-synchronous integrated flow of events in condition of its incomplete observability. Appl. Math. Sci. **9**(29), 1433–1451 (2015)

34. Gortsev, A.M., Nezhelskaya, L.A.: Estimation of the dead time period and intensity of synchronous doubly stochastic flow of events. Radio Eng. **10**, 8–16 (2004). (in Russian)

35. Vasileva, L.A., Gortsev, A.M.: Estimation of parameters of a double-stochastic flow of events under conditions of its incomplete observability. Autom. Remote Control **63**(3), 511–515 (2002). (in Russian)

36. Gortsev, A.M., Zavgorodnyaya, M.E.: States estimation of the alternating flow of events in condition of its incomplete observability. Atmos. Ocean Opt. **10**(3), 273–280 (1997). (in Russian)

37. Gortsev, A.M., Klimov, I.S.: Intensity estimation of the Poisson flow of events in condition of its incomplete observability. Radio Eng. **12**, 3–7 (1991). (in Russian)

38. Normey-Rico, J.E.: Control of dead-time processes (Advanced textbooks in control and signal processing). Springer, London (2007)

39. Gortsev, A.M., Kalyagin, A.A., Nezhelskaya, L.A.: Optimal states estimation of integrated semi-synchronous flow of events. Bull. Tomsk State Univ. Control Comput. Eng. Inf. **2**(11), 66–81 (2010). (in Russian)

40. Bakholdina, M., Gortsev, A.: Joint probability density of the intervals length of the modulated semi-synchronous integrated flow of events and its recurrence conditions. In: Dudin, A., Nazarov, A., Yakupov, R., Gortsev, A. (eds.) ITMM 2014. CCIS, vol. 487, pp. 18–25. Springer, Heidelberg (2014)

Mean-Field Analysis for Heterogeneous Work Stealing Models

Quan-Lin Li[⊠] and Feifei Yang

School of Economics and Management Sciences, Yanshan University,
Qinhuangdao, China
liquanlin@tsinghua.edu.cn

Abstract. In this paper, we provide a simple framework for applying the mean-field theory to dealing with a heterogeneous work stealing model of M clusters, each of which consists of N same servers and operates under two types of work stealing schemes: One within a cluster, and another between any two clusters. We first set up an infinite-dimensional system of mean-field equations, which is related to the M clusters. Then we use the martingale limit theory to prove the asymptotic independence of this heterogeneous work stealing model. Finally, we analyze and compute the fixed point, which can give performance analysis of this heterogeneous stealing model.

Keywords: Work stealing model · Cluster · Martingale limit theory · Asymptotic independence · Fixed point · Performance analysis

1 Introduction

Big networks, such as, computer networks, Internet of Things, manufacturing systems, transportation networks and healthcare systems, are becoming more complex, and analysis of their resource management is more difficult and challenging, e.g., see Li [10] and Benaim and Le Boudec [2]. For such a resource management, a basic issue is to continuously redistribute jobs and service ability among clusters or servers. Up to now, there have been two useful methods to do so: *Push strategies*, in which a processor that is overloaded will send work to the others; and *pull strategies*, in which an underloaded processor will ask for work from other processors with more workload. The push strategies are mainly used in the centralized systems, where a classical example is the *supermarket models*, e.g., see Vvedenskaya et al. [21], Mitzenmacher [15], Li et al. [11], and Li and Lui [12]. The pull strategies are more appropriate for non-centralized and relatively large-scale systems, where the *work stealing model* is one most representative scheme. Readers may refer to, such as, Blumofe and Papadopoulos [5], Blumofe and Leiserson [4], and Berenbrink et al. [3].

The queueing theory and Markov processes are applied to modeling and analysis for the work stealing models. Squillante and Nelson [20] and Squillante [19] provided the continuous time Markov models with work stealing scheme

© Springer International Publishing Switzerland 2015
A. Dudin et al. (Eds.): ITMM 2015, CCIS 564, pp. 28–40, 2015.
DOI: 10.1007/978-3-319-25861-4_3

over a small number of processors. Hendler and Shavit [9] studied the non-blocking steal-half work queues. Harchol-Balter et al. [8] and Osogami et al. [17] analyzed the queueing models with two independent processors under a work stealing scheme. Anselmi and Gaujal [1] gave a numerical method of the M/G/1 work stealing model. Lu et al. [13] provided a novel load balancing algorithm for dynamically scalable web services through a join-idle-queue. The mean-field theory is applied to analyzing the work stealing models. Important examples include Mitzenmacher [16], Gast and Gaujal [7], and Minnebo and Van Houdt [14].

The main contributions of this paper are twofold. The first one is to consider a heterogeneous work stealing model of M clusters, and to set up an infinite-dimensional system of mean-field equations, which is related to the M clusters. The second one is to apply the martingale limit theory to proving the asymptotic independence of this heterogeneous work stealing model. Based on this, we analyze and compute the fixed point, which can give performance analysis of this heterogeneous work stealing model.

The remainder of this paper is organized as follows. In Sect. 2, we describe a heterogeneous work stealing model of M clusters, and express the state of this heterogeneous work stealing model by means of an empirical measure process. In Sect. 3, we derive an infinite-dimensional system of mean-field equations. In Sect. 4, we apply the martingale limit theory to proving the asymptotic independence. Based on this, we analyze and compute the fixed point, which can give performance analysis of this heterogeneous work stealing model. Some computational remarks are given in Sect. 5.

2   A Work Stealing Model

In this section, we first describe a heterogeneous work stealing model of M clusters, each of which consists of N same servers and operates under two types of work stealing schemes. Then we express the states of this heterogeneous work stealing model by means of an empirical measure process.

2.1   Model Description

This heterogeneous work stealing model is composed of M clusters, each of which contains N same servers and operates under two types of work stealing schemes: One within a cluster, and another between any two clusters.

In this heterogeneous work stealing model, there are two types of work stealing schemes: (1) The work stealing scheme is operated within a cluster. If a server completes all its customers and enters an idle period, then the idle server steals the half of customers from a victim server with the longest queue length. Notice that the victim server and the idle server are within a cluster. (2) The work stealing scheme is operated between any two clusters. If a server completes all its customers and enters an idle period, then the idle server steals the half of customers from a victim server with the longest queue length in another cluster. Notice that the victim server and the idle server are in two different clusters,

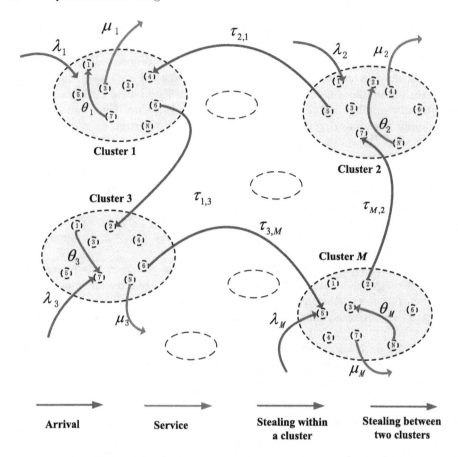

Fig. 1. A heterogeneous stealing model of M clusters

respectively. Figure 1 provides a physical illustration for this heterogeneous work stealing model of M clusters.

Based on Fig. 1, we provide some necessary parameters and notation as follows.

The Arriving Processes. Let C_r denote the rth cluster. Then customers arrive at cluster C_r as a Poisson process with arrival rate $N\lambda_r$ for $\lambda_r > 0$, this also indicates that customers arrive at each server in this cluster as a Poisson process with arrival rate λ_r. Upon arrival, each arriving customer chooses $d_r \geq 1$ servers independently and uniformly at random from the N servers, and joins the one whose queue length is the shortest among the d_r selected servers. If there is a tie, the servers with the shortest queue will be chosen randomly.

The Service Processes. The service times of each server in cluster C_r are i.i.d. and are exponential with service rate $\mu_r > 0$, and every server has an infinite waiting space. All customers in any server will be served in the first-come-first-served (FCFS) manner.

The Work Stealing Processes Within Cluster C_r**.** In cluster C_r, once one server finds no customers and enters an idle period, it chooses $b_r \geq 1$ servers independently and uniformly at random from the other $N - 1$ servers, and then steals the half of customers from one server whose queue length is the longest among the b_r selected servers. If there is a tie, the servers with the longest queue length will be chosen randomly. If each of the b_r selected servers has not enough customers, it continues to steal the half of customers from another group of b_r selected servers. We assume that the stealing time distribution from one server to another server within cluster C_r is exponential with stealing rate $\theta_r > 0$.

The Work Stealing Processes Between Any Two Clusters. Among the M clusters, each server in cluster C_s can steals the half of customers from one server in cluster C_r for $C_s \neq C_r$. When server j in cluster C_s finds no customers and enters an idle period, it first selects cluster C_r by probability $p_{r,s}$, then it chooses $a_r^{(s)} \geq 1$ servers independently and uniformly at random from the N servers in cluster C_r, and steals the half of customers from one server whose queue length is the longest among the $a_r^{(s)}$ selected servers. If each of the $a_r^{(s)}$ selected servers has not enough customers, it continue to steal the half of customers from another group of $a_r^{(s)}$ selected servers. We assume that the stealing time distribution from cluster C_r to cluster C_s is exponential with stealing rate $\tau_{r,s} > 0$.

We assume that the arrival, service and stealing processes are independent of each other.

2.2   An Empirical Measure Process

Now, we express the states of this heterogeneous work stealing model of M clusters by means of an empirical measure process.

In cluster C_r, we define $n_{r;k}^{(N)}(t)$ as the number of servers with at least k customers (the serving customer is also taken into account) at time $t \geq 0$ for $k \geq 0$ and $1 \leq r \leq M$. Clearly, $n_{r;0}^{(N)}(t) = N, 0 \leq n_{r;k}^{(N)}(t) \leq N$. Let

$$U_{r;k}^{(N)}(t) = \frac{n_{r;k}^{(N)}(t)}{NM}.$$

Then $U_{r;k}^{(N)}(t)$ is the fraction of servers with at least k customers at time t in the M clusters. Specifically, $n_{r;0}^{(N)}(t) \equiv N$ for all $t \geq 0$, this yields that $U_{r;0}^{(N)}(t) = n_{r;0}^{(N)}(t)/NM \equiv 1/M$ for all $t \geq 0$. Set

$$\mathbf{U}^{(N)}(t) = (\mathbf{U}_1^{(N)}(t), \mathbf{U}_2^{(N)}(t), \ldots, \mathbf{U}_M^{(N)}(t)),$$

where

$$\mathbf{U}_r^{(N)}(t) = (U_{r;0}^{(N)}(t), U_{r;1}^{(N)}(t), U_{r;2}^{(N)}(t), \ldots), 1 \leq r \leq M.$$

It is easy to see that for any given $t \geq 0$, $\mathbf{U}^{(N)}(t)$ is a random vector. From the exponential or Poisson assumption for the arrival, service and stealing processes,

$\{\mathbf{U}^{(N)}(t), t \geq 0\}$ is an infinite-dimensional Markov process whose state space is given by

$$\Omega^{(N)} = \{\mathbf{g} : \mathbf{g} = (\mathbf{g}_1, \mathbf{g}_2, \cdots \mathbf{g}_M), \mathbf{g}_r = (g_{r;0}, g_{r;1}, g_{r;2} \cdots),$$

$$\frac{1}{M} \geq g_{r;0} \geq g_{r;1} \geq g_{r;2} \cdots \geq 0, 1 \leq r \leq M,$$

$$1 = \sum_{r=1}^{M} g_{r;0} \geq \sum_{r=1}^{M} g_{r;1} \geq \sum_{r=1}^{M} g_{r;2} \geq \cdots \geq 0,$$

and $N g_{r;k}$ is a nonnegative integer for $k \geq 0\}$.

To analyze the Markov process $\{\mathbf{U}^{(N)}(t), t \geq 0\}$, we write

$$u_{r;k}^{(N)}(t) = E\left[U_{r;k}^{(N)}(t)\right], k \geq 0, 1 \leq r \leq M.$$

Let

$$\mathbf{u}^{(N)}(t) = \left(\mathbf{u}_1^{(N)}(t), \mathbf{u}_2^{(N)}(t), \ldots, \mathbf{u}_M^{(N)}(t)\right),$$

where

$$\mathbf{u}_r^{(N)}(t) = \left(u_{r;0}^{(N)}(t), u_{r;1}^{(N)}(t), u_{r;2}^{(N)}(t), \ldots\right), 1 \leq r \leq M.$$

3 The System of Mean-Field Equations

In this section, for this heterogeneous work stealing model of M clusters, we use a probability analysis to set up an infinite-dimensional system of mean-field equations satisfied by the expected fraction vector $\mathbf{u}^{(N)}(t)$ for $t \geq 0$.

Our computation contains the following four parts: (a) the arrival processes in cluster C_r, (b) the service processes in cluster C_r, (c) the work stealing processes in cluster C_r, and (d) the work stealing processes between any two clusters.

For (a) and (b), the computation is similar to that in Li et al. [11]. On the other hand, we only provide a detailed analysis for (c), while (d) can be discussed similarly.

Part (c). In cluster C_r, the rate that once one server finds no customers and enters an idle period, it chooses b_r servers independently and uniformly at random from the other $N - 1$ servers, and then steals the half of customers from one server whose queue length is the longest among the b_r selected servers is given by

$$(N-1)\theta_r \left[u_{r;0}^{(N)}(t) - u_{r;1}^{(N)}(t)\right] \left[u_{r;k}^{(N)}(t) - u_{r;2k}^{(N)}(t)\right]$$

$$\times \sum_{m=1}^{b_r} C_{b_r}^m \left[u_{r;2k}^{(N)}(t) - u_{r;2k+1}^{(N)}(t)\right]^m \left[1 - u_{r;2k}^{(N)}(t)\right]^{b_r - m} dt$$

$$+ (N-1)\theta_r \left[u_{r;0}^{(N)}(t) - u_{r;1}^{(N)}(t)\right] \left[u_{r;k}^{(N)}(t) - u_{r;2k-1}^{(N)}(t)\right]$$

$$\times \sum_{m=1}^{b_r} C_{b_r}^m \left[u_{r;2k-1}^{(N)}(t) - u_{r;2k}^{(N)}(t)\right]^m \left[1 - u_{r;2k-1}^{(N)}(t)\right]^{b_r - m} dt$$

$$= (N-1)\,\theta_r \left[u_{r;0}^{(N)}(t) - u_{r;1}^{(N)}(t) \right] \left[u_{r;k}^{(N)}(t) - u_{r;2k}^{(N)}(t) \right]$$

$$\times \left\{ \left[1 - u_{r;2k+1}^{(N)}(t) \right]^{b_r} - \left[1 - u_{r;2k}^{(N)}(t) \right]^{b_r} \right\} dt$$

$$+ (N-1)\,\theta_r \left[u_{r;0}^{(N)}(t) - u_{r;1}^{(N)}(t) \right] \left[u_{r;k}^{(N)}(t) - u_{r;2k-1}^{(N)}(t) \right]$$

$$\times \left\{ \left[1 - u_{r;2k}^{(N)}(t) \right]^{b_r} - \left[1 - u_{r;2k-1}^{(N)}(t) \right]^{b_r} \right\} dt.$$

The above rate computation uses a key factor that Equation $x - \langle x \rangle = k$ has only two different solutions: $x = 2k$ and $x = 2k - 1$, where $\langle x \rangle$ denotes the maximal integer part of the real number x.

Using a similar analysis to that in Sect. 2 of Li et al. [11], an infinite-dimensional system of mean-field equations is given by: For $k \geq 0, 1 \leq r \leq M$,

$$\frac{d}{dt} u_{r;k}^{(N)}(t) = \lambda_r \left\{ \left[u_{r;k-1}^{(N)}(t) \right]^{d_r} - \left[u_{r;k}^{(N)}(t) \right]^{d_r} \right\} - \mu_r \left[u_{r;k}^{(N)}(t) - u_{r;k+1}^{(N)}(t) \right]$$

$$- \frac{(N-1)}{N} \theta_r \left[u_{r;0}^{(N)}(t) - u_{r;1}^{(N)}(t) \right] \left[u_{r;k}^{(N)}(t) - u_{r;2k}^{(N)}(t) \right]$$

$$\times \left\{ \left[1 - u_{r;2k+1}^{(N)}(t) \right]^{b_r} - \left[1 - u_{r;2k}^{(N)}(t) \right]^{b_r} \right\}$$

$$- \frac{(N-1)}{N} \theta_r \left[u_{r;0}^{(N)}(t) - u_{r;1}^{(N)}(t) \right] \left[u_{r;k}^{(N)}(t) - u_{r;2k-1}^{(N)}(t) \right]$$

$$\times \left\{ \left[1 - u_{r;2k}^{(N)}(t) \right]^{b_r} - \left[1 - u_{r;2k-1}^{(N)}(t) \right]^{b_r} \right\}$$

$$- \sum_{s \neq r}^{M} p_{r,s} \tau_{r,s} \left[u_{s;0}^{(N)}(t) - u_{s;1}^{(N)}(t) \right] \left[u_{r;k}^{(N)}(t) - u_{r;2k}^{(N)}(t) \right]$$

$$\times \left\{ \left[1 - u_{r;2k+1}^{(N)}(t) \right]^{a_r^{(s)}} - \left[1 - u_{r;2k}^{(N)}(t) \right]^{a_r^{(s)}} \right\}$$

$$- \sum_{s \neq r}^{M} p_{r,s} \tau_{r,s} \left[u_{s;0}^{(N)}(t) - u_{s;1}^{(N)}(t) \right] \left[u_{r;k}^{(N)}(t) - u_{r;2k-1}^{(N)}(t) \right]$$

$$\times \left\{ \left[1 - u_{r;2k}^{(N)}(t) \right]^{a_r^{(s)}} - \left[1 - u_{r;2k-1}^{(N)}(t) \right]^{a_r^{(s)}} \right\}. \tag{1}$$

with the boundary condition

$$u_{r;0}^{(N)}(t) \equiv \frac{1}{M}, \sum_{r=1}^{M} u_{r;0}^{(N)}(t) \equiv 1, \tag{2}$$

and with the initial condition

$$\mathbf{u}^{(N)}(0) = \mathbf{g}, \tag{3}$$

where

$$\mathbf{g} \in \Omega^{(N)}.$$

4   The Fixed Point

In this section, we apply the martingale limit theory to proving the asymptotic independence of this heterogeneous work stealing model. Based on this, we analyze and compute the fixed point, which can give performance analysis of this heterogeneous work stealing model.

From the set $\Omega^{(N)}$, we write

$$\Omega_N = \left\{ \mathbf{g} \in \Omega^{(N)} : \mathbf{g}e < +\infty \right\},$$

where e is a column vector of ones with a suitable dimension in the context. Let

$$\widetilde{\Omega} = \{ \mathbf{g} : \mathbf{g} = (\mathbf{g}_1, \mathbf{g}_2, \dots \mathbf{g}_M), \mathbf{g}_r = (g_{r;0}, g_{r;1}, g_{r;2} \dots),$$

$$\frac{1}{M} \geq g_{r;0} \geq g_{r;1} \geq g_{r;2} \dots \geq 0, 1 \leq r \leq M,$$

$$1 = \sum_{r=1}^{M} g_{r;0} \geq \sum_{r=1}^{M} g_{r;1} \geq \sum_{r=1}^{M} g_{r;2} \geq \dots \geq 0 \right\}$$

and

$$\Omega = \left\{ \mathbf{g} \in \widetilde{\Omega} : \mathbf{g}e < +\infty \right\}.$$

In the vector space $\widetilde{\Omega}$, we take a metric

$$\rho\left(\mathbf{g}, \mathbf{g}'\right) = \max_{1 \leq r \leq M} \left\{ \sup_{k \geq 0} \left\{ \frac{|g_{r;k} - g'_{r;k}|}{k+1} \right\} \right\}, \mathbf{g}, \mathbf{g}' \in \widetilde{\Omega}.$$

Note that under the metric $\rho\left(\mathbf{g}, \mathbf{g}'\right)$, the vector space $\widetilde{\Omega}$ is separable and compact.

Now, we consider the Markov process $\{\mathbf{U}^{(N)}(t), t \geq 0\}$ on state space Ω_N (or $\Omega^{(N)}$ in a similar analysis). Note that the stochastic evolution of this heterogeneous work stealing model is described as the Markov process $\{\mathbf{U}^{(N)}(t), t \geq 0\}$, and

$$\frac{\mathrm{d}}{\mathrm{d}t}\mathbf{U}^{(N)}(t) = \mathbf{A}_N \, f(\mathbf{U}^{(N)}(t)),$$

where $\mathbf{A}_N$ acting on functions $f : \Omega_N \to \mathbf{C}^1$ is the generating operator of the Markov process $\{\mathbf{U}^{(N)}(t), t \geq 0\}$, and

$$\mathbf{A}_N = \sum_{r=1}^{M} \mathbf{A}_N^{(r)},$$

where

$$\mathbf{A}_N^{(r)} = \mathbf{A}_N^{(r);\text{Input}} + \mathbf{A}_N^{(r);\text{Output}} + \mathbf{A}_N^{(r);\text{WS-I}} + \mathbf{A}_N^{(r);\text{WS-II}},$$

$$\mathbf{A}_N^{(r);\text{Input}} f(\mathbf{g}) = N\lambda_r \sum_{k=1}^{\infty} \left(g_{r;k-1}^{d_r} - g_{r;k}^{d_r} \right) \left[f\left(\mathbf{g} + \frac{e_{r;k}}{N}\right) - f(\mathbf{g}) \right],$$

$$\mathbf{A}_N^{(r);\text{Output}} f(\mathbf{g}) = N\mu_r \sum_{k=1}^{\infty} \left(g_{r;k} - g_{r;k+1} \right) \left[f\left(\mathbf{g} - \frac{e_{r;k}}{N}\right) - f(\mathbf{g}) \right],$$

$$\mathbf{A}_N^{(r);\text{WS-I}} f(\mathbf{g}) = (N-1)\left(g_{r;0} - g_{r;1}\right)\theta_r \sum_{k=1}^{\infty} \left(g_{r;k} - g_{r;2k}\right)$$
$$\times \left[\left(1 - g_{r;2k+1}\right)^{b_r} - \left(1 - g_{r;2k}\right)^{b_r} \right]\left[f\left(\mathbf{g} - \frac{e_{r;2k}}{N} + \frac{e_{r;k}}{N}\right) - f(\mathbf{g}) \right]$$
$$+ (N-1)\left(g_{r;0} - g_{r;1}\right)\theta_r \sum_{k=1}^{\infty} \left(g_{r;k} - g_{r;2k-1}\right)$$
$$\times \left[\left(1 - g_{r;2k}\right)^{b_r} - \left(1 - g_{r;2k-1}\right)^{b_r} \right]\left[f\left(\mathbf{g} - \frac{e_{r;2k-1}}{N} + \frac{e_{r;k}}{N}\right) - f(\mathbf{g}) \right]$$

$$\mathbf{A}_N^{(r);\text{WS-II}} f(\mathbf{g}) = N\sum_{\substack{s \neq r}}^{M} p_{r,s} \mathcal{T}_{r,s} \left(g_{s;0} - g_{s;1}\right) \sum_{k=1}^{\infty} \left(g_{r;k} - g_{r;2k}\right)$$
$$\times \left[\left(1 - g_{r;2k+1}\right)^{a_r^{(s)}} - \left(1 - g_{r;2k}\right)^{a_r^{(s)}} \right]\left[f\left(\mathbf{g} - \frac{e_{r;2k}}{N} + \frac{e_{r;k}}{N}\right) - f(\mathbf{g}) \right]$$
$$+ N\sum_{\substack{s \neq r}}^{M} p_{r,s} \mathcal{T}_{r,s} \left(g_{s;0} - g_{s;1}\right) \sum_{k=1}^{\infty} \left(g_{r;k} - g_{r;2k-1}\right)$$
$$\times \left[\left(1 - g_{r;2k}\right)^{a_r^{(s)}} - \left(1 - g_{r;2k-1}\right)^{a_r^{(s)}} \right]\left[f\left(\mathbf{g} - \frac{e_{r;2k-1}}{N} + \frac{e_{r;k}}{N}\right) - f(\mathbf{g}) \right]$$

Let

$$\mathbf{A} = \lim_{N\to\infty} \mathbf{A}_N, \quad \mathbf{A}^{(r)} = \lim_{N\to\infty} \mathbf{A}_N^{(r)}; \quad \mathbf{A}^{(r);\clubsuit} = \lim_{N\to\infty} \mathbf{A}_N^{(r);\clubsuit}.$$

Then

$$\mathbf{A} f(\mathbf{g}) = \lambda_r \sum_{r=1}^{M} \sum_{k=1}^{\infty} \left(g_{r;k-1}^{d_r} - g_{r;k}^{d_r} \right) \frac{\partial}{\partial g_{r;k}} f(\mathbf{g})$$
$$- \mu_r \sum_{r=1}^{M} \sum_{k=1}^{\infty} \left(g_{r;k} - g_{r;k+1} \right) \frac{\partial}{\partial g_{r;k}} f(\mathbf{g})$$
$$- \sum_{r=1}^{M} \theta_r \left(g_{r;0} - g_{r;1} \right) \sum_{k=1}^{\infty} \left(g_{r;k} - g_{r;2k} \right)$$
$$\times \left[\left(1 - g_{r;2k+1}\right)^{b_r} - \left(1 - g_{r;2k}\right)^{b_r} \right]\left[\frac{\partial}{\partial g_{r;2k}} f(g_r) - \frac{\partial}{\partial g_{r;k}} f(g_r) \right]$$

$$- \sum_{r=1}^{M} \theta_r \left(g_{r;0} - g_{r;1} \right) \sum_{k=1}^{\infty} \left(g_{r;k} - g_{r;2k-1} \right)$$

$$\times \left[\left(1 - g_{r;2k} \right)^{b_r} - \left(1 - g_{r;2k-1} \right)^{b_r} \right] \left[\frac{\partial}{\partial g_{r;2k-1}} f\left(g_r \right) - \frac{\partial}{\partial g_{r;k}} f\left(g_r \right) \right]$$

$$- \sum_{r=1}^{M} \sum_{s \neq r}^{M} p_{r,s} \tau_{r,s} \left(g_{s;0} - g_{s;1} \right) \sum_{k=1}^{\infty} \left(g_{r;k} - g_{r;2k} \right)$$

$$\times \left[\left(1 - g_{r;2k+1} \right)^{a_r^{(s)}} - \left(1 - g_{r;2k} \right)^{a_r^{(s)}} \right] \left[\frac{\partial}{\partial g_{r;2k}} f\left(g_r \right) - \frac{\partial}{\partial g_{r;k}} f\left(g_r \right) \right]$$

$$- \sum_{r=1}^{M} \sum_{s \neq r}^{M} p_{r,s} \tau_{r,s} \left(g_{s;0} - g_{s;1} \right) \sum_{k=1}^{\infty} \left(g_{r;k} - g_{r;2k-1} \right)$$

$$\times \left[\left(1 - g_{r;2k} \right)^{a_r^{(s)}} - \left(1 - g_{r;2k-1} \right)^{a_r^{(s)}} \right] \left[\frac{\partial}{\partial g_{r;2k-1}} f\left(g_r \right) - \frac{\partial}{\partial g_{r;k}} f\left(g_r \right) \right].$$

Now, we consider the limiting behavior of the sequence $\{\mathbf{U}^{(N)}(t), t \geq 0\}$ of Markov processes. Let $\mathbf{U}(t) = \lim_{N \to \infty} \mathbf{U}^{(N)}(t)$ and $\mathbf{u}(t) = \lim_{N \to \infty} \mathbf{u}^{(N)}(t)$. Then the Markov process $\{\mathbf{U}^{(N)}(t), t \geq 0\}$ with generating operator $\mathbf{A}_N$ uniformly converges on any finite time interval to the limiting Markov process $\{\mathbf{U}(t), t \geq 0\}$ with generating operator $\mathbf{A}$.

As $N \to \infty$ it follows from the system of mean-field Eqs. (1) to (3) that $\mathbf{u}(t)$ is a solution to the following system of limiting mean-field equations: For $k \geq 0, 1 \leq r \leq M$,

$$\frac{\mathrm{d}}{\mathrm{d}t} u_{r;k}(t) = \lambda_r \left\{ [u_{r;k-1}(t)]^{d_r} - [u_{r;k}(t)]^{d_r} \right\} - \mu_r [u_{r;k}(t) - u_{r;k+1}(t)]$$

$$- \theta_r [u_{r;0}(t) - u_{r;1}(t)] [u_{r;k}(t) - u_{r;2k}(t)]$$

$$\times \left\{ [1 - u_{r;2k+1}(t)]^{b_r} - [1 - u_{r;2k}(t)]^{b_r} \right\}$$

$$- \theta_r [u_{r;0}(t) - u_{r;1}(t)] [u_{r;k}(t) - u_{r;2k-1}(t)]$$

$$\times \left\{ [1 - u_{r;2k}(t)]^{b_r} - [1 - u_{r;2k-1}(t)]^{b_r} \right\}$$

$$- \sum_{s \neq r}^{M} p_{r,s} \tau_{r,s} [u_{s;0}(t) - u_{s;1}(t)] [u_{r;k}(t) - u_{r;2k}(t)]$$

$$\times \left\{ [1 - u_{r;2k+1}(t)]^{a_r^{(s)}} - [1 - u_{r;2k}(t)]^{a_r^{(s)}} \right\}$$

$$- \sum_{s \neq r}^{M} p_{r,s} \tau_{r,sr,s} [u_{s;0}(t) - u_{s;1}(t)] [u_{r;k}(t) - u_{r;2k-1}(t)]$$

$$\times \left\{ [1 - u_{r;2k}(t)]^{a_r^{(s)}} - [1 - u_{r;2k-1}(t)]^{a_r^{(s)}} \right\}. \tag{4}$$

with the boundary condition

$$u_{r;0}(t) \equiv \frac{1}{M}, \quad \sum_{r=1}^{M} u_{r;0}(t) \equiv 1, \tag{5}$$

and with the initial condition

$$\mathbf{u}(0) = \mathbf{g}, \, \mathbf{g} \in \Omega. \tag{6}$$

Note that the convergence in the Skorohod topology means the convergence in distribution for the Skorohod topology on the space of trajectories. The following theorem applies the martingale limit theory to studying the weak convergence of the sequence $\{\mathbf{U}^{(N)}(t), t \geq 0\}$ of Markov processes as N tends to infinity.

Theorem 1. *If* $\mathbf{U}^{(N)}(0)$ *converges weakly to* $\mathbf{g} \in \Omega$ *as* N *tends to infinity, then the sequence* $\{\mathbf{U}^{(N)}(t), t \geq 0\}$ *of Markov processes converges in the Skorohod topology to the unique and global solution to the system of differential Eqs. (4) to (6).*

Proof: From the martingale characterization of the Markov jump process $\{\mathbf{U}^{(N)}(t), t \geq 0\}$, it follows from Rogers and Williams [18] that for $k \geq 0$ and $1 \leq r \leq M$,

$$M_{r;k}^{(N)}(t) = U_{r;k}^{(N)}(t) - U_{r;k}^{(N)}(0)$$

$$- \int_0^t \sum_{\Theta \in \Omega - \{\mathbf{U}^{(N)}(t)\}} \mathcal{Q}^{(N)}\left(\mathbf{U}^{(N)}(s), \Theta\right) \left[\Theta_{r;k} - U_{r;k}^{(N)}(s)\right] ds$$

is a martingale with respect to the natural filtration associated to the Poisson processes involved in the arrival, service and work stealing processes, where $\mathcal{Q}^{(N)}\left(\mathbf{U}^{(N)}(s), \Theta\right)$ is the Q-matrix of the Markov jump process $\{\mathbf{U}^{(N)}(t), t \geq 0\}$. Using a similar method to Darling and Norris [6], it is easy to see that if $\mathbf{U}^{(N)}(0)$ converges weakly to $\mathbf{g} \in \Omega$ as N tends to infinity, then the sequence $\{\mathbf{U}^{(N)}(t), t \geq 0\}$ of Markov processes is tight for the Skorohod topology, and any limit $\mathbf{U}(t)$ of $\{\mathbf{U}^{(N)}(t), t \geq 0\}$ asymptotically approaches a single trajectory identified by the unique and global solution to the system of differential Eqs. (4) to (6). This completes the proof. ∎

In what follows we analyze and compute the fixed point of the infinite-dimensional system of limiting mean-field Eqs. (4) to (6), and then can give performance analysis of this heterogeneous work stealing model of M clusters.

Let $\pi = (\pi_1, \pi_2, \ldots \pi_M)$ where $\pi_r = (\pi_{r;0}, \pi_{r;1}, \pi_{r;2}, \ldots)$. A row vector π is called a fixed point of the infinite-dimensional system of limiting mean-field Eqs. (4) to (6) if $\pi = \lim_{t \to +\infty} \mathbf{u}(t)$ or $\pi_{r;k} = \lim_{t \to +\infty} u_{r;k}(t)$ for $k \geq 0$ and $1 \leq r \leq M$. It is well-known that if π is a fixed point with respect to the vector $\mathbf{u}(t)$, then

$$\lim_{t \to +\infty} \left[\frac{d}{dt}\mathbf{u}(t)\right] = 0.$$

Taking $t \to +\infty$ in both sides of the system of limiting mean-field Eqs. (4) to (6), we obtain that for $k \geq 1, 1 \leq r \leq M$,

$$\lambda_r (\pi_{r;k-1}^{d_r} - \pi_{r;k}^{d_r}) - \mu_r \left(\pi_{r;k} - \pi_{r;k+1} \right)$$

$$- \theta_r \left(\pi_{r;0} - \pi_{r;1} \right) \left(\pi_{r;k} - \pi_{r;2k} \right) \left[(1 - \pi_{r;2k+1})^{b_r} - (1 - \pi_{r;2k})^{b_r} \right]$$

$$- \theta_r \left(\pi_{r;0} - \pi_{r;1} \right) \left(\pi_{r;k} - \pi_{r;2k-1} \right) \left[(1 - \pi_{r;2k})^{b_r} - (1 - \pi_{r;2k-1})^{b_r} \right]$$

$$- \sum_{s \neq r}^{M} p_{r,s} \tau_{r,s} \left(\pi_{s;0} - \pi_{s;1} \right) \left(\pi_{r;k} - \pi_{r;2k} \right) \left[(1 - \pi_{r;2k+1})^{a_r^{(s)}} - (1 - \pi_{r;2k})^{a_r^{(s)}} \right]$$

$$- \sum_{s \neq r}^{M} p_{r,s} \tau_{r,s} \left(\pi_{s;0} - \pi_{s;1} \right) \left(\pi_{r;k} - \pi_{r;2k-1} \right) \left[(1 - \pi_{r;2k})^{a_r^{(s)}} - (1 - \pi_{r;2k-1})^{a_r^{(s)}} \right] = 0, \quad (7)$$

with the boundary condition

$$\pi_{r;0} = \frac{1}{M}, \quad \sum_{r=1}^{M} \pi_{r;0} = 1. \tag{8}$$

In general, it is not easy to compute the fixed point π when considering the M clusters. Using the fixed point, now we can compute the mean of stationary queue length in any server in this heterogeneous work stealing model of M clusters.

Let Q_r and Q be the stationary queue lengths of any servers in cluster C_r and in this system, respectively. Then

$$E[Q_r] = \sum_{k=1}^{\infty} P\{Q_r \geq k\} = \sum_{k=1}^{\infty} \pi_{r,k}$$

and

$$E[Q] = \sum_{r=1}^{M} E[Q_r] = \sum_{r=1}^{M} \sum_{k=1}^{\infty} \pi_{r,k}.$$

5   Concluding Remarks

This paper discusses a heterogeneous work stealing model of M clusters under two types of work stealing schemes. We first set up an infinite-dimensional system of mean-field equations, which is related to the M clusters. Then we apply the martingale limit theory to proving the asymptotic independence of the heterogeneous work stealing model. Based on this, we analyze and compute the fixed point, which can give performance analysis of this heterogeneous work stealing model. Along such a line, there are a number of interesting directions for potential future research, for example:

- Providing effective algorithms for computing the fixed points of the heterogeneous work stealing model of M clusters;
- studying the stability or metastability of this heterogeneous work stealing model of M clusters; and
- analyzing influence of the M clusters on performance measures of this heterogeneous work stealing model.

Acknowledgement. This work is partly supported by the National Natural Science Foundation of China under grant (#71271187, #71471160), and the Fostering Plan of Innovation Team and Leading Talent in Hebei Universities under grant (# LJRC027).

References

1. Anselmi, J., Gaujal, B.: Performance evaluation of a work stealing algorithm for streaming applications. In: Abdelzaher, T., Raynal, M., Santoro, N. (eds.) OPODIS 2009. LNCS, vol. 5923, pp. 1–12. Springer, Heidelberg (2009)
2. Benaim, M., Le Boudec, J.Y.: A class of mean field interaction models for computer and communication systems. Perform. Eval. **65**(11), 823–838 (2008)
3. Berenbrink, P., Friedetzky, T., Goldberg, L.: The natural work-stealing algorithm is stable. SIAM J. Comput. **32**(5), 1260–1279 (2003)
4. Blumofe, R.D., Leiserson, C.E.: Scheduling multithreaded computations by work stealing. J. ACM **46**(5), 720–748 (1999)
5. Blumofe, R.D., Papadopoulos, D.: The performance of work stealing in multiprogrammed environments. ACM SIGMETRICS Perform. Eval. Rev. **26**(1), 266–267 (1998)
6. Darling, R.W.R., Norris, J.R.: Structure of large random hypergraphs. Ann. Appl. Probab. **15**(1), 125–152 (2005)
7. Gast, N., Gaujal, B.: A mean field model of work stealing in large-scale systems. ACM SIGMETRICS Perform. Eval. Rev. **38**(1), 13–24 (2010)
8. Harchol-Balter, M., Li, C., Osogami, T., Scheller-Wolf, A., Squillante, M.S.: Analysis of task assignment with cycle stealing under central queue. In: The 23rd International Conference of IEEE on Distributed Computing Systems, pp. 628–637 (2003)
9. Hendler, D., Shavit, N.: Non-blocking steal-half work queues. In: Proceedings of the Twenty-First Annual Symposium on Principles of Distributed Computing, ACM, pp. 280–289 (2002)
10. Li, Q.L.: Nonlinear Markov processes in big networks, pp. 1–23 (2015). arXiv:1504.07974
11. Li, Q.L., Dai, G., Lui, J.C.S., Wang, Y.: The mean-field computation in a supermarket model with server multiple vacations. Discrete Event Dyn. Syst. **24**(4), 473–522 (2014)
12. Li, Q.L., Lui, J.C.S.: Block-structured supermarket models. Discrete Event Dynamic Systems, 1–36, 29 June 2014
13. Lu, Y., Xie, Q., Kliot, G., Geller, A., Larus, J.R., Greenberg, A.: Join-Idle-Queue: a novel load balancing algorithm for dynamically scalable web services. Perform. Eval. **68**(11), 1056–1071 (2011)
14. Minnebo, W., Van Houdt, B.: A fair comparison of pull and push strategies in large distributed networks. IEEE/ACM Trans. Netw. **22**(3), 996–1006 (2014)
15. Mitzenmacher, M.D.: The power of two choices in randomized load balancing. Ph.D. thesis, Department of Computer Science, University of California at Berkeley, USA (1996)
16. Mitzenmacher, M.D.: Analyses of load stealing models based in di erential equations. In: The 10th ACM Symposium on Parallel Algorithms and Architectures, pp. 212–221 (1998)
17. Osogami, T., Harchol-Balter, M., Scheller-Wolf, A.: Analysis of cycle stealing with switching cost. J. ACM **31**(1), 184–195 (2003)
18. Rogers, L.C.G., Williams, D.: Diffusions, Markov Processes, and Martingales 2: Itô Calculus. Wiley, New York (1987)

19. Squillante, M.S.: Stochastic analysis of multiserver systems. ACM SIGMETRICS Perform. Eval. Rev. **34**(4), 44–51 (2007)
20. Squillante, M.S., Nelson, R.: Analysis of task migration in shared-memory multiprocessor scheduling. J. ACM **19**(1), 143–155 (1991)
21. Vvedenskaya, N.D., Dobrushin, R.L., Karpelevich, F.I.: Queueing system with selection of the shortest of two queues: an asymptotic approach. Probl. Inf. Transmissions **32**(1), 20–34 (1996)

Joint Probability Density Function of Modulated Synchronous Flow Interval Duration Under Conditions of Fixed Dead Time

Alexander Gortsev and Mariya Sirotina[✉]

National Research Tomsk State University, Tomsk, Russian Federation
mashuliagol@mail.ru

Abstract. A modulated synchronous doubly stochastic flow under conditions of a fixed dead time is considered. After each registered event there is a time of fixed duration T (dead time), during which another flow events are inaccessible for observation. When duration of the dead time period finishes, the first happened event creates the dead time period of duration T again and etc. An explicit form of a probability density function of interval duration between two adjacent events of modulated synchronous doubly stochastic flow under conditions of a fixed dead time is derived. Also an explicit form of a joint probability density function for modulated synchronous flow interval duration is obtained. A recurrent conditions for modulated synchronous flow as well as some probabilistic characteristics of the flow are obtained using the formula for a joint probability density function.

Keywords: Modulated synchronous doubly stochastic flow · Probability density function of interval duration · Joint probability density function of interval duration · Recurrent conditions of the flow

1 Introduction

This paper is a continuation of the modulated synchronous flow investigation which was started in paper [1].

Mathematical models of queueing theory are widely used when describing the real physical, technological and other processes and systems. In connection with rapid development of computer equipment and information technologies an important sphere of queueing theory applications appeared. This sphere was called as design and creation of data-processing networks, computer communication networks, satellite networks and telecommunication networks [2].

In the practice, an intensity of input flow varies along with time. Moreover, these variations are often of a random nature. This leads to consideration of a doubly stochastic flow of events [3–7]. An example of such flow is a modulated synchronous doubly stochastic flow [8,9].

© Springer International Publishing Switzerland 2015
A. Dudin et al. (Eds.): ITMM 2015, CCIS 564, pp. 41–52, 2015.
DOI: 10.1007/978-3-319-25861-4_4

2   Problem Statement

Let us consider the modulated synchronous doubly stochastic flow of events, whose rate is a piecewise constant random process $\lambda(t)$ with two states: λ_1, λ_2 ($\lambda_1 > \lambda_2$). The sojourn time of the process $\lambda(t)$ in state λ_i has exponential probability distribution function with the parameter $\alpha_i, i = 1, 2$. If at the moment t the process $\lambda(t)$ sojourns in the state λ_i than in the small half-interval $[t, t + \Delta t)$, with probability $\alpha_i \Delta t + o(\Delta t)$ the process finishes its stay in the state λ_i and moves to the state λ_j with probability is one $(i, j = 1, 2, i \neq j)$. During the time random interval when $\lambda(t) = \lambda_i$ Poisson flow with rate $\lambda_i, i = 1, 2$ arrives. A state transition of the process $\lambda(t)$ may also occur at the moment of Poisson flow event arrival. Moreover, transition from the state λ_1 to the state λ_2 is realized only at the moment of event occurrence with probability $p\,(0 < p \leq 1)$. With the complementary probability $1 - p$ the process remains at the state λ_1. Transition from the state λ_2 to the state λ_1 is also realized only at the moment of event occurrence with probability $q\,(0 < q \leq 1)$. With the complementary probability $1 - q$ the process remains at the state λ_2. In the described conditions $\lambda(t)$ is the Markovian process.

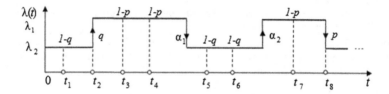

Fig. 1. Forming the modulated synchronous flow

An example of this situation is shown on the Fig. 1, where λ_1, λ_2 are the states of the process $\lambda(t)$, $t_1, t_2, \ldots$ are the moments of the flow events occurrence.

Block matrixes of infinitesimal coefficients are of the form:

$$D_1 = \begin{vmatrix} (1-p)\lambda_1 & p\lambda_1 \\ q\lambda_2 & (1-q)\lambda_2 \end{vmatrix}, D_0 = \begin{vmatrix} -(\lambda_1 + \alpha_1) & \alpha_1 \\ \alpha_2 & -(\lambda_2 + \alpha_2) \end{vmatrix}.$$

The elements of the matrix D_1 are intensities of the process $\lambda(t)$ transition from the state to the state with an event occurrence. Off-diagonal elements of matrix D_0 are intensities of the process $\lambda(t)$ transition from the state to the state without an event occurrence. Diagonal elements of matrix D_0 are the intensities of process $\lambda(t)$ leaving its states, which are taken with the opposite sign. We should note that if $\alpha_i = 0, i = 1, 2$ there is a usual synchronous flow of events [10].

Emphasize that we assume a priority of the event occurrence in the problem statement. Event occurs and after that the process $\lambda(t)$ makes a transition from the state to state. This circumstance is irrelevant when obtaining the analytical results because event occurrence and the process $\lambda(t)$ state transition happen immediately. When obtaining the numerical results using a simulation modeling, there is necessary to have the definiteness what is the first: the event occurrence or the state transition.

The process $\lambda(t)$ and event types (Poisson flow events of λ_1 and λ_2 intensities) are not observable in principle (in prepositions made $\lambda(t)$ is latent markovian process). Only the moments of observable events occurrence $t_1, t_2, \ldots t_k, \ldots$ are accessible for observation. A stationary mode of the flow is considered. A sequence of time moments of event occurrence $t_1, t_2, \ldots t_k, \ldots$ is an imbedded Markov chain $\{\lambda(t_k)\}$. So the flow has a markovian chain characteristic when its evolution is considered from the moment t_k, $k = 1, 2, \ldots$ (the moment of flow event occurrence).

Let $\tau_k = t_{k+1} - t_k$, $k = 1, 2, \ldots$, is a value of a k interval duration between the moments of an adjacent observable events occurrence. Since the flow functions in stationary mode then a probability density function of interval duration between the moments of adjacent observable events occurence is $p_T(\tau_k) = p_T(\tau)$, $\tau \geq 0$, for any k. So without the loss of generality the moment t_k of event occurence we can assume equal to null, that means $\tau = 0$.

Let $\tau_k = t_{k+1} - t_k, \tau_{k+1} = t_{k+2} - t_{k+1}$, $k = 1, 2, \ldots$, are a duration values of adjacent k and $k+1$ intervals between adjacent observable flow events occurrence. According to the flow stationarity we can assume $k = 1$ and consider $\tau_1 = t_2 - t_1, \tau_2 = t_3 - t_2, \tau_1 \geq 0, \tau_2 \geq 0$ interval durations. Then $\tau_1 = 0$ responds to the moment t_1 of observable flow event occurrence and $\tau_2 = 0$ responds to the moment t_2 of an observable flow event occurrence. Then corresponding joint probability density function is of the form $p_T(\tau_k, \tau_{k+1}) = p_T(\tau_1, \tau_2), \tau_1 \geq 0, \tau_2 \geq 0$.

The main purpose of the research work is to obtain an explicit form of a probability density function $p_T(\tau)$ as well as an explicit form of a joint probability density function $p_T(\tau_1, \tau_2)$. During the investigation recurrense conditions of observable flow are obtained as well as some probabilistic characteristic of the flow.

3   Probability Density Function of Modulated Synchronous Flow Interval Duration

Let τ is a value of the random variable of interval duration between the moments of two adjacent flow events occurrence. Then a probability density function of interval duration between the moments of modulated synchronous flow adjacent event occurrence is written in the form of:

$$p_T(\tau) = \begin{cases} 0, 0 \leq \tau < T, \\ \sum_{i=1}^{2} \pi_i(0|T) \sum_{j=1}^{2} q_{ij}(T) \sum_{k=1}^{2} \widetilde{p}_{jk}(\tau - T), \tau \geq T \end{cases} \tag{1}$$

where $\pi_i(0|T)$, $i = 1, 2$, is a conditional stationary probability that the process $\lambda(t)$ sojourns in the state λ_i at the moment $\tau = 0$ when the flow event occurred in the moment $\tau = 0$, and the dead time of T duration occurred, $i = 1, 2$, $(\pi_1(0|T) + \pi_2(0|T) = 1)$; $q_{ij}(T)$, $i, j = 1, 2$, is a probability that during the time of T duration the process $\lambda(t)$ moves from the state λ_i to the state λ_j, $i, j = 1, 2$;

$\widetilde{p}_{jk}(\tau - T)$ is a probability density function that on the interval $(\tau - T, \tau)$ there are no flow events and in the moment τ the process $\lambda(\tau) = \lambda_k$ when at the moment $\tau - T$ the process $\lambda(\tau - T) = \lambda_j$ $(j, k = 1, 2)$.

Let $t = \tau - T, \tau \geq T$. Let us introduce $p_{jk}(t)$, $j, k = 1, 2$, is a transition probability that there are no flow events on the interval $(0; t)$ and $\lambda(t) = \lambda_k$ at the moment t when $\lambda(0) = \lambda_j$ at the moment $t = 0$ $(j, k = 1, 2)$. Then for introduced probabilities $p_{jk}(t)$, $j, k = 1, 2$, we have the following system of differencial equations:

$$
\begin{aligned}
p'_{11}(t) &= -(\lambda_1 + \alpha_1)p_{11}(t) + \alpha_2 p_{12}(t), \\
p'_{12}(t) &= -(\lambda_2 + \alpha_2)p_{12}(t) + \alpha_1 p_{11}(t), \\
p'_{22}(t) &= -(\lambda_2 + \alpha_2)p_{22}(t) + \alpha_1 p_{21}(t), \\
p'_{21}(t) &= -(\lambda_1 + \alpha_1)p_{21}(t) + \alpha_2 p_{22}(t), \\
p_{11}(0) &= 1, \ p_{12}(0) = 0, \ p_{22}(0) = 1, \ p_{21}(0) = 0.
\end{aligned}
$$

Solving obtained system of differential equations we find out that

$$
\begin{aligned}
p_{11}(t) &= \tfrac{1}{z_2 - z_1}\left[(\lambda_2 + \alpha_2 - z_1)e^{-z_1 t} - (\lambda_2 + \alpha_2 - z_2)e^{-z_2 t}\right], \qquad (2) \\
p_{12}(t) &= \tfrac{\alpha_1}{z_2 - z_1}\left[e^{-z_1 t} - e^{-z_2 t}\right], \\
p_{22}(t) &= \tfrac{1}{z_2 - z_1}\left[(\lambda_1 + \alpha_1 - z_1)e^{-z_1 t} - (\lambda_1 + \alpha_1 - z_2)e^{-z_2 t}\right], \\
p_{21}(\tau) &= \tfrac{\alpha_2}{z_2 - z_1}\left[e^{-z_1 t} - e^{-z_2 t}\right], \\
z_{1,2} &= (\lambda_1 + \alpha_1 + \lambda_2 + \alpha_2) \mp \sqrt{(\lambda_1 + \alpha_1 - \lambda_2 - \alpha_2)^2 + 4\alpha_1\alpha_2}.
\end{aligned}
$$

Besides probability densities $\widetilde{p}_{jk}(t) = \widetilde{p}_{jk}(\tau - T)$, $j, k = 1, 2$, from the formula for the probability density $p_T(\tau)$ 1 are of the form:

$$
\begin{aligned}
\widetilde{p}_{11}(\tau) &= p_{11}(\tau)\lambda_1(1 - p) + p_{12}(\tau)\lambda_2 q, \qquad (3) \\
\widetilde{p}_{12}(\tau) &= p_{11}(\tau)\lambda_1 p + p_{12}(\tau)\lambda_2(1 - q), \\
\widetilde{p}_{22}(\tau) &= p_{22}(\tau)\lambda_2(1 - q) + p_{21}(\tau)\lambda_1 p, \\
\widetilde{p}_{21}(\tau) &= p_{22}(\tau)\lambda_2 q + p_{21}(\tau)\lambda_1(1 - p),
\end{aligned}
$$

where probabilities $p_{ij}(\tau)$, $i, j = 1, 2$, are defined in 2.

Inserting 2 into 3 then changing t on $\tau - T$ we can obtaining an explicit form of the probability densities $\widetilde{p}_{jk}(\tau - T)$, $j, k = 1, 2$:

$$
\begin{aligned}
\widetilde{p}_{11}(\tau - T) &= \tfrac{\lambda_1(1-p)}{z_2 - z_1}\left[(\lambda_2 + \alpha_2 - z_1)e^{-z_1(\tau - T)} - (\lambda_2 + \alpha_2 - z_2)e^{-z_2(\tau - T)}\right] \\
&\quad + \tfrac{\alpha_1 \lambda_2 q}{z_2 - z_1}\left[e^{-z_1(\tau - T)} - e^{-z_2(\tau - T)}\right], \\
\widetilde{p}_{12}(\tau - T) &= \tfrac{\lambda_1 p}{z_2 - z_1}\left[(\lambda_2 + \alpha_2 - z_1)e^{-z_1(\tau - T)} - (\lambda_2 + \alpha_2 - z_2)e^{-z_2(\tau - T)}\right] \\
&\quad + \tfrac{\alpha_1 \lambda_2 (1-q)}{z_2 - z_1}\left[e^{-z_1(\tau - T)} - e^{-z_2(\tau - T)}\right], \\
\widetilde{p}_{21}(\tau - T) &= \tfrac{\lambda_2 q}{z_2 - z_1}\left[(\lambda_1 + \alpha_1 - z_1)e^{-z_1(\tau - T)} - (\lambda_1 + \alpha_1 - z_2)e^{-z_2(\tau - T)}\right] \\
&\quad + \tfrac{\alpha_2 \lambda_1 (1-p)}{z_2 - z_1}\left[e^{-z_1(\tau - T)} - e^{-z_2(\tau - T)}\right], \\
\widetilde{p}_{22}(\tau - T) &= \tfrac{\lambda_2(1-q)}{z_2 - z_1}\left[(\lambda_1 + \alpha_1 - z_1)e^{-z_1(\tau - T)} - (\lambda_1 + \alpha_1 - z_2)e^{-z_2(\tau - T)}\right] \\
&\quad + \tfrac{\alpha_2 \lambda_1 p}{z_2 - z_1}\left[e^{-z_1(\tau - T)} - e^{-z_2(\tau - T)}\right],
\end{aligned}
$$

$$(4)$$

where $z_{1,2}$ are defined in 2.

Lets introduce probability $q_{ij}(\tau)$ is a probability that during the time of τ duration the process $\lambda(t)$ moves from the state λ_i to the state λ_j, $i, j = 1, 2$. For the introduced probabilities $q_{ij}(\tau)$, $i, j = 1, 2$, we have the following system of differential equations:

$$q'_{11}(\tau) = -(\alpha_1 + p\lambda_1)q_{11}(\tau) + (\alpha_2 + q\lambda_2)q_{12}(\tau),$$
$$q'_{12}(\tau) = -(\alpha_2 + q\lambda_2)q_{12}(\tau) + (\alpha_1 + p\lambda_1)q_{12}(\tau),$$
$$q'_{21}(\tau) = -(\alpha_2 + q\lambda_2)q_{22}(\tau) + (\alpha_1 + p\lambda_1)q_{21}(\tau),$$
$$q'_{22}(\tau) = -(\alpha_1 + p\lambda_1)q_{21}(\tau) + (\alpha_2 + q\lambda_2)q_{22}(\tau),$$

with a boundary conditions:

$$q_{11}(0) = 1, \ q_{12}(0) = 0, \ q_{21}(0) = 0, \ q_{22}(0) = 1.$$

Solving this system of differential equations and assuming $\tau = T$ in obtained solution we found out:

$$q_{11}(T) = \pi_2 + \pi_1 e^{-(\alpha_1 + p\lambda_1 + \alpha_2 + q\lambda_2)T}, \ q_{12}(T) = \pi_2 - \pi_2 e^{-(\alpha_1 + p\lambda_1 + \alpha_2 + q\lambda_2)T},$$
$$q_{21}(T) = \pi_1 - \pi_1 e^{-(\alpha_1 + p\lambda_1 + \alpha_2 + q\lambda_2)T}, \ q_{22}(T) = \pi_1 + \pi_2 e^{-(\alpha_1 + p\lambda_1 + \alpha_2 + q\lambda_2)T},$$
$$\pi_1 = \frac{\alpha_2 + q\lambda_2}{\alpha_1 + p\lambda_1 + \alpha_2 + q\lambda_2}, \ \pi_1 = \frac{\alpha_1 + p\lambda_1}{\alpha_1 + p\lambda_1 + \alpha_2 + q\lambda_2},$$

$$(5)$$

where π_i, $i = 1, 2$, is an apriori stationary probability that in the random time moment the flow sojourns in the state λ_i [2].

To find out probabilities $\pi_i(0|T)$, $i = 1, 2$, from the initial formula 1 let us introduce π_{ij}, $i, j = 1, 2$, is a probability that during the time period from the moment $\tau = 0$ to the moment of the next observable flow event occurrence and realizing the next flow state drawing the process $\lambda(t)$ moves from the state λ_i to the state λ_j, $i, j = 1, 2$.

Then for the introduced probabilities $\pi_i(0|T)$, $i = 1, 2$, and π_{ij}, $i, j = 1, 2$, according to the markovian property of the process $\lambda(t)$ it is correct to write the following system of linear equations:

$$\pi_1(0|T) = \pi_1(0|T)\pi_{11} + \pi_2(0|T)\pi_{21},$$
$$\pi_2(0|T) = \pi_2(0|T)\pi_{22} + \pi_1(0|T)\pi_{12},$$
$$\pi_1(0|T) + \pi_2(0|T) = 1,$$

where expressing probabilities $\pi_i(0|T)$, $i = 1, 2$, we obtain:

$$\pi_1(0|T) = \frac{\pi_{21}}{\pi_{12} + \pi_{21}}, \ \pi_2(0|T) = \frac{\pi_{12}}{\pi_{12} + \pi_{21}}. \qquad (6)$$

Besides for the probabilities π_{ij}, $i, j = 1, 2$, according to the markovian property of the process $\lambda(t)$, we can write the following system of linear equations:

$$\pi_{11} = q_{11}(T)p_{11} + q_{12}(T)p_{21}, \qquad (7)$$
$$\pi_{12} = q_{11}(T)p_{12} + q_{12}(T)p_{22},$$
$$\pi_{11} + \pi_{12} = 1,$$
$$\pi_{21} = q_{21}(T)p_{11} + q_{22}(T)p_{21},$$
$$\pi_{22} = q_{21}(T)p_{12} + q_{22}(T)p_{22},$$
$$\pi_{21} + \pi_{22} = 1,$$

where probabilities $q_{ij}(T)$, $i, j = 1, 2$, are defined in 5, p_{jk}, $j, k = 1, 2$, is a probability that during the time interval between the moment $t = 0$ (the moment of event occurrence) and the moment of the next event occurrence the process $\lambda(t)$ moves from the state λ_j to the state λ_k, $j, k = 1, 2$. The probability p_{ij} doesn't depend on time, it is a stationary probability of the process $\lambda(t)$ transition from the state λ_j to the state λ_k between the moments of two adjacent flow events. Then $p_{jk} = \int_0^\infty \widetilde{p}_{jk}(u)\Delta u$.

Integrating the probability density functions $\widetilde{p}_{jk}(t)$, $j, k = 1, 2$, obtained in 4, the following expressions for the transition probabilities p_{jk}, $j, k = 1, 2$, are derived:

$$p_{11} = \frac{\lambda_1(1-p)(\lambda_2+\alpha_2)+\lambda_2 q\alpha_1}{z_1 z_2}, \tag{8}$$
$$p_{12} = \frac{\lambda_1 p(\lambda_2+\alpha_2)+\lambda_2(1-q)\alpha_1}{z_1 z_2},$$
$$p_{21} = \frac{\lambda_2 q(\lambda_1+\alpha_1)+\lambda_1(1-p)\alpha_2}{z_1 z_2},$$
$$p_{12} = \frac{\lambda_2(1-q)(\lambda_1+\alpha_1)+\lambda_1 p\alpha_2}{z_1 z_2},$$
$$z_1 z_2 = \lambda_1\lambda_2 + \lambda_1\alpha_2 + \lambda_2\alpha_1,$$

Therefore, inserting formulas 8 and 5 into 7 the following expressions for probabilities π_{ij}, $i, j = 1, 2$, are obtained:

$$\pi_{11} = \frac{1}{z_1 z_2}((1-p)\lambda_1\alpha_2 + q\lambda_2(\lambda_1+\alpha_1)$$
$$+\lambda_1\lambda_2(1-p-q)(\pi_1 + \pi_2 e^{-(\alpha_1+p\lambda_1+\alpha_2+q\lambda_2)T})),$$
$$\pi_{12} = \frac{1}{z_1 z_2}((1-q)\lambda_2(\lambda_1+\alpha_1) + p\lambda_1\alpha_2$$
$$-\lambda_1\lambda_2(1-p-q)(\pi_1 + \pi_2 e^{-(\alpha_1+p\lambda_1+\alpha_2+q\lambda_2)T})),$$
$$\pi_{21} = \frac{1}{z_1 z_2}(q\lambda_2\alpha_1 + (1-p)\lambda_1(\lambda_2+\alpha_2) \tag{9}$$
$$-\lambda_1\lambda_2(1-p-q)(\pi_2 + \pi_1 e^{-(\alpha_1+p\lambda_1+\alpha_2+q\lambda_2)T})),$$
$$\pi_{22} = \frac{1}{z_1 z_2}(p\lambda_1(\lambda_2+\alpha_2) + q\lambda_2\alpha_1$$
$$+\lambda_1\lambda_2(1-p-q)(\pi_2 + \pi_1 e^{-(\alpha_1+p\lambda_1+\alpha_2+q\lambda_2)T})),$$

where π_i, $i = 1, 2$, and $z_1 z_2$ are defined in 8.

Inserting probabilities π_{ij}, $i, j = 1, 2$, defined in 9, into formulas 6 expressions for the probabilities $\pi_i(0|T)$, $i = 1, 2$, are written in the form:

$$\pi_1(0|T) = \frac{q\lambda_2\alpha_1+(1-p)\lambda_1(\lambda_2+\alpha_2)-\lambda_1\lambda_2(1-p-q)(\pi_2+\pi_1 e^{-(\alpha_1+p\lambda_1+\alpha_2+q\lambda_2)T})}{z_1 z_2-\lambda_1\lambda_2(1-p-q)(\pi_2+\pi_1 e^{-(\alpha_1+p\lambda_1+\alpha_2+q\lambda_2)T})}, \tag{10}$$
$$\pi_2(0|T) = \frac{(1-q)\lambda_2(\lambda_1+\alpha_1)+p\lambda_1\alpha_2-\lambda_1\lambda_2(1-p-q)(\pi_2+\pi_1 e^{-(\alpha_1+p\lambda_1+\alpha_2+q\lambda_2)T})}{z_1 z_2-\lambda_1\lambda_2(1-p-q)(\pi_2+\pi_1 e^{-(\alpha_1+p\lambda_1+\alpha_2+q\lambda_2)T})},$$

where π_i, $i = 1, 2$, and $z_1 z_2$ are defined in 8.

To get a probabilities $\pi_i(T)$, $i = 1, 2$, let introduce $\pi_i(\tau|T)$, $i = 1, 2$, is a probability that at the time moment τ the process $\lambda(\tau)$ sojourns in the state λ_i, $0 < \tau < T$. Then $\pi_i(\tau+\Delta\tau|T)$, $i = 1, 2$, is a probabilty that at the time moment $\tau+\Delta\tau$ the process $\lambda(\tau)$ sojourns in the state λ_i, $0 < \tau+\Delta\tau < T$. Considering all possible ways of process $\lambda(\tau)$ behaviour on the interval $(\tau, \tau+\Delta\tau)$ the following system of differential equations related on the probabilities $\pi_i(\tau|T)$, $i = 1, 2$ can be written out:

$$\pi_1'(\tau|T) = -(p\lambda_1 + \alpha_1)\pi_1(\tau|T) + (q\lambda_2 + \alpha_2)\pi_2(\tau|T),$$
$$\pi_2'(\tau|T) = (p\lambda_1 + \alpha_1)\pi_1(\tau|T) - (q\lambda_2 + \alpha_2)\pi_2(\tau|T),$$

with a boundary conditions $\pi_i(\tau|T)=\pi_i(0|T)$ for $\tau = 0$, $i = 1, 2$.

Solution of this system is written as follows:

$$\pi_1(\tau|T) = \pi_1 - (\pi_1 - \pi_1(0|T))e^{-(\alpha_1+p\lambda_1+\alpha_2+q\lambda_2)\tau},$$
$$\pi_2(\tau|T) = \pi_2 - (\pi_2 - \pi_2(0|T))e^{-(\alpha_1+p\lambda_1+\alpha_2+q\lambda_2)\tau},$$

where π_i, $i = 1, 2$, are defined in 5, $\pi_i(0|T)$, $i = 1, 2$, are defined in 10.

Inserting in the last formula for the $\pi_i(\tau|T)$, $i = 1, 2$, instead of τ the value T and denoting $\pi_i(T|T)$, $i = 1, 2$, as $\pi_i(T)$, $i = 1, 2$, we obtain:

$$\pi_1(T) = \pi_1 - (\pi_1 - \pi_1(0|T))e^{-(\alpha_1+p\lambda_1+\alpha_2+q\lambda_2)T},$$
$$\pi_2(T) = \pi_2 - (\pi_2 - \pi_2(0|T))e^{-(\alpha_1+p\lambda_1+\alpha_2+q\lambda_2)T}, \tag{11}$$

where π_i, $i = 1, 2$, are defined in 5, $\pi_i(0|T)$, $i = 1, 2$, are defined in 10.

It is easy to show that

$$\pi_1(T) = \pi_1(0|T)q_{11}(T) + \pi_2(0|T)q_{21}(T),$$
$$\pi_2(T) = \pi_2(0|T)q_{12}(T) + \pi_2(0|T)q_{22}(T). \tag{12}$$

Inserting 4, 5 and 10 into 1, using 12 and making a difficult enough calculations we derive an explicit form of a probability density function $p_T(\tau)$ for the modulated synchronous doubly stochastic flow in conditions of the fixed dead time:

$$p_T(\tau) = 0, \ 0 \le \tau < T,$$
$$p_T(\tau) = \gamma(T)z_1e^{-z_1(\tau-T)} + (1 - \gamma(T))z_2e^{-z_2(\tau-T)}, \ \tau \ge T, \tag{13}$$

where $\gamma(T) = \frac{1}{z_2-z_1}(z_2 - \pi_1(T)\lambda_1 - \pi_2(T)\lambda_2)$, $1-\gamma(T) = \frac{1}{z_2-z_1}(-z_1 + \pi_1(T)\lambda_1 + \pi_2(T)\lambda_2)$,

$z_{1,2}$ are defined in 2, $\pi_i(T)$, $i = 1, 2$, are defined in 11.

4 Joint Probability Density Function of Modulated Synchronous Flow Interval Duration

Let τ_1, τ_2 are a values of the random variable of duration of two adjacent intervals between the moments of adjacent flow events. Then a joint probability density function $p_T(\tau_1, \tau_2)$ is written in the form of:

$$p_T(\tau_1, \tau_2) = \begin{cases} 0, \ 0 \le \tau_1 < T, 0 \le \tau_2 < T, \\ \sum_{i=1}^{2} \pi_i(0|T) \sum_{j=1}^{2} q_{ij}(T) \sum_{k=1}^{2} \widetilde{p}_{jk}(\tau_1 - T) \\ \times \sum_{s=1}^{2} q_{ks}(T) \sum_{n=1}^{2} \widetilde{p}_{sn}(\tau_2 - T), \ \tau_1 \ge T, \tau_2 \ge T, \end{cases}$$

where $\pi_i(0|T), i = 1, 2$, are defined in 10, $q_{ij}(T), q_{ks}(T), i, j, k, s = 1, 2$, are defined in 5, $\widetilde{p}_{jk}(\tau_1-T), \widetilde{p}_{sn}(\tau_2-T), j, k, s, n = 1, 2$, are defined in 4 (in formulas 4 it is necessary to change τ on $\tau_1$or τ_2).

Using the formula 12 and making a changes $t_1 = \tau_1 - T$, $t_2 = \tau_2 - T$ it is not difficult to get a formula for a joint probability density function $p_T(\tau_1, \tau_2)$ in the form:

$$p_T(\tau_1, \tau_2) = \begin{cases} 0, \ 0 \le \tau_1 < T, 0 \le \tau_2 < T, \\ \sum_{i=1}^{2} \pi_i(T) \sum_{j=1}^{2} p_{ij}(t_1) \sum_{k=1}^{2} q_{jk}(T) \sum_{s=1}^{2} \widetilde{p}_{ks}(t_2), \ t_1 \ge 0, t_2 \ge 0, \end{cases} \tag{14}$$

Besides, according to the formulas 1 and 10 we can write the following:

$$p_T(\tau_1) = \sum_{i=1}^{2} \pi_i(T) \sum_{j=1}^{2} \widetilde{p}_{ij}(t_1), t_1 \geq 0,$$
$$p_T(\tau_2) = \sum_{i=1}^{2} \pi_i(T) \sum_{j=1}^{2} \widetilde{p}_{ij}(t_2), t_2 \geq 0.$$

Inserting the values of $q_{jk}(T)$, $j,k = 1,2$, from 5 to 14 and making a difficult enough manipulations the difference $p(\tau_1, \tau_2) - p(\tau_1)p(\tau_2)$ is derived as follows:

$$
\begin{aligned}
p(\tau_1, \tau_2) - p(\tau_1)p(\tau_2) = \\
e^{-(\alpha_1 + p\lambda_1 + \alpha_2 + q\lambda_2)T} \sum_{s=1}^{2} (\widetilde{p}_{1s}(t_2) - \widetilde{p}_{2s}(t_2)) \\
\times \{\pi_1(T)(\pi_2(0|T)\widetilde{p}_{11}(t_1) - \pi_1(0|T)\widetilde{p}_{12}(t_1)) \\
+\pi_2(T)(\pi_2(0|T)\widetilde{p}_{21}(t_1) - \pi_1(0|T)\widetilde{p}_{22}(t_1))\},
\end{aligned}
\tag{15}
$$

where $\pi_i(0|T), i = 1,2$, are defined in 10, $\pi_i(T), i = 1,2$, are defined in 11, $\widetilde{p}_{is}(t_2)$, $\widetilde{p}_{ij}(t_1)$, $i,j,s = 1,2$, are defined in 4. Then the union $\sum_{s=1}^{2}(\widetilde{p}_{1s}(t_2) - \widetilde{p}_{2s}(t_2))$ is of the form:

$$\sum_{s=1}^{2} (\widetilde{p}_{1s}(t_2) - \widetilde{p}_{2s}(t_2)) = -(\lambda_1 - \lambda_2)(z_1 e^{-z_1 \tau_2} - z_2 e^{-z_2 \tau_2})/(z_2 - z_1). \tag{16}$$

where $z_{1,2}$ are defined in 2.

Inserting $\pi_i(0|T), i = 1,2$, from 10, $\pi_i(T), i = 1,2$, from 11, $\widetilde{p}_{ij}(t_1), i,j = 1,2$, from 4 into expression in brace from the formula 15, then inserting 16 into 15, after that making a return changing $t_1 = \tau_1 - T$, $t_2 = \tau_2 - T$ the formula for the joint probability density function $p_T(\tau_1, \tau_2)$ is written as follows:

$$
\begin{aligned}
p_T(\tau_1, \tau_2) = 0,\ 0 \leq \tau_1 < T, 0 \leq \tau_2 < T, \\
p_T(\tau_1, \tau_2) = p(\tau_1)p(\tau_2) + e^{-(\alpha_1 + p\lambda_1 + \alpha_2 + q\lambda_2)T}(\lambda_1 - \lambda_2)\lambda_1\lambda_2(1 - p - q) \\
\times \frac{((\lambda_1 p + \alpha_1)\pi_1(0) - (\lambda_2 q + \alpha_2)\pi_2(0))((p+q)\lambda_1\lambda_2 + \lambda_2\alpha_1 + \lambda_1\alpha_2)}{(z_2 - z_2)^2(z_1 z_2 - \lambda_1\lambda_2(1 - p - q)e^{-(\alpha_1 + p\lambda_1 + \alpha_2 + q\lambda_2)T})^2(\alpha_1 + p\lambda_1 + \alpha_2 + q\lambda_2)^2} \times \{z_1 z_2 \\
+e^{-(\alpha_1 + p\lambda_1 + \alpha_2 + q\lambda_2)T}((\alpha_1 + p\lambda_1 + \alpha_2 + q\lambda_2)(\alpha_1 + \lambda_1 + \alpha_2 + \lambda_2) - 2z_1 z_2) \\
-e^{-2(\alpha_1 + p\lambda_1 + \alpha_2 + q\lambda_2)T}((\alpha_1 + p\lambda_1 + \alpha_2 + q\lambda_2)((1 - p)\lambda_1 + (1 - q)\lambda_2) - z_1 z_2)\} \\
\times(z_1 e^{-z_1(\tau_1 - T)} - z_2 e^{-z_2(\tau_1 - T)})(z_1 e^{-z_1(\tau_2 - T)} - z_2 e^{-z_2(\tau_2 - T)}),\ \tau_1 \geq T, \tau_2 \geq T,
\end{aligned}
$$

where $\pi_i(0), i = 1,2$, are defined in 10 for $T = 0$, $z_{1,2}$ are defined in 2, $z_1 z_2$ are defined in 8.

We can show that multiplication $\gamma(T)(1 - \gamma(T))$ is of the form:

$$
\begin{aligned}
\gamma(T)(1 - \gamma(T)) \\
= \frac{(\lambda_1 - \lambda_2)((\lambda_1 p + \alpha_1)\pi_1(0) - (\lambda_2 q + \alpha_2)\pi_2(0))((p+q)\lambda_1\lambda_2 + \lambda_2\alpha_1 + \lambda_1\alpha_2)}{(z_2 - z_2)^2(z_1 z_2 - \lambda_1\lambda_2(1 - p - q)e^{-(\alpha_1 + p\lambda_1 + \alpha_2 + q\lambda_2)T})^2(\alpha_1 + p\lambda_1 + \alpha_2 + q\lambda_2)^2} \times \{z_1 z_2 \\
+e^{-(\alpha_1 + p\lambda_1 + \alpha_2 + q\lambda_2)T}((\alpha_1 + p\lambda_1 + \alpha_2 + q\lambda_2)(\alpha_1 + \lambda_1 + \alpha_2 + \lambda_2) - 2z_1 z_2) \\
-e^{-2(\alpha_1 + p\lambda_1 + \alpha_2 + q\lambda_2)T}((\alpha_1 + p\lambda_1 + \alpha_2 + q\lambda_2)((1 - p)\lambda_1 + (1 - q)\lambda_2) - z_1 z_2)\} \\
\times(z_1 e^{-z_1(\tau_1 - T)} - z_2 e^{-z_2(\tau_1 - T)})(z_1 e^{-z_1(\tau_2 - T)} - z_2 e^{-z_2(\tau_2 - T)})z_1 z_2,
\end{aligned}
\tag{17}
$$

where $\pi_i(0), i = 1,2$, are defined in 10 for $T = 0$, $z_{1,2}$ are defined in 2, $z_1 z_2$ are defined in 5.

Lets denote the expression in brace from formula 17 as $f(T)$. After some manipulations $f(T)$ can be written as follows:

$$f(T) = z_1 z_2 (1 - e^{-(\alpha_1 + p\lambda_1 + \alpha_2 + q\lambda_2)T})^2$$
$$+ e^{-(\alpha_1 + p\lambda_1 + \alpha_2 + q\lambda_2)T}(\alpha_1 + p\lambda_1 + \alpha_2 + q\lambda_2)$$
$$+ e^{-(\alpha_1 + p\lambda_1 + \alpha_2 + q\lambda_2)T}(1 - e^{-(\alpha_1 + p\lambda_1 + \alpha_2 + q\lambda_2)T})$$
$$\times (\alpha_1 + p\lambda_1 + \alpha_2 + q\lambda_2)((1-p)\lambda_1 + (1-q)\lambda_2),$$

so that for any $T \geq 0$ we have $f(T) > 0$.

Finally the joint probability density function $p_T(\tau_1, \tau_2)$ is written as follows:

$$p_T(\tau_1, \tau_2) = 0, \ 0 \leq \tau_1 < T, \ 0 \leq \tau_2 < T,$$
$$p(\tau_1, \tau_2) = p(\tau_1)p(\tau_2) + e^{-(\alpha_1 + p\lambda_1 + \alpha_2 + q\lambda_2)T} \frac{\lambda_1 \lambda_2 (1-p-q)}{z_1 z_2} \gamma(T)(1 - \gamma(T))$$
$$\times (z_1 e^{-z_1(\tau_1-T)} - z_2 e^{-z_2(\tau_1-T)})(z_1 e^{-z_1(\tau_2-T)} - z_2 e^{-z_2(\tau_2-T)}), \ \tau_1 \geq T, \tau_2 \geq T,$$
$$(18)$$

where $z_{1,2}$ are defined in 2, $\gamma(T)$, $1 - \gamma(T)$ are defined in 13.

5   Recurrence Conditions for Observable Flow of Events

Let consider a particular cases when modulated synchronous doubly stochastic flow which functions under conditions of the fixed dead time is a recurrence flow.

1. From the formula 17 for the joint probability density function $p_T(\tau_1, \tau_2)$ the first recurrence condition follows. If $p + q = 1$ then the joint probability density 17 factors:

$$p_T(\tau_1, \tau_2) = p_T(\tau_1)p_T(\tau_2), \ \tau_1 \geq T, \ \tau_2 \geq T.$$

For this case the probability density function $p_T(\tau)$ is of the form:

$$p_T(\tau) = 0, \ 0 \leq \tau < T,$$
$$p_T(\tau) = \gamma(T)z_1 e^{-z_1(\tau-T)} + (1 - \gamma(T))z_2 e^{-z_2(\tau-T)}, \ \tau \geq T,$$
$$\gamma(T) = \frac{1}{z_2 - z_1}(z_2 - \lambda_1\pi_1 - \lambda_2\pi_2$$
$$+ (\lambda_1(\pi_1 - q) + \lambda_2(\pi_2 - p))e^{-(\alpha_1 + p\lambda_1 + \alpha_2 + q\lambda_2)T}),$$
$$1 - \gamma(T) = \frac{1}{z_2 - z_1}(-z_1 + \lambda_1\pi_1 + \lambda_2\pi_2$$
$$- (\lambda_1(\pi_1 - q) + \lambda_2(\pi_2 - p))e^{-(\alpha_1 + p\lambda_1 + \alpha_2 + q\lambda_2)T}),$$

where $z_{1,2}$ are defined in 2, π_i, $i = 1, 2$, are defined in 5.

2. From the formula 18 for the multiplication $\gamma(T)(1 - \gamma(T))$ the second recurrence flow condition follows. If $(\lambda_1 p + \alpha_1)\pi_1(0) = (\lambda_2 q + \alpha_2)\pi_2(0)$ then the joint probability density 17 factors:

$$p_T(\tau_1, \tau_2) = p_T(\tau_1)p_T(\tau_2), \ \tau_1 \geq T, \ \tau_2 \geq T.$$

Besides $\gamma(T) = 1$. For this case the probability density function $p_T(\tau)$ is of the form:

$$p_T(\tau) = 0, \ 0 \leq \tau < T,$$
$$p_T(\tau) = z_1 e^{-z_1(\tau-T)}, \ \tau \geq T,$$

where z_1 are defined in 2.

Because a sequence of the moments of observable flow events occurrence $t_1, t_2, \ldots,$ $t_k, \ldots$ is an imbedded Markov chain $\{\lambda(t_k)\}$ then if the one or both of the flow recurrence conditions work it is easy to show using the method of mathematical induction that the joint probability density function $p(\tau_1, \tau_2, \ldots \tau_k)$ factors for any k: $p(\tau_1, \tau_2, \ldots \tau_k) = p(\tau_1)p(\tau_2)\ldots p(\tau_k)$. Therefore an observable event flow becomes a recurrence flow.

When considering a flow recurrence conditions it is necessary to use results obtained in [1].

For the first factorization condition $p+q = 1$ a posteriori probability $w(\lambda_1|t_k + 0)$ of the flow state λ_1 in the time moment t_k is of the form:

$$w(\lambda_1|t_k + 0) = q, \ k = 1, 2, \ldots .$$

herefore a posteriori probability $w(\lambda_1|t)$ doesn't depend on prehistory and only defines by it meaning in the moment of the observable flow event occurrence. In this case there is some proximity of the considered flow to the simple stream in sense that posteriori probability of the state λ_1 of the process $\lambda(t)$ in the moments of the observable flow events occurence has a constant value equal to q.

For the second factorization condition $(\lambda_1 p + \alpha_1)\pi_1(0) = (\lambda_2 q + \alpha_2)\pi_2(0)$ a posteriori probability $w(\lambda_1|t_k + 0)$ of the flow state λ_1 in the time moment t_k is of the form:

$$w(\lambda_1|t_k + 0) = \frac{q\lambda_2 + [(1-p)\lambda_1 - q\lambda_2]\, w(\lambda_1|t_k - 0)}{\lambda_2 + [(\lambda_1 - \lambda_2]\, w(\lambda_1|t_k - 0)}, \ k = 1, 2, \ldots .$$

So a posteriori probability $w(\lambda_1|t)$ depends on prehistory in spite of the flow is recurrent and the probability density $p_T(\tau)$ has an exponential distribution $\cdot p_T(\tau) = z_1 e^{-z_1(\tau - T)}, \ \tau \geq T.$

6   Probabilistic Characteristics and Probabilities of the Observable Flow Event Types

It is not difficult to obtain probabilistic characteristics of the observable flow such as a mean value of interval duration between adjacent events, variance and covariance.

$$M(\tau) = T + \frac{\gamma(T)}{z_1} + \frac{1-\gamma(T)}{z_2}, \ D(\tau) = 2\left(\frac{\gamma(T)}{z_1^2} + \frac{1-\gamma(T)}{z_2^2}\right) - \left(\frac{\gamma(T)}{z_1} + \frac{1-\gamma(T)}{z_2}\right)^2,$$
$$cov(\tau_1, \tau_2) = e^{-(\alpha_1 + p\lambda_1 + \alpha_2 + q\lambda_2)T} \gamma(T)(1 - \gamma(T))\lambda_1\lambda_2(1 - p - q)\frac{(z_1 - z_2)^2}{(z_1 z_2)^3}.$$

There are two types of events in the flow on the question: (1) Poisson flow events of the λ_1 intensity; (2) Poisson flow events of the λ_2 intensity.

Let $q_1^{(i)}(T)$ is a stationary probability that the event occurred is a Poisson flow event of the λ_1 intensity and the process $\lambda(t)$ moves from the state λ_1 to the state λ_i, $i = 1, 2$, $q_2^{(i)}(T)$ is a stationary probability that the event occurred is a Poisson flow event of the λ_2 intensity and the process λ_2 moves from the state λ_2 to the state λ_i, $i = 1, 2$.

Then for the introduced probabilities we can derive the following explicit expressions:

$$q_1^{(1)}(T) = (1-p)\lambda_1 \frac{\alpha_2+\lambda_2\pi_1+\lambda_2(q-\pi_1)e^{-(\alpha_1+p\lambda_1+\alpha_2+q\lambda_2)T}}{z_1z_2-\lambda_1\lambda_2(1-p-q)e^{-(\alpha_1+p\lambda_1+\alpha_2+q\lambda_2)T}},$$

$$q_1^{(2)}(T) = p\lambda_1 \frac{\alpha_2+\lambda_2\pi_1+\lambda_2(q-\pi_1)e^{-(\alpha_1+p\lambda_1+\alpha_2+q\lambda_2)T}}{z_1z_2-\lambda_1\lambda_2(1-p-q)e^{-(\alpha_1+p\lambda_1+\alpha_2+q\lambda_2)T}},$$

$$q_2^{(1)}(T) = q\lambda_2 \frac{\alpha_1+\lambda_1\pi_2+\lambda_1(p-\pi_2)e^{-(\alpha_1+p\lambda_1+\alpha_2+q\lambda_2)T}}{z_1z_2-\lambda_1\lambda_2(1-p-q)e^{-(\alpha_1+p\lambda_1+\alpha_2+q\lambda_2)T}},$$

$$q_2^{(2)}(T) = (1-q)\lambda_2 \frac{\alpha_1+\lambda_1\pi_2+\lambda_1(p-\pi_2)e^{-(\alpha_1+p\lambda_1+\alpha_2+q\lambda_2)T}}{z_1z_2-\lambda_1\lambda_2(1-p-q)e^{-(\alpha_1+p\lambda_1+\alpha_2+q\lambda_2)T}},$$

where $\pi_i, i = 1,2$, z_1z_2 are defined in 8.

Then a stationary probability $q_1(T)$ that an occurred event is a Poisson flow event of the λ_1 intensity can be written in the form:

$$q_1(T) = q_1^{(1)}(T) + q_1^{(2)}(T) = \lambda_1 \frac{\alpha_2+\lambda_2\pi_1+\lambda_2(q-\pi_1)e^{-(\alpha_1+p\lambda_1+\alpha_2+q\lambda_2)T}}{z_1z_2-\lambda_1\lambda_2(1-p-q)e^{-(\alpha_1+p\lambda_1+\alpha_2+q\lambda_2)T}},$$

where $\pi_i, i = 1,2$, z_1z_2 are defined in 8.

A stationary probability $q_2(T)$ that an occurred event is a Poisson flow event of the λ_2 intensity can be written in the form:

$$q_2(T) = q_2^{(1)}(T) + q_2^{(2)}(T) = \lambda_2 \frac{\alpha_1+\lambda_1\pi_2+\lambda_1(p-\pi_2)e^{-(\alpha_1+p\lambda_1+\alpha_2+q\lambda_2)T}}{z_1z_2-\lambda_1\lambda_2(1-p-q)e^{-(\alpha_1+p\lambda_1+\alpha_2+q\lambda_2)T}},$$

where $\pi_i, i = 1,2$, z_1z_2 are defined in 8.

Note that $\pi_1(0|T) = q_1^{(1)}(T) + q_2^{(1)}(T)$, $\pi_1(0|T) = q_1^{(2)}(T) + q_2^{(2)}(T)$.

7 Conclusion and Future Research

During this research the explicit form of the joint probability density function $p_T(\tau_1, \tau_2)$ of the interval duration between an adjacent events of the modulated synchronous doubly stochastic flow in conditions of the fixed dead time is derived. There are shown recurrence flow conditions as well as some probabilistic characteristics of the flow. Also there are shown the explicit form of the such characteristics as flow event types. The formulas obtained allow us to carry out an estimation of the flow parameters using the maximum likelihood method or method of matching moments.

Acknowledgments. The work is supported by Tomsk State University Competitiveness Improvement Program.

References

1. Gortsev, A., Sirotina, M.: Joint probability density function of modulated synchronous flow interval duration. In: Dudin, A., Nazarov, A., Yakupov, R., Gortsev, A. (eds.) ITMM 2014. CCIS, vol. 487, pp. 145–152. Springer, Heidelberg (2014)

2. Dudin, A.N., Klimenuk, V.N.: Queue Systems with Correlated Flows. Belorussian State University, Minsk (2000)
3. Kingman, J.F.C.: On doubly stochastic Poisson process. Proc. Cambridge Phylosophical Soc. **60**(4), 923–930 (1964)
4. Basharin, G.P., Kokotushkin, V.A., Naumov, V.A.: About the method of renewals of subnetwork computation. AN USSR Techn. Kibernetics **6**, 92–99 (1979)
5. Neuts, M.F.: A versatile Markov point process. J. Appl. Probab. **16**, 764–779 (1979)
6. Lucantoni, D.M.: New results on the single server queue with a batch markovian arrival process. Commun. Stat. Stochast. Models **7**, 1–46 (1991)
7. Card, H.C.: Doubly stochastic Poisson processes in artifical neural learning. IEEE Trans. Neural Netw. **9**(1), 229–231 (1998)
8. Gortsev, A.M., Golofastova, M.N.: Optimal state estimation of modulated synchronous doubly stochastic flow of events. Control Comput. Inf. Tomsk State Univ. J. **23**(2), 42–53 (2013)
9. Sirotina, M.N.: Optimal state estimation of modulated synchronous doubly stochastic flow of events in conditions of fixed dead time. Control Comput. Inf. Tomsk State Univ. J. **26**(1), 63–72 (2014)
10. Bushlanov, I.V., Gortsev, A.M., Nezhel' skaya, L.A.: Estimating parameters of the synchronous twofold-stochastic flow of events. Autom. Remote Control **69**(9), 1517–1533 (2008)

Stationary Distribution of the Queueing Networks with Batch Negative Customer Arrivals

Yury Malinkovsky[1,2](✉)

[1] Francisk Skorina Gomel State University, Gomel, Belarus
malinkovsky@gsu.by, yury.malinkovsky@gmail.com
[2] National Research Tomsk State University, Tomsk, Russia

Abstract. Stationary functioning of a queueing network with batch negative customer arrivals is analyzed. Necessary and sufficient condition for ergodisity of the isolated node is established. Stationary product-form distribution of network states is found. Given network model is generalization of classic G-network model on the case of several types of negative customers.

Keywords: Queueing network · Positive customers · Negative customers · Ergodicity · Stationary distribution · Product-form

1 Isolated Node

We consider queueing system with exponential single server and $T+1$ mutually independent arriving Poisson flows: positive customers with intensity λ^+ and negative customers of T types with intensities $\lambda_1^-, \ldots, \lambda_T^-$ respectively. Arriving negative customer of flow with number l instantly deletes (kills) exactly l positive customers if there are such quantity in the system and deletes all positive customers if there are less than l customers in the system ($l = \overline{1, T}$). Negative customer and deleted positive customers instantly leave the system and don't exert influence on the system's behavior. System state $n(t)$ at moment t is quantity of the positive customers in the system. Obviously $n(t)$ is Markov chain with continuous time and state space Z_+. If its stationary distribution $\{p(n), n = 0, 1, \ldots\}$ exists then satisfies the system of equilibrium equations for vertical sections in transition graph:

$$\lambda^+ p(n) = (\mu + \lambda_1^- + \ldots + \lambda_T^-)p(n+1) + (\lambda_2^- + \ldots + \lambda_T^-)p(n+2)$$
$$+(\lambda_3^- + \ldots + \lambda_T^-)p(n+3) + \ldots + \lambda_T^- p(n+T), \qquad n = 0, 1, \ldots. \tag{1}$$

This is homogenous linear difference equation of order T. Partial solution of (1) we are looking for in the form $p(n) = z^n$. Substituting one in Eq. (1) we obtain the characteristic equation

$$g(z) = \sum_{l=1}^{T} z^l \sum_{s=l}^{T} \lambda_s^- + \mu z - \lambda^+ = \sum_{s=1}^{T} \lambda_s^- \sum_{l=1}^{s} z^l + \mu z - \lambda^+ = 0. \tag{2}$$

© Springer International Publishing Switzerland 2015
A. Dudin et al. (Eds.): ITMM 2015, CCIS 564, pp. 53–63, 2015.
DOI: 10.1007/978-3-319-25861-4_5

We will prove sufficiency of condition

$$\rho = \frac{\lambda^+}{\mu + \sum\limits_{t=1}^{T} t\lambda_t^-} < 1 \tag{3}$$

for ergodicity of the process $n(t)$. At first we shall use Descartes theorem [1]. It is only one reversal of sign on transition from μ to $-\lambda^+$. Therefore Eq. (2) has only one strong positive root. By this $g(0) = -\lambda^+ < 0$ and $g(1) = \sum\limits_{t=1}^{T} t\lambda_t^- + \mu - \lambda > 0$ on the strength (3). So this root $z_0 \in (0, 1)$. Hence equilibrium Eq. (1) has the solution $p(n) = Cz_0^n$. From normalization condition one coincides with geometric distribution:

$$p(n) = (1 - z_0)z_0^n, \quad n = 0, 1, \ldots. \tag{4}$$

We will use Foster ergodic theorem [2]. For irredusible conservative regular Markov chain with continuous time to be ergodic it is necessary and sufficient the system of equilibrium equations has nonzero solution that $\sum\limits_{n=0}^{\infty} |p(n)| < \infty$. Equation (2) as we seen has root $z_0 \in (0, 1)$ when condition (3) is satisfied and it is being known that (4) is the partial solution of equilibrium Eq. (1). The series $\sum\limits_{n=0}^{\infty} |p(n)|$ converges as sum of the geometric progression members with ratio less than one. Obviously chain is irreducible and conservative. Regularity follows from leaving rate $q(n)$ of process $n(t)$ from the state n is bounded [3]. Hence the condition (3) is sufficient for ergodicity $n(t)$ and when (3) holds then ergodic distribution has the form (4).

Lemma 1. *1. All the roots of characteristic Eq. (2) on modulo strong more than one if $\rho > 1$.*
2. Characteristic Eq. (2) has only root $z = 1$ (simple root) and rest roots on modulo strong more than one if $\rho = 1$.

Proof. Introduce complex variable functions

$$\phi(z) = \sum_{s=1}^{T} \lambda_s^- \sum_{l=1}^{s} z^l + \mu z, \qquad f(z) = -\lambda^+,$$

then characteristic Eq. (2) becomes $g(z) = \phi(z) + f(z) = 0$.

1. Let be $\rho > 1$. Introduced functions $\phi(z)$ and $f(z)$ are analitical in closed disk $|z| \le 1$. Inequality

$$|\phi(z)| \le \sum_{s=1}^{T} s\lambda_s^- + \mu < \lambda^+ = |f(z)|$$

is true on the boundary $|z| = 1$. By Rouche's theorem the functions $\phi(z) + f(z)$ and $f(z)$ have the same quantity of zeros in the open disk $|z| < 1$ so function

$g(z) = \phi(z) + f(z)$ hasn't zeros in disk $|z| < 1$. On the boundary $|z| = 1$ characteristic Eq. (2) also hasn't zeros. Indeed if $z = e^{i\varphi}$, $0 \leq \varphi < 2\pi$, is root of (2) then

$$Re\, g(z) = \sum_{s=1}^{T} \lambda_s^- \sum_{l=1}^{s} \cos l\varphi + \mu \cos \varphi - \lambda^+ = 0.$$

But inequality

$$\left| \sum_{s=1}^{T} \lambda_s^- \sum_{l=1}^{s} \cos l\varphi + \mu \cos \varphi \right| \leq \sum_{s=1}^{T} s\lambda_s^- + \mu < \lambda^+$$

contradicts above equality. So all the roots of (2) have moduluses strong greater than one.

2. Let be $\rho = 1$. The proof is based on modification of Rouche's theorem proposed by V. Klimenok [4] and very useful for ergodic conditions research of queueing processes:

Let the functions $\phi(z)$ $f(z)$ be analytic in the open disk $|z| < 1$ and continuous on the boundary $z = 1$ and the following relations hold:

$$|f(z)|_{|z|=1,\, z\neq 1} > |\phi(z)|_{|z|=1,\, z\neq 1},$$

$$f(1) = -\phi(1) \neq 0.$$

Let also the functions $f(z)$ and $\phi(z)$ have the derivatives at the point $z = 1$ and the following inequality holds

$$\frac{f'(1) + \phi'(1)}{f(1)} < 0,$$

then functions $f + \phi$ and f have the same quantity of zeros in open disk $|z| < 1$.

$z = 1$ is the root of characteristic Eq. (3). This root is simple because $g'(1) \geq \mu > 0$. We prove there are no other roots on $|z| = 1$. As $g(-1) \leq -\mu - \lambda^+ < 0$, then $z = -1$ isn't the root of (2). Let $z = e^{i\varphi}$ be the root of (2) then

$$Re\, g(z) = \sum_{s=1}^{T} \lambda_s^- \sum_{l=1}^{s} \cos l\varphi + \mu \cos \varphi - \lambda^+ = 0.$$

As on cycle $0 \leq \varphi < 2\pi$ for $\varphi \neq 0, \varphi \neq \pi$

$$\left| \sum_{s=1}^{T} \lambda_s^- \sum_{l=1}^{s} \cos l\varphi + \mu \cos \varphi \right| \leq \sum_{s=1}^{T} \lambda_s^- \sum_{l=1}^{s} |\cos l\varphi| + \mu |\cos \varphi| < \sum_{s=1}^{T} s\lambda_s^- + \mu = \lambda^+,$$

then above equality is not true. So $z = 1$ is unique and simple root of characteristic Eq. (2) on the disk $|z| = 1$.

As $z = e^{i\varphi}$, $\varphi \neq 0$, then

$$|\phi(z)| = \left| \sum_{s=1}^{T} \lambda_s^- \sum_{l=1}^{s} e^{il\varphi} + \mu e^{i\varphi} \right| < \sum_{s=1}^{T} s\lambda_s^- + \mu = \lambda^+ = |f(z)|,$$

because modulus of the sum of complex numbers with multiple $\varphi \neq 0$ arguments strong less than sum of its moduluses. Next

$$f(1) = -\lambda^+ = -\sum_{s=1}^{T} s\lambda_s^- - \mu = -\phi(1) < 0,$$

$$\frac{f'(1) + \phi'(1)}{f(1)} = \frac{\phi'(1)}{f(1)} = -\frac{1}{\lambda^+}\left(\sum_{s=1}^{T} \frac{s(s+1)}{2}\lambda_s^- + \mu\right) < 0.$$

Thereby all conditions of modification V. Klimenok of Rouche's theorem are satisfied. So characteristic Eq. (2) has only root $z = 1$ and simple on the boundary $|z| = 1$, rest roots on modulo are strong more than one if $\rho = 1$. ☐

Lemma 2. *Let $Q_j(n)$ are some polinomials on variable n, $0 \leq \varphi_j < 2\pi$, $\varphi_j \neq \varphi_m$ if $j \neq m$ $(j, m = 1, \ldots, k)$ and*

$$\sum_{j=l}^{k} Q_j(n)e^{in\varphi_j} \rightarrow 0 \qquad when \quad n \rightarrow \infty,$$

then $Q_j(n) \equiv 0$ $(j = 1, \ldots, k)$.

Proof. Let

$$A = \max_{1 \leq j \leq k} \deg Q_j(n),$$

then

$$n^A \sum_{j=1}^{k} \frac{Q_j(n)}{n^A}e^{in\varphi_j} \rightarrow 0 \quad when \quad n \rightarrow \infty.$$

For $j = 1, \ldots, k$ constant C_j exists such that $\frac{Q_j(n)}{n^A} \rightarrow C_j$. Then $n^A\left(\sum_{j=1}^{k} C_j e^{in\varphi_j} + o(1)\right) \rightarrow 0$, and so $\sum_{j=1}^{k} C_j e^{in\varphi_j} + o(1) \rightarrow 0$ because $A \geq 0$. Hence $\varepsilon_n = \sum_{j=1}^{k} C_j e^{in\varphi_j} \rightarrow 0$ when $n \rightarrow \infty$. Therefore for every $\varepsilon > 0$ there exists a number $N = N_\varepsilon$ then for all $n \geq N$ there is $|\varepsilon_n| < \varepsilon$. By introduced notation for ε_n one has

$$\begin{cases} \varepsilon_N = \sum_{j=1}^{k} C_j e^{iN\varphi_j} \\ \varepsilon_{N+1} = \sum_{j=1}^{k} C_j e^{i(N+1)\varphi_j} \\ \cdots\cdots\cdots\cdots\cdots \\ \varepsilon_{N+k-1} = \sum_{j=1}^{k} C_j e^{i(N+k-1)\varphi_j} \end{cases}.$$

Absolute magnitude of the determinant Δ of this system of the linear equations (relative to $C_1, \ldots, C_k$) coincides with absolute magnitude of Vandermonde

determinant, that is $|\Delta| = \prod\limits_{j<k} \left| e^{i\varphi_j} - e^{i\varphi_k} \right| \neq 0.$ Let Δ_m is determinant differs
from Δ by substitution of the column of free terms instead of column m of determinant Δ. Factorizing Δ_m on this column we have $|\Delta_m| < C\varepsilon$ where C – some constant independent on ε. By Cramer rule

$$|C_m| = \frac{|\Delta_m|}{|\Delta|} < \frac{C}{|\Delta|}\varepsilon.$$

Hence $C_m = 0$, $m = 1, 2, \ldots, k$ on the strength arbitrariness $\varepsilon > 0$. So $Q_j(n) \to 0$ when $n \to \infty$ which means that $Q_j(n) \equiv 0$, $j = 1, 2, \ldots, k$ because $Q_j(n)$ is polinomial.   □

Theorem 1. *Markov chain $n(t)$ is regular. It is ergodic if and only if inequality (3) holds. The stationary distribution of chain has geometric distribution form (4) in this case.*

Proof. 1. Let $\lambda^+ > \mu + \sum_{s=1}^{T} s\lambda_s^-$. All the roots of characteristic Eq. (2) modulus strong more than one by Lemma 1. We prove that stationary distribution doesn't exist, hence Markov chain $n(t)$ is not ergodic. The total solution of Eq. (1) has form

$$p(n) = \sum_{j=1}^{l} Q_j(n) z_j^n, \tag{5}$$

where $z_1, z_2, \ldots, z_l$ are all different roots of characteristic Eq. (2), $Q_1(n)$, $Q_2(n), \ldots, Q_l(n)$ are polinomials of n with degrees one less than orders of roots $z_1, z_2, \ldots, z_l$ respectively. Without loss of generality $1 < |z_1| \leq |z_2| \leq \ldots \leq |z_l|$. We divide all the roots into groups of roots with equal magnitudes:

$$|z_1| = |z_2| = \ldots = |z_{l_1}| = r_1 < |z_{l_1+1}| = |z_{l_1+2}| = \ldots = |z_{l_2}| = r_2 < |z_{l_2+1}| =$$
$$|z_{l_2+2}| = \ldots = |z_{l_3}| = r_3 < \ldots < |z_{l_p+1}| = |z_{l_p+2}| = \ldots = |z_l| = r_{p+1}.$$

Then (5) overwrites as

$$p(n) = \sum_{j=1}^{l_1} Q_j(n) z_j^n + \sum_{j=l_1+1}^{l_2} Q_j(n) z_j^n + \ldots + \sum_{j=l_p+1}^{l} Q_j(n) z_j^n. \tag{6}$$

Representating the roots of characteristic equanion in exponential form $z_k = |z_k| e^{i\varphi_k}$, $0 \leq \varphi_k < 2\pi$, we will rewrite (6) as

$$p(n) = r_1^n \sum_{j=1}^{l_1} Q_j(n) e^{in\varphi_j} + r_2^n \sum_{j=l_1+1}^{l_2} Q_j(n) e^{in\varphi_j} + \ldots + r_{p+1}^n \sum_{j=l_p+1}^{l} Q_j(n) e^{in\varphi_j} =$$

$$r_{p+1}^n \left[\left(\frac{r_1}{r_{p+1}}\right)^n \sum_{j=1}^{l_1} Q_j(n) e^{in\varphi_j} + \left(\frac{r_2}{r_{p+1}}\right)^n \sum_{j=l_1+1}^{l_2} Q_j(n) e^{in\varphi_j} + \ldots \right.$$

$$+ \left(\frac{r_p}{r_{p+1}}\right)^n \sum_{j=l_{p-1}+1}^{l_p} Q_j(n)e^{in\varphi_j} + \sum_{j=l_p+1}^{l} Q_j(n)e^{in\varphi_j}\Bigg], \tag{7}$$

where $1 < r_1 < r_2 < \ldots < r_{p+1}$. If the stationary distribution exists then $p(n) \to 0$ when $n \to \infty$. Then the expression in square brackets of (7) morewhere tends to zero, because $r_{p+1} > 1$. But all terms of this expression except last term go to 0 because $\left|e^{in\varphi_j}\right| = 1$ and the polinomials $Q_j(n)$, if their degrees are not zeros, go to ∞ considerably slowly than $\left(\frac{r_j}{r_{p+1}}\right)^n \to 0$ when $n \to \infty$ (this statement obviously if $Q_j(n)$ is polinomial of zero degree). Therefore the last term of expression in square brackets of (7) also tends zero:

$$\sum_{j=l_p+1}^{l} Q_j(n)e^{in\varphi_j} \to 0 \qquad n \to \infty.$$

Without loss of generality account that all $\varphi_j \in [0, 2\pi)$ and $\varphi_j \neq \varphi_m$ when $j \neq m$. By Lemma 2 $Q_j(n) \equiv 0$ when $l_{p+1} + 1 \leq j \leq l$: Thus (6) take the next form:

$$p(n) = \sum_{j=1}^{l_l} Q_j(n)z_j^n + \sum_{j=l_1+1}^{l_2} Q_j(n)z_j^n + \ldots + \sum_{j=l_{p-1}+1}^{l_p} Q_j(n)z_j^n. \tag{8}$$

Factor out r_p^n in (8) from brackets in similar mode we will prove that all $Q_j(n) \equiv 0$ in last sum of (8). By induction $p(n) \equiv 0$. Hence nontrivial solution of equilibrium Eq. (1) such that $p(n) \to 0$ $n \to \infty$ doesn't exist, that is stationary distribution doesn't exist. So Markov chain $n(t)$ is not ergodic.

2. Let $\lambda^+ = \mu + \sum_{s=1}^{T} s\lambda_s^-$. The characteristic Eq. (2) has only root $z = 1$ (simple) and rest roots modulus strong more than one by Lemma 1. The total solution of difference equation differs from (5) by availability constant term:

$$p(n) = C + \sum_{j=1}^{l} Q_j(n)z_j^n,$$

where $z_1, z_2, \ldots, z_l$ are all different roots of characteristic Eq. (2) apart from simple root $z = 1$, $Q_1(n), Q_2(n), \ldots, Q_l(n)$ are polinomials of n with degrees one less than orders of roots $z_1, z_2, \ldots, z_l$ respectively. The proof that if $p(n) \to 0$ when $n \to \infty$ then $p(n) \equiv 0$ in full repeats the proof of point 1. At the end we get $p(n) \equiv C$. But $p(n) \to 0$ hence $p(n) \equiv 0$. So Markov chain $n(t)$ is not ergodic. ☐

2   Queueing Network

We consider queueing network consisting of N single-line exponential nodes with service rate μ_i for the server of node i $(i = \overline{1, N})$. $(T+1)N$ mutually independent

Poisson flows arrive from without to the network. More specifically positive (usual) customer flow with parameter Λ_i and T negative customer flows with parameters λ_{il} arrive in node i ($i = \overline{1,N}, l = \overline{1,T}$). The negative customers don't demand service. Arriving negative customer of flow l instantly deletes (kills) exactly l positive customers if there are such quantity in the node i and deletes all positive customers if there are less than l customers in node i ($i = \overline{1,N}, l = \overline{1,T}$). Negative customer and de ed positive customers instantly leave the network and don't exert influence on the network behavior. The positive customer served in node i instantly and independently on other customers moves to node j as positive with probability p_{ij}^+, as negative customer of flow with number l with probability p_{ijl}^-, or arrives the network with probability p_{i0} ($i, j = \overline{1,N}, l = \overline{1,T}$) and $\sum_{j=1}^{N} \left(p_{ij}^+ + \sum_{l=1}^{T} p_{ijl}^- \right) + p_{i0} = 1$ ($i = \overline{1,N}$). Quantity of places for waiting of positive customers is unbounded. For distinctness we supposed the positive customers are served in order of their arrival moments.

We will describe the state of network by random vector

$$\mathbf{n}(t) = (n_1(t), n_2(t), \ldots, n_N(t)),$$

where $n_i(t)$ is a quantity of positive customers in node i at time t. Because primitive assumptions about entering flows and service times distributions $\mathbf{n}(t)$ is multidimensional Markov chain with continuous time and state spase $X = Z_+^N$ where $Z_+ = \{0, 1, \ldots\}$. Assume $\mathbf{n}(t)$ is irreducible. For example we can assume all $\Lambda_i > 0$ and for every i exists l such that $\lambda_{il} > 0$. Our purpose is to establish the ergodic condition and to determine the stationary distribution.

We consider isolated node believeing customer flows arrive with rates like those rates of corresponding flows in the network (which isn't Poisson). We add index i as first index corresponding to node number to all notations for isolated node of Sect. 1. Characteristic Eq. (2) with substituting of root z_{i0} of (2) become identity

$$\sum_{l=1}^{T} \sum_{s=l}^{T} \lambda_{is}^- z_{0i}^l + \mu_i z_{i0} - \lambda_i^+ = 0. \tag{9}$$

If ergodic condition

$$\rho_i = \frac{\lambda_i^+}{\mu_i + \sum_{t=1}^{T} t\lambda_{it}^-} < 1, \quad i = \overline{1,N}, \tag{10}$$

holds it follows by results of Sect. 1 that the stationary distribution of isolated node has form

$$p_i(n_i) = (1 - z_{i0})z_{i0}^{n_i}, \qquad n_i = 0, 1, \ldots. \tag{11}$$

Hence the probability of full server employment in steady-state is z_{i0}. So flow intensities of positive and negative customers in network satisfies the next traffic equations system:

$$\lambda_i^+ = \Lambda_i + \sum_{j=1}^{N} \mu_j z_{j0} p_{ji}^+, \qquad i = \overline{1, N}, \tag{12}$$

$$\lambda_{il}^- = \lambda_{il} + \sum_{j=1}^{N} \mu_j z_{j0} p_{jil}^-, \qquad l = \overline{1, T}, \quad i = \overline{1, N}. \tag{13}$$

By the continuity theorem of implicit function and Brauer fixed point theorem we can prove positive solution of the traffic equations system (12), (13) exists.

If the steady-state distribution $\{p(\mathbf{n}\}$ of Markov chain $\mathbf{n}(t)$ exists then one satisfies to the global balance equations

$$p(\mathbf{n}) \sum_{i=1}^{N} \left[\Lambda_i + (\mu_i + \lambda_{i1} + \ldots + \lambda_{iT}) I_{\{n_i \neq 0\}} \right] =$$

$$\sum_{i=1}^{N} \left\{ p(\mathbf{n} - \mathbf{e}_i) \Lambda_i I_{\{n_i \neq 0\}} p(\mathbf{n} + \mathbf{e}_i) \left[\mu_i p_{i0} + \lambda_{i1} + (\lambda_{i2} + \ldots + \lambda_{iT}) I_{\{n_i = 0\}} \right] \right.$$

$$+ \sum_{l=2}^{T} p(\mathbf{n} + l\mathbf{e}_i) \left[\lambda_{il} + (\lambda_{il+1} + \ldots + \lambda_{iT}) I_{\{n_i = 0\}} \right] + \sum_{j=1}^{N} \left[p(\mathbf{n} + \mathbf{e}_j - \mathbf{e}_i) \mu_j p_{ji}^+ I_{\{n_i \neq 0\}} \right.$$

$$+ \sum_{l=1}^{T} p(\mathbf{n} + \mathbf{e}_j + l\mathbf{e}_i) \mu_j (p_{jil}^- + (p_{jil+1}^- + \ldots + p_{jiT}^-) I_{\{n_i = 0\}})$$

$$\left. \left. + p(\mathbf{n} + \mathbf{e}_j) \mu_j (p_{ji1}^- + p_{ji2}^- + \ldots + p_{jiT}^-) I_{\{n_i = 0\}} \right] \right\}, \qquad \mathbf{n} \in Z_+^N. \tag{14}$$

Here $\mathbf{e}_i$ is a unit vector of direction i and I_A is an indicator of event A equal to 1 if event A occurs and to 0 if event A doesn't occur.

The main result has the next form.

Theorem 2. *Markov process* $\mathbf{n}(t)$ *is regular and if inequalities (10) hold then it is ergodic. Its stationary distribution* $\{p(\mathbf{n}\}$ *is defined by*

$$p(\mathbf{n}) = \prod_{i=1}^{n} p_i(n_i), \qquad \mathbf{n} \in Z_+^N,$$

where $p_i(n_i)$ *and* ρ_i *are defined by equalities (11) and (10) respectively,* z_{i0}, $i = \overline{1, N}$, *are the roots of Eq. (9) belonging to segment* $[0, 1]$.

Proof. We have $I_{\{n_i = 0\}} = 1 - I_{\{n_i \neq 0\}}$, so (14) shapes

$$p(\mathbf{n}) \sum_{i=1}^{N} \left[\Lambda_i + (\mu_i + \lambda_{i1} + \ldots + \lambda_{iT}) I_{\{n_i \neq 0\}} \right] =$$

$$\sum_{i=1}^{N} \left\{ p(\mathbf{n} - \mathbf{e}_i) \Lambda_i I_{\{n_i \neq 0\}} + p(\mathbf{n} + \mathbf{e}_i) (\mu_i p_{i0} + \lambda_{i1} + \lambda_{i2} + \ldots + \lambda_{iT}) - \right.$$

$$p(\mathbf{n} + \mathbf{e}_i)(\lambda_{i2} + \ldots + \lambda_{iT})I_{\{n_i \neq 0\}} + \sum_{l=2}^{T} p(\mathbf{n} + l\mathbf{e}_i)(\lambda_{il} + \lambda_{il+1} + \ldots + \lambda_{iT})$$

$$- \sum_{l=2}^{T} p(\mathbf{n} + l\mathbf{e}_i)(\lambda_{il+1} + \ldots + \lambda_{iT})I_{\{n_i \neq 0\}}$$

$$+ \sum_{j=1}^{N} \Big[p(\mathbf{n} + \mathbf{e}_j - \mathbf{e}_i)\mu_j p_{ji}^+ I_{\{n_i \neq 0\}} + \sum_{l=1}^{T} p(\mathbf{n} + \mathbf{e}_j + l\mathbf{e}_i)\mu_j(p_{jil}^- + \ldots + p_{jiT}^-)$$

$$- \sum_{l=1}^{T} p(\mathbf{n} + \mathbf{e}_j + l\mathbf{e}_i)\mu_j(p_{jil+1}^- + \ldots + p_{jiT}^-)I_{\{n_i \neq 0\}} + p(\mathbf{n} + \mathbf{e}_j)\mu_j(p_{ji1}^- + \ldots + p_{jiT}^-)$$

$$- p(\mathbf{n} + \mathbf{e}_j)\mu_j(p_{ji1}^- + p_{ji2}^- + \ldots + p_{jiT}^-)I_{\{n_i \neq 0\}}\Big]\Big\}, \qquad \mathbf{n} \in Z_+^N. \qquad (15)$$

We partition this equation into local balance equations. The sum of terms in the left side of (15) including factor $I_{\{n_i \neq 0\}}$ equates to the same sum in the right side of (15). After the sum of terms in the left side of (15) doesn't containing factor $I_{\{n_i \neq 0\}}$ equates to the same sum in the right side of (15):

$$p(\mathbf{n})\sum_{i=1}^{N} \Lambda_i = \sum_{i=1}^{N} \Big\{ p(\mathbf{n}+\mathbf{e}_i)(\mu_i p_{i0} + \lambda_{i1} + \ldots + \lambda_{iT}) + \sum_{l=2}^{T} p(\mathbf{n}+l\mathbf{e}_i)(\lambda_{il} + \ldots + \lambda_{iT})$$

$$+ \sum_{j=1}^{N} \Big[\sum_{l=1}^{T} p(\mathbf{n}+\mathbf{e}_j+l\mathbf{e}_i)\mu_j(p_{jil}^- + \ldots + p_{jiT}^-) + p(\mathbf{n}+\mathbf{e}_j)\mu_j(p_{ji1}^- + \ldots + p_{jiT}^-)\Big]\Big\}. \qquad (16)$$

$$p(\mathbf{n})\sum_{i=1}^{N}(\mu_i + \lambda_{i1} + \ldots + \lambda_{iT}) = \sum_{i=1}^{N} \Big\{ p(\mathbf{n} - \mathbf{e}_i)\Lambda_i - p(\mathbf{n} + \mathbf{e}_i)(\lambda_{i2} + \ldots + \lambda_{iT})$$

$$- \sum_{l=2}^{T} p(\mathbf{n} + l\mathbf{e}_i)(\lambda_{il+1} + \ldots + \lambda_{iT}) + \sum_{j=1}^{N} \Big[p(\mathbf{n} + \mathbf{e}_j - \mathbf{e}_i)\mu_j p_{ji}^+$$

$$\sum_{l=1}^{T} p(\mathbf{n}+\mathbf{e}_j+l\mathbf{e}_i)\mu_j(p_{jil+1}^- + \ldots + p_{jiT}^-) - p(\mathbf{n}+\mathbf{e}_j)\mu_j(p_{ji1}^- + p_{ji2}^- + \ldots + p_{jiT}^-)\Big]\Big\}.$$

We partition previous equations into more detail balance equations:

$$p(\mathbf{n})(\mu_i + \lambda_{i1} + \ldots + \lambda_{iT}) = p(\mathbf{n} - \mathbf{e}_i)\Lambda_i - p(\mathbf{n} + \mathbf{e}_i)(\lambda_{i2} + \ldots + \lambda_{iT})$$

$$- \sum_{l=2}^{T} p(\mathbf{n} + l\mathbf{e}_i)(\lambda_{il+1} + \ldots + \lambda_{iT}) + \sum_{j=1}^{N} \Big[p(\mathbf{n} + \mathbf{e}_j - \mathbf{e}_i)\mu_j p_{ji}^+ -$$

$$\sum_{l=1}^{T} p(\mathbf{n}+\mathbf{e}_j+l\mathbf{e}_i)\mu_j(p_{jil+1}^- + \ldots + p_{jiT}^-) - p(\mathbf{n}+\mathbf{e}_j)\mu_j(p_{ji1}^- + p_{ji2}^- + \ldots + p_{jiT}^-)\Big]. \qquad (17)$$

Let probabilities $p_i(n_i)$ are defined by equalities (13). We will prove that

$$p(\mathbf{n}) = p_1(n_1)p_2(n_2)\ldots p_N(n_N), \qquad \mathbf{n} \in Z_+^N \qquad (18)$$

is solution of local balance Eqs. (16) and (17), that is global balance Eq. (14). We will devide both sides of (16) on $p(\mathbf{n})$ and use (10), (11), (13) – (15):

$$\sum_{i=1}^{N} \Lambda_i = \sum_{i=1}^{N} \left\{ z_{i0}(\mu_i p_{i0} + \lambda_{i1} + \lambda_{i2} + \ldots + \lambda_{iT}) + \sum_{l=2}^{T} z_{i0}^l (\lambda_{il} + \lambda_{il+1} + \ldots + \lambda_{iT}) \right.$$

$$\left. + \sum_{j=1}^{N} \left[\sum_{l=1}^{T} z_{j0} z_{i0}^l \mu_j (p_{jil}^- + p_{jil+1}^- + \ldots + p_{jiT}^-) + z_{j0}\mu_j (p_{ji1}^- + p_{ji2}^- + \ldots + p_{jiT}^-) \right] \right\} =$$

$$\sum_{i=1}^{N} \left[z_{i0}\mu_i p_{i0} + \sum_{l=1}^{T} z_{i0}^l (\lambda_{il} + \lambda_{il+1} + \ldots + \lambda_{iT}) \right.$$

$$\left. + \sum_{l=1}^{T} z_{i0}^l (\lambda_{il}^- - \lambda_{il} + \lambda_{il+1}^- - \ldots + \lambda_{iT}^- - \lambda_{iT}) + \lambda_{i1}^- - \lambda_{i1} + \lambda_{i2}^- - \ldots + \lambda_{iT}^- - \lambda_{iT} \right] =$$

$$\sum_{i=1}^{N} \left[z_{i0}\mu_i p_{i0} + \sum_{l=1}^{T} z_{i0}^l (\lambda_{il} + \ldots + \lambda_{iT}) + \lambda_{i1}^- - \lambda_{i1} + \lambda_{i2}^- - \lambda_{i2} + \ldots + \lambda_{iT}^- - \lambda_{iT} \right] =$$

$$\sum_{i=1}^{N} \left[z_{i0}\mu_i p_{i0} + \lambda_i^+ - \mu_i z_{i0} + \lambda_{i1}^- - \lambda_{i1} + \lambda_{i2}^- - \lambda_{i2} + \ldots + \lambda_{iT}^- - \lambda_{iT} \right] =$$

$$\sum_{i=1}^{N} \left[z_{i0}\mu_i p_{i0} + \Lambda_i + \sum_{j=1}^{N} z_{j0}\mu_j p_{ji}^+ - \mu_i z_{i0} + \lambda_{i1}^- - \lambda_{i1} + \lambda_{i2}^- - \lambda_{i2} + \ldots + \lambda_{iT}^- - \lambda_{iT} \right] =$$

$$\sum_{i=1}^{N} \left[z_{i0}\mu_i p_{i0} + \Lambda_i + \sum_{j=1}^{N} z_{j0}\mu_j \left(1 - \sum_{l=1}^{T} p_{jil}^- - p_{j0}\right) - \mu_i z_{i0} + \lambda_{i1}^- - \lambda_{i1} + \ldots + \lambda_{iT}^- - \lambda_{iT} \right]$$

$$\sum_{i=1}^{N} \left[\Lambda_i + \lambda_{i1}^- - \lambda_{i1} + \lambda_{i2}^- - \lambda_{i2} + \ldots + \lambda_{iT}^- - \lambda_{iT} - \sum_{l=1}^{T}\sum_{j=1}^{N} z_{j0}\mu_j p_{jil}^- \right] =$$

$$\sum_{i=1}^{N} \left[\Lambda_i + \lambda_{i1}^- - \lambda_{i1} + \lambda_{i2}^- - \lambda_{i2} + \ldots + \lambda_{iT}^- - \lambda_{iT} - \sum_{l=1}^{T} (\lambda_{il}^- - \lambda_{il}) \right] = \sum_{i=1}^{N} \Lambda_i,$$

that is (17) becomes identity. In much the same way we check implementation of local balance Eq. (18):

$$\mu_i + \lambda_{i1} + \ldots + \lambda_{iT} = \frac{\Lambda_i}{z_{i0}} - z_{i0}(\lambda_{i2} + \ldots + \lambda_{iT}) - \sum_{l=2}^{T} z_{i0}^l (\lambda_{il+1} + \ldots + \lambda_{iT})$$

$$+ \sum_{j=1}^{N} \left[\frac{z_{j0}}{z_{i0}} \mu_j p_{ji}^{+} - \sum_{l=1}^{T} z_{j0} z_{i0}^{l} \mu_j (p_{jil+1}^{-} + \ldots + p_{jiT}^{-}) - z_{j0} \mu_j (p_{ji1}^{-} + \ldots + p_{jiT}^{-}) \right]$$

$$= \frac{\Lambda_i}{z_{i0}} - \sum_{l=1}^{T} z_{i0}^{l} (\lambda_{il+1} + \lambda_{il+2} + \ldots + \lambda_{iT}) + \frac{\lambda_i^{+} - \Lambda_i}{z_{i0}}$$

$$- \sum_{l=1}^{T} z_{i0}^{l} (\lambda_{il+1}^{-} - \lambda_{il+1} + \ldots + \lambda_{iT}^{-} - \lambda_{iT}) - (\lambda_{i1}^{-} - \lambda_{i1} + \ldots + \lambda_{iT}^{-} - \lambda_{iT}) =$$

$$\frac{\lambda_i^{+}}{z_{i0}} - \sum_{l=0}^{T} z_{i0}^{l} (\lambda_{il+1} + \lambda_{il+2} + \ldots + \lambda_{iT}) + \lambda_{i1} + \lambda_{i2} + \ldots + \lambda_{iT} =$$

$$\mu_i + \frac{1}{z_{i0}} \sum_{l=1}^{T} \sum_{s=l}^{T} \lambda_{is}^{-} z_{i0}^{l} - \sum_{l=0}^{T} \sum_{s=l+1}^{T} \lambda_{is}^{-} z_{i0}^{l} + \lambda_{i1} + \lambda_{i2} + \ldots + \lambda_{iT} = \mu_i + \lambda_{i1} + \ldots + \lambda_{iT}.$$

Using Foster ergodic theorem [2] completes the proof. □

3 Conclusion

We have considered stationary functioning of an open queueing network with batch arrivals of negative customers. Expression for stationary distribution has been derived in product form. Given network model is generalization of classic G-network model on the case of several types of negative customers. Research results have practical importance and may be used for real networks investigation.

References

1. Kurosh, A.G.: Course of Higher Algebra. Nauka, Moscow (1971). (in Russian)
2. Bocharov, P.P., Pechinkin, A.V.: Queueing Theory. RUDN, Moskow (1995). (in Russian)
3. Gikhman, I.I., Skorohod, A.V.: Introduction in Theory of Stochastic Processes. Nauka, Moskow (1977). (in Russian)
4. Klimenok, V.: On the Modification of Rouche's Theorem for the Queueing Theory Problems. Queueing Syst. **38**, 431–434 (2001)
5. Gelenbe, E.: Product-form queueing networks with negative and positive customers. J. Appl. Prob. **28**, 656–663 (1991)

Sojourn Time Analysis of Finite Source Markov Retrial Queuing System with Collision

Anna Kvach$^{(\boxtimes)}$ and Anatoly Nazarov

Tomsk State University, Tomsk, Russia
kvach_as@mail.ru, nazarov.tsu@gmail.com

Abstract. This paper deals with a finite source retrial queueing system of type M/M/1//N with collision of the customers. This means that the system has one server and N sources. Analysis of the sojourn time in the system is presented. The analysis is performed under an asymptotic condition of infinitely increasing number of sources. The approximation of the distribution of the total sojourn time in the system is derived.

Keywords: Finite source queueing system · Retrial queue · Collision · Asymptotic analysis · Sojourn time

1 Introduction

Retrial queue [1–3] is a queuing system characterized by the following basic assumption: a customer who cannot get service goes to the orbit and, after some random period of time, returns to the system and tries to get service again. It is assumed that the orbit is infinitely large and every call repeats his attempts until he is satisfied. Retrial queueing systems are important to study computer and telephone systems, digital communication networks with random access protocols, engineering cellular mobile radio networks, computer networks and other technical systems. For a comprehensive review of retrial queues and a summary of many results and literature, the reader is directed to the works by Falin and Templeton [4], Artalejo and Gomez-Corral [5], and references therein.

In many practical situations, it is important to take into consideration the fact that the rate of generation of a primary calls degreases as the number of customers in the system increases. This can be done with the help of finite source models where each source generates its own flow of a primary customers.

Finite source retrial model can be applied for researching magnetic disk memory systems, local area networks with CSMA/CD protocols with star topology, ets. The seminal papers of this area are [6–9]. Dragieva V. in [10] considered a single server unreliable finite source retrial model in which breakdowns occur only when the server is busy and after breakdown the server is immediately sent for repair. A various types of unreliable system with finite numbers of sources are investigated by Almási B., Sztrick J., Roszik J., for example, in [11,12]. In this works authors used the software tool MOSEL (Modeling, Specification, and Evaluation Language) to formulate the model and to calculate and display the main performance measures.

A. Dudin et al. (Eds.): ITMM 2015, CCIS 564, pp. 64–72, 2015.
DOI: 10.1007/978-3-319-25861-4_6

In present paper we consider the M/M/1//N retrial queue with collision. In the main model it is assumed that if an arriving customer finds the server busy, then the arriving customer collides with a customer in service and they both goes to the orbit and the server becomes idle immediately. Choi et al. [13] considered retrial queues with collision arising from the specific communication protocol CSMA/CD. In the papers Nazarov A., Lyubina T. are considered the various open retrial queuing systems with collision of customers [14,15].

In our previous paper [16] we considered a closed retrial queueing system M/M/1//N with collision. Using method of asymptotic analysis under conditions of infinitely increasing number of sources, we obtained a distribution of the number of sources in "waiting" state.

In this paper we propose method of asymptotic analysis under conditions of infinitely increasing number of sources to research the sojourn time in finite source Markov retrial queueing system with collision.

2   Model Description

We consider a finite source retrial queuing system of type M/M/1//N in Kendals notation with collision of the customers. This mean that the system has one server and N sources. Each one of them generates a primary customers according to a Poisson flow with rate λ/N. We assume that sources can be in two states: generating a primary customers and waiting for the end of successful service. Source which send the customer for service, moves into the "waiting" state and stays in this state till the end of the service of this customer. If a primary customer finds server idle, he enters into service immediately, during service time, which distributed exponentially with parameter μ. Otherwise, if server is busy, arriving customer involves into collision with servicing customer and they both moves into the orbit. Retrial customer repeat his demand for service with an exponential distribution with rate σ/N. We assume that primary customers, retrial customers and service time are mutually independent.

Lets select a random customer from the system and shall call him the observed customer. Let us first consider the time between the moment, when a primary customer enters service for the first time and the time point on which this customer successfully ends his service. This time period is called the sojourn time. In the system occur of a situation of the conflict (collision of the customers) is possible, this feature is necessary to consider in the study of the sojourn time in the system. Therefore, the sojourn time consist of the total time, which customer spend on the orbit and the total time of the service. Total service time includes all period of time in which the observed customer tried to get service, but it was interrupted by arriving customer and the service time in which observed customer successfully finished his service.

At time t let $i(t)$ be the number of sources locating in "waiting" state and $k(t)$ determines the server state

$$k(t) = \begin{cases} 0, & \text{if the server is free,} \\ 1, & \text{if the server is busy (not by observed customer),} \\ 2, & \text{if the server is busy by observed customer.} \end{cases}$$

Introduce $T(t)$ - the residual sojourn time of the observed customer in the system at time t.

Assuming that the observed customer locates in the orbit, lets denote by $G_k(u, i, t) = M\{e^{juT(t)}|k(t) = k, i(t) = i\}$ the joint conditional characteristic function.

For the functions $G_k(u, i, t)$, we can write the system of the finite-difference equation:

$$G_0(u, i, t - \Delta t) = \left(1 - \lambda \frac{N-i}{N} \Delta t\right)\left(1 - \sigma \frac{i}{N} \Delta t\right) e^{ju\Delta t} G_0(u, i, t)$$
$$+ \lambda \frac{N-i}{N} \Delta t G_1(u, i+1, t) + \sigma \frac{i-1}{N} \Delta t G_1(u, i, t)$$
$$+ \frac{\sigma}{N} \Delta t G_2(u, i, t) + o(\Delta t),$$

$$G_1(u, i, t - \Delta t) = \left(1 - \lambda \frac{N-i}{N} \Delta t\right)\left(1 - \sigma \frac{i-1}{N} \Delta t\right)(1 - \mu \Delta t) e^{ju\Delta t} G_1(u, i, t)$$
$$+ \lambda \frac{N-i}{N} \Delta t G_0(u, i+1, t) + \sigma \frac{i-1}{N} \Delta t G_0(u, i, t)$$
$$+ \mu \Delta t G_0(u, i-1, t) + o(\Delta t),$$

$$G_2(u, i, t - \Delta t) = \left(1 - \lambda \frac{N-i}{N} \Delta t\right)\left(1 - \sigma \frac{i-1}{N} \Delta t\right)(1 - \mu \Delta t) e^{ju\Delta t} G_2(u, i, t)$$
$$+ \lambda \frac{N-i}{N} \Delta t G_0(u, i+1, t) + \sigma \frac{i-1}{N} \Delta t G_0(u, i, t) + \mu \Delta t + o(\Delta t).$$

The Kolmogorov backward differential equations are

$$-\frac{\partial G_0(u, i, t)}{\partial t} = \left[ju - \lambda \frac{N-i}{N} - \sigma \frac{i}{N}\right] G_0(u, i, t) + \lambda \frac{N-i}{N} G_1(u, i+1, t)$$
$$+ \sigma \frac{i-1}{N} G_1(u, i, t) + \frac{\sigma}{N} G_2(u, i, t),$$
$$-\frac{\partial G_1(u, i, t)}{\partial t} = \left[ju - \lambda \frac{N-i}{N} - \sigma \frac{i-1}{N} - \mu\right] G_1(u, i, t) + \lambda \frac{N-i}{N} G_0(u, i+1, t)$$
$$+ \sigma \frac{i-1}{N} G_0(u, i, t) + \mu G_0(u, i-1, t),$$
$$-\frac{\partial G_2(u, i, t)}{\partial t} = \left[ju - \lambda \frac{N-i}{N} - \sigma \frac{i-1}{N} - \mu\right] G_2(u, i, t) + \lambda \frac{N-i}{N} G_0(u, i+1, t)$$
$$+ \sigma \frac{i-1}{N} G_0(u, i, t) + \mu.$$

Note this system in steady state

$$\left[ju - \lambda \frac{N-i}{N} - \sigma \frac{i}{N}\right] G_0(u, i) + \lambda \frac{N-i}{N} G_1(u, i+1)$$
$$+ \sigma \frac{i-1}{N} G_1(u, i) + \frac{\sigma}{N} G_2(u, i) = 0,$$
$$\left[ju - \lambda \frac{N-i}{N} - \sigma \frac{i-1}{N} - \mu\right] G_1(u, i) + \lambda \frac{N-i}{N} G_0(u, i+1)$$
$$+ \sigma \frac{i-1}{N} G_0(u, i) + \mu G_0(u, i-1) = 0, \quad (1)$$
$$\left[ju - \lambda \frac{N-i}{N} - \sigma \frac{i-1}{N} - \mu\right] G_2(u, i) + \lambda \frac{N-i}{N} G_0(u, i+1)$$
$$+ \sigma \frac{i-1}{N} G_0(u, i) + \mu = 0.$$

In order to solve this system, we use method of asymptotic analysis [17] under conditions of infinitely increasing number of sources ($N \to \infty$).

3   Method of Asymptotic Analysis

Let us denote $\dfrac{1}{N} = \varepsilon$.

Introducing following substitute

$$i\varepsilon = x, \qquad u = \varepsilon w, \qquad G_k(u, i) = F_k(w, x, \varepsilon), \qquad (2)$$

we can transform system (1) to the form:

$$[j\varepsilon w - \lambda(1 - x) - \sigma x] F_0(w, x, \varepsilon) + \lambda(1 - x) F_1(w, x + \varepsilon, \varepsilon)$$

$$+ \sigma(x - \varepsilon) F_1(w, x, \varepsilon) + \sigma \varepsilon F_2(w, x, \varepsilon) = 0,$$

$$[j\varepsilon w - \lambda(1 - x) - \sigma(x - \varepsilon) - \mu] F_1(w, x, \varepsilon) + \lambda(1 - x) F_0(w, x + \varepsilon, \varepsilon)$$
$$\tag{3}$$
$$+ \sigma(x - \varepsilon) F_0(w, x, \varepsilon) + \mu F_0(w, x - \varepsilon, \varepsilon) = 0,$$

$$[j\varepsilon w - \lambda(1 - x) - \sigma(x - \varepsilon) - \mu] F_2(w, x, \varepsilon) + \lambda(1 - x) F_0(w, x + \varepsilon, \varepsilon)$$

$$+ \sigma(x - \varepsilon) F_0(w, x, \varepsilon) + \mu = 0.$$

Theorem 1. *The limiting value* $F_0(w, x)$, $F_1(w, x)$, $F_2(w, x)$ *of function* $F_0(w, x, \varepsilon)$, $F_1(w, x, \varepsilon)$, $F_2(w, x, \varepsilon)$ *(the solutions of the system (3)), can be represented in the following form*

$$F_0(w, x) = F_1(w, x) = F(w, x) = \frac{d}{d - jw},$$

$$F_2(w, x) = \frac{\mu + a(\kappa_1) F(w, x)}{b(\kappa_1)},$$

where

$$d = \frac{\sigma \mu}{2a(\kappa_1) + \mu},$$

$$a(\kappa_1) = \lambda(1 - \kappa_1) + \sigma \kappa_1,$$

$$b(\kappa_1) = \lambda(1 - \kappa_1) + \sigma \kappa_1 + \mu,$$

$$\kappa_1 = \frac{2\mu R_1^2}{\sigma(1 - 2R_1)},$$

$$R_1 = \frac{\sigma(2\lambda + \mu) - \sqrt{\sigma^2(2\lambda - \mu)^2 + 8\sigma\mu\lambda^2}}{4\mu(\sigma - \lambda)}.$$

Proof. There are two stages of proving.

Stage 1. Using the following denotation $\lim_{\varepsilon \to 0} F_k(w, x, \varepsilon) = F_k(w, x)$ as $\varepsilon \to 0$, the system (3) has the form

$$- [\lambda (1 - x) + \sigma x] F_0(w, x) + [\lambda (1 - x) + \sigma x] F_1(w, x) = 0 \ ,$$

$$- [\lambda (1 - x) + \sigma x + \mu] F_1(w, x) + [\lambda (1 - x) + \sigma x + \mu] F_0(w, x) = 0 \ , \qquad (4)$$

$$- [\lambda (1 - x) + \sigma x + \mu] F_2(w, x) + [\lambda (1 - x) + \sigma x] F_0(w, x) + \mu = 0 \ .$$

From system (4) we obtain that the functions $F_0(w, x)$ and $F_1(w, x)$ is equal and function $F_2(w, x)$ can be represented as

$$F_0(w, x) = F_1(w, x) \doteq F(w, x),$$

$$F_2(w, x) = \frac{[\lambda(1 - x) + \sigma x] F(w, x) + \mu}{\lambda(1 - x) + \sigma x + \mu}. \qquad (5)$$

Stage 2. Lets consider the system (3). Using the expansion into a Taylor series of the first order of smallness about a point x, we get

$$[j\varepsilon w - \lambda (1 - x) - \sigma x] F_0(w, x, \varepsilon) + [\lambda (1 - x) + \sigma (x - \varepsilon)] F_1(w, x, \varepsilon)$$

$$+ \sigma \varepsilon F_2(w, x, \varepsilon) + \lambda (1 - x) \varepsilon \frac{\partial F_1(w, x, \varepsilon)}{\partial x} = 0 \ ,$$

$$[j\varepsilon w - \lambda (1 - x) - \sigma (x - \varepsilon) - \mu] F_1(w, x, \varepsilon) + [\lambda (1 - x) + \sigma (x - \varepsilon)$$

$$+ \mu] F_0(w, x, \varepsilon) + [\lambda (1 - x) - \mu] \varepsilon \frac{\partial F_0(w, x, \varepsilon)}{\partial x} = 0 \ , \qquad (6)$$

$$[j\varepsilon w - \lambda (1 - x) - \sigma (x - \varepsilon) - \mu] F_2(w, x, \varepsilon) + [\lambda (1 - x) + \sigma (x - \varepsilon)]$$

$$\cdot F_0(w, x, \varepsilon) + \lambda (1 - x) \varepsilon \frac{\partial F_0(w, x, \varepsilon)}{\partial x} + \mu = 0 \ .$$

Denote the solution of the system (6) as follows

$$F_k(w, x, \varepsilon) = F_k(w, x) + \varepsilon f_k(w, x) + o(\varepsilon), \quad k = 0, 1, 2. \qquad (7)$$

Substituting (7) to the system (6) we obtain

$$\varepsilon \left\{ jw F_0(w, x) - \sigma F_1(w, x) + \sigma F_2(w, x) + \lambda (1 - x) \frac{\partial F_1(w, x)}{\partial x} \right.$$

$$+ \left[\lambda (1 - x) + \sigma x \right] \cdot \left(f_1(w, x) - f_0(w, x) \right) \bigg\}$$

$$+ \left[\lambda (1 - x) + \sigma x \right] \left(F_1(w, x) - F_0(w, x) \right) = O(\varepsilon^2),$$

$$\varepsilon \Big\{ (jw + \sigma) F_1(w, x) - \sigma F_0(w, x) + \Big[\lambda (1 - x) - \mu \Big] \frac{\partial F_0(w, x)}{\partial x}$$
$$+ \Big[\lambda (1 - x) + \sigma x + \mu \Big] \cdot \Big(f_0(w, x) - f_1(w, x) \Big) \Big\}$$
$$+ \Big[\lambda (1 - x) + \sigma x + \mu \Big] \Big(F_0(w, x) - F_1(w, x) \Big) = O(\varepsilon^2),$$

$$\varepsilon \Big\{ (jw + \sigma) F_2(w, x) - \sigma F_0(w, x) + \lambda (1 - x) \frac{\partial F_0(w, x)}{\partial x}$$
$$+ \Big[\lambda (1 - x) + \sigma x \Big] \cdot \Big(f_0(w, x) - f_2(w, x) \Big) - \mu f_2(w, x) \Big\}$$
$$+ \Big[\lambda (1 - x) + \sigma x \Big] F_0(w, x) - \Big[\lambda (1 - x) + \sigma x + \mu \Big] F_2(w, x) + \mu = O(\varepsilon^2).$$

Considering expressions (5) for the functions $F_0(w, x)$, $F_1(w, x)$ and $F_2(w, x)$ the system rewrite as

$$\varepsilon \Big\{ (jw - \sigma) F(w, x) + \sigma F_2(w, x) + \lambda (1 - x) \frac{\partial F(w, x)}{\partial x}$$
$$+ \Big[\lambda (1 - x) + \sigma x \Big] \cdot \Big(f_1(w, x) - f_0(w, x) \Big) \Big\} = O(\varepsilon^2),$$

$$\varepsilon \Big\{ jw F(w, x) + \Big[\lambda (1 - x) - \mu \Big] \frac{\partial F(w, x)}{\partial x}$$
$$+ \Big[\lambda (1 - x) + \sigma x + \mu \Big] \cdot \Big(f_0(w, x) - f_1(w, x) \Big) \Big\} = O(\varepsilon^2), \qquad (8)$$

$$\varepsilon \Big\{ (jw + \sigma) F_2(w, x) - \sigma F(w, x) + \lambda (1 - x) \frac{\partial F(w, x)}{\partial x} +$$
$$+ \Big[\lambda (1 - x) + \sigma x \Big] \cdot \Big(f_0(w, x) - f_2(w, x) \Big) - \mu f_2(w, x) \Big\} = O(\varepsilon^2).$$

Dividing each part of the equation of the system (8) and executing an asymptotic transition as $\varepsilon \to 0$, we obtain the following system

$$\Big[\lambda (1 - x) + \sigma x \Big] \cdot \Big(f_0(w, x) - f_1(w, x) \Big) = (jw - \sigma) F(w, x)$$
$$+ \sigma F_2(w, x) + \lambda (1 - x) \frac{\partial F(w, x)}{\partial x},$$

$$-\Big[\lambda (1 - x) + \sigma x + \mu \Big] \cdot \Big(f_0(w, x) - f_1(w, x) \Big) = jw F(w, x)$$
$$+ \Big[\lambda (1 - x) - \mu \Big] \frac{\partial F(w, x)}{\partial x}, \qquad (9)$$

$$\Big[\lambda (1 - x) + \sigma x \Big] \cdot \Big(f_2(w, x) - f_0(w, x) \Big) + \mu f_2(w, x) =$$
$$(jw + \sigma) F_2(w, x) - \sigma F(w, x) + \lambda (1 - x) \frac{\partial F(w, x)}{\partial x}.$$

Using the following denotation

$$a(x) = \lambda(1 - x) + \sigma x,$$
$$b(x) = \lambda(1 - x) + \sigma x + \mu, \tag{10}$$

lets multiply the first equation of (9) by $b(x)$, the second equation by $a(x)$ and add the resulting equation together:

$$-\left\{\lambda(1 - x)b(x) + \left[\lambda(1 - x) - \mu\right]a(x)\right\}\frac{\partial F(w, x)}{\partial x} = \left[(jw - \sigma)b(x) + jwa(x)\right]F(w, x) + \sigma b(x)F_2(w, x). \tag{11}$$

Taking into account the entered denotation (10), expression (5) can be rewritten as

$$b(x)F_2(w, x) = a(x)F(w, x) + \mu.$$

Substituting this expression to the Eq. (11) we obtain

$$-\left\{\lambda(1 - x)b(x) + \left[\lambda(1 - x) - \mu\right]a(x)\right\}\frac{\partial F(w, x)}{\partial x} = \left[(jw - \sigma)b(x) + (jw + \sigma)a(x)\right]F(w, x) + \sigma\mu. \tag{12}$$

In our previous paper [16] we investigated the closed M/M/1//N retrial queueing system with collision. In this article it was shown that the number of sources in "waiting" state $i(t)\varepsilon$ asymptoticaly converge to the deterministic quantity κ_1. Therefore, taking into account the denotation (2) $x = i\varepsilon$, we obtain that $x = \kappa_1$.

Putting $x = \kappa_1$ in the Eq. (12), the multiplier before partial derivative $\frac{\partial F(w, x)}{\partial x}$ becomes equal to zero and Eq. (12) can be rewritten as

$$\left[(jw - \sigma)b(\kappa_1) + (jw + \sigma)a(\kappa_1)\right]F(w, \kappa_1) + \sigma\mu = 0.$$

Performing this equation and entering denotation $d = \dfrac{\sigma\mu}{2 \cdot a(\kappa_1) + \mu}$, we obtain the following expression for the function $F(w, \kappa_1)$

$$F(w, \kappa_1) = \frac{d}{d - jw}.$$

Note, that function $F(w, \kappa_1)$ does not depend on argument x. Taking into account this fact and (5), we can write

$$F(w) = \frac{d}{d - jw},$$

$$F_2(w) = \frac{\mu}{b(\kappa_1)} + \frac{a(\kappa_1)}{b(\kappa_1)}F(w).$$

The theorem is proved.   □

We obtain the characteristic functions of the distribution of the residual sojourn time. Using the formula of total probability, we can write the following expression for the characteristic functions of the distribution of the total sojourn time:

$$H(w) = R_0 F_2(w) + (1 - R_0)F(w) = \frac{\mu}{b(\kappa_1)}R_0 + \left(1 - \frac{\mu}{b(\kappa_1)}R_0\right)\frac{d}{d - jw}, \quad (13)$$

where R_0 was previously obtained in [16].

Lets perform the inverse substitutions (2) in the formula (13):

$$H(u) \approx \frac{\mu}{b(\kappa_1)}R_0 + \left(1 - \frac{\mu}{b(\kappa_1)}R_0\right)\frac{d/N}{d/N - ju}.$$

Using denotation $q = \frac{\mu}{b(\kappa_1)}R_0$, we can write the following expression for the approximation $h(u)$ of the characteristic function $H(u)$:

$$h(u) = q + \left(1 - q\right)\frac{d/N}{d/N - ju}.$$

Knowing $h(u)$, it is easy to show that the approximation of distribution of the total sojourn time can be written as

$$A(x) = 1 - (1 - q)e^{-\frac{d}{N}x}.$$

4   Conclusion

In this paper, we have considered a finite source retrial queuing system $M/M/1//N$ with collision of the customers. We obtain the equations for conditional characteristic function of the distribution of the residual sojourn time. This equation was solved under an asymptotic condition of infinitely increasing number of sources. As the result, we obtain the approximation of the distribution of the total sojourn time in the system.

Acknowledgments. The work is supported by Tomsk State University Competitiveness Improvement Program.

References

1. Nazarov, A.A., Terpugov, A.F.: The Queuing Theory. NTL, Tomsk (2004). (In Russian)
2. Gnedenko, B.V., Kovalenko, I.N.: Introduction to Queuing Theory. KomKniga, MOSCOW (2007). (In Russian)
3. Koening, D., Shtoyan, D.: Methods of the Queuing Theory. Radio and Communications, Moscow (1981). (In Russian)
4. Falin, G.I., Templeton, J.G.C.: Retrial Queues. Chapman & Hall, London (1997)

5. Artalejo, J.R., Gomez-Corral, A.: Retrial Queueing Systems: A Computational Approach. Springer, Heidelberg (2008)
6. Artalejo, J.R.: Retrial queues with a finite number of sources. J. Korean Math. Soc. **35**, 503–525 (1998)
7. Falin, G.I., Artalejo, J.R.: A finite source retrial queue. Eur. J. Oper. Res. **108**, 409–424 (1998)
8. Falin, G.I.: A multiserver retrial queue with a finite number of sources of primary calls. Math. Comput. Model. **30**, 33–49 (1999)
9. Dragieva, V.I.: Single-line queue with finite source and repeated calls. Prob. Inf. Trans. **30**, 283–289 (1994)
10. Dragieva, V.I.: System State Distributions In One Finite Source Unreliable Retrial Queue. http://elib.bsu.by/handle/123456789/35903
11. Almási, B., Roszik, J., Sztrik, J.: Homogeneous finite-source retrial queues with server subject to breakdowns and repairs. Math. Comput. Model. **42**, 673–682 (2005)
12. Sztrik, J., Almási, B., Roszik, J.: Heterogeneous finite-source retrial queues with server subject to breakdowns and repairs. J. Math. Sci. **132**, 677–685 (2006)
13. Choi, B.D., Shinand, Y.W., Ahn, W.C.: Retrial queues with collision arising from unslotted CSMA/CD protocol. Queueing Syst. **11**, 335–356 (1992)
14. Lyubina, T.V., Nazarov, A.A.: Research of the Markov dynamic retrial queue system with collision. Tomsk State Univ. Bull. J. Control Comput. Sci. **12**(3), 73–84 (2010). (In Russian)
15. Lyubina, T.V., Nazarov, A.A.: Research of the non-Markov dynamic retrial queue system with collision. Kemerovo State Univ. Bull. **49**(1), 38–44 (2012). (In Russian)
16. Nazarov, A., Kvach, A., Yampolsky, V.: Asymptotic analysis of closed Markov retrial queuing system with collision. In: Dudin, A., Nazarov, A., Yakupov, R., Gortsev, A. (eds.) ITMM 2014. CCIS, vol. 487, pp. 334–341. Springer, Heidelberg (2014)
17. Nazarov, A.A., Moiseeva, S.P.: Methods of Asymptotic Analysis in a Queuing Theory. NTL, Tomsk (2006). (In Russian)

Asymptotic Analysis of the Queueing Network $SM - (GI/\infty)^K$

Alexander Moiseev$^{(\boxtimes)}$

Tomsk State University, Tomsk, Russia
`moiseev.tsu@gmail.com`

Abstract. We consider the infinite-server queueing network with semi-Markov arrivals. The system of differential equations for characteristic function of customers number at the network nodes is derived. The system is solved under asymptotic condition of high-rate arrivals. It is shown that probability distribution of customers at the network nodes can be approximated by multi-dimensional Gaussian distribution which parameters are obtained in the paper. Presented results of numerical experiments allow to determine the approximation applicability.

Keywords: Queueing network · Semi-Markov process · Asymptotic analysis

1 Introduction

Queueing networks [1] are used for modelling of modern telecommunications and other systems in many fields where we deal with transmission any objects from one point to another. In the queueing theory such systems are named as networks and points are named as the network nodes. Most investigations consider the networks with Poisson arrivals but there are results [2] which prove that the Poisson model can be adequate only in a few cases of modern telecommunication streams. Therefore, many researches use more complex models of the streams such as Markovian arrival process [3] or semi-Markov process [4]. For more information about investigations in queues with semi-Markov arrivals see [5–8].

In the paper, we consider the queueing network with infinite number of servers at every node. Usually, such models can not be directly applied to real systems but results of their analysis can be used to obtain some characteristics of the evolution of systems with limited number of servers. More information about infinite-server models see [9]. We have considered a simple infinite-server queueing system with semi-Markov arrivals in the paper [4].

The mathematical model under study is presented in the Sect. 2.

In the Sect. 3, the main method of the investigation – method of the multi-dimensional screening is described. Kolmogorov equations for the screened processes are derived in the Sect. 4.

© Springer International Publishing Switzerland 2015
A. Dudin et al. (Eds.): ITMM 2015, CCIS 564, pp. 73–84, 2015.
DOI: 10.1007/978-3-319-25861-4_7

Asymptotic analysis of the obtained equations is performed in the Sects. 5 and 6. In the Sect. 7, the approximation for the multi-dimensional stationary probability distribution of customers number at the network nodes is obtained. Numerical results (Sect. 8) prove the applicability of the approximation and determine its accuracy.

2   Mathematical Model

Consider a queueing network with K nodes, semi-Markov arrivals and infinite number of servers at every node. A customer that arrives at the network enters into the k-th node with probability v_k, where $k = 1, \ldots, K$. All servers at one network node have i.i.d. service times with a cumulative distribution function $B_k(x)$ where k is a number of the node. When the service is completed at the k-th node the customer moves to the node ν with probability $r_{k\nu}$ or leaves the network with probability r_{k0}. Note that

$$r_{k0} = 1 - \sum_{\nu=1}^{K} r_{k\nu}.$$

The epochs of customers' arrivals are equal to transition times $t_0, t_1, \ldots, t_n, \ldots$ of the high-rate semi-Markov process which is determined by semi-Markov matrix $\boldsymbol{A}(x)$ with entries $A_{lm}(x)$ by the following way:

$$A_{lm}(x) = \mathrm{P}\left\{ \xi_{n+1} = m, \tau_{n+1} < \frac{x}{N} \middle| \xi_n = l \right\}.$$

Here $l, m = 1, \ldots, L$, $\{\xi_n, \tau_n\}$ is a stationary Markovian process, N is a parameter which determines the high rate of the arrivals and has a large value (theoretically, we suppose that $N \to \infty$). Values τ_n determine inter-transition intervals: $t_{n+1} = t_n + \tau_{n+1}$ for $n \geq 0$.

Let's consider semi-Markov process $l(t)$ which is defined as follows [4]:

$$l(t) = \xi_{n+1}, \ t_n \leq t < t_{n+1}.$$

Denote by $z(t)$ the residual time from the moment t to the next transition epoch of the semi-Markov process. So, the process $\{l(t), z(t)\}$ is Markovian and we can write the following matrix differential equation for its stationary probability distribution $r_l(z) = \mathrm{P}\{l(t) = l, z(t) < \frac{z}{N}\}$ for $l = 1, \ldots, L$ [4]:

$$r'(z) = r'(0)\left[\boldsymbol{I} - A(z)\right] \tag{1}$$

where $r(z) = \{r_1(z), \ldots, r_L(z)\}$ is a row vector and $\boldsymbol{I}$ is identity matrix.

In the paper [4], we obtained that $r'(0) = \lambda r$ where row vector r is a stationary probability distribution of states of the embedded Markov chain ξ_n. This vector satisfies the Kolmogorov equation $r = r \cdot \boldsymbol{P}$ where $\boldsymbol{P} = \lim_{z \to \infty} \boldsymbol{A}(z)$ is a stochastic matrix of transitions probabilities of the embedded chain. Parameter λ is determined as follows

$$\lambda = \frac{1}{r\,\boldsymbol{A}\boldsymbol{e}},$$

here $\boldsymbol{A} = \int\limits_0^\infty [\boldsymbol{P} - \boldsymbol{A}(x)]\,dx$ and $\boldsymbol{e}$ is a column vector with entries all equal to 1.

Denote by $i_k(t)$ a number of customers at the k-th node of the network at the time moment t, $k = 1,\ldots,K$. The goal of the study is to find the stationary multi-dimensional probability distribution of the customers number $\boldsymbol{i}(t) = \{i_1(t),\ldots,i_K(t)\}$ at the network nodes.

3   Method of Multi-dimensional Dynamic Screening

A direct investigation of the process $\boldsymbol{i}(t)$ doesn't seem possible. So, we use the method of the multi-dimensional dynamic screening [10] in the study. Let's briefly describe it here.

Let's fix some arbitrary time moment T. Denote by $S_k(t)$ the probability that the customer which arrives at the network at the moment $t \le T$ will be served at the k-th node at the moment T ($k = 1,\ldots,K$). Denote by $S_0(t) = 1 - \sum\limits_{k=1}^{K} S_k(t)$ the probability that the customer leave the network before the moment T.

Consider K so called screened point processes numbered from 1 to K which are formed as follows. The customer which comes in the network at the time moment t generates a point in the k-th process with probability $S_k(t)$ for all $k = 1,\ldots,K$ and doesn't generate a point in any of the processes with probability $S_0(t)$. Let the network be empty at some time moment $t_0 < T$. Denote by $n_k(t)$ a number of points generated in the k-th process before the moment t and let's use vector notation $\boldsymbol{n}(t) = \{n_1(t),\ldots,n_K(t)\}$. It's obvious that

$$\mathrm{P}\{\boldsymbol{i}(T) = \boldsymbol{i}\} = \mathrm{P}\{\boldsymbol{n}(T) = \boldsymbol{i}\} \tag{2}$$

for every values of vector $\boldsymbol{i}$. So, if we find the probability distribution for the process $\boldsymbol{n}(t)$ we obtain the probability distribution of the process $\boldsymbol{i}(t)$ at the arbitrary time moment T by using the property (2) and substituting $t = T$.

In the paper [10], the following expression for a vector of the screening probabilities $\boldsymbol{s}(t) = \{S_1(t),\ldots,S_K(t)\}$ were obtained:

$$\boldsymbol{s}(t) = \frac{1}{2\pi}\boldsymbol{v} \int\limits_{-\infty}^{\infty} e^{-j\alpha(T-t)}\left(\boldsymbol{I} - \boldsymbol{B}^*(\alpha)\boldsymbol{R}\right)^{-1}\left(\boldsymbol{B}^*(\alpha) - \boldsymbol{I}\right)\frac{1}{j\alpha}\,d\alpha$$

where $\boldsymbol{v} = \{v_1,\ldots,v_K\}$, $\boldsymbol{R} = \{r_{k\nu}\}_{k,\nu=1,\ldots,K}$ is a routing matrix, $j = \sqrt{-1}$ is imaginary unit, $\boldsymbol{B}^*(\alpha)$ is a diagonal matrix which diagonal entries given by the Fourier–Stieltjes transforms $B_k^*(\alpha)$ of the distribution functions $B_k(x)$:

$$B_k^*(\alpha) = \int\limits_0^\infty e^{j\alpha x}\,dB_k(x),\quad \text{for } k = 1,\ldots,K.$$

4   Kolmogorov Equations

Obviously the process $\{n(t), l(t), z(t)\}$ is Markovian. Denote by $P(n, l, z, t) = P\{n(t) = n, l(t) = l, z(t) < \frac{z}{N}\}$ its probability distribution. We can write the following Kolmogorov equations:

$$\frac{1}{N}\frac{\partial P(n, l, z, t)}{\partial t} = \frac{\partial P(n, l, z, t)}{\partial z} - \frac{\partial P(n, l, 0, t)}{\partial z}$$

$$+ \sum_{k=1}^{K}\sum_{m=1}^{L}\frac{\partial P(n - e_k, m, 0, t)}{\partial z}A_{ml}(z)S_k(t) + \sum_{m=1}^{L}\frac{\partial P(n, m, 0, t)}{\partial z}A_{ml}(z)S_0(t) \quad (3)$$

for all values $n \geq 0$, $l = 1, \ldots, L$ and $z > 0$. Here e_k is a vector with entries all equal to 0 except the k-th one which equals to 1.

For partial characteristic functions

$$H(u, l, z, t) = \sum_{n_1=0}^{\infty} \cdots \sum_{n_K=0}^{\infty} e^{ju n_1 + \cdots + ju n_K} P(n, l, z, t), \quad l = 1, \ldots, M,$$

using vector notation $h(u, z, t) = \{H(u, 1, z, t), \ldots, H(u, K, z, t)\}$, we can rewrite the Eq. (3) in the following way

$$\frac{1}{N}\frac{\partial h(u, z, t)}{\partial t} = \frac{\partial h(u, z, t)}{\partial z} + \frac{\partial h(u, 0, t)}{\partial z}\left\{A(z)\left[1 + \sum_{k=1}^{K}\left(e^{ju_k} - 1\right)S_k(t)\right] - I\right\}. \quad (4)$$

The initial condition for the problem is the following:

$$h(u, z, t_0) = r(z). \quad (5)$$

In the paper we will solve the problem (4)–(5) under the asymptotic condition of arrivals' high rate: $N \to \infty$.

5   The First-Order Asymptotic Analysis

Let's make the following changes of variables in the problem (4)–(5):

$$\frac{1}{N} = \varepsilon, \quad u = \varepsilon w, \quad h(u, t) = f_1(w, t, \varepsilon).$$

So, the problem can be written in the following form

$$\varepsilon\frac{\partial f_1(w, z, t, \varepsilon)}{\partial t} = \frac{\partial f_1(w, z, t, \varepsilon)}{\partial z}$$

$$+ \frac{\partial f_1(w, 0, t, \varepsilon)}{\partial z}\left\{A(z)\left[1 + \sum_{k=1}^{K}\left(e^{j\varepsilon w_k} - 1\right)S_k(t)\right] - I\right\}, \quad (6)$$

$$f_1(w, z, t_0, \varepsilon) = r(z). \quad (7)$$

Let's prove the following statement about the asymptotic solution of this problem
$$f_1(w, z, t) = \lim_{\varepsilon \to 0} f_1(w, z, t, \varepsilon).$$

Theorem 1. *The asymptotic solution $f_1(w, z, t)$ of the problem (6)–(7) under a condition $\varepsilon \to 0$ is the following:*

$$f_1(w, z, t) = r(z) \exp \left\{ \lambda \sum_{k=1}^{K} jw_k \int_{t_0}^{t} S_k(x)\, dx \right\}. \tag{8}$$

Proof. We perform the proof by two stages.

Stage 1. Let $\varepsilon \to 0$ in the Eq. (6). We obtain the following equation

$$\frac{\partial f_1(w, z, t)}{\partial z} + \frac{\partial f_1(w, 0, t)}{\partial z} [A(z) - I] = 0.$$

This equation has a form similar to the Eq. (1). So, we can make a conclusion that the function $f_1(w, z, t)$ can be represented as

$$f_1(w, z, t) = r(z)\Phi_1(w, t) \tag{9}$$

where $\Phi_1(w, t)$ is some scalar function.

Stage 2. Let's perform an asymptotic transition $z \to \infty$ in the Eq. (6):

$$\varepsilon \frac{\partial f_1(w, \infty, t, \varepsilon)}{\partial t} = \frac{\partial f_1(w, 0, t, \varepsilon)}{\partial z} \left\{ P \left[1 + \sum_{k=1}^{K} \left(e^{j\varepsilon w_k} - 1 \right) S_k(t) \right] - I \right\}.$$

Summing up all rows of this matrix equation by multiplying it by the vector e, we obtain

$$\varepsilon \frac{\partial f_1(w, \infty, t, \varepsilon)}{\partial t} e = \frac{\partial f_1(w, 0, t, \varepsilon)}{\partial z} e \sum_{k=1}^{K} \left(e^{j\varepsilon w_k} - 1 \right) S_k(t).$$

Dividing this expression by ε and performing the transition $\varepsilon \to 0$, we obtain the following equation:

$$\frac{\partial f_1(w, \infty, t)}{\partial t} e = \frac{\partial f_1(w, 0, t)}{\partial z} e \sum_{k=1}^{K} jw_k S_k(t).$$

Substituting here the expression (9) and taking into account that $r(\infty)e = 1$ and $r'(0)e = \lambda re = \lambda$, we obtain the following differential equation for the unknown function $\Phi_1(w, t)$

$$\frac{\partial \Phi_1(w, t)}{\partial t} = \Phi_1(w, t)\lambda \sum_{k=1}^{K} jw_k S_k(t).$$

Solution of the equation with initial condition $\Phi_1(w, t_0) = 1$ is the following

$$\Phi_1(w, t) = \exp \left\{ \lambda \sum_{k=1}^{K} jw_k \int_{t_0}^{t} S_k(x)dx \right\}.$$

Finally, substituting this expression into the formula (9), we obtain the final form of the function $F_1(w, z, t)$ as the expression (8). The proof is completed.

6   The Second-Order Asymptotic Analysis

Let $h_2(u, z, t)$ be the function that is defined by the expression

$$h(u, z, t) = h_2(u, z, t) \exp \left\{ N\lambda \sum_{k=1}^{K} ju_k \int_{t_0}^{t} S_k(x) dx \right\}. \tag{10}$$

Substituting this expression into the formulas (4)–(5), we obtain the following boundary value problem:

$$\frac{1}{N} \frac{\partial h_2(u, z, t)}{\partial t} + \lambda h_2(u, z, t) \sum_{k=1}^{K} ju_k S_k(t) = \frac{\partial h_2(u, z, t)}{\partial z}$$

$$+ \frac{\partial h_2(u, 0, t)}{\partial z} \left\{ A(z) \left[1 + \sum_{k=1}^{K} \left(e^{ju_k} - 1 \right) S_k(t) \right] - I \right\} \tag{11}$$

with initial condition

$$h_2(u, z, t_0) = r(z). \tag{12}$$

Let's make the following changes of variables

$$\frac{1}{N} = \varepsilon^2, \quad u = \varepsilon w, \quad h_2(u, z, t) = f_2(w, z, t, \varepsilon). \tag{13}$$

Using new variables, the problem (11)–(12) can be written in the form

$$\varepsilon^2 \frac{\partial f_2(w, z, t, \varepsilon)}{\partial t} + \lambda f_2(w, z, t, \varepsilon) \sum_{k=1}^{K} j\varepsilon w_k S_k(t) = \frac{\partial f_2(w, z, t, \varepsilon)}{\partial z}$$

$$+ \frac{\partial f_2(w, 0, t, \varepsilon)}{\partial z} \left\{ A(z) \left[1 + \sum_{k=1}^{K} \left(e^{j\varepsilon w_k} - 1 \right) S_k(t) \right] - I \right\}, \tag{14}$$

$$f_2(w, z, t_0, \varepsilon) = r(z). \tag{15}$$

Let's prove the following statement.

Theorem 2. *The asymptotic solution* $f_2(w, z, t) = \lim_{\varepsilon \to 0} f_2(w, z, t, \varepsilon)$ *of the problem (14)–(15) under a condition* $\varepsilon \to 0$ *is the following:*

$$f_2(w, z, t) = r(z)$$

$$\times \exp \left\{ \lambda \sum_{k=1}^{K} \frac{(jw_k)^2}{2} \int_{t_0}^{t} S_k(x) dx + \kappa \sum_{k=1}^{K} \sum_{\nu=1}^{K} \frac{jw_k jw_\nu}{2} \int_{t_0}^{t} S_k(x) S_\nu(x) dx \right\} \tag{16}$$

where

$$\kappa = 2f_0' e$$

and row vector f_0' satisfies the system of linear equations

$$\begin{cases} f_0' [I - P] = \lambda [rP - r(\infty)], \\ f_0' Ae = \dfrac{\lambda^2 a_2}{2} - 1. \end{cases}$$

Here $a_2 = rA_2 e$, $A_2 = \int\limits_0^\infty x^2 dA(x)$.

Proof. We perform the proof by three stages.

Stage 1. Let $\varepsilon \to 0$ in the Eq. (14). We obtain the following equation

$$\frac{\partial f_2(w, z, t)}{\partial z} + \frac{\partial f_2(w, 0, t)}{\partial z} [A(z) - I] = 0.$$

This expression has a similar form as the Eq. (1). So, we can make the conclusion that the function $f_2(w, z, t)$ can be represented in the form

$$f_2(w, z, t) = r(z)\Phi_2(w, t) \tag{17}$$

where $\Phi_2(w, t)$ is some scalar function.

Stage 2. Taking into account the formula (17), we can write the function $F_2(w, z, t, \varepsilon)$ in the following expansion form

$$f_2(w, z, t, \varepsilon) = \Phi_2(w, t) \left[r(z) + f(z) \sum_{k=1}^K j\varepsilon w_k S_k(t) \right] + O\left(\varepsilon^2\right) \tag{18}$$

where $f(z)$ is some row vector function which has the property $f(\infty) = 0$ due to normalization condition.

Let's substitute the expansion (18) and the expansion $e^{j\varepsilon w_k} = 1 + j\varepsilon w_k + O\left(\varepsilon^2\right)$ into the Eq. (14). We obtain the following formula

$$\Phi_2(w, t)r(z)\lambda \sum_{k=1}^K j\varepsilon w_k S_k(t) = \Phi_2(w, t) \left[r'(z) + f'(z) \sum_{k=1}^K j\varepsilon w_k S_k(t) \right] +$$

$$\Phi_2(w, t) \left\{ r'(0) + f'(0) \sum_{k=1}^K j\varepsilon w_k S_k(t) \right\} \left\{ A(z) \left[1 + \sum_{k=1}^K j\varepsilon w_k S_k(t) \right] - I \right\} + O\left(\varepsilon^2\right). \tag{19}$$

Taking into account the Eq. (1), after some transforms, we reduce (19) to the following form

$$\lambda r(z) = f'(z) + r'(0)A(z) + f'(0)[A(z) - I] + O(\varepsilon).$$

Performing the asymptotic transition $\varepsilon \to 0$, we obtain the following matrix differential equation for the function $f(z)$

$$f'(z) = f'(0)[I - A(z)] - \lambda[rA(z) - r(z)].$$

Using notation $f'_0 = f'(0)$, we can obtain the following linear system for the vector f'_0 (see [4]):

$$\begin{cases} f'_0\,[I - P] = \lambda\,[rP - r(\infty)], \\ f'_0 Ae = \dfrac{\lambda^2 a_2}{2} - 1. \end{cases}$$

Stage 3. Let's substitute the expression (18) and the expansion

$$e^{j\varepsilon w_k} = 1 + j\varepsilon w_k + \frac{(j\varepsilon w_k)^2}{2} + O\left(\varepsilon^3\right)$$

into the Eq. (14) and perform the asymptotic transition $z \to \infty$. We obtain the following equation

$$\varepsilon^2 \frac{\partial \Phi_2(w,t)}{\partial t} r(\infty) + \Phi_2(w,t) r(\infty)\lambda \sum_{k=1}^{K} j\varepsilon w_k S_k(t) + \lambda \Phi_2(w,t) f(\infty)\left[\sum_{k=1}^{K} j\varepsilon w_k S_k(t)\right]^2$$

$$= \Phi_2(w,t)\left\{r'(0) + f'(0)\sum_{k=1}^{K} j\varepsilon w_k S_k(t)\right\}$$

$$\times \left\{P - I + P\sum_{k=1}^{K} j\varepsilon w_k S_k(t) + P\sum_{k=1}^{K} \frac{(j\varepsilon w_k)^2}{2} S_k(t)\right\} + O\left(\varepsilon^3\right).$$

Multiplying both parts of this matrix equation by the vector e and taking into account that $r(\infty)e = 1$, $Pe = e$, $r'(0)e = \lambda$ and $f(\infty) = 0$, we can reduce this equation to the following form

$$\varepsilon^2 \frac{\partial \Phi_2(w,t)}{\partial t} + \lambda \Phi_2(w,t)\sum_{k=1}^{K} j\varepsilon w_k S_k(t) + O\left(\varepsilon^3\right) =$$

$$\Phi_2(w,t)\left\{\lambda\sum_{k=1}^{K} j\varepsilon w_k S_k(t) + \lambda\sum_{k=1}^{K} \frac{(j\varepsilon w_k)^2}{2} S_k(t) + f'(0)e\left[\sum_{k=1}^{K} j\varepsilon w_k S_k(t)\right]^2\right\}.$$

Dividing both parts of this equation by ε^2 and performing the transition $\varepsilon \to 0$, we obtain the following differential equation for the unknown function $\Phi_2(w,t)$

$$\frac{\partial \Phi_2(w,t)}{\partial t} = \Phi_2(w,t)\left\{\lambda\sum_{k=1}^{K} \frac{(jw_k)^2}{2} S_k(t) + 2f'_0 e\sum_{k=1}^{K}\sum_{\nu=1}^{K} \frac{jw_k jw_\nu}{2} S_k(t)S_\nu(t)\right\}.$$

Using the notation $\kappa = 2f'_0 e$ and taking into account the initial condition $\Phi_2(w,t_0) = 1$, we obtain the following solution:

$$\Phi_2(w,t) = \exp\left\{\lambda\sum_{k=1}^{K} \frac{(jw_k)^2}{2}\int_{t_0}^{t} S_k(x)dx + \kappa\sum_{k=1}^{K}\sum_{\nu=1}^{K} \frac{jw_k jw\nu}{2}\int_{t_0}^{t} S_k(x)S_\nu(x)dx\right\}.$$

Substituting this expression into the formula (17), we obtain the final form of the function $F_2(w,z,t)$ as the expression (16). The theorem is proved.

7 Stationary Probability Distribution of Customers Number in the Network

Finally, let's make in the formula (14) substitutions that are inverse to the changes of variables (13). Using the formula (10), we obtain the following expression for the vector characteristic function $\boldsymbol{h}(\boldsymbol{u}, z, t)$ of the number of points have been generated in the multi-dimensional screened process $\boldsymbol{n}(t)$ inside the interval $[t_0, t]$:

$$
\boldsymbol{h}(\boldsymbol{u}, z, t) = \boldsymbol{r}(z) \exp \left\{ N\lambda \sum_{k=1}^{K} ju_k \int_{t_0}^{t} S_k(x)dx \right.
$$

$$
\left. + N\lambda \sum_{k=1}^{K} \frac{(ju_k)^2}{2} \int_{t_0}^{t} S_k(x)dx + N\kappa \sum_{k=1}^{K} \sum_{\nu=1}^{K} \frac{ju_k ju_\nu}{2} \int_{t_0}^{t} S_k(x) S_\nu(x)dx \right\}. \quad (20)
$$

We have derived this expression under the asymptotic condition $N \to \infty$, therefore, the formula (20) defines the approximation of the actual distribution for enough large values of the parameter N.

Let's make the transition $z \to \infty$ and let's multiply both parts of the expression (20) by vector $\boldsymbol{e}$. Using the main formula of the multi-dimensional screening method (2), we perform a transition to the characteristic function $h(\boldsymbol{u}, T)$ of the process under study $\boldsymbol{i}(t)$ at the time moment $t = T$. So, we can write the following approximation for the function $h(\boldsymbol{u}, T)$ under the condition that N is large enough:

$$
h(\boldsymbol{u}, T) = \exp \left\{ N\lambda \sum_{k=1}^{K} ju_k \int_{t_0}^{T} S_k(x)dx \right.
$$

$$
\left. + N\lambda \sum_{k=1}^{K} \frac{(ju_k)^2}{2} \int_{t_0}^{T} S_k(x)dx + N\kappa \sum_{k=1}^{K} \sum_{\nu=1}^{K} \frac{ju_k ju_\nu}{2} \int_{t_0}^{T} S_k(x) S_\nu(x)dx \right\}.
$$

If we set here $t_0 \to \infty$ and $T = 0$, we obtain the following expression for the approximation of the characteristic function of the customers number at the network nodes:

$$
h(\boldsymbol{u}) = \exp \left\{ N\lambda j \boldsymbol{u} \boldsymbol{S} \boldsymbol{e} + \frac{1}{2} Nj\boldsymbol{u} \left[\lambda \boldsymbol{S} + \kappa \boldsymbol{V} \right] j\boldsymbol{u}^T \right\}. \quad (21)
$$

Here $\boldsymbol{S}$ is a diagonal matrix with diagonal entries equal to $S_k = \int_{-\infty}^{0} S_k(x)dx$ and $\boldsymbol{V}$ is a matrix with entries $V_{k\nu} = \int_{-\infty}^{0} S_k(x) S_\nu(x)dx$.

So, the stationary probability distribution of the number of the customers at the network nodes for the queueing network $SM - (GI/\infty)^K$ under the condition of the high-rate arrivals can be approximated by the multi-dimensional Gaussian distribution with vector of means $N\lambda \boldsymbol{S} \boldsymbol{e}$ and covariance matrix $N[\lambda \boldsymbol{S} + \kappa \boldsymbol{V}]$.

8 Numerical Results

An accuracy of the approximation (21) we have checked for various numerical examples by comparing values of the approximating distribution with results of simulations. We introduce here one example which help us to demonstrate the approximation applicability.

Consider the queueing network $SM - (GI/\infty)^K$ with semi-Markov matrix $\boldsymbol{A}(x)$ of the arrival process written in the form $\boldsymbol{A}(x) = \boldsymbol{P} \circ \boldsymbol{G}(x)$ where $\boldsymbol{P}$ is a stochastic matrix and $\boldsymbol{G}(x)$ is a matrix with the entries equal to some cumulative distribution functions. The symbol "∘" here is the Hadamard (entrywise) matrix production. Consider the example where all entries of the matrix $\boldsymbol{G}(x)$ are cumulative distribution functions of the gamma distribution in the form $G_{k\nu}(x) = \frac{\gamma(\alpha_{k\nu}, \beta_{k\nu} x)}{\Gamma(\alpha_{k\nu})}$. Let parameters of the arrival process be the following:

$$
\boldsymbol{P} = \begin{bmatrix} 0.5\&0.4\&0.1 \\ 0.3\&0.2\&0.5 \\ 0.4\&0.1\&0.5 \end{bmatrix}, \quad \alpha = \begin{bmatrix} 0.5\&0.2\&0.8 \\ 0.5\&1.5\&1.5 \\ 0.3\&0.1\&0.4 \end{bmatrix}, \quad \beta = N \begin{bmatrix} 1\&0.2\&0.4 \\ 0.5\&1.5\&2 \\ 0.1\&0.2\&0.4 \end{bmatrix}.
$$

The network consists of four nodes. The service times at nodes also have gamma distributions with the following values of the shape α_k and the rate β_k parameters:

$$
\begin{aligned}
\alpha_1 &= 1.5, & \beta_1 &= 1.0, \\
\alpha_2 &= 0.5, & \beta_2 &= 0.5, \\
\alpha_3 &= 0.4, & \beta_3 &= 0.2, \\
\alpha_4 &= 1.5, & \beta_4 &= 1.5
\end{aligned}
$$

where k is a node number.

We concentrate an attention on the node number 3 because the average number of customers at this node at the stationary regime is equal to N. So, we can demonstrate how a value of the parameter N affects on the accuracy of the approximation (21). We will characterize the approximation accuracy by the Kolmogorov distance [11]

$$
d = \sup_x \left| \tilde{F}(x) - F(x) \right|
$$

where $\tilde{F}(x)$ is the cumulative distribution function of the Gaussian approximation (21) and $F(x)$ is the cumulative distribution function constructed based on the results of the simulation.

Figure 1 represent probability distributions of the number of customers at the node that are based on the both simulation results and analytical formulas of Gaussian approximation (21) for $N = 1, 5, 10, 50$. The Kolmogorov distance (d) for various values of the parameter N are presented in the Table 1. It is easy to see that the growth of the arrivals rate (parameter N) implies that the Gaussian approximation (21) becomes more accurate. We assume that the approximation is good enough when the Kolmogorov distance $d \leq 0.03$. So, using the results of our numerical experiments, we draw a conclusion that the

Gaussian approximation (21) can be applicable when the value of the parameter N (or, in other words, the average number of customers at the network node) is about 10 or greater.

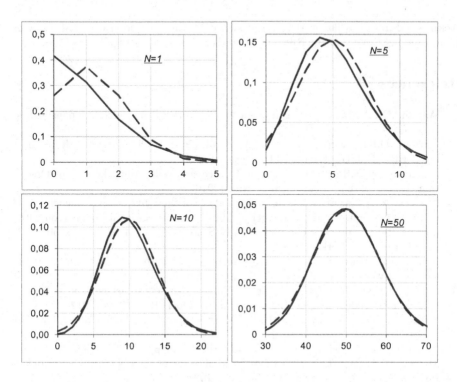

Fig. 1. Probability distributions of the number of customers at one network node for various values of the parameter N: analytical approximation (dashed line), simulation results (solid line).

Table 1. The Kolmogorov distance d between approximation and simulation distributions in relation to values N.

N	1	5	10	50	100
d	0.155	0.050	0.027	0.015	0.010

9   Conclusions

The asymptotic analysis of the queueing network with high-rate semi-Markov arrivals is presented in the paper. It is obtained that the multi-dimensional stationary probability distribution of the customers number at the network nodes can be approximated by multi-dimensional Gaussian distribution. Parameters of the approximation is also derived in the paper.

Numerical examples show that the approximation is enough accurate for values of the high-rate arrivals parameter N (or, in other words, the average number of customers at the network node) is about 10 or more.

Acknowledgments. The work is performed under the state order of the Ministry of Education and Science of the Russian Federation (No. 1.511.2014/K).

References

1. Walrand, J.: An Introduction to Queueing Networks. Prentice-Hall, Upper Saddle River (1988)
2. Heyman, D.P., Lucantoni, D.: Modelling multiple IP traffic streams with rate limits. IEEE/ACM Trans. Netw. **11**, 948–958 (2003)
3. Chakravarthy, S.R.: Markovian arrival processes. Wiley Encyclopedia of Operations Research and Management Science (2010)
4. Moiseev, A., Nazarov, A.: Asymptotic analysis of the infinite-server queueing system with high-rate semi-markov arrivals. In: IEEE International Congress on Ultra Modern Telecommunications and Control Systems (ICUMT 2014), pp. 507–513. IEEE Press, New York (2014)
5. Neuts, M.F., Chen, S.Z.: The infinite server queue with semi-Markovian arrivals and negative exponential services. J. Appl. Prob. **9**, 178–184 (1972)
6. Smith, W.: The infinitely-many-server queue with semi-Markovian arrivals and customer-dependent exponential service times. Oper. Res. **22**, 907–912 (1972)
7. Takács, L.: A storage process with semi-Markov input. Adv. Appl. Prob. **7**, 830–844 (1975)
8. Çinlar, E.: Queues with semi-Markovian arrivals. J. Appl. Prob. **4**, 365–379 (1967)
9. Massey, W.A., Whitt, W.: Networks of infinite-server queues with nonstationary Poisson input. Queueing Sys. **13**, 183–250 (1993)
10. Nazarov, A., Moiseev, A.: Analysis of an open non-Markovian $GI - (GI|\infty)^K$ queueing network with high-rate renewal arrival process. Prob. Inf. Transm. **49**(2), 167–178 (2013)
11. Kolmogorov, A.: Sulla determinazione empirica di una legge di distribuzione. Giornale dell' Intituto Italiano degli Attuari **4**, 83–91 (1933)

Number of Lost Calls During the Busy Period in an M/G/1//N Queue with Inactive Orbit

Velika Dragieva[✉]

Sofia University of Forestry, 10 Kliment Ohridski Blvd, Sofia, Bulgaria
dragievav@yahoo.com
http://www.ltu.bg/

Abstract. The main objective of the paper is to investigate the distribution of the number of lost calls, made during the busy period in one single server finite queueing system. It is assumed that the customers, failed to get service are temporarily blocked in the orbit of inactive customers for an exponentially distributed time interval. This model and its variants have many applications, especially for optimization of the corresponding models with retrials. Using the discrete transformations method we derive formulas for computing the mean value of the number of lost calls made during the busy period.

Keywords: Finite queues · Queues with losses · Inactive orbit · Busy period · Number of lost calls

1 Introduction

The queueing system under consideration has one server which serves N customers. Each of these customers in its free state (not being under service or blocked) produces a Poisson process of demands (calls) of the same rate λ. If the server is busy at the moment of a call arrival, the customer is blocked for an exponentially distributed (with intensity μ) time interval. During this interval the customer is not allowed to do any attempts for service and is said to be blocked (to be in inactive state, or in the orbit of inactive customers). When the blocking time is over, the customer moves again in free state and can produce new demands. The service times have probability distribution function $G(x)$, with $G(0) = 0$, Laplace-Stieltjes transform - $g(s)$, first moment - ν^{-1} and hazard rate function

$$\gamma(x) = \frac{G'(x)}{1 - G(x)}.$$

In fact, this system is a queueing model with finite number of sources and losses, a particular case of the Engset models, and has many applications both, in itself and for optimization of the finite retrial queues. We may find finite queues with lost or returning customers in our daily activities as well as in many telephone, computer and communication systems.

© Springer International Publishing Switzerland 2015
A. Dudin et al. (Eds.): ITMM 2015, CCIS 564, pp. 85–98, 2015.
DOI: 10.1007/978-3-319-25861-4_8

The generalized Engset models have been studied in a number of papers but to the best of our knowledge they are mainly concerned with the blocking probability in the cases of multiserver system with exponential service times ([11,13,14]). In these models it is assumed that the customers are ready to produce new calls immediately after the failures. In many real situations this is not realistic. The customers are lost for the system, but they have to satisfy their demands by another way, for example by another firm, or operator and so on. They will hardly have new requests before the completion of the previous one. During this time the customers should be considered as inactive or missing from the system. In the finite queues these missing customers change the input flow of demands and have to be taken into account, as it is in our model with inactive orbit.

In the queueing models with retrials it is assumed that a customer, unable to get service at a moment of his/her arrival (because of a busy server/servers, server vacation or repair and so on) repeats his/her attempts in pre-determined or random intervals, until receives service. There is a large literature devoted to retrial queues in the past twenty years. The reader can find a detailed review of main results, methods of analysis and the literature on retrial queues in the monographs [2,8]. The simplest example of such model is the situation when a telephone subscriber gets a busy signal and repeats the call until the demand is satisfied. Let us now consider the same example, but with the assumption that the system operator may temporarily prohibit access to the system for all unsuccessful subscribers. In other words, in some situations (high intensity of the input flow or of the server utilization and others), all subscribers which obtain a busy signal are "blocked" for pre-determined or random time interval, during which they are not allowed to make new calls, neither repeated, nor primary. But will it improve the system performance? And when exactly? This is an interesting problem, a particular case of the optimization problems (like optimal control, cost functions and others), which are ones of the main motivations and tools for constructing and analyzing competitive queueing models (see for example [3,6,10] and the references therein). It is namely one of the main motivations for the analysis presented in this paper.

The steady state analysis of this system is carried out in [4], distribution of the busy period and the number of successful calls, made during it are studied in [5]. Here we extend this investigation and consider the number of lost calls, made during the busy period. The method of analysis is similar to those in the finite systems with retrials (see ([1,7,12]).

We assume that the busy period starts at time $t_0 = 0$, at which there are no blocked customers and one of them generates a call. It ends at the first epoch at which the server is free and there are no blocked customers. Denote:

- by ζ the length of the busy period, its distribution function, $P\{\zeta \leq x\}$ – by $H(x)$ and its Laplace – Stieltjes transform – by $\eta(s)$;
- by $N^L(t)$ - the number of lost calls made during the time interval $(0, t), t \geq 0$;
- by N^{LBP} - the number of lost calls made during the busy period;
- by ζ_k^L - the length of the busy period, during which k lost calls occur.

In Sect. 2 we derive formulas for successively computing the probabilities that during the busy period exactly k lost calls occur, Sect. 3 is devoted to moments of the lost calls distribution, especially to the computation of its mean value. Conclusion closes the paper.

2 Distribution of the Lost Calls Made During the Busy Period

Let us introduce the following probabilities (densities)

$$P_{1nk}^L(t,x)dx = P\left\{\zeta > t, N^L(t) = k, C(t) = 1, R(t) = n, x \leq z(t) < x + dx\right\},$$

$$P_{ink}^L(t) = P\left\{\zeta > t, N^L(t) = k, C(t) = i, R(t) = n\right\}, i = 0, 1,$$

$$P_k^{LBP} = P\left\{N^{LBP} = k\right\},$$

$$h_k^L(t) = \frac{dP\left\{\zeta_k^L \leq t\right\}}{dt} = \frac{dP\left\{\zeta \leq t, N^{LBP} = k\right\}}{dt},$$

with initial conditions

$$P_{0nk}^L(0) = 0, P_{1nk}^L(0,x) = \delta(x)\delta_{(n,k)(0,0)}$$

and Laplace transforms $\overline{P}_{ink}^L(s), \overline{P}_{1nk}^L(s,x), \overline{h}_k^L(s)$. In terms of these quantities, for the distribution of the lost calls, made during the busy period we have

$$P_k^{LBP} = P\left\{N^{LBP} = k\right\} = \int_0^\infty h_k^L(t)dt = \overline{h}_k^L(0),$$

and for the distribution function of the busy period length -

$$H(x) = \sum_{k=0}^{\infty} \int_0^x h_k^L(t)dt$$

Here $C(t)$ is the number of busy servers at instant t (i.e. $C(t)$ is 0 or 1 according to whether the server is free or busy at time t), $R(t)$ is the number of inactive (blocked) customers at the instant t, $z(t)$ is equal to the elapsed service time in the case of busy server, $\delta(x)$ is Dirac delta and δ_{ij} is Kronecker's delta.

To calculate $\overline{h}_k^L(0)$ we first derive Kolmogorov's equations for the probabilities (densities) $P_{1nk}^L(t,x)$ and $P_{0nk}^L(t)$,

$$\frac{d}{dt}P_{0nk}^L(t) = -[(N-n)\lambda + n\mu]P_{0nk}^L(t) + (n+1)\mu P_{0,n+1,k}^L(t) + \int_0^t P_{1nk}^L(t,x)\gamma(x)dx,$$

$$\frac{\partial}{\partial t}P_{1nk}^L(t,x) = -\left[(N-n-1)\lambda + n\mu + \gamma(x) + \frac{\partial}{\partial x}\right]P_{1nk}^L(t,x)$$

$$+(n+1)\mu P_{1,n+1,k}^L(t,x) + (N-n)\lambda P_{1,n-1,k-1}^L(t,x), 0 \leq n \leq N-1, k \geq 0,$$

$$P^L_{1nk}(t,0) = (N-n)\lambda P^L_{0nk}(t), 1 \le n \le N-1, k \ge 0,$$

with

$$P^L_{0Nk}(t) = P^L_{1Nk}(t,x) = P^L_{1,-1,k}(t,x) = P^L_{1,n,-1}(t,x) = 0,$$

and

$$P^L_{0nk}(0) = 0, P^L_{1nk}(0,x) = \delta(x)\delta_{(n,k)(0,0)}.$$

Besides these equations, the following relations hold

$$h^L_k(t) = \int_0^\infty P^L_{10k}(t,x)\gamma(x)dx + \mu P^L_{01k}(t),$$

$$\sum_{n=0}^{N-1} P^L_{0nk}(t) + \sum_{n=0}^{N-1} P^L_{1nk}(t) = \sum_{q=k}^\infty \int_t^\infty h^L_q(x)dx.$$

Applying Laplace transforms we get

$$[(N-n)\lambda + n\mu + s]\overline{P}^L_{0nk}(s)$$

$$= (1-\delta_{n,N-1})(n+1)\mu\overline{P}^L_{0,n+1,k}(s) + \int_0^\infty \overline{P}^L_{1nk}(s,x)\gamma(x)dx, \qquad (1)$$

$$[(N-n-1)\lambda + n\mu + \gamma(x) + s + \frac{\partial}{\partial x}]\overline{P}^L_{1nk}(s,x) = \delta(x)\delta_{(n,k)(0,0)}$$

$$+ (1-\delta_{n,N-1})(n+1)\mu\overline{P}^L_{1,n+1,k}(s,x)$$

$$+ (1-\delta_{k0})(1-\delta_{n0})(N-n)\lambda\overline{P}^L_{1,n-1,k-1}(s,x), \ 0 \le n \le N-1, k \ge 0, \qquad (2)$$

$$0 \le n \le N-1, k \ge 0,$$

$$\overline{P}^L_{1nk}(s,0) = (N-n)\lambda\overline{P}^L_{0nk}(s), \qquad (3)$$

$$1 \le n \le N-1, k \ge 0,$$

$$\overline{h}^L_k(s) = \int_0^\infty \overline{P}^L_{10k}(s,x)\gamma(x)dx + \mu\overline{P}^L_{01k}(s), \qquad (4)$$

$$\sum_{n=0}^{N-1} \overline{P}^L_{0nk}(s) + \sum_{n=0}^{N-1} \overline{P}^L_{1nk}(s) = \frac{1}{s}\left\{\sum_{q=k}^\infty \left[\overline{h}^L_q(0) - \overline{h}^L_q(s)\right]\right\} s \ne 0. \qquad (5)$$

The systems (1)-(5) can be solved successively for $k = 0, 1, \ldots$ with the help of the discrete transformations method (see for example ([7,9,12]). According to this method we rewrite Eq. (2) in a matrix form,

$$[\theta I - A]\overline{P}^L_{1k}(s,x) = B_k,$$

and find the matrices Y and Λ, such that $Y^{-1}AY = \Lambda$, where Λ is a diagonal matrix. Thus, applying in (2) the transformations

$$\overline{P}^L_{1k}(s,x) = Y\overline{Q}^L_{1k}(s,x) \qquad (6)$$

we get it in the simpler form

$$[\theta I - \Lambda]\overline{Q}^L_{1k}(s, x) = Y^{-1}B_k. \tag{7}$$

Here

$$\theta = \gamma(x) + s + \frac{\partial}{\partial x},$$

I is the identity matrix of order N, Λ is constructed from (2) in the usual way, $\overline{P}^L_{1k}(s, x)$ is the vector of unknown quantities,

$$\overline{P}^L_{1k}(s, x) = \left(\overline{P}^L_{10k}(s, x), \ldots, \overline{P}^L_{1,N-1,k}(s, x)\right)^T,$$

and

$$B_0 = (\delta(x), 0, \ldots, 0)^T,$$

$$B_k = \left(0, (N-1)\lambda\overline{P}^L_{1,0,k-1}(s, x), \ldots, \lambda\overline{P}^L_{1,N-2,k-1}(s, x)\right)^T, k = 1, 2, \ldots$$

The matrices Y and Λ are found in [4], where the following proposition is proved.

Proposition 1. *The diagonal matrix Λ is equal to*

$$\Lambda = diag\{0, -(\mu + \lambda), \ldots, -(N-1)(\mu + \lambda)\},$$

and the entries of the kth column of Y, $(y_{0k}, \ldots, y_{N-1,k})^T$, $k = 0, 1, \ldots, N-1$ can be calculated by the recursive relations

$$y_{0k} = 1, \tag{8}$$

$$y_{nk} = \frac{-k(\lambda + \mu)}{n\mu}(y_{0k} + \cdots + y_{n-1,k}) + \frac{(N-n)\lambda}{n\mu}y_{n-1,k}, \tag{9}$$

$$n = 1, \ldots, N-1,$$

or by their equivalent formulas

$$y_{nk} = \sum_{i=0}^{n}(-1)^{n-i}\left(\frac{\lambda}{\mu}\right)^i\binom{N-k-1}{i}\binom{k}{n-i}, \tag{10}$$

with

$$\binom{j}{l} = 0 \text{ if } l > j.$$

Furthermore, for the sum of the first n coordinates of the kth column we have

$$\sum_{i=0}^{n}y_{ik} = \begin{cases} \sum_{i=0}^{n}\left(\frac{\lambda}{\mu}\right)^i\binom{N-1}{i} \text{ for } k = 0, \\ \sum_{i=0}^{n}(-1)^{n-i}\left(\frac{\lambda}{\mu}\right)^i\binom{N-k-1}{i}\binom{k-1}{n-i}, \\ \text{ for } k = 1, \ldots, N-1 \end{cases}$$

and therefore

$$\sum_{i=0}^{N-1} y_{ik} = \begin{cases} \left(1 + \frac{\lambda}{\mu}\right)^{N-1} & for\ k = 0 \\ 0\ for\ k = 1, \dots, N-1. \end{cases} \tag{11}$$

The results of this proposition are sufficient to solve the equations of statistical equilibrium as well as the equations determining the busy period length distribution (see [4]), but here, as Eq. (7) shows, we need the inverse matrix of Y. To this end we prove the next Theorem 1. The matrix Y depends on the system paremeters λ, μ and N. In the proof of Theorem 1 we use the dependence on N and, only in this theorem we will extend notations: the matrix Y will be denoted as $Y^{(N)}$, its entries - as $y_{nk}^{(N)}$ and the adjoint quantity of $y_{nk}^{(N)}$ - as $Y_{nk}^{(N)}$, $k = 0, 1, \dots, N-1, 0, \dots, N-1$.

Theorem 1. *The determinant of the matrix $Y^{(N)}$, defined in Proposition 1 is equal to*

$$\det\left(Y^{(N)}\right) = \left[-\left(1 + \frac{\lambda}{\mu}\right)\right]^{\frac{N(N-1)}{2}}, \tag{12}$$

and its inverse matrix, $\left(Y^{(N)}\right)^{-1}$ is equal to

$$\left(Y^{(N)}\right)^{-1} = \left(\frac{\mu}{\lambda + \mu}\right)^{N-1} Y^{(N)}. \tag{13}$$

Proof. It is easy to verify Eqs. (12) and (13) for $N = 2,\ 3$, i.e. that

$$\left(Y^{(2)}\right)^{-1} = \frac{\mu}{\lambda + \mu} Y^{(2)}, \quad \det(Y^{(2)}) = -\left(1 + \frac{\lambda}{\mu}\right),$$

$$\left(Y^{(3)}\right)^{-1} = \frac{\mu^2}{(\lambda + \mu)^2} Y^{(3)}, \det\left(Y^{(3)}\right) = -\left(1 + \frac{\lambda}{\mu}\right)^3.$$

Further, let us suppose that they hold for $N - 1$,

$$\left(Y^{(N-1)}\right)^{-1} = \frac{\mu^{N-2}}{(\lambda + \mu)^{N-2}} Y^{(N-1)},$$

and that

$$\det\left(Y^{(N-1)}\right) = \left[-\left(1 + \frac{\lambda}{\mu}\right)\right]^{\frac{(N-1)(N-2)}{2}}.$$

The first of these equations means that the adjoint quantity $Y_{ij}^{(N-1)}$ of the matrix $Y^{(N-1)}$ is equal to

$$Y_{ij}^{(N-1)} = \det(Y^{(N-1)}) \frac{\mu^{N-2}}{(\lambda + \mu)^{N-2}} y_{ji}^{(N-1)}. \tag{14}$$

To prove that these relations hold and for the matrix $Y^{(N)}$, we establish relations between the entries of $Y^{(N)}$ and $Y^{(N-1)}$. For $n = 1, 2, \ldots, N - 1$ it holds:

$$y_{ni}^{(N)} = y_{n,i-1}^{(N-1)} - y_{n-1,i-1}^{(N-1)}, i = 1, \ldots, N - 1, \tag{15}$$

$$y_{n0}^{(N)} = y_{n0}^{(N-1)} + \frac{\lambda}{\mu} y_{n-1,0}^{(N-1)}. \tag{16}$$

The first of these equations can be proved with the help of formula (10), the second follows from (9) for $k = 0$ and the binomial formula

$$\binom{N-1}{k} = \binom{N-2}{k} + \binom{N-2}{k-1}.$$

Using (15), (16) and induction on N, it is easy to prove that for all $n = 1, 2, \ldots$ it holds:

$$y_{ni}^{(N)} - y_{n,i+1}^{(N)} = \left(1 + \frac{\lambda}{\mu}\right) y_{n-1,i}^{(N-1)}, i = 0, \ldots, N - 2. \tag{17}$$

Now, if in $Y^{(N)}$ we subtract successively the kth column from the $(k-1)^{th}$, $k = 1, \ldots, N - 1$, then from (17), we get it in the form

$$Y^{(N)} = \begin{pmatrix} 0 & 0 & \cdots & 0 & 1 \\ \left(1+\frac{\lambda}{\mu}\right) y_{0,0}^{(N-1)} & \left(1+\frac{\lambda}{\mu}\right) y_{01}^{(N-1)} & \cdots & \left(1+\frac{\lambda}{\mu}\right) y_{0,N-2}^{(N-1)} & y_{1,N-1}^{(N)} \\ \left(1+\frac{\lambda}{\mu}\right) y_{1,0}^{(N-1)} & \left(1+\frac{\lambda}{\mu}\right) y_{11}^{(N-1)} & \cdots & \left(1+\frac{\lambda}{\mu}\right) y_{1,N-2}^{(N-1)} & y_{2,N-1}^{(N)} \\ & & \cdot & & \cdot \\ \left(1+\frac{\lambda}{\mu}\right) y_{N-2,0}^{(N-1)} & \left(1+\frac{\lambda}{\mu}\right) y_{N-1,1}^{(N-1)} & \cdots & \left(1+\frac{\lambda}{\mu}\right) y_{N-2,N-2}^{(N-1)} & y_{N-1,N-1}^{(N)} \end{pmatrix}$$

This means that

$$\det(Y^{(N)}) = (-1)^{N-1} \left(1 + \frac{\lambda}{\mu}\right)^{N-1} \det(Y^{(N-1)}), \tag{18}$$

which proves formula (12).

Further, applying the same procedure in the adjoint quantity $Y_{nk}^{(N)}$, $k = 1, \ldots, N - 1$, we obtain that

$$Y_{nk}^{(N)} = (-1)^{N-1} \left(1 + \frac{\lambda}{\mu}\right)^{N-2} \left[Y_{n-1,k}^{(N-1)} - Y_{n-1,k-1}^{(N-1)}\right], k = 1, \ldots, N - 1,$$

which, according to (14), (15) and (18) gives

$$Y_{nk}^{(N)} = (-1)^{N-1} \det(Y^{(N-1)}) \left[y_{k,n-1}^{(N-1)} - y_{k-1,n-1}^{(N-1)}\right]$$

$$= \left(\frac{\mu}{\lambda + \mu}\right)^{N-1} \det(Y^{(N)}) y_{kn}^{(N)}.$$

For $k = 0$, if we add all rows of $Y^{(N)}$ to the nth one and apply formulas (11), we have

$$\det(Y^{(N)}) = \left(1 + \frac{\lambda}{\mu}\right)^{N-1} Y_{n0}^{(N)}.$$

Regarding to the fact that $y_{0j}^{(N)} = 1$, the last equation completes the proof of (13) and of the Theorem 1.

Now we are ready to calculate successively the probabilities P_k^{LBP} that exactly k lost calls occur during the busy period. For $k = 0$ the matrix Eq. (7) gives N linear differential equations,

$$\frac{\partial}{\partial x}\overline{Q}_{1nk}^L(s, x) + [n(\lambda + \mu) + \gamma(x) + s]\overline{Q}_{1n0}^L(s, x)$$

$$= \left(\frac{\mu}{\lambda + \mu}\right)^{N-1} \delta(x)y_{n0}, \ \ n = 0, 1, \ldots, N - 1. \qquad (19)$$

These equations allow to express the quantities $\overline{Q}_{1n0}^L(s, x)$ in terms of their initial conditions, $\overline{Q}_{1n0}^L(s, 0)$, and from the relation (6) between $\overline{Q}_{1n0}^L(s, x)$ and $\overline{P}_{1n0}^L(s, x)$ - to express $\overline{P}_{1n0}^L(s, x)$ in terms of $\overline{Q}_{1n0}^L(s, 0)$. Then, from the relations (3) between $\overline{P}_{0n0}^L(s)$ and $\overline{P}_{1n0}^L(s, 0)$, and the relations (4) between $\overline{P}_{100}^L(s, x), \overline{P}_{010}^L(s)$ and $\overline{h}_0^L(s)$ we express $\overline{P}_{0n0}^L(s)$ and $\overline{h}_0^L(s)$ by $\overline{Q}_{1n0}^L(s, 0)$. At the end, substituting in (1) and in the normalizing conditions (5) we obtain a system of linear equations for the quantities $\overline{Q}_{1n0}^L(s, 0)$. The Laplace – Stieltjes transform of the busy period length, $\eta(s)$ should be calculated by the formulas, given in [5]. Here these formulas are presented in the next Section, Theorem 2.

For $k > 0$ the procedure is the same, but in the right hand side of (19) participate the Laplace transforms $\overline{P}_{1,0,k-1}^L(s, x), \ldots, \overline{P}_{1,N-2,k-1}^L(s, x)$ which we already know. Here we will not go to details about the formulas for computing the distribution of the number of lost customers, made during the busy period. We will turn our attention to the moments of this distribution.

3   Moments of the Number of Lost Calls Made During the Busy Period

We define

$$ML_{in}^{(j)}(t) = \sum_{k=0}^{\infty} k^j P_{ink}^L(t),$$

$$M^{(j)}(t) = \sum_{k=0}^{\infty} k^j h_k^L(t)$$

with Laplace transforms $\overline{ML}_{in}^{(j)}(s)$, $\overline{M}^{(j)}(s)$ and

$$E\left[(N^{LBP})^j\right] = \sum_{k=0}^{\infty} k^j P\left\{N^{SBP} = k\right\} = \sum_{k=0}^{\infty} k^j \overline{h}_k^L(0) = \overline{M}^{(j)}(0).$$

For $j = 0$ we have

$$\overline{ML}_{1n}^{(0)}(s,x) = \sum_{k=0}^{\infty} \overline{P}_{1nk}^{L}(s,x) = \int_{0}^{\infty} e^{-st} P_{1n}(t,x)dt = \overline{P}_{1n}(s,x),$$

$$\overline{ML}_{in}^{(0)}(s) = \sum_{k=0}^{\infty} \overline{P}_{ink}^{L}(s) = \int_{0}^{\infty} e^{-st} P_{in}(t)dt = \overline{P}_{in}(s),$$

where

$$P_{1n}(t,x) = P\{\zeta > t, C(t) = 1, N(t) = n, 0 < z(t) \le x\},$$

$$P_{in}(t) = P\{\zeta > t, C(t) = i, N(t) = n\}.$$

Formulas for computing the Laplace transforms $\overline{P}_{1n}(s,x), \overline{P}_{in}(s)$ as well as the Laplace – Stieltjes transform, $\eta(s)$, of the busy period distribution function are obtained in [5], where the following theorem is proved.

Theorem 2. *The Laplace transforms* $\overline{P}_{1n}(s,x), \overline{P}_{in}(s)$ *of the probabilities* $P_{1n}(t,x), P_{in}(i),$ $i = 0,1$ *and the Laplace – Stieltjes transform,* $\eta(s)$, *of the busy period distribution function can be calculated by the formulas*

$$\overline{P}_{1n}(s,x) = [1 - G(x)] \sum_{k=0}^{N-1} y_{nk} e^{-[k(\lambda+\mu)+s]x} \left[\overline{Q}_{1k}(s,0) \right.$$

$$\left. + \left(\frac{\mu}{\lambda+\mu} \right)^{N-1} y_{k0} \right], \tag{20}$$

$$\overline{P}_{1n}(s) = \int_{0}^{\infty} \overline{P}_{1n}(s,x)dx$$

$$= \sum_{k=0}^{N-1} y_{nk} \frac{1 - g_k(s)}{k(\lambda+\mu) + s} \left[\overline{Q}_{1k}(s,0) + \left(\frac{\mu}{\lambda+\mu} \right)^{N-1} y_{k0} \right],$$

$$0 \le n \le N - 1,$$

$$\overline{P}_{0n}(s) = \frac{1}{(N-n)\lambda} \sum_{k=0}^{N-1} y_{nk} \overline{Q}_{1k}(s,0), 1 \le n \le N - 1.$$

$$\eta(s) = \left\{ \sum_{k=0}^{N-1} \overline{Q}_{1k}(s,0) \left[1 + g_k(s) - \frac{k(\lambda+\mu)}{(N-1)\lambda} \right] \right.$$

$$\left. + \sum_{k=0}^{N-1} g_k(s) \left(\frac{\mu}{\lambda+\mu} \right)^{N-1} y_{k0} \right]$$

where the initial conditions $\overline{Q}_{1k}(s,0)$ *satisfy the following system of linear equations*

$$\sum_{k=0}^{N-1} \overline{Q}_{1k}(s,0) \left\{ y_{nk} \left[\delta_{n,N-1} + \frac{n\mu + s}{(N-n)\lambda} - g_k(s) \right] \right.$$

$$\left. + (1 - \delta_{n,N-1}) \frac{k(\lambda+\mu)}{(N-n-1)\lambda} (y_{0k} + \cdots + y_{nk}) \right\}$$

$$= \left(\frac{\mu}{\lambda + \mu}\right)^{N-1} {}_{k=0}^{N-1} g_k(s) y_{nk} y_{k0} \tag{21}$$

$$1 \leq n \leq N - 1,$$

$$\overline{Q}_{10}(s,0) \left\{ 1 + g_0(s) + \sum_{n=1}^{N-1} \frac{s y_n^{(0)}}{(N-n)\lambda} + [1 - g_0(s)] \left(\frac{\lambda + \mu}{\mu}\right)^{N-1} \right\}$$

$$+ \sum_{k=1}^{N-1} \overline{Q}_{1k}(s,0) \left\{ 1 + g_k(s) + \sum_{n=1}^{N-1} \frac{s y_{nk}}{(N-n)\lambda} - \frac{k(\lambda + \mu)}{(N-1)\lambda} \right\}$$

$$= g_0(s) - \left(\frac{\lambda + \mu}{\mu}\right)^{N-1} \sum_{k=0}^{N-1} g_k(s) y_{k0}. \tag{22}$$

Here y_{nk} are given in Proposition 1, $g_k(s) = g(k(\lambda + \mu) + s)$.

Further, for $j = 1$ we derive equations for $\overline{ML}_{in}^{(1)}(s)$, $\overline{M}^{(1)}(s)$ multiplying each of the Eqs. (1)-(4) by k and summing over $k = 1, 2, \ldots$:

$$[(N-n)\lambda + n\mu + s]\overline{ML}_{0n}^{(1)}(s)$$

$$= (1 - \delta_{n,N-1})(n+1)\mu\overline{ML}_{0,n+1}^{(1)}(s) + \int_0^\infty \overline{ML}_{1n}^{(1)}(s,x)\gamma(x)dx, \tag{23}$$

$$[(N-n-1)\lambda + n\mu + \gamma(x) + s + \frac{\partial}{\partial x}]\overline{ML}_{1n}^{(1)}(s,x)$$

$$= (1 - \delta_{n,N-1})(n+1)\mu\overline{ML}_{1,n+1}^{(1)}(s,x) + (1 - \delta_{n0})(N-n)\lambda\overline{ML}_{1,n-1}^{(1)}(s,x)$$

$$+ (1 - \delta_{n0})(N-n)\lambda\overline{ML}_{1,n-1}^{(0)}(s,x), \tag{24}$$

$$0 \leq n \leq N - 1,$$

$$\overline{ML}_{1n}^{(1)}(s,0) = (N-n)\lambda\overline{ML}_{0n}^{(1)}(s), \tag{25}$$

$$1 \leq n \leq N - 1,$$

$$\overline{M}^{(1)}(s) = \sum_0^\infty \overline{ML}_{10}^{(1)}(s,x)\gamma(x)dx + \mu\overline{ML}_{01}^{(1)}(s). \tag{26}$$

In the same way, summing the normalizing condition (5) over k we have:

$$\frac{1 - \eta(s)}{s} = \frac{\overline{M}^{(1)}(0) - \overline{M}^{(1)}(s)}{s}, \quad s \neq 0. \tag{27}$$

Solving the system (23)-(27) we can find the mean number of lost calls, made during the busy period, $E\left[N^{LBP}\right] = \overline{M}^{(1)}(0)$. We use again the method of discrete transformations and rewrite Eq. (24) in a matrix form,

$$[\theta I - A]\overline{ML}_1^{(1)}(s,x) = D_1. \tag{28}$$

Here θ, I and A are the same as in the matrix form of (2), $\overline{ML}_1^{(1)}(s,x)$ is the column vector of the unknown quantities,

$$\overline{ML}_1^{(1)}(s,x) = \left(\overline{ML}_{10}^{(1)}(s,x),\ldots,\overline{ML}_{1,N-1}^{(1)}(s,x)\right)^T,$$

and

$$D_1 = \left(0,(N-1)\lambda\overline{ML}_{10}^{(0)}(s,x),\ldots,\lambda\overline{ML}_{1,N-2}^{(0)}(s,x)\right)^T,$$

This mean that the transformation

$$\overline{ML}_1^{(1)}(s,x) = Y\overline{Q}_1^{(1)}(s,x) \tag{29}$$

will simplify (28) to the form

$$[\theta I - \Lambda]\,\overline{Q}_1^{(1)}(s,x) = Y^{-1}D_1. \tag{30}$$

Now, to obtain formulas for calculating the mean number of lost calls, made during the busy period we follow the already discribed procedure. From the differential equations, equivalent to (30) we express the functions $\overline{Q}_1^{(1)}(s,x)$ in terms of their initial conditions, $\overline{Q}_1^{(1)}(s,0)$. Then we express by $\overline{Q}_1^{(1)}(s,0)$ all Laplace transforms $\overline{ML}_{1n}^{(1)}(s,x)$, $\overline{ML}_{1n}^{(1)}(s)$, $\overline{ML}_{0n}^{(1)}(s)$ and substituting in (23) and (27) obtain system of linear equations for $\overline{Q}_1^{(1)}(s,0)$. At the end, with the help of (26) we find $\overline{M}^{(1)}(s)$ and $E\left[N^{LBP}\right] = \overline{M}^{(1)}(0)$. Thus we prove the next Theorem 3.

Theorem 3. *The Laplace transforms $\overline{ML}_{in}^{(1)}(s)$, $\overline{M}^{(1)}(s)$ can be calculated by the formulas*

$$\overline{ML}_{1i}^{(1)}(s) = \sum_{n=0}^{N-1} y_{in}\frac{1-g_n(s)}{n(\lambda+\mu)+s}\overline{Q}_{1n}^{(1)}(s,0)$$

$$+\left(\frac{\mu}{\lambda+\mu}\right)^{N-1}\sum_{n=0}^{N-1}\sum_{k=1}^{N-1}N-1(N-k)\lambda y_{nk}\sum_{q=0}^{N-1}y_{k-1,q}\overline{f}_{qn}$$

$$\times\left[\overline{Q}_{1q}(s,0)+\left(\frac{\mu}{\lambda+\mu}\right)^{N-1}y_{q0}\right], \tag{31}$$

where

$$\overline{f}_{qn} = \begin{cases} \frac{1}{(n-q)(\lambda+\mu)}\frac{1-g_q(s)}{q(\lambda+\mu)+s} & \text{for } q\neq n \\ \frac{-g_n(s)}{[n(\lambda+\mu)+s]^2}-\frac{g_n'(s)}{n(\lambda+\mu)+s} & \text{for } q=n \end{cases}$$

$$\overline{ML}_{0i}^{(1)}(s) = \frac{1}{(N-i)\lambda}\sum_{n=0}^{N-1} y_{in}\left\{\overline{Q}_{1n}^{(1)}(s,0)+\left(\frac{\mu}{\lambda+\mu}\right)^{N-1}\right.$$

$$\times\sum_{k=1}^{N-1}(N-k)\lambda y_{nk}\sum_{q=0}^{N-1}y_{k-1,q}\left[\overline{Q}_{1q}(s,0)+\left(\frac{\mu}{\lambda+\mu}\right)^{N-1}y_{q0}\right]\widetilde{f}_{qn}\right\}, \tag{32}$$

with

$$\tilde{f}_{qn} = \begin{cases} \frac{1}{(n-q)(\lambda+\mu)} & \text{for } q \neq n \\ 0 & \text{for } q = n \end{cases}$$

$$n = 0, 1, \ldots, N-1,$$

$$\overline{M}^{(1)}(s) = \sum_{n=0}^{N-1} y_{0n} g_n(s) \left[\overline{Q}_{1n}^{(1)}(s,0) + \left(\frac{\mu}{\lambda+\mu}\right)^{N-1} \right.$$

$$\times \sum_{k=1}^{N-1} (N-k)\lambda y_{nk} \sum_{q=0}^{N-1} y_{k-1,q} \left[\overline{Q}_{1q}(s,0) + \left(\frac{\mu}{\lambda+\mu}\right)^{N-1} y_{q0} \right] F_{qn} \right]$$

$$+ \frac{\mu}{(N-1)\lambda} \sum_{n=0}^{N-1} y_{1n} \left[\overline{Q}_{1n}^{(1)}(s,0) + \left(\frac{\mu}{\lambda+\mu}\right)^{N-1} \right.$$

$$\times \sum_{k=1}^{N-1} (N-k)\lambda y_{nk} \sum_{q=0}^{N-1} y_{k-1,q} \left[\overline{Q}_{1q}(s,0) + \left(\frac{\mu}{\lambda+\mu}\right)^{N-1} y_{q0} \right] \tilde{f}_{qn} \right], \quad (33)$$

$$F_{qn} = \begin{cases} \frac{1}{(n-q)(\lambda+\mu)} g_{n-k}(0) & \text{for } q \neq n \\ \frac{1}{\nu} & \text{for } q = n \end{cases} \quad (34)$$

The quantities $\overline{Q}_{1k}(s,0)$ are solutions of the system (21), (22) of Theorem 2, and $\overline{Q}_{1k}^{(1)}(s,0)$ satisfy the following system of linear equations:

$$\sum_{n=0}^{N-1} \overline{Q}_{1n}(s,0) \left\{ y_{in} \left[\delta_{i,N-1} + \frac{i\mu+s}{(N-i)\lambda} - g_i(s) \right] \right.$$

$$\left. + (1-\delta_{i,N-1}) \frac{n(\lambda+\mu)}{(N-i-1)\lambda} (y_{0n} + \cdots + y_{in}) \right\}$$

$$= \left(\frac{\mu}{\lambda+\mu}\right)^{N-1} \left\{ \frac{(1-\delta_{i,N-1})(i+1)\mu}{(N-i-1)\lambda} \sum_{n=0}^{N-1} y_{i+1,n} \sum_{k=1}^{N-1} (N-k)\lambda y_{nk} \sum_{q=0}^{N-1} y_{k-1,q} \right.$$

$$\times \left[\overline{Q}_{1q}(s,0) + \left(\frac{\mu}{\lambda+\mu}\right)^{N-1} y_{q0} \right] \tilde{f}_{qn}$$

$$+ \sum_{n=0}^{N-1} y_{in} e^{-[n(\lambda+\mu)+s]x} \sum_{k=1}^{N-1} (N-k)\lambda y_{nk} \sum_{q=0}^{N-1} y_{k-1,q}$$

$$\times \left[\overline{Q}_{1q}(s,0) + \left(\frac{\mu}{\lambda+\mu}\right)^{N-1} y_{q0} \right] R_{qn}(s)$$

$$- \frac{[(N-i)\lambda+i\mu+s]}{(N-i)\lambda} \sum_{n=0}^{N-1} y_{in} \sum_{k=1}^{N-1} (N-k)\lambda y_{nk} \sum_{q=0}^{N-1} y_{k-1,q}$$

$$\times \left[\overline{Q}_{1q}(s,0) + \left(\frac{\mu}{\lambda+\mu}\right)^{N-1} y_{q0} \right] \tilde{f}_{qn}, \quad (35)$$

$$i = 1, \ldots, N - 1,$$

$$R_{qn}(s) = \begin{cases} \frac{1}{(n-q)(\lambda+\mu)} g_k(s) \; for \; q \neq n \\ g_n'(s) \; for \; q = n \end{cases}, \tag{36}$$

$$A(s) - A(0) = 1 - \eta(s) - [B(s) - B(0)], \tag{37}$$

where

$$A(s) = \sum_{n=0}^{N-1} \overline{Q}_{1n}^{(1)}(s, 0) \left[y_{0n} g_n(s) - \frac{\mu}{(N-1)\lambda} y_{1n} \right],$$

$$B(s) = \sum_{n=0}^{N-1} y_{0n} g_n(s) \left(\frac{\mu}{\lambda + \mu} \right)^{N-1}$$

$$\times \sum_{k=1}^{N-1} (N-k)\lambda y_{nk} \sum_{q=0}^{N-1} y_{k-1,q} \left[\overline{Q}_{1q}(s, 0) + \left(\frac{\mu}{\lambda + \mu} \right)^{N-1} y_{q0} \right] F_{qn} \right]$$

$$+ \frac{\mu}{(N-1)\lambda} \left(\frac{\mu}{\lambda + \mu} \right)_{n=0}^{N-1} y_{1n} \sum_{k=1}^{N-1} (N-k)\lambda y_{nk} \sum_{q=0}^{N-1} y_{k-1,q}$$

$$\times \left[\overline{Q}_{1q}(s, 0) + \left(\frac{\mu}{\lambda + \mu} \right)^{N-1} y_{q0} \right] \widetilde{f}_{qn}.$$

Here, $\overline{Q}_{1n}^{(1)}(0,0)$ *are solutions of (35)-(37) for* $s = 0$ *(*$s \to 0$*),* $g_k(s)$*, as in Theorem 2, is equal to the Laplace-Stieltjes transform,* $g(s)$ *of the service times in the point* $s + k(\lambda + \mu), y_{in}$ *can be calculated according to the formulas of Proposition 1.*

In a similar way we can deal with the moments of higher order.

4   Conclusion

In this paper we consider a finite source queueing system of M/G/1 type in which the failed customers are not allowed neither to queue nor to do repetitions. Instead, they are temporarily blocked in the orbit of inactive customers. We investigate a descriptor of the system functioning, connected with its busy period: the number of lost calls made during the busy period. Formulas for computing the mean value of this descriptor are derived. The discrete transformations, which are established here for analysis of the lost calls distribution can be also applied for transient analysis of the system, as a possible future work.

References

1. Amador, J.: On the distribution of the successful and blocked events in retrial queues with finite number of sources. In: Proceedings of the 5th International Conference on Queueing Theory and Network Applications, pp. 15–22 (2010)
2. Artalejo, J.R., Gomez-Corral, A.: Retrial Queueing Systems: A Computational Approach. Springer, Berlin Heidelberg (2008)

3. Artalejo, J.R., Phung-Duc, T.: Markovian retrial queues with two way communications. J. Ind. Manage. Optim. **8**(4), 781–806 (2012)
4. Dragieva, V.I.: Steady state analysis of the M/G/1//N queue with orbit of blocked customers. Annals of Operations Research (accepted)
5. Dragieva, V.I.: On the busy period in one finite queue of M/G/1 type with inactive orbit. Serdica J. Comput. **8**(3), 291–308 (2014)
6. Efrosinin, D., Breuer, L.: Threshold policies for controlled retrial queues with heterogeneous servers. Ann. Oper. Res. **141**, 139–162 (2006)
7. Falin, G.I., Artalejo, J.R.: A finite source retrial queue. Eur. J. Oper. Res. **108**, 409–424 (1998)
8. Falin, G.I., Templeton, J.G.C.: Retrial Queues. Chapman and Hall, London (1997)
9. Jaiswal, N.: Priority Queues. Academic press, New York (1969)
10. Kim, C., Klimenok, V., Birukov, A., Dudin, A.: Optimal multi-threshold control by the BMAP/SM/1 retrial system. Ann. Oper. Res. **141**, 193–210 (2006)
11. Moscholios, L., Logothetis, M.: Engset multi - rate state - dependent loss models with Qos guarantee. Int. J. Commun. Syst. **19**(1), 67–93 (2006)
12. Wang, J., Zhao, L., Zhang, F.: Analysis of the finite source retrial queues with server breakdowns and repairs. J. Ind. Manage. Optim. **7**, 655–676 (2011)
13. Wong, E.W.M., Zalesky, A., Zukerman, M.: On generalization of the Engset model. IEEE Commun. Lett. **11**(4), 360–362 (2007)
14. Zhang, J., Peng, Y., Wong, E.W.M., Zukerman, M.: Sensitivity of blocking probability in the generalized Engset model for OBS. IEEE Commun. Lett. **15**(11), 1243–1245 (2011)

Performance of the DCF Access Method in 802.11 Wireless LANs

Pavel Mikheev$^{(\boxtimes)}$ and Sergey Suschenko

Tomsk State University, Lenina Street, 36, 634050 Tomsk, Russia
{doka.patrick,ssp.inf.tsu}@gmail.com

Abstract. A mathematical model of access method "carrier sense multiple access with collision avoidance" for two active stations was proposed. The effect of carrier capture and unimodal dependence of the operating characteristics from the initial width of the contention window was detected. Measures of preventing the effect of carrier capture, based on the modifications of the standard protocol were proposed. For research into the capture effect with a large number of rivals, a simulation model of the competition process was developed. The efficiency of prevention measures ensuring fair distribution of a jointly used time resource within a shared communication medium with an insignificant decrease in the general throughput of the access method has been shown.

Keywords: 802.11 wireless networks · Contention · A random delay timer · Positive acknowledgment · The effect of carrier capture · Throughput

1 Random Multiple Access Method in 802.11 Wireless Networks

Let us analyze the wireless local area network (LAN) based on the IEEE 802.11 standard. The fundamental access method of such LANs is called DCF (Distributed Coordination Function) [1,2] known as *carrier sense multiple access with collision avoidance* (CSMA/CA) [2–4]. This mechanism is based upon the fact that the transmitting station checks whether the carrier signal is present in the medium, and, before starting transmission of a data frame, expects release of the communication medium. IEEE 802.11 stations, in contrast to wired Ethernet, are not capable of detecting collisions in a communication medium [1,5]. Due to this fact, detection of collisions and non-conflict transmissions of protocol-based data units is based on the time-outs mechanism and on the algorithm of positive decision feedback.

Let us analyze the cycle of a data frame transmission from the sending station to the recipient station. First and foremost, the sending station senses the medium to determine if another station is transmitting. Thereafter, at the end of the inter-frame interval, the random delaying algorithm is initiated to select

A. Dudin et al. (Eds.): ITMM 2015, CCIS 564, pp. 99–113, 2015.
DOI: 10.1007/978-3-319-25861-4_9

a random backoff interval (the number of a slot in which the data transmission may be started). The slot number is selected with equal probability from the interval $[0, S_n - 1]$, where S_n is the size of the contention window measured in slot intervals t_c and determined by the relation $S_n = 2^{N_0+m}$, $m = n$ if $n \leq 10 - N_0$ and $m = 10 - N_0$ if $n \geq 10 - N_0$. Here $N_0 = \overline{1, 10}$ is the initial value predetermining the width of the contention window during the first attempt of a sender to transfer data, and $n \geq 0$ is the number of retransmission. The width of the contention window may not exceed the maximum value established by the standard. For all physical layers and methods of modulation, the IEEE 802.11 standard has established the maximum width of the contention window equal to $S_{max} = 1024$ [2]. The number of a selected slot shall be assigned to the backoff interval counter t_o, after which the countdown of slot intervals begins. At the end of each slot interval, the backoff interval counter shall decrement as long as medium is idle. If the medium is determined to be busy at any time during a backoff slot, then the backoff procedure is suspended. Decrementing is resumed when the medium is idle again. Transmission shall commence when the backoff interval counter reaches zero ($t_o = 0$). When the transmission is completed, the sender waits for a acknowledgement during the time t_{out}, after which it is considered that a conflict has occurred, and stations having got into such conflict increase the n value by one, and the actions targeted at data transmission are repeated. The width of the contention window is doubled with each attempt of data frame transmission, until the maximum value is achieved; and the width of the contention window remains equal to S_{max} with each subsequent attempt of data frame transmission. After successful transmission, the window width obtains the initial value S_0.

Thus, the wireless access technology, due to lack of possibility to detect collisions in a communication medium, has three significant differences from the random access method implemented in the wired medium. Firstly, the wireless transmission method employs the mechanism of positive feedback (positive acknowledgements). Secondly, in contrast to the random access method, in wired networks the WiFi technology employs the random delay mechanism as early as during the first transmission. And at last, the wireless access protocol employs the mechanism of "suspension" of the delaying timer from the time of detection of the medium occupation until expiration of the random delay timer.

2   Mathematic Modelling of 802.11 Wireless LAN

Let us analyze the operation of a wireless local area network until the first error-free data frame transmission with obtained acknowledgement on successful delivery of data. Let us suppose that the wireless LAN contains K stations which are data sources. Consider that all the sources are independent and equal, and always have data frames for sending, and all interval spaces are expressed in slot intervals t_c. Let all the stations exchange frames of equal sizes. Then, according to the sequence of protocol actions, the elementary cycle of data frame transfer to the recipient will be determined by the size of the interframe space t_m, random delay period t_o, duration of "suspension" of the random delay timer t_z, time of data frame transmission t_k, and the value of time-out for expecting a positive acknowledgement t_{out}, which consists of a short interframe space plus the time of transmission of a positive acknowledgement [2,4]. The average time of data frame transmission $T(K, N_0)$ consists of the weighted sum of average periods of waiting for failed transmissions and the time of successful transmission [6]:

$$T(K, N_0) = d + \sum_{N=0}^{\infty} \left[Nd + \sum_{n=0}^{N-1} t(n, K, N_0) + \tau(N, K, N_0) \right] f(N, K, N_0). \quad (1)$$

Here $d = t_m + t_k + t_{out}$, $t(n, K, N_0)$ and $\tau(N, K, N_0)$ are the average conditional times until failed and successful N-th repeated attempts to send a data frame by a subscriber, and $f(N, K, N_0)$ is the function of probability [7] of the duration of competition between subscribers for the medium, which is determined by the probability of successful data frame transmission on the N-th repeated step after $N - 1$ failures [6]:

$$f(N, K, N_0) = P(N, K, N_0) \prod_{n=0}^{N-1} \pi(n, K, N_0). \quad (2)$$

Along with the average time of data frame transmission, one of the main indicators showing the efficiency of functioning the data transfer network is the throughput performance. In the case under analysis, we will look for an individual throughput performance, the standardized value of which shall be determined as a ratio between the time necessary for data frame transmission t_k and the average time of data frame transmission $T(K, N_0)$:

$$C(K, N_0) = \frac{t_k}{T(K, N_0)}. \quad (3)$$

Let us analyze the competition of two wireless stations $(K = 2)$ of a local area network. We denote the competing (conflicting) stations through A and B. Let us find the probability timing characteristics of the data transmission process executed by the A station. Let us denote via $p_n(i)$ the probability of selection of random backoff interval with a duration equal to i slot intervals on the n-th repeated transmission by the A station, and via $f_n(j)$ the probability of selection of random backoff interval with a duration equal to j slot intervals on the n-th repeated transmission by the B station. Then the conditional probability of a conflict on the n-th repeated transmission for the A station is determined by the relation

$$\pi(n, 2, N_0) = \begin{cases} \sum_{i=0}^{S_0-1} p_0(i) \sum_{j=0}^{i} f_0(j) L_{i-j}, & n = 0; \\ \sum_{k=1}^{n} E_k(n) \left[\sum_{i=0}^{S_k-1} p_n(i) \sum_{j=0}^{i} f_k(j) L_{i-j} \right. \\ \qquad \left. + \sum_{i=S_k}^{S_n-1} p_n(i) \sum_{j=0}^{S_k-1} f_k(j) L_{i-j} \right], & n \geq 1. \end{cases} \quad (4)$$

Here L_k represents recurrent probabilities of movement of the B station "bottom-up" from originally selected slot interval j to a conflict slot interval i selected by the A station (k is a difference between j-th and i-th slots), for many steps with successful transmissions:

$$
L_k = \begin{cases} \sum\limits_{i=0}^{\infty} f_0^i(0) \sum\limits_{i=1}^{k} f_0(i) L_{k-i}, & k = \overline{1, S_0 - 1}, \ L_0 = 1; \\ \sum\limits_{i=0}^{\infty} f_0^i(0) \sum\limits_{i=1}^{S_0-1} f_0(i) L_{k-i}, & k = \overline{S_0, S_n - 1}. \end{cases} \tag{5}
$$

In other words elements L_k include probabilities of all possible actions of the B station before collision with the A station, if the B station originally selected slot interval j and the A station selected slot interval i. From this point, it is not difficult to see that, before the conflict with the A rival, the competing B station may carry out an unlimited number of successful transmissions in case of "fallout" of random delay having zero duration. Using the relations for the arithmetic-geometrical progression [8] for L_k with $k = \overline{1, S_0 - 1}$, we obtain the final relation:

$$
L_k = \frac{S_0^{k-1}}{(S_0 - 1)^k}, \quad k = \overline{1, S_0 - 1}. \tag{6}
$$

Inserting (6) into (4), we find the probability of a conflict on the first attempt of data frame transmission:

$$
\pi(0, 2, N_0) = \frac{S_0 - 1}{S_0^2} \left[\left(\frac{S_0}{S_0 - 1} \right)^{S_0} - 1 \right]. \tag{7}
$$

The coefficients $E_k(n)$ in the relation (4) are the probabilities that on the n-th repeated transmission by the A station, the B station will be in the condition of the k-th repeated transmission:

$$
E_1(1) = 1;
$$

$$
E_1(n) = \sum_{k=1}^{n-1} \frac{E_k(n-1)}{\pi(n-1, 2, N_0)} \left[\sum_{i=1}^{S_k-1} p_{n-1}(i) \sum_{j=0}^{i-1} f_k(j) L_{i-j} \right.
$$

$$
\left. + \sum_{i=S_k}^{S_{n-1}-1} p_{n-1}(i) \sum_{j=0}^{S_k-1} f_k(j) L_{i-j} \right], \quad n \geq 2; \tag{8}
$$

$$
E_k(n) = \frac{E_{k-1}(n-1) \sum\limits_{i=0}^{S_{k-1}-1} p_{n-1}(i) f_{k-1}(i)}{\pi(n-1, 2, N_0)}, \quad n \geq 2, \ k = \overline{2, n}.
$$

The average conditional times until failed and successful n-th attempt of data transmission $t(N, K, N_0)$ and $\tau(N, K, N_0)$ consist of the average duration of random delay $N_s(n)$ (average number of slots until the start of transmission) and the average number of suspensions caused by medium capture by the B station, $Z_t(n, N_0)$ in case of failure and $Z_\tau(n, N_0)$ in case of success, respectively:

$$
t(n, 2, N_0) = N_s(n) + Z_t(n, N_0) d, \quad \tau(n, 2, N_0) = N_s(n) + Z_\tau(n, N_0) d.
$$

Here

$$N_s = \sum_{i=0}^{S_n-1} i p_n(i) = \frac{S_n - 1}{2}, \tag{9}$$

and the average numbers of suspensions $Z_t(n, N_0)$ and $Z_\tau(n, N_0)$ look similar:

$$Z_t(n, N_0) = \begin{cases} \sum_{i=1}^{S_0-1} p_0(i) \sum_{j=0}^{i-1} f_0(j) M_{i-j}, & n = 0; \\ \sum_{k=1}^{n} E_k(n) \left[\sum_{i=1}^{S_k-1} p_n(i) \sum_{j=0}^{i-1} f_k(j) M_{i-j} \right. \\ \left. + \sum_{i=S_k}^{S_n-1} p_n(i) \sum_{j=0}^{S_k-1} f_k(j) M_{i-j} \right], & n \geq 1; \end{cases} \tag{10}$$

$$Z_\tau(n, N_0) = \begin{cases} \sum_{i=1}^{S_0-1} p_0(i) \sum_{j=0}^{i-1} f_0(j) V_{i-j}, & n = 0; \\ \sum_{k=1}^{n} E_k(n) \left[\sum_{i=1}^{S_k-1} p_n(i) \sum_{j=0}^{i-1} f_k(j) V_{i-j} \right. \\ \left. + \sum_{i=S_k}^{S_n-1} p_n(i) \sum_{j=0}^{S_k-1} f_k(j) V_{i-j} \right], & n \geq 1; \end{cases} \tag{11}$$

The elements M_k and V_k are indicators of the average number of suspensions of the delaying timer for the A station after selection of random delay with the duration i on the n-th repeated transmission upon selection of the j-th slot preceding to the i-th one by the competing B station (k is a difference between j-th and i-th slots):

$$M_k = \begin{cases} 0, & k = 0; \\ \sum_{m=1}^{k} f_0(m) \sum_{i=0}^{\infty} (i + 1 + M_{k-m}) f_0^i(0), & k = \overline{1, S_0 - 1}; \\ \sum_{m=1}^{S_0-1} f_0(m) \sum_{i=0}^{\infty} (i + 1 + M_{k-m}) f_0^i(0), & k = \overline{S_0, S_n - 1}; \end{cases}$$

$$V_k = \begin{cases} \sum_{i=0}^{\infty} (i + 1) f_0^i(0) \sum_{m=k+1}^{S_0-1} f_0(m) \\ + \sum_{m=1}^{k-1} f_0(m) \sum_{i=0}^{\infty} (i + 1 + V_{k-m}) f_0^i(0), & k = \overline{1, S_0 - 1}; \\ \sum_{m=1}^{S_0-1} f_0(m) \sum_{i=0}^{\infty} (i + 1 + V_{k-m}) f_0^i(0), & k = \overline{S_0, S_n - 1}. \end{cases}$$

After inserting here the probabilities of fallout of delay duration $f_0(m)$, we obtain the following relations:

$$M_k = \begin{cases} \dfrac{S_0}{S_0 - 1} \left[\left(\dfrac{S_0}{S_0 - 1} \right)^k - 1 \right], & k = \overline{1, S_0 - 1}; \\ \dfrac{S_0}{S_0 - 1} + \dfrac{\sum_{m=1}^{S_0-1} M_{k-m}}{S_0 - 1}, & k = \overline{S_0, S_n - 1}. \end{cases} \tag{12}$$

$$V_k = \begin{cases} \dfrac{S_0 - 2}{S_0 - 1}\left(\dfrac{S_0}{S_0 - 1}\right)^k, & k = \overline{1, S_0 - 1}; \\ \dfrac{S_0}{S_0 - 1} + \dfrac{\sum_{m=1}^{S_0-1} V_{k-m}}{S_0 - 1}, & k = \overline{S_0, S_n - 1}. \end{cases} \tag{13}$$

The indicator of the general throughput performance can be found by analogy with individual operational speed (3), therewith the numerator of such relation should be adjusted not only for the package successfully transferred by the A station, but also for the average number of packages transferred by the B station for the concerned period:

$$C_g(2, N_0) = \frac{(G(N_0) + 1)t_k}{T(2, N_0)},$$

where $G(N_0)$ will be determined by the weighted amount of the average number of suspensions of the delaying timer of the A station in expectation of failed and successful transmissions, which are determined by the relations (10) and (11):

$$G(N_0) = \sum_{N=0}^{\infty} \left[\sum_{n=0}^{N-1} Z_t(n, N_0) + Z_\tau(N, N_0)\right] f(N, 2, N_0).$$

The numeric research into the average time of data frame transmission by the A station shows that the function (1) has a strongly manifested minimum at the coordinate N_0 (see Fig. 1) determining the initial size of the competition window and, subsequently, the degree of scattering of stations by durations of delays before the start of the competition procedure. For two competing stations, the minimum is reached at $N_0 = 4$. It is obvious that the value N_0 minimizing the average time of data frame transmission maximizes the individual throughput (see Fig. 1). Moreover, as early as at the stage of formalization of the task, the probability of capture of the communication medium by one of the subscribers mentioned in [9,10] has become obvious. This effect manifests itself especially strongly with small values N_0. The effect of capturing the communication medium causes discrimination-related individual indicators against a good level of the general throughput performance of the network (see Fig. 1).

As early as at the first attempt of competition between two stations, capture of the communication medium becomes possible (e.g. by the B station), and its probability will be determined by the probabilities that for one of the stations (B) the delay duration will turn out to be shorter than the duration of delay of the other station (A); then the "succeeded" station (B) will have fallout of zero duration, which will alternate with shorter delays than the residual value of the station's A delaying timer:

$$P_z(0, 2, N_0) = \sum_{i=1}^{S_0-1} p_0(i)L_{i-1} \sum_{k=1}^{\infty} f_0^k(0) = \frac{1}{S_0^2}\left(\frac{S_0}{S_0 - 1}\right)^{S_0-1}.$$

From this point, it is not difficult to see that the probability of capture is considerably determined by the initial width of the contention window S_0 (see Fig. 2).

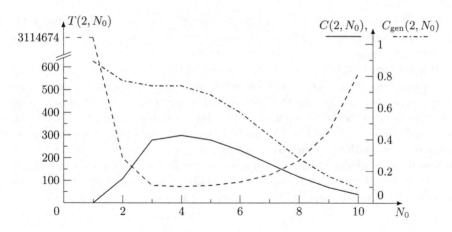

Fig. 1. Average time of data frame transmission, and individual and general throughput performances

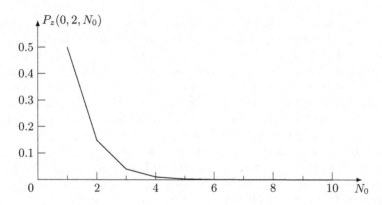

Fig. 2. Probability of the medium capture by one of the stations

After several conflicts, the possibility of capture for the "succeeded" station becomes yet more probable.

The main reason for the effect of capturing the communication medium is the protocol action — "suspension of delay", because this results in a fact that after a non-conflict transmission the station may capture the communication medium for an infinitely long time, getting into the delay interval from 0 to the residual value of the delay of other stations.

Another reason for an increase in the probability of capturing the communication medium by one of the subscribers after several conflicts, consists in various sizes of the contention window for stations withdrawn from the conflict and stations continuing resolution of the conflict in the condition of waiting for expiration of delay time and suspension periods. After a positive resolution of the conflict by one of the stations (or by several stations), the size of its contention window is reduced in multiples down to the initial value $S_0 < S_n$, which gives this station a priority right in subsequent competition for the medium with "conflicting" stations, because the shorter duration of an occasional delay for such station has a significantly higher probability as compared with the similar operational indicator of the "conflicting" station.

It is obvious that to reduce the probability of the effect of medium capturing for an infinitely long time, it is possible to offer, on one hand, to fix the size of the contention window for the first and all subsequent transmissions, and on the other hand – the duration of random delay t_o should be selected within the interval from 1 to $2^{N_0} - 1$ of slot periods t_c, thus excluding the delay of the zero size. Therewith, medium capturing by one station will never exceed $2^{N_0} - 2$ successful transmissions until the subsequent conflict or its resolution. Further we will analyze the measures proposed for modification of the procedure of access to the communication medium for prevention of the "medium capturing effect".

3   Preventing the Effect of Capturing Communication Medium

Let us start considering the methods for prevention of capturing the medium from the method based upon narrowing the range of random delay values. We will consider that during the process of competition for the medium, wireless communication stations make selection of random delay values from the interval $\overline{1, S_n - 1}$ of slot intervals. In such a case, any possibility of selection of zero delay is excluded, and no station will be able to capture the communication medium for an infinite time. Let us denote this variant of preventing the effect of capturing the access to the communication medium as the "modified method".

In case of a fixed width of the contention window, the probability timing characteristics of the system will be equal for each attempt of data transmission by the A and B stations. Then the conditional probability of a conflict on the n-th repeated transmission for the A station will be determined by the relation (7).

For the procedure of preventing capture effect, based upon a fixed width of the competition window, and by excluding any random delay with zero. duration, the probability timing characteristics of the communications system will be invariant to the number of the repeated transmission. It should be noted that this method of access to the communication medium should be used with $N_0 \geq 2$, because for $N_0 = 1$, the stations will select one and the same slot, i.e. will always get into a conflict, and none of the stations will be able to transfer data.

Figure 3 represents comparative curves showing the average time of data frame transmission by the subscriber in the network from two stations for all methods of access to the communication medium analyzed by us. The figure shows that (in case of 100 % loading) in a network of two stations the basic method of access to the communication medium with a fixed width of a competitive window ($BMFW$) is preferential as compared with the three other methods of struggle for the medium. On one hand, this method ensures less time of data frame transmission than the time provided for the basic and modified

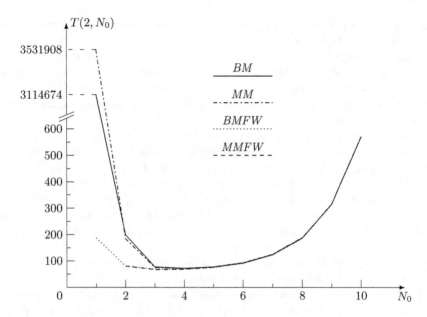

Fig. 3. Average time of data frame transmission for the basic (*BM*) and modifying (*MM*) access methods, and for the methods with fixed contention window width — (*BMFW*) and (*MMFW*)

methods, — with $N_0 = \overline{1,4}$, and on the other hand, it is not inferior to any of the methods along the entire fragment of acceptable values of the primary degree of the contention windows width. Generally, with $N_0 = \overline{5,10}$, the difference between the efficiency of the methods is offset.

4   Simulation Modeling of 802.11 Wireless LAN

With a view to studying the behavior of a wireless LAN and to analyzing its operation speed indices, a program simulating the logic of protocol actions has been developed for the basic method of access to the shared communication medium, as well as its suggested modifications to prevent any capture effect when the physical speed of transmission is equal to 54 Mb/s [2]. As for the programming language, the choice between C++ [11] and Python [12] was made in favour of the latter to save time for development. For modeling of protocol operations of access to the shared communication medium of the wireless LAN, the operation of a network with an arbitrary number of stations (subscribers) is imitated until the general number of packages successfully transferred via the network reaches the given value before modeling the value. It is supposed that all stations are always ready to transfer data in the form of packages having equal length. The operational characteristics under analysis are the average time of data frame transmission and individual throughput of each station. Moreover, the quantity of collisions and the number of stations involved in the conflict are recorded. The individual average time of data frame transmission is found as the ratio between the time consumed for transmission of a predetermined number of packages by all stations and the number of packages transferred by a definite station. The integral throughput of the method of access to the shared communication medium is determined by the ratio between the number of transferred packages given before modeling and the time actually consumed for delivery of all information to receivers.

Before initiation of the imitation algorithm, an array of stations is created, whereof each station is an instance of wireless station class. When such class instance is created, the range of variables such as the degree of width of the contention window (N_0), the number of successfully transferred packages, the number of collisions, as well as the Boolean variable registering the activity of the station within the current slot interval is initialized. Moreover, each station possesses a variable — a delaying timer and variables in which the values of average time of data frame transmission and individual operational speed are stored. Additionally, a variable accumulating the total (program or model [13,14]) time and other auxiliary values has been determined.

After initialization, a cycle of non-conflict package transmission is executed, which is completed with obtaining an acknowledgement. The total time of package transmission is immediately complemented with the inter-frame space. Then all the stations having transferred the package (initially all), regardless of success or failure, pass the procedure of initialization of the delaying timer (selection of slot for package transmission). For determination of random numbers, an embedded random number generator is used, which in the Python language implements the Mersenne Twister algorithm [15].

Further, all stations are monitored in each slot interval for registration of subscribers with the delaying timer value equal to zero. For each station with the zero delaying timer in this slot interval, the Boolean sign of activity is set, and a variable containing the number of stations, which have transferred the package increases by one. Upon completion of scanning, the value of this variable is analyzed: if its value is equal to one, the transmission has been successful, if the values are more than one — a collision has occurred. For stations having got into a conflict, the value of the degree of width of the competition window is increased by one. If there has been no transmitting stations in this slot interval, the control is transferred to the cycle which is carried out until emergence of at least one active station (station with zero delaying timer) in the next slot interval. During this cycle, successively, the variable of the total time of package transmission increases by one, and the values of the delaying timers decrease by one. If the stations, the delaying timer of which has reached zero, are detected, the total time of package transmission is complemented by the time necessary for transmission of one package plus the time-out for acknowledgement expectation. If the activity of only one station is recorded, the value of its window width degree takes the initial value N_0, and the number of packages successfully transferred by it increases by one. If there are two or more transmitting stations, the variable containing the number of conflicts increases by one.

On the basis of the proposed simulation model, the indices of operational speed of the 802.11 wireless LAN has been analyzed for various numbers of network subscribers. For each set of parameters K, N_0 (the number of stations and initial width degree of the contention window), ten cycles to implement the competition process were executed; each of these cycles included successful transmission of one million packages by subscribers. The results of modeling the activity of two, three, five and ten stations averaged by the basic competition protocol are given in the Table 1 containing the distribution of various operation parameters of the network starting from the initial width degree of the contention window N_0.

The most informative values are the minimum ($C_{\min}$) and maximum ($C_{\max}$) values of individual operational speed (for the network of two stations, the individual operational speed of each station C_i is given), and the total throughput of the system (C_g). Moreover, Table 1 represents the dynamics of changing the number of collisions attributed to the number of successfully transferred packages (Q) and distribution of collisions by the number of stations involved in the conflict $q(I)$, where I is the number of conflicting stations. The standard deviation (σ) of the total throughput is given as a measure of adequacy of obtained results.

Figure 4 for the standard method of access to the shared medium shows the comparative curves of the minimum ($C_{\min}$) and maximum ($C_{\max}$) individual operational speeds, as well as the curves of the total throughput of the system (C_{gen}) as functions of the initial width degree of the contention window. The assemblage of curves showing the total throughput also includes a curve corresponding to the analytical resolution for the network with two active stations (C_{teor}). The represented numerical results make it obvious that with small values of the N_0, there is an effect of capturing the shared communication medium by one of the stations for any number of competing subscribers in the wireless local network. At the same time, due to capturing the shared medium by any station and, as a consequence, non-conflict transfer of any number of packages, there are good indicators of the total throughput of the network and a strong imbalance in the indicators of individual operational speed and average time of data frame transmission of various stations. With an increase in the width of the contention window, the values of individual characteristics of the stations become aligned, with a maximum observed in the total throughput against the parameter of the initial width degree of the contention window. With an increase in the number of stations in the network, this maximum shifts towards the maximum possible size of the contention window, and while there is no extremum as such for the network of two stations, — for the major number of active subscribers the maximum is strongly manifested: for three stations — $N_0 = 4$, for five — $N_0 = 5$, for ten — $N_0 = 6$, and for twenty — $N_0 = 7$. With such initial parameters, collision transmissions are decreasing, and individual indicators of stations are equaled against the peak of the total throughput of the wireless network. Moreover, for small N_0 the value of total throughput is higher for a network with the small number of stations, and as the window width increases, the inverse dependence occurs — the more stations, the higher is the throughput. Comparison

Table 1. Characteristics of the wireless LAN with K stations

$K = 2$	$N_0 = 1$	$N_0 = 2$	$N_0 = 3$	$N_0 = 4$	$N_0 = 5$	$N_0 = 6$	$N_0 = 7$
C_1	0.000004	0.41525	0.38040	0.37676	0.35161	0.30000	0.22920
C_2	0.893831	0.38517	0.37831	0.37701	0.35134	0.29988	0.22951
Q	0.000976	0.08718	0.11254	0.06184	0.03124	0.01573	0.00768
C_g	0.893834	0.80042	0.75871	0.75377	0.70295	0.59988	0.45871
σ	0.000016	0.00092	0.00028	0.00011	0.00010	0.00024	0.00013
$K = 3$	$N_0 = 1$	$N_0 = 2$	$N_0 = 3$	$N_0 = 4$	$N_0 = 5$	$N_0 = 6$	$N_0 = 7$
$C_{\min}$	0.000004	0.25109	0.24296	0.24690	0.24087	0.21869	0.17974
$C_{\max}$	0.690570	0.26090	0.24437	0.24736	0.24181	0.21915	0.17992
Q	0.001984	0.13956	0.16529	0.10749	0.05873	0.03059	0.01537
$q(2)$	0.998	0.968	0.945	0.963	0.979	0.99	0.995
$q(3)$	0.002	0.032	0.055	0.037	0.021	0.01	0.005
C_g	0.892914	0.76442	0.73057	0.74149	0.72433	0.65670	0.53950
σ	0.000038	0.00126	0.00026	0.00012	0.00015	0.00011	0.00022
$K = 5$	$N_0 = 1$	$N_0 = 2$	$N_0 = 3$	$N_0 = 4$	$N_0 = 5$	$N_0 = 6$	$N_0 = 7$
$C_{\min}$	0.002372	0.13886	0.13904	0.14280	0.14463	0.14002	0.12450
$C_{\max}$	0.319206	0.14928	0.13963	0.14317	0.14580	0.14057	0.12501
Q	0.004215	0.20618	0.23916	0.17296	0.10604	0.05792	0.03008
$q(2)$	0.99478	0.92317	0.86557	0.91493	0.94892	0.97155	0.9857
$q(3)$	0.00498	0.07483	0.09659	0.07882	0.04970	0.02804	0.0142
$q(4)$	0.00024	0.00198	0.03773	0.00617	0.00138	0.00041	0.0001
$q(5)$	0	0.00002	0.00010	0.00008	0.00001	0	0
C_g	0.890868	0.72341	0.69665	0.71472	0.72653	0.70126	0.62367
σ	0.000053	0.00027	0.00039	0.00014	0.00016	0.00023	0.00005
$K = 10$	$N_0 = 1$	$N_0 = 2$	$N_0 = 3$	$N_0 = 4$	$N_0 = 5$	$N_0 = 6$	$N_0 = 7$
$C_{\min}$	0.020412	0.06339	0.06364	0.06621	0.06940	0.07105	0.06902
$C_{\max}$	0.147836	0.07023	0.06706	0.06831	0.07079	0.07221	0.06970
Q	0.00987	0.29624	0.32358	0.27000	0.19061	0.11598	0.06418
$q(2)$	0.98936	0.86016	0.83915	0.86135	0.90057	0.93400	0.96348
$q(3)$	0.01013	0.12944	0.14509	0.12644	0.09282	0.06327	0.03574
$q(4)$	0.00020	0.00997	0.01481	0.01144	0.00630	0.00268	0.00078
$q(5)$	0	0.00043	0.00091	0.00073	0.00030	0.00004	0
$q(6)$	0.00030	0.00001	0.00004	0.00004	0.00001	0	0
$q(7)$	0	0	0	0.00001	0	0	0
C_g	0.885725	0.67481	0.65277	0.67070	0.69965	0.71598	0.69360
σ	0.000057	0.00057	0.00033	0.00025	0.00023	0.00017	0.00022

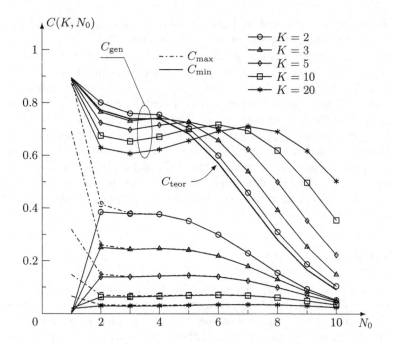

Fig. 4. Individual and general throughput performance

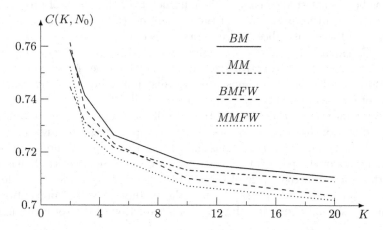

Fig. 5. General throughput performance for the basic (*BM*) and modifying (*MM*) access methods, and for the methods with fixed contention window width — (*BMFW*) and (*MMFW*)

of results of analytical and simulation modeling of a network consisting of two stations shows that the values of total throughput for various sizes N_0 differ by less than 4 %.

Apart from the basic method of access to a communication medium, the imitation modeling was conducted by taking into account the measures proposed in

Sect. 3 to reduce the "effect of capture" of the communication medium. Figure 5 represents comparative curves showing the transmitted flow for various methods of access to the communication medium from the number of active stations of the network with an optimal value of the contention window width. Figure 5 shows that in networks with six or more active subscribers, exclusion of zero delay from the interval for selection of random delay before transmission should be recognized as the best measure to prevent the capture effect. Such measure completely prevents any capture of communication medium for an infinitely long time (capture of medium by one subscriber is always limited on top), with a minimum payment for fair distribution of the communication medium. Therewith, the properties of adaptation of a standard protocol to the number of competing subscribers are preserved due to the function of doubling of the competition window size after each conflict.

5   Conclusion

The performed analysis is targeted at studying the method of carrier sense multiple access with collision avoidance. Analytic correlations have been obtained for probability timing characteristics of the competition process between two stations. The "medium capture effect" and the extreme dependence of operational parameters on the initial contention window size have been revealed.

It has been suggested to change the parameters of the protocol procedure of competition, ensuring prevention of the capture effect by saving high values of individual and integral indices of operational speed.

It has been shown that the optimal initial width of the contention window (S_0) is determined by the active size of the network (the number of competing stations), and it ensures almost uniform distribution of a jointly used time resource of the medium between competing subscribers.

Research into the measures preventing the effect of capturing the communication medium in the local area network containing up to 20 competing subscribers conducted on the imitation model has shown that the proposed modifications of the parameters of the protocol competition procedure are effective. It has been shown that for a nontrivial size of the active network $K \geq 6$, the most effective measure to prevent the capture effect is the modified method preventing the delay of zero size and ensuring the minimum reduction of the total throughput capacity.

Acknowledgments. The work is performed under the state order of the Ministry of Education and Science of the Russian Federation (No. 1.511.2014/K).

References

1. Tanenbaum, A.S., Wetherall, D.J.: Computer Networks, 5th edn., p. 960. Prentice Hall, Boston (2010)
2. IEEE Std 802.11 – 2007, Revision of IEEE Std 802.11 – 1999: Part 11: Wireless LAN Medium Access Control (MAC) and Physical Layer (PHY) Specifications, p. 1184. IEEE Computer Society (2007)
3. Geier, J.: Wireless Networks first-step. Cisco Press, Indianapolis (2005)
4. Roshan, P., Leary, J.: 802.11 Wireless LAN Fundamentals. Cisco Press, Indianapolis (2004)
5. Bianchi, G.: Performance Analysis of the IEEE 802.11 Distributed Coordination Function. IEEE J. Sel. Areas Commun. **18**(3), 535–547 (2000)
6. Kustov, N.T., Suschenko, S.P.: Capacity of the random multiple access method. Autom. Remote Control. **62**(1), 76–85 (2001)
7. Hastings, N.A.J., Peacock, J.B.: Statistical Distributions in Scientific Work Series: A Handbook for Students and Practitioners, p. 130. Butterworth, London (1975)
8. Prudnikov, A.P., Brychkov, I.A., Marichev, O.I.: Integrals and Series: Special functions. Overseas Publishers Association, Amsterdam (1986)
9. Bononi, L., Conti, M., Donatiello, L.: Design and performance evaluation of distributed contention control (DCC) mechanism for IEEE 802.11 wireless local area network. J. Parallel Distrib. Comput. **60**(4), 407–430 (2000)
10. Vishnevsky, V.M., Lyakhov, A.I.: IEEE 802.11 wireless LAN: saturation throughput analysis with seizing effect consideration. Cluster Comput. **5**, 133–144 (2002)
11. Stroustrup, B.: The C++ Programming Language. 923 p. AT&T Labs, New Jersey (1997)
12. Lutz, M.: Learning Python. 845 p. O'Reilly Media Inc, Sebastopol (2009)
13. Shannon, R.E.: Systems Simulation, The Art and Science, 387 p. Prentice-Hall, New Jersey (1975)
14. Kelton, D.W., Law, A.M.: Simulation modeling and analysis, 3rd edn. McGraw-Hill, New York (2000)
15. Matsumoto, M., Nishimura, T.: mersenne twister: a 623-dimensionally equidistributed uniform pseudo-random number generator. ACM Trans. Model. Comput. Simul. (TOMACS) **8**(1), 3–30 (1998). Special issue uniform random number generation

On a Flow of Repeated Customers in Stable Tandem Cyclic Queueing Systems

Andrei V. Zorine$^{(\boxtimes)}$ and Vladimir A. Zorin

N.I. Lobachevsky State University of Nizhni Novgorod, 23 Prospekt Gagarina,
603950 Nizhni Novgorod, Russia
zoav1602@gmail.com, zor.vaz264@yandex.ru

Abstract. We investigate tandem queueing systems with control of conflicting input flows using cyclic algorithms with readjustments. Input flows are modulated by a finite-state synchronous Markov chain. Customers arrive in Poisson flows of batches with intensities and batch size distributions determined by the environment. All serviced customers from the first input flow and randomly selected serviced customers from the second conflicting input flow in the first system are transferred with random speeds to the second queueing system. We develop a numerical algorithm to evaluate the stationary probability distribution for the number of customers joining the transfer queue at each stage of servers' operation.

Keywords: Tandem queueing systems · Retrial customers · Non-local description of output flows · Stationary probability distribution · Censoring of Markov chain

1 Introduction

Study of output flows from queueing systems is a known problem in queueing theory. For instance, in a network the output from some nodes is fed into another nodes. Although many approaches to model flows exist [1], quite a few results are available so far in queueing problems. For the sake of mathematical tractability, researcher uses an extended state space to be able to watch for the output at any time instant. Frequent extension of the state space consists in adding elapsed/remained time variables to a classical discrete-state number-in-the-queue model [2]. This leads to investigation of functional equations for sequences of functions and their partial derivatives. On the other hand, in [3] a non-classical approach was suggested. It consists in choosing a discrete time-scale and considering the integral characteristics of the flow over intervals between

A.V. Zorine—This work was fulfilled as a part of State Budget Research and Development program No. 01201456585 "Mathematical modeling and analysis of stochastic evolutionary systems and decision processes" of N.I. Lobachevsky State University of Nizhni Novgorod and was supported by State Program "Promoting the competitiveness among world's leading research and educational centers".

A. Dudin et al. (Eds.): ITMM 2015, CCIS 564, pp. 114–127, 2015.
DOI: 10.1007/978-3-319-25861-4_10

successive observation epochs. This approach has been already proven fruitful in statistical study of several data present in literature [4], in stability study of output flows from an intersection under control of a fixed-duration cyclic traffic signal [5,6], and in a study of an intersection with dependent non-identically distributed interarrival times [7]. In the present paper we develop a numerical algorithm to evaluate the stationary probability distribution of the flow of retrial customers in tandem queueing system from [8–10].

2   Retrial Customers in Tandem Queueing Systems

Although this paper is a sequel to [8–10] it will be useful to recall the following. Two queueing systems in tandem are considered. The first system accepts conflicting input flows Π_1, Π_2, and the second system accepts conflicting input flows Π_3, Π_4. Essentially, the conflict of the flows means that simultaneous servicing of the customers of different flows is prohibited, that they cannot be summarized, and that the study cannot be reduced to queueing system with fewer input flows. The flows Π_1, Π_2, and Π_4 are formed in a random environment with a finite number d of states $e^{(1)}$, $e^{(2)}$, ..., $e^{(d)}$. A change of the environmental state may occur only at instants of the changing of the servers' states. The probability of transition from $e^{(k)}$ to $e^{(l)}$ in one step equals $a_{k,l}$. It is assumed throughout the paper that the states of the random environment make an aperiodic irreducible Markov chain. In state of the environment $e^{(k)}$ customers in the flow Π_j, $j = 1$, 2, 4 arrive in groups so that the flow of groups is Poisson with intensity $\lambda_j^{(k)}$, and a number b in the group appears with probability $f(b; j, k)$. After service in the first queueing system, each customer from the flow Π_1 is directed into the second queueing system and joins a flow Π_5 of repeated customers. After service in the first queueing system and independently of each other, each customer from the flow Π_2 with probability α becomes a repeated customer joining the flow Π_5, or with probability $(1 - \alpha)$ leaves the tandem systems. Customers of the flow Π_j where $j = 1, 2 \ldots, 5$, arrive into a queue O_j of unlimited capacity.

We will consider the pair of servers in both systems as a single server with n cyclically changing states $\Gamma^{(1)}$, $\Gamma^{(2)}$, ..., $\Gamma^{(n)}$ and a fixed duration T_r of the state $\Gamma^{(r)}$, $r = 1, 2, \ldots, n$. In addition to the traditional functions of servicing the conflicting flows Π_1, Π_2, Π_3, and Π_4, the server also serves the demands of the flow Π_5. The set $\Gamma = \{\Gamma^{(1)}, \Gamma^{(2)}, \ldots, \Gamma^{(n)}\}$ is partitioned into nine mutually disjoint classes named Γ^{I}, Γ^{II}, ..., Γ^{IX}. Customers from O_5 are serviced in every state $\Gamma^{(r)} \in \Gamma$. Further, at $\Gamma^{(r)} \in \Gamma^{\mathrm{II}}$ customers in O_1 are serviced. At $\Gamma^{(r)} \in \Gamma^{\mathrm{III}}$ customers in O_2 are serviced, while, at $\Gamma^{(r)} \in \Gamma^{\mathrm{IV}}$ customers in O_3 are serviced. At $\Gamma^{(r)} \in \Gamma^{\mathrm{V}}$ customers in O_4 are serviced, while, at $\Gamma^{(r)} \in \Gamma^{\mathrm{VI}}$ customers in O_1 and O_3 are serviced. At $\Gamma^{(r)} \in \Gamma^{\mathrm{VII}}$ customers in O_2 and O_3 are serviced, while, at $\Gamma^{(r)} \in \Gamma^{\mathrm{VIII}}$ customers in O_1 and O_4 are serviced. Finally, at $\Gamma^{(r)} \in \Gamma^{\mathrm{IX}}$ customers in O_2 and O_4 are serviced. For $j = 1$ let us set $^j\Gamma = \Gamma^{\mathrm{II}} \cup \Gamma^{\mathrm{VI}} \cup \Gamma^{\mathrm{VIII}}$, for $j = 2$ set $^j\Gamma = \Gamma^{\mathrm{III}} \cup \Gamma^{\mathrm{VII}} \cup \Gamma^{\mathrm{IX}}$. The service process will be defined by means of saturation flows Π_j^{sat}, $j = 1, 2, \ldots, 5$. Let the saturation flow Π_j^{sat} contain a nonrandom number $\ell_{r,j} \geqslant 1$ of customers during the time T_r at $\Gamma^{(r)} \in {}^j\Gamma$ and 0 customers during the same time at $\Gamma^{(r)} \notin {}^j\Gamma$, $j = 1, 2, 3$, and 4. To define the saturation flow Π_5^{sat} we assume that each customer in O_5 during the time T_r with probability p_r completes service, leaves O_5 and joins Π_3, or with probability $(1 - p_r)$ it remains in the queue until the next tact.

All the random objects considered in the paper can be constructed on a probability space $(\Omega, \mathfrak{F}, \mathbf{P})$. Let the observation instants $\tau_0, \tau_1, \tau_2, \ldots$ coincide with the moments of the server's state changes. Denote by $\xi_i(\omega) \in E = \{e^{(1)}, e^{(2)}, \ldots, e^{(d)}\}$ the environmental state during the interval $(\tau_i, \tau_{i+1}]$, by $\kappa_{j,i}(\omega)$ the number in the queue O_j at time τ_i, by $\Gamma_i(\omega) \in \Gamma$ the server state during the time interval $(\tau_{i-1}, \tau_i]$, by $\eta_{j,i}(\omega)$ the number of arrivals in Π_j during the time interval $(\tau_i, \tau_{i+1}]$, by $\xi_{j,i}(\omega)$ the number of customers in Π_j^{sat} during the time interval $(\tau_i, \tau_{i+1}]$, by $\bar{\xi}_{j,i}(\omega)$ the number of customers in Π_j^{out}, the j-th output flow, during the time interval $(\tau_i, \tau_{i+1}]$. Let us define a function $\psi(b; x, u) = \binom{x}{b} u^x (1 - u)^{b-x}$ for $0 \leqslant u \leqslant 1$, $b = 0, 1, \ldots$, and $x = 0, 1, \ldots, b$ and functions $\varphi_{j(x;k,t)}$, $t > 0$, $k = 1, 2, \ldots, d$, from series expansions

$$\sum_{x=0}^{\infty} z^x \varphi_{j(x;k,t)} = \exp\left\{\lambda_j^{(k)} t \left(\sum_{b=1}^{\infty} z^b f(b; j, k) - 1\right)\right\}, \quad j = 1, 2.$$

Here $\varphi_{j(x;k,t)}$ determines the probability of x arrivals in Π_j during time t in the environment state $e^{(k)}$. For $r = 1, 2, \ldots, n-1$ set $r \oplus 1 = r+1$, and set $n \oplus 1 = 1$. The assumptions on the queueing and service process imply relations

$$\kappa_{j,i+1}(\omega) = \kappa_{j,i}(\omega) + \eta_{j,i}(\omega) - \bar{\xi}_{j,i}(\omega), \ \bar{\xi}_{j,i}(\omega) = \min\{\kappa_{j,i}(\omega) + \eta_{j,i}(\omega), \xi_{j,i}(\omega)\},$$
$$\Gamma_{i+1}(\omega) = u(\Gamma_i(\omega)) \qquad \text{where } u(\Gamma^{(r)}) = \Gamma^{(r \oplus 1)}.$$

Moreover, the conditional probability

$$\mathbf{P}(\{\omega \colon \eta_{j,i}(\omega) = b_j, \xi_{j,i}(\omega) = y_j, \ j = 1, 2; \eta_{5,i}(\omega) = b_5, \chi_{i+1}(\omega) = e^{(l)}\} \mid \mathfrak{F}_i)(\omega)$$

w.r.t a σ-algebra $\mathfrak{F}_i = \sigma(\Gamma_t, \chi_t, \kappa_{1,t}, \kappa_{2,t} \colon t = 0, 1, \ldots, i)$ is equal on the event $\{\omega \colon \Gamma_i(\omega) = \Gamma^{(r)}, \chi_i(\omega) = e^{(k)}, \kappa_{1,i}(\omega) = x_1^i, \kappa_{2,i}(\omega) = x_2^i\}$ to

$$a_{k,l} \varphi_1(b_1; e^{(k)}, T_{r \oplus 1}) \varphi_2(b_1; e^{(k)}, T_{r \oplus 1}) \delta_{y_1, \ell_{r \oplus 1,1}} \delta_{y_2, 0} \delta_{b_5, \min\{x_1 + b_1, \ell_{r \oplus 1,1}\}}$$

for $\Gamma^{(r \oplus 1)} \in {}^1\Gamma$, equal to

$$a_{k,l} \varphi_1(b_1; e^{(k)}, T_{r \oplus 1}) \varphi_2(b_1; e^{(k)}, T_{r \oplus 1}) \delta_{y_1, 0} \delta_{y_2, \ell_{r \oplus 1,1}} \psi(b_5; \min\{x_2 + b_2, \ell_{r \oplus 1,2}\}, \alpha)$$

for $\Gamma^{(r \oplus 1)} \in {}^2\Gamma$, and equal to

$$a_{k,l} \varphi_1(b_1; e^{(k)}, T_{r \oplus 1}) \varphi_2(b_1; e^{(k)}, T_{r \oplus 1}) \delta_{y_1, 0} \delta_{y_2, 0} \delta_{b_5, 0}$$

for $\Gamma^{(r \oplus 1)} \notin {}^1\Gamma \cup {}^2\Gamma$.

Put $\bar{\xi}_{j,-1}(\omega) = 0$ and

$$\hat{\nu}_i(\omega) = (\Gamma_i(\omega), \chi_i(\omega), \kappa_{1,i}(\omega), \kappa_{2,i}(\omega), \bar{\xi}_{1,i-1}(\omega), \bar{\xi}_{2,i-1}(\omega)).$$

Then a marked point process

$$\{(\tau_i(\omega), \hat{\nu}_i(\omega), \eta_{5,i-1}(\omega)); i = 1, 2, \ldots\}$$

with the mark $\hat{\nu}_i$ of customers arrived in Π_5 during the time interval $(\tau_{i-1}, \tau_i]$ is a nonlocal description of the flow Π_5. The stochastic properties of the marks are presented in the next theorem.

Denote by a_k the stationary probability of the state $e^{(k)}$ and set

$$T = T_1 + T_2 + \ldots + T_n,$$

$$\ell_j = \sum_{r \in {}^j\Gamma} \ell_{r,j},$$

$$\bar{\lambda}_j^{(k)} = \lambda_j^{(k)} \sum_{b=1}^{\infty} b \cdot f(b; j, k).$$

Theorem 1. *Stochastic sequences*

$$\{(\Gamma_i(\omega), \kappa_{j,i}(\omega), \chi_i(\omega), \bar{\xi}_{j,i-1}(\omega)); i = 0, 1, \ldots\}, \qquad j = 1, 2; \qquad (1)$$

$$\{(\Gamma_i(\omega), \kappa_{1,i}(\omega), \kappa_{2,i}(\omega), \chi_i(\omega), \bar{\xi}_{1,i-1}(\omega), \bar{\xi}_{2,i-1}(\omega)); i = 0, 1, \ldots\} \qquad (2)$$

with a fixed probability distribution of a random vector

$$(\Gamma_0(\omega), \kappa_{1,0}(\omega), \kappa_{2,0}(\omega), \chi_0(\omega))$$

are time-homogeneous Markov chains with period n. The stationary probability distribution of the Markov chain (1), j fixed, exists if

$$T \sum_{k=1}^{d} \mathsf{a}_k \bar{\lambda}_j^{(k)} - \ell_j < 0; \qquad (3)$$

if it exists then

$$T \sum_{k=1}^{d} \mathsf{a}_k \bar{\lambda}_j^{(k)} - \ell_j \leqslant 0. \qquad (4)$$

Both inequalities (3) with $j = 1, 2$ are sufficient and both inequalities (4) with $j = 1, 2$ are necessary for the existence of the probability distribution of the Markov chain (2).

In the remainder of the paper we assume that inequality (3) holds for $j = 1, 2$. The Lemma below implies that random variables $\eta_{5,i}(\omega)$, $i = 0, 1, \ldots$ are identically distributed when sequences in (1) are strictly stationary, and their common law of probability distribution is entirely determined by the stationary probability distribution of the Markov chains (1).

Lemma 1. *One has*

$$\mathbf{P}(\{\omega \colon \eta_{5,i}(\omega) = b\})$$

$$= \sum_{l=1}^{d} \sum_{\Gamma^{(r)} \in {}^1\Gamma} \mathbf{P}(\{\omega \colon \Gamma_{i+1}(\omega) = \Gamma^{(r)}, \chi_{i+1}(\omega) = e^{(l)}, \bar{\xi}_{1,i}(\omega) = b\})$$

$$+\sum_{l=1}^{d}\sum_{\Gamma^{(r)}\in {}^2\Gamma}\sum_{x=b}^{\ell_{r,2}}\mathbf{P}(\{\omega\colon \Gamma_{i+1}(\omega)=\Gamma^{(r)},\chi_{i+1}(\omega)=e^{(l)},\bar{\xi}_{2,i}(\omega)=x\})\psi(b;x,\alpha)$$

$$+\sum_{l=1}^{d}\sum_{\Gamma^{(r)}\in\Gamma\setminus({}^1\Gamma\cup{}^2\Gamma)}\delta_{0,b}\mathbf{P}(\{\omega\colon \Gamma_{i+1}(\omega)=\Gamma^{(r)},\chi_{i+1}(\omega)=e^{(l)}\}).$$

Now we are able to find explicitly the transition probabilities of the Markov chain (1). For the sake of brevity, let us introduce events

$$A_{j,i}(r,k,x_j^1,b_j^1)=\{\omega\colon \Gamma_i(\omega)=\Gamma^{(r)},\chi_i(\omega)=e^{(k)},\kappa_{j,i}(\omega)=x_j^1,\bar{\xi}_{j,i-1}(\omega)=b_j^1\},$$

$$D_{j,i}(x_j^1,x_j^2,b_j^2)$$
$$=\{\omega\colon x_j^2=\max\{0,x_j^1+\eta_{j,i}(\omega)-\xi_{j,i}(\omega)\},b_j^2=\min\{x_j^1+\eta_{j,i}(\omega),\xi_{j,i}(\omega)\}\}$$

for $x_j^1=0,1,\ldots,$ $x_j^2=0,1,\ldots,$ $b_j^2=0,1,\ldots.$ Then the transition probability of the Markov chain (1) has the form:

$$\mathbf{P}(A_{j,i+1}(r\oplus 1,l,x_j^2,b_j^2)\mid A_{j,i}(r,k,x_j^1,b_j^1))$$
$$=a_{k,l}\mathbf{P}(D_{j,i}(x_j^1,x_j^2,b_j^2)\mid A_{j,i}(r,k,x_j^1,b_j^1)).$$

For $\Gamma^{(r\oplus 1)}\notin {}^j\Gamma$ we have

$$\mathbf{P}(D_{j,i}(x_j^1,x_j^2,b_j^2)\mid A_{j,i}(r,k,x_j^1,b_j^1))$$
$$=\mathbf{P}(\{\omega\colon x_j^2=x_j^1+\eta_{j,i}(\omega),b_j^2=0\}\mid A_{j,i}(r,k,x_j^1,b_j^1))$$
$$=\delta_{0,b_j^2}\varphi(x_j^2-x_j^1;j,k,T_{r\oplus 1}).$$

Let us consider now the case $\Gamma^{(r\oplus 1)}\in {}^j\Gamma$. Assume $x_j^2=0$. Then for $x_j^1\leqslant \ell_{r\oplus 1,j}$ and $b_j^2\leqslant \ell_{r\oplus 1,j}$ we have

$$\mathbf{P}(D_{j,i}(x_j^1,x_j^2,b_j^2)\mid A_{j,i}(r,k,x_j^1,b_j^1))$$
$$=\mathbf{P}(\{\omega\colon x_j^1+\eta_{j,i}(\omega)\leqslant \ell_{r\oplus 1,j},b_j^2=x_j^1+\eta_{j,i}(\omega)\}\mid A_{j,i}(r,k,x_j^1,b_j^1))$$
$$=\varphi(b_j^2-x_j^1;j,k,T_{r\oplus 1}),$$

when $x_j^1\leqslant \ell_{r\oplus 1,j}$ and $b_j^2 > \ell_{r\oplus 1,j}$

$$\mathbf{P}(D_{j,i}(x_j^1,x_j^2,b_j^2)\mid A_{j,i}(r,k,x_j^1,b_j^1))=0,$$

but when $x_j^1 > \ell_{r\oplus 1,j}$ we have

$$\mathbf{P}(D_{j,i}(x_j^1,x_j^2,b_j^2)\mid A_{j,i}(r,k,x_j^1,b_j^1))=0.$$

Finally, for $x_j^2 > 0$ we have

$$\mathbf{P}(D_{j,i}(x_j^1,x_j^2,b_j^2)\mid A_{j,i}(r,k,x_j^1,b_j^1))$$
$$=\mathbf{P}(\{\omega\colon x_j^1+\eta_{j,i}(\omega)\leqslant \ell_{r\oplus 1,j},b_j^2=x_j^1+\eta_{j,i}(\omega)\}\mid A_{j,i}(r,k,x_j^1,b_j^1))$$
$$=\delta_{\ell_{r\oplus 1,j},b_j^2}\varphi(b_j^2+\ell_{r\oplus 1,j}-x_j^1;j,k,T_{r\oplus 1}).$$

In effect, this proves the following.

Lemma 2. *The transition probability*

$$\mathbf{P}(A_{j,i+1}(r \oplus 1, l, x_j^2, b_j^2) \mid A_{j,i}(r, k, x_j^1, b_j^1)) \tag{5}$$

equals

$$a_{k,l}\varphi(x_j^2 - x_j^1; j, k, T_{r \oplus 1})$$

for $\Gamma^{(r \oplus 1)} \notin {}^j\Gamma$ and $b_j^2 = 0$; it equals

$$a_{k,l}\varphi(b_j^2 - x_j^1; j, k, T_{r \oplus 1})$$

for $\Gamma^{(r \oplus 1)} \in {}^j\Gamma$, $x_j^2 = 0$, $x_j^1 \leqslant \ell_{r \oplus 1, j}$, and $b_j^2 \leqslant \ell_{r \oplus 1, j}$; it equals

$$a_{k,l}\varphi(x_j^2 + \ell_{r \oplus 1, j} - x_j^1; j, k, T_{r \oplus 1})$$

for $\Gamma^{(r \oplus 1)} \in {}^j\Gamma$, $x_j^2 > 0$, $x_j^1 \leqslant x_j^2 + \ell_{r \oplus 1, j}$, and $b_j^2 = \ell_{r \oplus 1, j}$; it equals 0 in the remaining cases. Hence, the states of the Markov chain (1) of the form $(\Gamma^{(r)}, e^{(k)}, x_j, b_j)$ with either $\Gamma^{(r)} \notin {}^j\Gamma$, $b_j > 0$, or $\Gamma^{(r)} \in {}^j\Gamma$, $x_j = 0$, $b_j > \ell_{r \oplus 1, j}$, or $\Gamma^{(r)} \in {}^j\Gamma$, $x_j > 0$, $b_j \neq \ell_{r \oplus 1, j}$ are inessential. The remaining states belong to a single class of communicating states.

Denote by $Q_{x,j}(r, k, b)$ the stationary probability for the state $(\Gamma^{(r)}, x, e^{(k)}, b)$ of the Markov chain (1) with fixed $j = 1, 2$. Let the essential states be ordered so that the value of k is shifted first, then the value of b, and, finally, that of x. This way they make series (to be read row-wise):

$$(\Gamma^{(1)}, 0, e^{(1)}, 0), \quad (\Gamma^{(1)}, 0, e^{(2)}, 0), \quad \ldots, \quad (\Gamma^{(1)}, 0, e^{(d)}, 0),$$
$$(\Gamma^{(1)}, 0, e^{(1)}, 1), \quad (\Gamma^{(1)}, 0, e^{(2)}, 1), \quad \ldots, \quad (\Gamma^{(1)}, 0, e^{(d)}, 1),$$
$$\ldots,$$
$$(\Gamma^{(1)}, 0, e^{(1)}, \tilde{\ell}_{1,j}), \quad (\Gamma^{(1)}, 0, e^{(2)}, \tilde{\ell}_{1,j}), \quad \ldots, \quad (\Gamma^{(1)}, 0, e^{(d)}, \tilde{\ell}_{1,j}),$$
$$(\Gamma^{(2)}, 0, e^{(1)}, 0), \quad (\Gamma^{(2)}, 0, e^{(2)}, 0), \quad \ldots, \quad (\Gamma^{(2)}, 0, e^{(d)}, 0),$$
$$\ldots,$$
$$(\Gamma^{(n)}, 0, e^{(1)}, \tilde{\ell}_{n,j}), \quad (\Gamma^{(n)}, 0, e^{(2)}, \tilde{\ell}_{n,j}), \quad \ldots, \quad (\Gamma^{(n)}, 0, e^{(d)}, \tilde{\ell}_{n,j}),$$
$$(\Gamma^{(1)}, 1, e^{(1)}, \tilde{\ell}_{1,j}), \quad (\Gamma^{(1)}, 1, e^{(2)}, \tilde{\ell}_{1,j}), \quad \ldots, \quad (\Gamma^{(1)}, 1, e^{(d)}, \tilde{\ell}_{1,j}),$$
$$(\Gamma^{(2)}, 1, e^{(1)}, \tilde{\ell}_{2,j}), \quad (\Gamma^{(2)}, 1, e^{(2)}, \tilde{\ell}_{2,j}), \quad \ldots, \quad (\Gamma^{(2)}, 1, e^{(d)}, \tilde{\ell}_{2,j}),$$
$$\ldots,$$
$$(\Gamma^{(n)}, 1, e^{(1)}, \tilde{\ell}_{n,j}), \quad (\Gamma^{(n)}, 1, e^{(2)}, \tilde{\ell}_{n,j}), \quad \ldots, \quad (\Gamma^{(n)}, 1, e^{(d)}, \tilde{\ell}_{n,j}),$$
$$(\Gamma^{(1)}, 2, e^{(1)}, \tilde{\ell}_{1,j}), (\Gamma^{(1)}, 2, e^{(2)}, \tilde{\ell}_{1,j}), \ldots, (\Gamma^{(1)}, 2, e^{(d)}, \tilde{\ell}_{1,j}),$$
$$\ldots$$

Then introduce the row vectors

$$Q_{0,j}(r, b) = (Q_{0,j}(r, 1, b), Q_{0,j}(r, 2, b), \ldots, Q_{0,j}(r, d, b)),$$
$$Q_{0,j}(r) = (Q_{0,j}(r, 1), Q_{0,j}(r, 2), \ldots, Q_{0,j}(r, \tilde{\ell}_{r,j})),$$

$$Q_{x,j}(r) = (Q_{x,j}(r,1,\tilde{\ell}_{r,j}), Q_{x,j}(r,2,\tilde{\ell}_{r,j}), \ldots, Q_{x,j}(r,k,\tilde{\ell}_{r,j})), \quad x \geqslant 1,$$
$$Q_{x,j} = (Q_{x,j}(1), Q_{x,j}(2), \ldots, Q_{x,j}(n)), \quad x \geqslant 0,$$
$$Q_j = (Q_{0,j}, Q_{1,j}, \ldots)$$

with stationary probabilities as its elements in accordance to the selected ordering. Notice that the vector $Q_{0,j}(r,k)$ contains one element for those r with $\Gamma^{(r)} \notin {}^j\Gamma$. Our nearest goal now is to exhibit a matrix structure of the transition probabilities (5) which is induced by the block-wise partition of the stationary probability distribution vector Q_j. Since the index j remains unchanged in course of this operation it will be omitted in notations $Q_{x,j}$, Q_j. Let

$$\Phi_r(x) = \begin{pmatrix} a_{1,1}\varphi(x;j,1,T_r), & a_{1,2}\varphi(x;j,1,T_r), & \ldots, & a_{1,d}\varphi(x;j,1,T_r) \\ a_{2,1}\varphi(x;j,2,T_r), & a_{2,2}\varphi(x;j,2,T_r), & \ldots, & a_{2,d}\varphi(x;j,2,T_r) \\ \vdots & \vdots & \ddots & \vdots \\ a_{d,1}\varphi(x;j,d,T_r), & a_{d,2}\varphi(x;j,d,T_r), & \ldots, & a_{d,d}\varphi(x;j,d,T_r) \end{pmatrix}, \quad x = 0, 1, \ldots,$$
$$\Phi_r(x) = 0, \quad x < 0.$$

Then the $d \times d$ matrix consisting of probabilities (5) for $k = 1, 2, \ldots, d$, $l = 1, 2, \ldots, d$ with fixed values for r, x_j^1, x_j^2, b_j^1, and b_j^2 can be expressed as $\Phi_{r\oplus 1}(x_j^2 - x_j^2)$ for $\Gamma^{(r\oplus 1)} \notin {}^j\Gamma$ and $b_j^2 = 0$, as $\Phi_r(b_j^2 - x_j^1)$ for $\Gamma^{(r\oplus 1)} \in {}^j\Gamma$, $x_j^2 = 0$, $x_j^2 \leqslant \ell_{r\oplus 1,j}$, and $b_j^2 \leqslant \ell_{r\oplus 1,j}$, as $\Phi_{r\oplus 1}(\ell_{r\oplus 1,j} + x_j^2 - x_j^1)$ for $\Gamma^{(r\oplus 1)} \in {}^j\Gamma$, $x_j^2 > 0$, $x_j^1 \leqslant x_j^2 + \ell_{r\oplus 1,j}$, and $b_j^2 = \ell_{r\oplus 1,j}$, and as the zero matrix in the remaining cases. The first and the third cases can be combined if we put $\tilde{\ell}_{r\oplus 1,j} = \ell_{r\oplus 1,j}$ for $\Gamma^{(r\oplus 1)} \in {}^j\Gamma$ and $\tilde{\ell}_{r\oplus 1,j} = 0$ and $\Gamma^{(r\oplus 1)} \notin {}^j\Gamma$: $\Phi_{r\oplus 1}(\tilde{\ell}_{r\oplus 1,j} + x_j^2 - x_j^1)$. At the next grouping step we fix values $x_j^1 \geqslant 1$ and $x_j^2 \geqslant 1$ implying $b_j^1 = \tilde{\ell}_{r,j}$ and $b_j^2 = \tilde{\ell}_{r\oplus 1,j}$, and find the square matrices of transition probabilities for $r = 1, 2, \ldots, n$, this corresponds to series Q_x of stationary probabilities with $x \geqslant 1$. Set

$$\Phi(x) = \begin{pmatrix} 0 & \Phi_2(\tilde{\ell}_{2,j} + x) & 0 & \ldots & 0 \\ 0 & 0 & \Phi_3(\tilde{\ell}_{3,j} + x) & \ldots & 0 \\ \vdots & \vdots & \vdots & \ddots & \vdots \\ 0 & 0 & 0 & \ldots & \Phi_n(\tilde{\ell}_{n,j} + x) \\ \Phi_1(\tilde{\ell}_{1,j} + x) & 0 & 0 & \ldots & 0 \end{pmatrix}.$$

Then the interesting part of the transition probability matrix is $\Phi(x_j^2 - x_j^1)$. Now assume that $x_j^1 = x_j^2 = 0$. Then b_j^1 varies from 0 through $\tilde{\ell}_{r,j}$, and b_j^2 varies from 0 through $\tilde{\ell}_{r\oplus 1,j}$. In this regard let us introduce a matrix of size $(\tilde{\ell}_{r,j} + 1) \times (\tilde{\ell}_{r\oplus 1,j} + 1)$ with blocks

$$\tilde{\Phi}_{r\oplus 1} = \begin{pmatrix} \Phi_{r\oplus 1}(0) & \Phi_{r\oplus 1}(1) & \ldots & \Phi_{r\oplus 1}(\tilde{\ell}_{r\oplus 1,j}) \\ \Phi_{r\oplus 1}(0) & \Phi_{r\oplus 1}(1) & \ldots & \Phi_{r\oplus 1}(\tilde{\ell}_{r\oplus 1,j}) \\ \vdots & \vdots & \ddots & \vdots \\ \Phi_{r\oplus 1}(0) & \Phi_{r\oplus 1}(1) & \ldots & \Phi_{r\oplus 1}(\tilde{\ell}_{r\oplus 1,j}) \end{pmatrix}$$

and a block matrix

$$\tilde{\Phi} = \begin{pmatrix} 0 & \tilde{\Phi}_2 & 0 & \dots & 0 \\ 0 & 0 & \tilde{\Phi}_3 & \dots & 0 \\ \vdots & \vdots & \vdots & \ddots & \vdots \\ 0 & 0 & 0 & \dots & \tilde{\Phi}_n \\ \tilde{\Phi}_1 & 0 & 0 & \dots & 0 \end{pmatrix}.$$

The repeated rows in the matrix $\tilde{\Phi}_r$ are explained by the fact that the value of the transition probability (5) doesn't depend on b_j^1. Then the block of transition probabilities corresponding to the sets of states with $x_j^1 = x_j^2 = 0$, is $\tilde{\Phi}$. Next, assume that $x_j^1 = 0$ and $x_j^2 \geqslant 1$. Then $b_j^2 = \tilde{\ell}_{r\oplus 1,j}$ and $b_j^1 = 0, 1, \dots, \tilde{\ell}_{r,j}$. Introduce the column vectors of size $(\tilde{\ell}_{r,j} + 1) \times 1$

$$(\tilde{\tilde{\Phi}}_{r\oplus 1}(x))^T = \left(\Phi_{r\oplus 1}(\tilde{\ell}_{r\oplus 1,j} + x)\ \Phi_{r\oplus 1}(\tilde{\ell}_{r\oplus 1,j} + x) \ \dots \ \Phi_{r\oplus 1}(\tilde{\ell}_{r\oplus 1,j} + x)\right)^T$$

and matrices

$$\tilde{\tilde{\Phi}}(x) = \begin{pmatrix} 0 & \tilde{\tilde{\Phi}}_2(x) & 0 & \dots & 0 \\ 0 & 0 & \tilde{\tilde{\Phi}}_3(x) & \dots & 0 \\ \vdots & \vdots & \vdots & \ddots & \vdots \\ 0 & 0 & 0 & \dots & \tilde{\tilde{\Phi}}_n(x) \\ \tilde{\tilde{\Phi}}_1(x) & 0 & 0 & \dots & 0 \end{pmatrix}.$$

The block of transition probabilities corresponding to $x_j^1 = 0$ and $x_j^2 \geqslant 1$ is $\tilde{\tilde{\Phi}}(x_j^2)$. Finally, assume that $x_j^1 \geqslant 1$ and $x_j^2 = 0$. Then $b_j^1 = \tilde{\ell}_{r,j}$, $b_j^2 = 0, 1, \dots, \tilde{\ell}_{r\oplus 1,j}$. Let E_d stand for the $d \times d$ identity matrix. Introduce an auxiliary block matrix Λ with $d \times d$ blocks. Let its blocks make n columns, the first $\tilde{\ell}_{1,j} + 1$ block rows have the form $(E_d, 0, 0, \dots, 0)$, then there are $\tilde{\ell}_{2,j} + 1$ rows of the form $(0, E_d, 0, \dots, 0)$, and so on, and, finally, $\tilde{\ell}_{n,j} + 1$ rows of the form $(0, 0, 0, \dots, E_d)$ in conclusion. It's not hard to prove that $\tilde{\tilde{\Phi}}(x) = \Lambda \Phi(x)$. Finally, we will need $1 \times (\tilde{\ell}_{r\oplus 1,j} + 1)$ matrices

$$\tilde{\tilde{\tilde{\Phi}}}_{r\oplus 1}(x) = \left(\Phi_{r\oplus 1}(-x)\ \Phi_{r\oplus 1}(1 - x) \ \dots \ \Phi_{r\oplus 1}(\tilde{\ell}_{r\oplus 1,j} - x)\right)$$

and matrices

$$\tilde{\tilde{\tilde{\Phi}}}(x) = \begin{pmatrix} 0 & \tilde{\tilde{\tilde{\Phi}}}_2(x) & 0 & \dots & 0 \\ 0 & 0 & \tilde{\tilde{\tilde{\Phi}}}_3(x) & \dots & 0 \\ \vdots & \vdots & \vdots & \ddots & \vdots \\ 0 & 0 & 0 & \dots & \tilde{\tilde{\tilde{\Phi}}}_n(x) \\ \tilde{\tilde{\tilde{\Phi}}}_1(x) & 0 & 0 & \dots & 0 \end{pmatrix}.$$

Then the block of transition probabilities with $x_j^1 \geqslant 1$ and $x_j^2 = 0$ is $\tilde{\tilde{\tilde{\Phi}}}(x_j^2)$. Put $\ell^* = \max\{\ell_{r,j} : r = 1, 2, \dots, n\}$. In effect, we have found that the transition probability matrix of the Markov chain (1) has the following block form:

$$
\begin{pmatrix}
\tilde{\tilde{\Phi}} & \tilde{\tilde{\Phi}}(1) & \tilde{\tilde{\Phi}}(2) & \tilde{\tilde{\Phi}}(3) & \tilde{\tilde{\Phi}}(4) & \cdots \\
\tilde{\tilde{\Phi}}(1) & \Phi(0) & \Phi(1) & \Phi(2) & \Phi(3) & \cdots \\
\tilde{\tilde{\Phi}}(2) & \Phi(-1) & \Phi(0) & \Phi(1) & \Phi(2) & \cdots \\
\vdots & \vdots & \vdots & \vdots & \vdots & \ddots \\
\tilde{\tilde{\Phi}}(\ell^*) & \Phi(1-\ell^*) & \Phi(2-\ell^*) & \Phi(3-\ell^*) & \Phi(4-\ell^*) & \cdots \\
0 & \Phi(-\ell^*) & \Phi(1-\ell^*) & \Phi(2-\ell^*) & \Phi(3-\ell^*) & \cdots \\
0 & 0 & \Phi(-\ell^*) & \Phi(1-\ell^*) & \Phi(2-\ell^*) & \cdots \\
\vdots & \vdots & \vdots & \vdots & & \ddots & \ddots
\end{pmatrix} .
$$

So, the stationary probabilities satisfy the following matrix equalities:

$$
Q_0 = \left(Q_0 \tilde{\Phi}(0) + \sum_{x=1}^{\ell^*} Q_x \tilde{\tilde{\Phi}}(x) \right), \tag{6}
$$

$$
Q_w = \left(Q_0 \Lambda\Phi(w) + \sum_{x=1}^{\ell^*+w} Q_x \Phi(w-x) \right), \qquad w \geqslant 1. \tag{7}
$$

Because of the specific structure of the transition probability matrix we use censored Markov chains [12] to obtain an effective numerical algorithm to evaluate the stationary probability distribution of the Markov chain (1). Fix any integer $a > \ell^*$ and observe the Markov chain (1) at passages of the set

$$
S_a = \{ (\Gamma^{(r)}, x, e^{(k)}, b) : r = 1, \ldots, n; k = 1, \ldots, d;
$$
$$
(x = 0, b = 0, \ldots, \tilde{\ell}_{r,j}) \vee (1 \leqslant x \leqslant a, b = \tilde{\ell}_{r,j}) \}.
$$

Let

$$
\{ (\hat{\Gamma}_i(\omega), \hat{\kappa}_{j,i}(\omega), \hat{\chi}_i(\omega), \hat{\xi}_{j,i}(\omega)); i = 0, 1, \ldots \} \tag{8}
$$

be the embedded Markov chain. Denote by $\mathfrak{A}(r', u, l; r, v, k)$ the conditional probability to first enter the set S_a through state $(\Gamma^{(r')}, a-u, l, \tilde{\ell}_{r',j}) \in S_a$ starting in a state $(\Gamma^{(r)}, a + v, e^{(k)}, \tilde{\ell}_{r,j}) \notin S_a$. Further, denote by $\mathfrak{B}(r', u, l; r, k)$ the conditional probability that the state of the Markov chain (1) is $(\Gamma^{(r')}, a-u, e^{(l)}, \tilde{\ell}_{r',j})$ at the first strong descending ladder time [13] of a sequence $\{\kappa_{j,i}(\omega); i = 0, 1, \ldots\}$ given that the initial state is $(\Gamma^{(r)}, a, e^{(k)}, \tilde{\ell}_{r,j})$. Introduce matrices

$$
\mathfrak{A}(r', u; r, v) = \begin{pmatrix}
\mathfrak{A}(r', 1, u; r, 1, v) & \cdots & \mathfrak{A}(r', d, u; r, 1, v) \\
\vdots & \ddots & \vdots \\
\mathfrak{A}(r', 1, u; r, d, v) & \cdots & \mathfrak{A}(r', d, u; r, d, v)
\end{pmatrix},
$$

$$
\mathfrak{A}(u; v) = \begin{pmatrix}
\mathfrak{A}(1, u; 1, v) & \cdots & \mathfrak{A}(n, u; 1, v) \\
\vdots & \ddots & \vdots \\
\mathfrak{A}(1, u; n, v) & \cdots & \mathfrak{A}(n, u; n, v)
\end{pmatrix},
$$

$$\mathfrak{B}(r', u; r) = \begin{pmatrix} \mathfrak{B}(r', u, 1; r, 1) & \cdots & \mathfrak{B}(r', u, d; r, 1) \\ \vdots & \ddots & \vdots \\ \mathfrak{B}(r', u, 1; r, d) & \cdots & \mathfrak{B}(r', u, d; r, d) \end{pmatrix},$$

$$\mathfrak{B}(u) = \begin{pmatrix} \mathfrak{B}(1, u; 1) & \cdots & \mathfrak{B}(n, u; 1) \\ \vdots & \ddots & \vdots \\ \mathfrak{B}(1, u; n) & \cdots & \mathfrak{B}(n, u; n) \end{pmatrix}.$$

Then the block matrix of transition probabilities of the censored Markov chain (8) corresponding to transitions from $(\Gamma^{(r)}, x_j^1, e^{(k)}, b_j^1)$ to $(\Gamma^{(r')}, x_j^2, e^{(l)}, b_j^2)$ with fixed queue levels x_j^1 and x_j^2, and packed the same way as matrix $\mathfrak{A}(u; v)$ is, equals

$$\tilde{\Phi} \qquad \qquad \text{for } x_j^1 = x_j^2 = 0,$$

$$\Lambda\Phi(x_j^2) + \sum_{y=1}^{\infty} \Lambda\Phi(a + y)\mathfrak{A}(a - x_j^2; y) \qquad \text{for } x_j^1 = 0, x_j^2 > 0,$$

$$\tilde{\tilde{\Phi}}(x_j^1) \qquad \qquad \text{for } x_j^1 > 0, x_j^2 = 0,$$

$$\Phi(x_j^2 - x_j^1) + \sum_{y=1}^{\infty} \Phi(a + y - x_j^1)\mathfrak{A}(a - x_j^2; y) \qquad \text{for } x_j^1 > 0, x_j^2 > 0.$$

Lemma 3. *One has*

$$\mathfrak{B}(u) = \Phi(-u) + \sum_{y=0}^{\infty} \Phi(y)\mathfrak{A}_{a-1}(u - 1; y + 1), \qquad u = 1, 2, \ldots, \ell^*, \qquad (9)$$

$$\mathfrak{A}(u; v) = \sum_{x=1}^{\min\{\ell^* - u, v - 1\}} \mathfrak{A}_{a+x}(0; v - x)\mathfrak{B}_{a+x}(u + x)$$

$$+ \mathfrak{B}(u + v), \qquad u = 0, 1, \ldots, \ell^* - 1, v = 1, 2, \ldots \quad (10)$$

To prove (9) one should apply the law of total probability to the definition of the probability $\mathfrak{B}(r', u, l; r, k)$ conditioned on the state after one step of the Markov chain (1). Then, equation (10) is proven by the law of total probability applied to the definition of $\mathfrak{A}(r', u, l; r, v, k)$ conditioned on the state of the censored chain (1) at the first strong descending ladder time of the sequence $\{\kappa_{j,i}(\omega); i = 0, 1, \ldots\}$.

Define, for convenience, by $\mathfrak{A}(0; 0) = E_{dn}$ the $dn \times dn$ identity matrix.

Theorem 2. *Assume that*

$$E_{dn} - \sum_{y=0}^{\infty} \Phi(y)\mathfrak{A}(0; y)$$

is an invertible matrix. Then the stationary probabilities $Q_x(r, k, b)$ of essential states $(\Gamma^{(r)}, x, e^{(k)}, b)$ satisfy the following matrix equations

$$Q_0 = Q_0 \tilde{\Phi} + \sum_{x=1}^{\ell^*} Q_x \tilde{\tilde{\Phi}}(x), \tag{11}$$

$$Q_w = Q_0 \Lambda \bigg(\Phi(w) + \sum_{y=1}^{\infty} \Phi(\ell^* + 1 + y) \mathfrak{A}(\ell^* + 1 - w; y)$$

$$+ \bigg(\sum_{y=0}^{\infty} \Phi(\ell^* + y + 1) \mathfrak{A}(0; y) \bigg) \bigg(E_{dn} - \sum_{y=0}^{\infty} \Phi(y) \mathfrak{A}(0; y) \bigg)^{-1}$$

$$\times \bigg(\Phi(w - \ell^* - 1) + \sum_{y=1}^{\infty} \Phi(y) \mathfrak{A}(\ell^* + 1 - w; y) \bigg) \bigg)$$

$$+ \sum_{x=1}^{\ell^*} Q_x \bigg(\Phi(w - x) + \sum_{y=1}^{\infty} \Phi(\ell^* + 1 + y - x) \mathfrak{A}(\ell^* + 1 - w; y)$$

$$+ \bigg(\sum_{y=0}^{\infty} \Phi(\ell^* + 1 + y - x) \mathfrak{A}(0; y) \bigg) \bigg(E_{dn} - \sum_{y=0}^{\infty} \Phi(y) \mathfrak{A}(0; y) \bigg)^{-1}$$

$$\times \bigg(\Phi(w - \ell^* - 1) + \sum_{y=1}^{\infty} \Phi(y) \mathfrak{A}(\ell^* + 1 - w; y) \bigg) \bigg), \tag{12}$$

$$Q_0 \Lambda \bigg(E_{dn} - \sum_{y=0}^{\infty} \bigg(\sum_{a=0}^{\ell^*} \Phi(y + a) \bigg) \mathfrak{A}(0; y) \bigg)$$

$$+ \sum_{x=1}^{\ell^*} Q_x \bigg(E_{dn} - \sum_{y=0}^{\infty} \bigg(\sum_{a=0}^{\ell^* - x} \Phi(y + a) \bigg) \mathfrak{A}(0; y) \bigg)$$

$$= \frac{1}{n} (\mathsf{a}_1, \mathsf{a}_2, \dots, \mathsf{a}_d, \mathsf{a}_1, \mathsf{a}_2, \dots, \mathsf{a}_d, \dots, \mathsf{a}_1, \mathsf{a}_2, \dots, \mathsf{a}_d)$$

$$\times \bigg(E_{dn} - \sum_{y=0}^{\infty} \bigg(\sum_{a=0}^{\infty} \Phi(a + y) \bigg) \mathfrak{A}(0; y) \bigg) \tag{13}$$

$$Q_a = \bigg\{ \sum_{x=1}^{a-1} Q_x \bigg(\sum_{y=0}^{\infty} \Phi(a + y - x) \mathfrak{A}(0; y) \bigg)$$

$$+ Q_0 \Lambda \bigg(\sum_{y=0}^{\infty} \Phi(a + y) \mathfrak{A}(0; y) \bigg) \bigg\} \bigg(E_{dn} - \sum_{y=0}^{\infty} \Phi(y) \mathfrak{A}(0; y) \bigg)^{-1}, \tag{14}$$

for $w = 1, 2, \dots, \ell^*$, and $a = \ell^* + 1, \ell^* + 2, \dots$.

Here Eqs. (11) and (12) are the Chapman–Kolmogorov stationarity equations for the censoring set S_{ℓ^*}. Equation (14) follows from the Chapman–Kolmogorov stationarity equations for the censoring set S_a with $a \geqslant \ell^* + 1$. Finally, Eq. (13) is obtained by substituting Eqs. (11), (12), and (14) into a normalization condition taking into account equal probabilities of all server states and independence of the server and the environment.

The invertibility condition in Theorem 2 holds when $\ell_{r,j} \geqslant 2$ for all $\Gamma^{(r)} \in {}^j\Gamma$. Indeed, a sufficient condition of invertibility consists in the localization of eigenvalues of the matrix

$$\sum_{y=0}^{\infty} \Phi(y)\mathfrak{A}(0; y),\tag{15}$$

those should be less than unity in absolute value [11, c. 118]. The matrix (15) has only non-negative elements. It is known that among eigenvalues of a non-negative matrix there is a simple non-negative one which is the largest in absolute value [11]. Call it ρ. To locate the extreme eigenvalue ρ of the matrix (15) we will use Gershgorin Circle theorem [11, p. 415]. Denote by $\phi_{s,t} \geqslant 0$ the element in row s, column t of the matrix (15). Then ρ satisfies one of equalities

$$|\phi_{s,s} - \rho| \leqslant \sum_{\substack{t=1 \\ t \neq s}}^{nd} \phi_{s,t}.$$

Hence

$$0 \leqslant \rho \leqslant \sum_{t=1}^{nd} \phi_{s,t}.$$

Now, let $\hat{\phi}_{s,t}(y)$ be the element in row s, column t of $\Phi(y)$, $\hat{\hat{\phi}}_{s,t}(y)$ the element in row s, column t of the matrix $\mathcal{A}(0; y)$. According to the rule of matrix multiplication,

$$\sum_{t=1}^{nd} \phi_{s,t} = \sum_{y=0}^{\infty} \sum_{t=1}^{nd} \sum_{q=1}^{nq} \hat{\phi}_{s,q}(y)\hat{\hat{\phi}}_{d,t}(y) = \sum_{y=0}^{\infty} \sum_{q=1}^{nq} \hat{\phi}_{s,q}(y) \sum_{t=1}^{nd} \hat{\hat{\phi}}_{q,t}(y).$$

Here the total of $\hat{\hat{\phi}}_{q,t}(y)$ for $t = 1, 2, \ldots, nd$ is the probability that at the first passage time of the set S_a the chain hits the boundary of it. When the assumption on $\ell_{r,j}$ is true this probability must be strictly less than unity. Thus

$$\sum_{y=0}^{\infty} \sum_{q=1}^{nq} \hat{\phi}_{s,q}(y) \sum_{t=1}^{nd} \hat{\hat{\phi}}_{q,t}(y) < \sum_{y=0}^{\infty} \sum_{q=1}^{nq} \hat{\phi}_{s,q}(y) \leqslant 1.$$

The second inequality has a probabilistic interpretation as well: summation w.r.t. $q = 1, 2, \ldots, nd$ gives the probability that the queue O_j grows by y in one step.

The stationary probabilities numerically evaluated by Theorem 2 are to be used to obtain the marginal stationary probability distributions of the sequence

$\{\eta_{5,i}(\omega); i = 0, 1, \ldots\}$. From now on we show the index j again. So, for $\Gamma^{(r)} \in {}^j\Gamma$, $j = 1, 2$ in the steady state we have

$$\mathbf{P}(\{\omega \colon \Gamma_{i+1}(\omega) = \Gamma^{(r)}, \chi_{i+1}(\omega) = e^{(l)}, \bar{\xi}_{1,i}(\omega) = b\})$$
$$= Q_0(r, l, b), \qquad b = 0, 1, \ldots, \ell_{r,1} - 1,$$
$$\mathbf{P}(\{\omega \colon \Gamma_{i+1}(\omega) = \Gamma^{(r)}, \chi_{i+1}(\omega) = e^{(l)}, \bar{\xi}_{1,i}(\omega) = \ell_{r,1}\})$$
$$= \sum_{x_j=0}^{\infty} Q_{x,j}(r, l, \ell_{r,1}) = \frac{a_l}{n} - \sum_{b=0}^{\ell_{r,1}-1} Q_0(r, l, b).$$

A program has been developed in Octave programming environment [14]. As an example consider the following settings. Suppose we have an environment with $d = 2$ internal states and transition probabilities $a_{1,1} = 0.9$, $a_{1,2} = 0.1$, $a_{2,1} = 0.4$, $a_{2,2} = 0.6$. In the state $e^{(1)}$ the flows Π_1 and Π_2 are Poissonian with rates $\lambda_1^{(1)} = 0.5$, $\lambda_2^{(1)} = 0.4$. In the state $e^{(2)}$ the flows Π_1 and Π_2 are Bartlett with parameters $\lambda_1^{(2)} = 0.25$, $\rho_1 = 0.5$, $q_1 = 0.5$, $\lambda_2^{(2)} = 0.2$, $\rho_2 = 0.5$, $q_2 = 0.5$ and probability generating functions

$$\sum_{x=0}^{\infty} z^x \varphi_j(x; 2, t) = \exp\left\{\lambda^{(2)} t\left(\frac{\rho_j(1 - q_j)z^2}{1 - q_j z} + (1 - \rho_j)z - 1\right)\right\}.$$

With this choice of parameters the input rates for each flow are constant over time. Then, put $n = 4$, $T_1 = 40$, $T_2 = T_4 = 4$, $T_3 = 30$, the flow Π_1 is services only in the state $\Gamma^{(1)}$ and $\ell_{1,1} = 50$, the flow Π_2 is serviced only in the state $\Gamma^{(3)}$ and $\ell_{1,1} = 40$. Using successively Lemma 1, Theorem 2, and Lemma 1 one finds:

$$\mathbf{E}\,\eta_{5,i}(\omega) = 14, \qquad \mathbf{var}\,\eta_{5,i}(\omega) = 283.44.$$

Now, if the flows in the state $e^{(2)}$ are the same as in the state $e^{(1)}$, then

$$\mathbf{E}\,\eta_{5,i}(\omega) = 14, \qquad \mathbf{var}\,\eta_{5,i}(\omega) = 280.64.$$

The growth in the variance of a batch size leads to increase in the variance of the number of repeated customers.

References

1. Jagerman, D.L., Melamed, B., Willinger, W.: Stochastic modeling of traffic process. In: Dshalalov, J.H. (ed.) Frontiers in Queueing: Models and Applications in Science and Engineering, pp. 271–320. CRC Press, Boca Raton (1997)
2. Cox, D.: The analysis of non-Markovian stochastic processes by the inclusion of supplementary variables. Math. Proc. Camb. Phil. Soc. 3(51), 433–441 (1955)
3. Fedotkin, M.A.: Service processes and control systems. Math. Probl. Cybersecur. 6, 333–344 (1998). Nauka. Fizmatlit, Moscow
4. Anisimova, L.N., Fedotkin, M.A.: Reliability of a control system and statistical analysis of failures of its elements. Vestn. Nizhegorodskogo Univ. Ser. Math. Model. Optim. Cont. 1, 14–22 (2000)

5. Fedotkin, M.A., Fedotkin, A.M.: Analysis and optimization of output processes of conflicting Gnedenko-Kovalenko traffic streams under cyclic control. Autom. Remote Control. **70**(12), 2024–2038 (2010)
6. Proidakova, E.V., Fedotkin, M.A.: Control of output flows in the system with cyclic servicing and readjustments. Autom. Remote Control. **6**(69), 993–1002 (2009)
7. Zorine, A.V.: A cybernetic model of cyclic control of conflicting flows with an after-effect. Proc. Kazan Univ. **156**(3), 66–75 (2014)
8. Zorine, A.: Stochastic model for communicating retrial queuing systems with cyclic control in random environment. Cybern. Syst. Anal. **6**(49), 890–897 (2013)
9. Zorine, A.V.: On the conditions of existence of a stationary mode in a tandem of queueing systems with cyclic control in a random environment. Autom. Control Comput. Sci. **47**(4), 183–191 (2013)
10. Zorine, A.V., Kuznetsov, N., Kuznetsov, I.N.: Analysis of a stochastic model of communicating retrial queueing systems with a cyclic control algorithm in a random environment. Vestn. Lobachevsky State Univ. Nizhni Novgorod **5**, 217–223 (2013)
11. Gantmacher, F.R.: The Theory of Matrices. AMS Chelsea Publishing, New York (1959)
12. Grassmann, W.K., Heyman, D.P.: Equilibrium distribution of block-structured Markov chains with repeating rows. J. of Appl. Probab. **27**, 557–576 (1990)
13. Feller, W.: An Introduction to Probability Theory And Its Applications, 2nd edn. Wiley, New York (1971)
14. Eaton, J.W., Bateman, D., Hauberg, S.: GNU Octave Manual Version 3. Network theory, Ltd., Bristol (2008)

The $M/GI/\infty$ System Subject to Semi-Markovian Random Environment

Anatoly Nazarov$^{(\boxtimes)}$ and Galina Baymeeva

Tomsk State University, Tomsk, Russia
{nazarov.tsu,baymeevag}@gmail.com

Abstract. In this paper we consider an $M/GI/\infty$ queueing system operating in a semi-Markovian random environment. That is, the arrival rate and service-time distribution change according to the external semi-Markov process state transitions. The service policy subject to environment transitions is as follows: the service-time distribution of the present customers does not change until their service is finished. The purpose of our study is to obtain the probability distribution of the number of customers in the system under asymptotic condition of high arrival rate and frequent environment transitions. To do this, we first apply the method of supplementary variable and the original method of dynamic screening to our system. We then conduct the asymptotic analysis of the system to obtain the discrete probability distribution.

Keywords: Queueing theory · Random environment · Semi-Markov process · Method of dynamic screening · Method of asymptotic analysis

1 Introduction

Queueing systems subject to external stochastic influences such as Markov modulation and random environment are of considerate interest in scientific literature. Such influences are often represented as system breakdowns or arrivals of priority customers (including batch arrivals) that force the system to behave at a different mode. Namely, several cases of Markov-modulated single-server queues are studied in [1]. Represented as a road subject to traffic incidents, the $M/M/\infty$ system operating under batch partial failures is considered in [2]. The random environment is assumed to have only two states: when there is an incident and when there are no incidents on the road. In [4], authors consider a more general case of $M/M/\infty$ queue in Markovian random environment with arbitrary finite number of states. The expression for steady-state factorial moments is obtained. In [5], the analysis of $M/G/\infty$ system in Markovian random environment is given. As a result, transient mean and stationary variance of the number of customers present in the system are obtained; a deeper analysis of exponential service case is conducted; the asymptotic normality of the number of customers probability distribution is shown under conditions of high arrival rate and frequent environment transitions due to the central limit theorem. In turn, the

© Springer International Publishing Switzerland 2015
A. Dudin et al. (Eds.): ITMM 2015, CCIS 564, pp. 128–140, 2015.
DOI: 10.1007/978-3-319-25861-4_11

steady-state mean number of customers in $M/M/\infty$ in semi-Markovian random environment is obtained in [7] as well as the steady-state distribution of the number of customers for the environment with 2 states.

Depending on the model application, different service policies of present customers with respect to environment transition may be considered. For instance, in one of the earliest papers [3] a single-server queue with general service-time is considered subject to interruptions with generally distributed durations. Two different cases of customer behavior after interruption clearance are studied: first, the resume policy, when the service is continued as it was left before the interruption; second, the repeat policy, when the service starts over. In [10,11], the $M/G/\infty$ system in semi-Markovian random environment is studied. These papers cover three cases of the present customers' reaction to environment state transitions. The first one is considered in the present paper — service-time distribution stays the same while the customer is in the system. This policy is also assumed in [5]. In the second case all customers are immediately cleared from the queue as environment state transition happens. The last case considers customers in service moving to a secondary queue which is an infinite-server system with bulk arrivals. This case is specifically analyzed in [10], and as a result the steady-state mean number of customers in the secondary queue is obtained.

Infinite-server queues are often used to approximate the behavior of systems with sufficiently large number of servers, such as banks, call-centers, supermarkets or digital distribution platforms. Such objects in reality are often affected by extraneous factors of stochastic nature which affect their performance. For instance, the change of bank rate set by the Central bank affects the conditions under which commercial banks give loans to their clients. These, in turn, significantly influence the intensity of clients' arrival. In this article we consider a mathematical model of such situation as an $M/GI/\infty$ queue operating in a random environment, for which the underlying process is a semi-Markov process with finite number of states. The arrival rate and service-time distribution change according to the environment state. Note that distribution of service-time customers which are currently being served does not change until the service-time is finished. Say the bank provided a credit to the client on certain conditions and during the repayment period there was a change of bank rate. The client will continue to repay his debt on those initial conditions — as mentioned in a loan agreement.

2   Problem Statement

We consider an $M/GI/\infty$ queueing system operating in semi-Markovian random environment. The system under discussion is an infinite-server queue with one stationary Poisson arrival process with parameter $\lambda_s N$ and the unlimited number of servers each having service-time distribution function $B_s(x)$, $s = \overline{1,K}$. We use a large parameter N that represents the condition of high arrival rate. Here $s = \overline{1,K}$ is the current state of a semi-Markov stochastic process $s(t)$ defined by the matrix product $\mathbf{P} \cdot \mathbf{A}(x)$. The matrix $\mathbf{P}$ here is a probability matrix of $s(t)$

state transitions and $\mathbf{A}(x)$ is a diagonal matrix with conditional sojourn time cdfs for every state $s = \overline{1,K}$ of $s(t)$ on its main diagonal. As there is always a free server in the system, there is no queue or loss option and each arriving customer is immediately placed at any free server and stays there for random time with distribution function $B_s(x)$. Note that we study the case when service-time distribution of a customer which is currently being served does not change until its service is finished.

That being said, for considered model we define a two-component process $\{i(t), s(t)\}$, where $i(t)$ with values $i \geq 0$ is the number of customers in the system at time t. Apparently, this process is non-Markovian. To deal with it, we first apply the original method of dynamic screening and the method of supplementary variable.

3   Method of Dynamic Screening

The method of dynamic screening can be used for the analysis of both queueing systems and networks. Further applications may be found in [14–16]. We apply this method to our system in the following way.

Given that at a certain time t_0 the system is empty, we pick a moment T and track the customer arrivals during the time interval (t_0, T). The customer will be referred to as "screened" at time t with probability

$$S_s(t) = 1 - B_s(T - t), s = \overline{1,K}, t_0 < t < T,$$

if it arrived at the system at time $t < T$ and was not fully serviced until the time T. Thus, the screened customers will be in the system taking up its servers at time T.

Let us denote by $n(t)$ the number of customers that were screened until time t. Stochastic process $n(t)$ is a screened point process with its points being the screened customers. The following identity always takes place:

$$i(T) = n(T). \tag{1}$$

We need to choose time t_0 so that at all times $t < t_0$ there are no screened customers, i.e.

$$S_s(t) = 1 - B_s(T - t) = 0, s = \overline{1,K}, t < t_0.$$

Since $B_s(x)$ is a cumulative distribution function, it is obvious enough to put $t_0 = -\infty$.

We write the possible state transitions of $n(t)$ and their probabilities assuming $n(t) = n, n \geq 0$ as follows:

$$n(t + \Delta t) = \begin{cases} n + 1, & \text{with prob. } \lambda_s \Delta t S_s(t) + o(\Delta t), \\ n, & \text{with prob. } 1 - \lambda_s \Delta t S_s(t) + o(\Delta t), \end{cases} s = \overline{1,K}$$

Equality (1) allows us to analyze a point process $n(t)$ instead of $i(t)$. Characteristics of the process $n(t)$ at time T coincide with the characteristics of value $i(T)$.

4   Kolmogorov Differential Equations

In order to deal with semi-Markovian process $s(t)$, we first need to apply the method of supplementary variable. We define $z(t)$ as residual sojourn time of $s(t)$ process in the current state, i.e. the interval from t until the next environment transition. It follows that the three-dimensional process $\{s(t), n(t), z(t)\}$ is a Markovian one. Therefore, we define the probabilities of system and environment state at time t as follows:

$$P(s, n, z, t) = P\left\{s(t) = s, n(t) = n, z(t) < \frac{z}{N}\right\}, s = \overline{1, K}, n \geq 0. \qquad (2)$$

Here the big parameter N justifies the condition of frequent environment transitions that compensates high arrival rate. The matrices that define the process $s(t)$ are determined as follows:

$$\mathbf{P} = \begin{pmatrix} p_{11} & p_{12} & \cdots & p_{1K} \\ p_{21} & p_{22} & \cdots & p_{2K} \\ \vdots & \vdots & \ddots & \vdots \\ p_{K1} & p_{K2} & \cdots & p_{KK} \end{pmatrix}, \mathbf{A}(x) = \begin{pmatrix} A_1(x) & 0 & \cdots & 0 \\ 0 & A_2(x) & \cdots & 0 \\ \vdots & \vdots & \ddots & \vdots \\ 0 & 0 & \cdots & A_K(x) \end{pmatrix}.$$

Let τ_s be the sojourn time of $s(t)$ in state $s = \overline{1, K}$. Then functions $A_s(x)$ are defined in the following way:

$$A_s(x) = P\left\{\frac{\tau_s}{N} < x\right\} = P\{\tau_s < Nx\}, s = \overline{1, K},$$

which means that $A_s(x)$ are the distribution functions of N-fold sojourn time of $s(t)$ in state $s = \overline{1, K}$.

The system of Kolmogorov differential equations that defines the probabilities (2) is written as follows:

$$\frac{1}{N}\frac{\partial P(s, n, z, t)}{\partial t} - \frac{\partial P(s, n, z, t)}{\partial z} + \frac{\partial P(s, n, 0, t)}{\partial z} =$$
$$\lambda_s S_s(t)\{P(s, n-1, z, t) - P(s, n, z, t)\} + \qquad (3)$$
$$A_s(z)\sum_{k=1}^{K} p_{ks}\frac{\partial P(k, n, 0, t)}{\partial z}, s = \overline{1, K}, n \geq 0$$

Here we use the denotation

$$\frac{\partial P(s, n, 0, t)}{\partial z} = \frac{\partial P(s, n, z, t)}{\partial z}\bigg|_{z=\infty}.$$

Provided $z \to \infty$, the initial condition to such system's solution is defined as follows:

$$P(s, n, t_0) = \begin{cases} r(s), & \text{if } n = 0, \\ 0, & \text{if } n > 0, \end{cases} s = \overline{1, K} \qquad (4)$$

Here $r(s)$ are the stationary probabilities of embedded Markov chain states of $s(t)$, $s = \overline{1,K}$. The partial characteristic functions of the process $\{s(t), n(t), z(t)\}$ are defined as follows:

$$H(s, u, z, t) = \sum_{n=0}^{\infty} e^{jun} P(s, n, z, t), s = \overline{1, K}$$

Here $j = \sqrt{-1}$ is the imaginary unit. We rewrite the system (4) using partial characteristic functions in the following way:

$$\frac{1}{N} \frac{\partial H(s, u, z, t)}{\partial t} - \frac{\partial H(s, u, z, t)}{\partial z} + \frac{\partial H(s, u, 0, t)}{\partial z} =$$
$$\lambda_s S_s(t)(e^{ju} - 1)H(s, u, z, t) + \tag{5}$$
$$A_s(z) \sum_{k=1}^{K} p_{ks} \frac{\partial H(k, u, 0, t)}{\partial z}, s = \overline{1, K}$$

We then use the following vector and matrix denotations:

$$\mathbf{H}(u, z, t) = \big(H(1, u, z, t)\ H(2, u, z, t) \cdots H(K, u, z, t) \big),$$

$$\Lambda = \begin{pmatrix} \lambda_1 & 0 & \cdots & 0 \\ 0 & \lambda_2 & \cdots & 0 \\ \vdots & \vdots & \ddots & \vdots \\ 0 & 0 & \cdots & \lambda_K \end{pmatrix}, \mathbf{S}(t) = \begin{pmatrix} S_1(t) & 0 & \cdots & 0 \\ 0 & S_2(t) & \cdots & 0 \\ \vdots & \vdots & \ddots & \vdots \\ 0 & 0 & \cdots & S_K(t) \end{pmatrix},$$

to rewrite the system (4) as follows:

$$\frac{1}{N} \frac{\partial \mathbf{H}(u, z, t)}{\partial t} - \frac{\partial \mathbf{H}(u, z, t)}{\partial z} +$$
$$\frac{\partial \mathbf{H}(u, 0, t)}{\partial z} [\mathbf{I} - \mathbf{P}\mathbf{A}(z)] = (e^{ju} - 1)\mathbf{H}(u, z, t)\Lambda\mathbf{S}(t). \tag{6}$$

Here $\mathbf{I}$ is the identity matrix. Our goal is to obtain the solution to system (6) as $z \to \infty$ that satisfies the initial condition derived from (4):

$$\mathbf{H}(u, t_0) = \mathbf{r}. \tag{7}$$

The row vector $\mathbf{r}$ here is the stationary probability distribution of the embedded Markov chain of the process $s(t)$ and solves the following system of matrix-vector equations:

$$\begin{cases} \mathbf{rP} = \mathbf{r}, \\ \mathbf{re} = 1. \end{cases} \tag{8}$$

5   Method of Asymptotic Analysis

Method of asymptotic analysis for queueing systems is the analysis of equations that define any of the system's characteristics or parameters [13]. It allows us to obtain the explicit distribution, parameters and moments under certain asymptotic conditions.

We obtain the solution to system (6) under asymptotic conditions of high arrival rate and frequent environment transitions, that is, as $N \to \infty$.

5.1   First-Order Asymptotic Analysis

Let us define substitutions for system (6) as follows:

$$\varepsilon = \frac{1}{N}, u = \varepsilon w, \mathbf{H}(u, z, t) = \mathbf{F}_1(w, z, t, \varepsilon).$$

Then (6) can be rewritten as

$$\varepsilon \frac{\partial \mathbf{F}_1(w, z, t, \varepsilon)}{\partial t} - \frac{\partial \mathbf{F}_1(w, z, t, \varepsilon)}{\partial z}$$
$$+ \frac{\partial \mathbf{F}_1(w, 0, t, \varepsilon)}{\partial z} [\mathbf{I} - \mathbf{PA}(z)] \qquad (9)$$
$$= (e^{j\varepsilon w} - 1)\mathbf{F}_1(w, z, t, \varepsilon)\mathbf{\Lambda S}(t).$$

As $\varepsilon \to 0$, the following equality holds:

$$\frac{\partial \mathbf{F}_1(w, z, t)}{\partial z} = \frac{\partial \mathbf{F}_1(w, 0, t)}{\partial z} [\mathbf{I} - \mathbf{PA}(z)]. \qquad (10)$$

We then represent the function $\mathbf{F}_1(w, z, t)$ as a product

$$\mathbf{F}_1(w, z, t) = \mathbf{r}(z)\Phi_1(w, t). \qquad (11)$$

Substitution (11) applied to (10) gives the following equation that defines row-vector $\mathbf{r}(z)$:

$$\mathbf{r}(z) = \int_0^z \mathbf{r}'(0) [\mathbf{I} - \mathbf{PA}(x)] \, dx \qquad (12)$$

To determine the value $\mathbf{r}'(0)$, we make the following substitution

$$\mathbf{r}'(0) = C\mathbf{r}, \ C = const. \qquad (13)$$

Note that according to (8)

$$\lim_{z \to \infty} \mathbf{r}(z) = C \int_0^\infty \mathbf{r} [\mathbf{I} - \mathbf{PA}(x)] \, dx$$
$$= C \int_0^\infty \mathbf{r} [\mathbf{I} - \mathbf{A}(x)] \, dx = C\mathbf{r}\mathbf{A}.$$

Apparently, the matrix $\mathbf{A}$ here is the diagonal matrix containing means A_s, $s = \overline{1, K}$ of distribution functions from $\mathbf{A}(x)$ on its main diagonal. According to (8), the constant C is derived as follows:

$$C = \frac{1}{\mathbf{rAe}} = \frac{1}{a}.$$

Finally, we write the expression for $\mathbf{r}(z)$:

$$\mathbf{r}(z) = \frac{1}{a} \int_0^z \mathbf{r} \left[\mathbf{I} - \mathbf{PA}(x) \right] dx.$$

Note that

$$\mathbf{r}(z) \Big|_{z=\infty} = \frac{\mathbf{rA}}{a}.$$

Now we set $z = \infty$ in (5.1) and make substitution (11):

$$\varepsilon \frac{1}{a} \mathbf{rA} \frac{\partial \Phi_1(w, t)}{\partial t} + \Phi_1(w, t)\mathbf{r} \left[\mathbf{I} - \mathbf{P} \right]$$
$$= (e^{j\varepsilon w} - 1)\Phi_1(w, t)\frac{1}{a}\mathbf{rA\Lambda S}(t).$$

Post-multiplication by $\mathbf{e}$ of both parts of the latter equation gives us the following first-order ordinary differential equation:

$$\frac{\partial \Phi_1(w, t)}{\partial t} = \frac{1}{a} \frac{e^{j\varepsilon w} - 1}{\varepsilon} \Phi_1(w, t)\mathbf{rA\Lambda S}(t)\mathbf{e}. \qquad (14)$$

As $\varepsilon \to 0$, the function $\Phi_1(w, t)$ that solves the equation above and satisfies the initial condition derived from (7) is as follows:

$$\Phi_1(w, t) = exp\left\{ jw\kappa_1(t) \right\},$$

$$\kappa_1(t) = \frac{1}{a} \int_{-\infty}^t \mathbf{rA\Lambda S}(\tau)\mathbf{e} d\tau.$$

Finally, we can write

$$\mathbf{H}(u, t) = \mathbf{F}_1(w, t, \varepsilon) \approx \mathbf{F}_1(w, t) = \frac{\mathbf{rA}}{a}\Phi_1(w, t) = \frac{\mathbf{rA}}{a}exp\{jw\kappa_1(t)\},$$

where $w = Nu$. It follows that

$$M\{e^{jun(t)}\} = \mathbf{H}(u, t)\mathbf{e} \approx h_1(u, t) = exp\{ju\kappa_1(t)N\}.$$

Since (1) takes place, we can finally conclude:

$$M\{e^{jui(T)}\} = M\{e^{jun(T)}\} = \mathbf{H}(u, T)\mathbf{e}$$
$$\approx h_1(u, T) = exp\{ju\kappa_1(T)N\}.$$

Let us calculate the value $\kappa_1(T)$:

$$\kappa_1(T) = \int_{-\infty}^{T} \mathbf{r}\mathbf{A}\mathbf{\Lambda}\mathbf{S}(t)\mathbf{e}\,dt = \int_{-\infty}^{T} \sum_{s=1}^{K} \mathbf{r}(s)A_s\lambda_s S_s(t)\,dt$$

$$= \sum_{s=1}^{K} \mathbf{r}(s)A_s\lambda_s \int_{-\infty}^{T} \{1 - B_s(T - t)\}\,dt$$

$$= \sum_{s=1}^{K} \mathbf{r}(s)A_s\lambda_s \int_{0}^{\infty} \{1 - B_s(\tau)\}\,d\tau = \sum_{s=1}^{K} \mathbf{r}(s)A_s\lambda_s b_s,$$

where b_s are the service-time means, $s = \overline{1, K}$. Thus

$$\kappa_1(T) = \sum_{s=1}^{K} \mathbf{r}(s)\lambda_s \int_{0}^{\infty} \{1 - B_s(\tau)\}\,d\tau = \sum_{s=1}^{K} \mathbf{r}(s)A_s\lambda_s b_s = \mathbf{r}\mathbf{A}\mathbf{\Lambda}\mathbf{B}\mathbf{e},$$

where $\mathbf{B}$ is a diagonal matrix containing service-time means b_s.

5.2   Second-Order Asymptotic Analysis

In the equation (6) we make a substitution

$$\mathbf{H}(u, z, t) = \mathbf{H}_2(u, z, t)e\{ju\kappa_1(t)N\}. \tag{15}$$

The function $\mathbf{H}_2(u, z, t)$ here is the centered characteristic function as the following relation takes place:

$$\mathbf{H}_2(u, z, t)\mathbf{e} = \mathbf{H}(u, z, t)e^{-ju\kappa_1(t)N}\mathbf{e}$$
$$= M\left\{exp\left[ju(n(t) - \kappa_1(t)N)\right]\right\}.$$

The substitution (15) yields an equation which defines $\mathbf{H}_2(u, z, t)$:

$$\frac{1}{N}\frac{\partial \mathbf{H}_2(u, z, t)}{\partial t} - \frac{\partial \mathbf{H}_2(u, z, t)}{\partial z}$$
$$+ \frac{\partial \mathbf{H}_2(u, 0, t)}{\partial z}[\mathbf{I} - \mathbf{P}\mathbf{A}(z)] \tag{16}$$
$$= \mathbf{H}_2(u, z, t)\left\{(e^{ju} - 1)\mathbf{\Lambda}\mathbf{S}(t) - ju\kappa_1'(t)\mathbf{I}\right\}$$

We rewrite the latter system using substitutions

$$\varepsilon^2 = \frac{1}{N},\, u = \varepsilon w,\, \mathbf{H}_2(u, t) = \mathbf{F}_2(w, z, t, \varepsilon)$$

in the following way:

$$\varepsilon^2 \frac{\partial \mathbf{F}_2(w, z, t, \varepsilon)}{\partial t} - \frac{\partial \mathbf{F}_2(w, z, t, \varepsilon)}{\partial z}$$
$$+ \frac{\partial \mathbf{F}_2(w, 0, t, \varepsilon)}{\partial t}[\mathbf{I} - \mathbf{P}\mathbf{A}(z)] \tag{17}$$
$$= \mathbf{F}_2(w, z, t, \varepsilon)[(e^{j\varepsilon w} - 1)\mathbf{\Lambda}\mathbf{S}(t) - j\varepsilon w\kappa_1'(t)\mathbf{I}]$$

As $\varepsilon \to 0$, the following relation takes place:

$$\frac{\partial \mathbf{F}_2(w, z, t)}{\partial z} = \frac{\partial \mathbf{F}_2(w, 0, t)}{\partial z} \left[\mathbf{I} - \mathbf{PA}(z)\right].$$

It follows that the function $\mathbf{F}_2(w, z, t)$ may be represented as follows:

$$\mathbf{F}_2(w, z, t) = \mathbf{r}(z)\Phi_2(w, t). \tag{18}$$

In turn, the function $\mathbf{F}_2(w, z, t, \varepsilon)$ may be approximated with the following expression:

$$\mathbf{F}_2(w, z, t, \varepsilon) = \Phi_2(w, t) \left\{\mathbf{r}(z) + j\varepsilon w \mathbf{f}_2(z, t)\right\} + O(\varepsilon^2). \tag{19}$$

The row-vector function $\mathbf{f}_2(z, t)$ is to be defined. To do this, first we make a substitution (19) in the system (17). We also make the following approximation in (17):

$$e^{j\varepsilon w} - 1 = j\varepsilon w + O(\varepsilon^2),$$

and then set $\varepsilon \to 0$. These manipulations yield us the following equation that defines $\mathbf{f}_2(z, t)$:

$$\frac{\partial \mathbf{f}_2(z, t)}{\partial z} - \frac{\partial \mathbf{f}_2(0, t)}{\partial z}\left[\mathbf{I} - \mathbf{PA}(z)\right] + \Phi_2(w, t)\mathbf{r}(z)\left[\mathbf{\Lambda S}(t) - \kappa_1'(t)\mathbf{I}\right] = \mathbf{0}. \tag{20}$$

Here $\mathbf{0}$ is the row-vector filled with zeros. It follows that as $z \to \infty$, we have the relation

$$\mathbf{f}_2(t) = \int\limits_0^\infty \left\{\frac{\partial \mathbf{f}_2(0, t)}{\partial z}\left[\mathbf{I} - \mathbf{PA}(x)\right] - \mathbf{r}(x)\left[\mathbf{\Lambda S}(t) - \kappa_1'(t)\mathbf{I}\right]\right\} dx. \tag{21}$$

The right part of the latter relation is the improper integral. In order for it to converge, it is necessary that the integrand function converges to 0 as the variable of integration approaches ∞. That is, the following relation stands for $\mathbf{f}_2(0, t)$:

$$\frac{\partial \mathbf{f}_2(0, t)}{\partial z}\left[\mathbf{I} - \mathbf{P}\right] = \frac{\mathbf{rA}}{a}\left[\mathbf{\Lambda S}(t) - \kappa_1'(t)\mathbf{I}\right]. \tag{22}$$

The equation above is the non-homogeneous underdetermined system of linear equations. We represent its solution as a sum of general solution to homogeneous system and a partial solution to non-homogeneous system:

$$\frac{\partial \mathbf{f}_2(0, t)}{\partial z} = c(t)\mathbf{r} + \mathbf{g}(t), \tag{23}$$

where $c(t)$ is an arbitrary scalar function of t. We write the additional condition for the function $\mathbf{g}(t)$ as follows:

$$\mathbf{g}(t)\mathbf{e} = 0, \tag{24}$$

Let us now define the explicit expression for (21):

$$\mathbf{f}_2(t) = \int_0^\infty \left\{ \frac{\partial \mathbf{f}_2(0,t)}{\partial z} \left[\mathbf{I} - \mathbf{P} + \mathbf{P}(\mathbf{I} - \mathbf{A}(z)) \right] - \mathbf{r}(z) \left[\mathbf{\Lambda S}(t) - \kappa_1'(t)\mathbf{I} \right] \right\} dz$$

$$= \int_0^\infty \left[\frac{1}{a} \mathbf{rA} - \mathbf{r}(z) \right] \left[\mathbf{\Lambda S}(t) - \kappa_1'(t)\mathbf{I} \right] dz$$

$$+ \int_0^\infty \frac{\partial \mathbf{f}_2(0,t)}{\partial z} \mathbf{P} \left[\mathbf{I} - \mathbf{A}(z) \right] dz$$

$$= \frac{1}{a} \mathbf{rA} \int_0^\infty \left\{ \mathbf{I} - \mathbf{A}^{-1} \int_0^z \left[\mathbf{I} - \mathbf{A}(x) \right] dx \right\} dz \left[\mathbf{\Lambda S}(t) - \kappa_1'(t)\mathbf{I} \right] + \frac{\partial \mathbf{f}_2(0,t)}{\partial z} \mathbf{PA}.$$

Note that $\mathbf{A}^{-1} \int_0^z [\mathbf{I} - \mathbf{A}(x)]\, dx$ is a diagonal matrix that contains distribution functions of both elapsed and residual sojourn time of $s(t)$ at each of its states. Then denoted by $\overline{\mathbf{A}}$ is the diagonal matrix that contains means of such cdfs respectively. Finally, we rewrite the expression for $\mathbf{f}_2(t)$ as follows:

$$\mathbf{f}_2(t) = \frac{1}{a} \mathbf{rA}\overline{\mathbf{A}} \left[\mathbf{\Lambda S}(t) - \kappa_1'(t)\mathbf{I} \right] + \frac{\partial \mathbf{f}_2(0,t)}{\partial z} \mathbf{PA}. \qquad (25)$$

Now we show that the row-vector function $\mathbf{f}_2(t)$ does not actually depend on the arbitrary scalar function $c(t)$ that is present in (23). To do that, we consider the following term that is present in (25):

$$\frac{\partial \mathbf{f}_2(0,t)}{\partial z} \mathbf{PA} \left[\mathbf{\Lambda S}(t) - \kappa_1'(t)\mathbf{I} \right] \mathbf{e}$$

$$= \left[c(t)\mathbf{r} + \mathbf{g}(t) \right] \mathbf{PA} \left[\mathbf{I} - \frac{1}{a} \mathbf{erA} \right] \mathbf{\Lambda S}(t)\mathbf{e}$$

$$= \mathbf{g}(t)\mathbf{PA} \left[\mathbf{I} - \frac{1}{a} \mathbf{erA} \right] \mathbf{\Lambda S}(t)\mathbf{e}$$

$$+ c(t)\mathbf{rPA\Lambda S}(t)\mathbf{e} - c(t)\frac{1}{a}\mathbf{rPAerA\Lambda S}(t)\mathbf{e}.$$

With (8) and $a = \mathbf{rAe}$ in mind, we conclude that the two latter terms cancel each other. Thus, the function $c(t)$ is not present in (25).

Now let us determine the function $\Phi_2(w,t)$. For this purpose, we again make substitution (19) in (17) and also the following approximation:

$$e^{j\varepsilon w} - 1 = j\varepsilon w + \frac{(j\varepsilon w)^2}{2} + O(\varepsilon^3).$$

As $\varepsilon \to 0$ and $z \to \infty$, this yields us the first-order ODE that defines $\Phi_2(w,t)$:

$$\frac{\partial \Phi_2(w,t)}{\partial t} = \frac{(jw)^2}{2} \Phi_2(w,t) \left\{ \kappa_1'(t) + 2\mathbf{f}_2(t) \left[\mathbf{\Lambda S}(t) - \kappa_1'(t)I \right] \mathbf{e} \right\}. \qquad (26)$$

Its solution that satisfies the initial condition derived from (7) is of the following form:

$$\Phi_2(w,t) = exp\left\{\frac{(jw)^2}{2}\kappa_2(t)\right\},\tag{27}$$

where

$$\kappa_2(t) = \kappa_1(t) + 2\int_{-\infty}^{t} \mathbf{f}_2(\tau)\left[\mathbf{\Lambda S}(\tau) - \kappa_1'(\tau)\mathbf{I}\right]\mathbf{e}\,d\tau.\tag{28}$$

Thus, the expression for the centered characteristic function $\mathbf{H}_2(u,t)$ is obtained and is written as follows:

$$\mathbf{H}_2(u,t) = \mathbf{F}_2(w,t,\varepsilon) \approx \mathbf{F}_2(w,t) = \frac{\mathbf{rA}}{a}\Phi_2(w,t)$$
$$= \frac{\mathbf{rA}}{a}exp\{\frac{(jw)^2}{2}\kappa_2(t)\} = \frac{\mathbf{rA}}{a}exp\{\frac{(ju)^2}{2}\kappa_2(t)N\}.$$

It follows that

$$\mathbf{H}(u,t) = \mathbf{H}_2(u,t)e^{ju\kappa_1(t)N} \approx \frac{\mathbf{rA}}{a}exp\left\{ju\kappa_1(t)N + \frac{(ju)^2}{2}\kappa_2(t)N\right\},\tag{29}$$

$$M\{e^{jun(t)}\} = \mathbf{H}(u,t)\mathbf{e} \approx h_2(u,t) = exp\left\{ju\kappa_1(t)N + \frac{(ju)^2}{2}\kappa_2(t)N\right\}.\tag{30}$$

Considering (1), the following identities are true:

$$M\{e^{jui(T)}\} = M\{e^{jun(T)}\} = \mathbf{H}(u,T)\mathbf{e} \approx h_2(u,T)$$
$$= exp\{ju\kappa_1(T)N + \frac{(ju)^2}{2}\kappa_2(T)N\},\tag{31}$$

where $\kappa_2(T)$ is of the following form:

$$\kappa_2(T) = \kappa_1(T) + 2\int_{-\infty}^{T} \mathbf{f}_2(t)\left[\mathbf{\Lambda S}(t) - \kappa_1'(t)\mathbf{I}\right]\mathbf{e}\,dt\tag{32}$$

According to the definition of functions $S_s(t) = 1 - B_s(T - t)$ it is clear that $\lim_{t\to\infty} S_s(t) = 0, s = \overline{1,K}$. Therefore, it is clear that the improper integral (32) is converging and thus can be calculated numerically given specific system and environment parameters.

Obviously, the asymptotic steady-state probability distribution of the number of customers in the system defined by (31) is normal with first and second cumulants $\kappa_1(t)N$ and $\kappa_2(t)N$ respectively. It is known that

$$M\{i(T)\} \approx \kappa_1(T)N, D\{i(T)\} \approx \kappa_2(T)N.\tag{33}$$

Inverse Fourier transform of (31) gives the probability density function of the normally distributed random variable:

$$p(x) = \frac{1}{\sqrt{2\pi\kappa_2(T)N}}exp\left\{-\frac{(x-\kappa_1(T)N)^2}{2\kappa_2(T)N}\right\}.$$ (34)

It is necessary to switch from this continuous distribution to discrete as follows:

$$P(i) = Cp(i), i \geq 0,$$ (35)

where the constant value C is defined considering the normalizing condition:

$$\sum_{i=0}^{\infty} P(i) = C\sum_{i=0}^{\infty} p(i) = 1.$$ (36)

Due to (36), C is given as follows:

$$C = 1/\sum_{i=0}^{\infty} p(i)$$ (37)

6 Conclusion

Thus, the Gaussian approximation of the probability distribution of the number of customers in the system $M(\lambda_s)/G(B_s(x))/\infty$ is obtained during the asymptotic analysis under conditions of high arrival rate and frequent environment transitions. Using the method of dynamic screening, we considered a non-stationary Markov point process $n(t)$ instead of non-Markovian $i(t)$ which is the number of customers in the system. Then, according to the method of supplementary variable we defined the residual sojourn time $z(t)$ in the present state of the environment process $s(t)$ to be able to analyze it with theory of Markov processes tools. After deriving the system of differential equations in terms of vector characteristic functions of the number of customers in the system, we conducted the asymptotic analysis of the system in question.

Earlier we considered a problem of $M/G/\infty$ queue operating in Markovian random environment with the same service policy when service-time distribution does not change while the customer is in the system. Similarly, we obtained the steady-state probability distribution of the number of customers in the system. However, the Markov case narrows down the application area significantly. Thus, in this paper we considered a more general case with random environment being semi-Markovian.

Acknowledgments. The work is supported by Tomsk State University Competitiveness Improvement Program.

References

1. Prabhu, N.U., Zhu, Y.: Markov-modulated queueing systems. Queueing Syst. **5**, 215–245 (1989)
2. Baykal-Gursoy, M., Xiao, W.: Stochastic decomposition in $M/M/\infty$ queues with Markov modulated service rates. Queueing Syst. **48**, 75–88 (2004)
3. Keilson, J.: Queues subject to service interruption. Ann. Math. Statist. **33**, 1314–1322 (1962)
4. O'Cinneide, C.A., Purdue, P.: The $M/M/\infty$ queue in a random environment. J. Appl. Prob. **23**, 175–184 (1986)
5. Blom, J., Kella, O., Mandjes, M., Thorsdottir, H.: Markov-modulated infinite-server queues with general service times. Queueing Syst. **76**, 403–424 (2014)
6. D'Auria, B.: $M/M/\infty$ queues in semi-Markovian random environment. Queueing Syst. **58**, 221–237 (2008)
7. Falin, G.: The $M/M/\infty$ queue in random environment. Queueing Syst. **58**, 65–76 (2008)
8. Fralix, B.H., Adan, I.J.B.F.: An infinite-server queue influenced by a semi-Markovian environment. Queueing Syst. **61**, 65–84 (2009)
9. D'Auria, B.: Stochastic decomposition of the $M/G/\infty$ queue in a random environment. Oper. Res. Lett. **35**, 805–812 (2007)
10. Purdue, P., Linton, D.: An infinite-server queue subject to an extraneous phase process and related models. J. Appl. Prob. **18**, 236–244 (1981)
11. Linton, D., Purdue, P.: An $M/G/\infty$ queue with m customer types subject to periodic clearing. Opsearch **16**, 80–88 (1979)
12. Nazarov, A.A., Baymeeva, G.V.: The study of $M/G/\infty$ in random environment
13. Nazarov, A.A., Moiseeva, S.P.: Method of Asymptotic Analysis in Queueing Theory. NTL, Tomsk (2006) (in Russian)
14. Nazarov, A.A., Moiseev, A.N.: Analysis of an open non-Markovian $GI - (GI|\infty)^K$ queueing network with high-rate renewal arrival process. Prob. Inf. Transm. **49**, 167–178 (2013)
15. Moiseev, A.N., Nazarov, A.A.: Asymptotic analysis of a multistage queuing system with a high-rate renewal arrival process Optoelectronics. Instrum. Data Process. **50**(2), 163–171 (2014)
16. Moiseev, A., Nazarov, A.: Asymptotic analysis of the infinite-server queueing system with high-rate semi-Markov arrivals. In: IEEE International Congress on Ultra Modern Telecommunications and Control Systems (ICUMT 2014), pp. 507–513. IEEE Press (2014)

Probability Density Function for Modulated MAP Event Flows with Unextendable Dead Time

Luydmila Nezhel'skaya[✉]

National Research Tomsk State University, Lenina Avenue 36,
634050 Tomsk, Russia
ludne@mail.ru

Abstract. We consider a modulated MAP flow of events with two states; it is one of the mathematical models for an incoming stream of claims (events) in digital integral servicing networks. The observation conditions for this flow are such that each event generates a period of dead time during which other events from the flow are inaccessible for observation and do not extend the dead time period (unextendable dead time). We find an explicit form for a probability density of interval duration between neighboring events in the observed flow. We consider the stationary operation mode for the observed flow, so we disregard the transient processes. The duration estimation of unextendable dead time is essential to determine the number of the lost events which have the useful information.

Keywords: Modulated map event flows · Unextendable dead time · Probability density function of interval duration · Transition probability

1 Introduction

Mathematical models of queueing theory are widely used to describe real physical, technical, and other systems and processes. Thanks to the fast development of computer hardware and information technologies, another important field of queueing theory applications has arisen, namely the design and creation of informational and computational networks, computer communication networks, satellite networks, telecommunication networks, etc. In practice, parameters that determine the incoming flow of events change in time, and the changes are often random; the latter has led researchers to consider doubly stochastic flows of events. One of the first works in this direction was probably the paper [1] in which a doubly stochastic flow is defined as a flow whose intensity is a random process. Doubly stochastic flows can be divided into two classes: flows whose intensity is a continuous random process and flows whose intensity is a piecewise constant random process with a finite number of states. We emphasize that flows of the second class were introduced virtually at the same time in 1979, in [2–4]. In [2,3], these flows were called MC (Markov Chain) flows; in [4], MVP

© Springer International Publishing Switzerland 2015
A. Dudin et al. (Eds.): ITMM 2015, CCIS 564, pp. 141–151, 2015.
DOI: 10.1007/978-3-319-25861-4_12

(Markov Versatile Processes) flows. Starting from the end of the 1980s, the latter, especially after [5], have usually been called MAP (Markovian Arrival Process) event flows. We note that MAP-flows of events are especially characteristic for real telecommunication networks [6].

This article is an immediate development of the studies which were carried out in [7–15]. In the studies of event flows, we can distinguish two classes of problems: (1) estimating the states of an event flow [7–9]; (2) estimating flow parameters [10–15].

One of the distorting factors in our estimates of event flow states and parameters is the dead time of sensing devices [16] which results from a detected event. Other events that occur during a dead time period are inaccessible for observation (simply speaking, they are lost). We can assume that this period lasts for some fixed time (unextendable dead time). One example of such flows is given by the CSMA/CD protocol, a random multiple access protocol with conflict detection which is widely used in computer networks. At the moment a conflict is registered (detected) on the input of a certain network node, the "stub" ("plug") signal is broadcast in the network; while the "stub" signal is being sent out, claims arriving to this network node are refused service and are forwarded to callback source. Here the time during which the network node is closed for servicing claims that arrive there after a conflict is found can be treated as the dead time of the device that registers conflicts in the network node.

In this paper an explicit form of a probability density function of interval duration between neighboring events in modulated MAP event flow with unextendable dead time is derived. The explicit form of the probability density provides the possibility to solve the problem of the flow parameters estimation and of the dead time duration.

2 Problem Setting

We consider a modulated MAP flow of events with intensity represented by a piecewise constant random process $\lambda(t)$ with two states: $\lambda(t) = \lambda_1$ or $\lambda(t) = \lambda_2$ ($\lambda_1 > \lambda_2 \geq 0$). The time during which process $\lambda(t)$ remains at the ith, $i = 1, 2$, state is determining by two random values: (1) the first random value has exponential distribution function $F_i^{(1)}(t) = 1 - e^{\alpha_i t}, i = 1, 2$; when the ith state ends process $\lambda(t)$ transits with probability equal one from the ith state to the jth, $i, j = 1, 2$ ($i \neq j$); (2) the second random value has exponential distribution function $F_2^{(i)}(t) = 1 - e^{\lambda_i t}, i = 1, 2$; when the ith state ends process $\lambda(t)$ transits with probability $P_1(\lambda_j|\lambda_i)$ from the ith state to the jth ($i \neq j$) and a flow event occurs or process $\lambda(t)$ transits with probability $P_0(\lambda_j|\lambda_i)$ from the ith state to the jth ($i \neq j$), but the flow event does not occur, or process $\lambda(t)$ transits from the ith state to the ith with probability $P_1(\lambda_i|\lambda_i)$ and a flow event occurs. Here $P_1(\lambda_j|\lambda_i) + P_0(\lambda_j|\lambda_i) + P_1(\lambda_i|\lambda_i) = 1; i, j = 1, 2; i \neq j$. The first and the second random values are independent from each other. Under these assumptions, $\lambda(t)$ is a Markov process. The infinitesimal characteristics matrices for the process $\lambda(t)$ are as follows [6]:

$$D_0 = \left\| \begin{matrix} -(\alpha_1 + \lambda_1) & \alpha_1 + \lambda_1 P_0(\lambda_2|\lambda_1) \\ \alpha_2 + \lambda_2 P_0(\lambda_1|\lambda_2) & -(\alpha_2 + \lambda_2) \end{matrix} \right\|,$$

$$D_1 = \left\| \begin{matrix} \lambda_1 P_1(\lambda_1|\lambda_1) & \lambda_1 P_1(\lambda_2|\lambda_1) \\ \lambda_2 P_1(\lambda_1|\lambda_2) & \lambda_1 P_2(\lambda_2|\lambda_2) \end{matrix} \right\|.$$

The elements of the matrix D_1 are intensities of the process $\lambda(t)$ passing from the state to the state with an event occurrence. Diagonal elements of matrix D_0 are the intensities of process $\lambda(t)$ leaving its states, which are taken with the opposite sign. Off-diagonal elements of matrix D_0 are intensities of the process $\lambda(t)$ passing from the state to the state without an event occurrence. We should note that if $\alpha_i = 0, i = 1, 2$, there is a usual MAP flow of events [8].

After each event registered at time t_k, there begins a time of fixed duration T (dead time) during which other events from the original modulated MAP flow are inaccessible for observation. When dead time is over, the first new event again gives rise to a period of dead time of duration T and so on. One possible scenario of the resulting situation is shown on Fig. 1, where $t_1, t_2, \ldots$ denote the moments when events occur in the observed flow; 1 and 2 are states of the random process $\lambda(t)$; black circles denote modulated MAP flow events inaccessible for observation; dashed lines denote dead time durations.

The process $\lambda(t)$ is unobservable in principle, i.e. latent Markov process, and we can only observe time moments when events in the observed flow $t_1, t_2, \ldots$ occur.

We consider the stationary operation mode the observed flow. Under the made assumptions the sequence $\{\lambda(t_k)\}$ at the time moments $t_1, t_2, \ldots, t_k$ of the

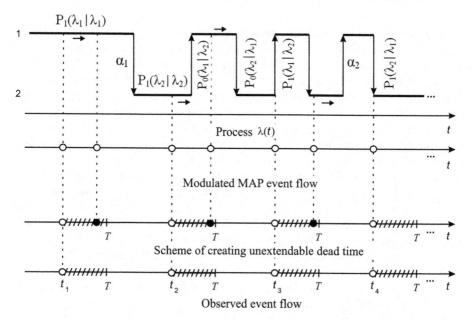

Fig. 1. Forming the observed flow

events occurrence is an embedded Markov chain. So the observed flow has a Markovian property if its evolution is considered from the moment $t_k, k = 1, 2, ...,$ of the event occurrence.

Denote by $\tau_k = t_{k+1} - t_k, k = 1, 2, ...$ the value of intervals k duration between neighboring events in the observed flow. In a steady-state conditions we may take that the probability density of the interval k duration is $p(\tau_k) = p(\tau), \tau \geq 0$, for any k. Then we can let $t_k = 0$ without loss of generality, i.e. the moment of the event occurrence is $\tau = 0$. On the other hand, since the event observed at time moment $\tau = 0$ gives rise to a dead time period of duration T, we get that $\tau = T + t$, where t is the duration value of the interval between the end of dead time $\tau = T$ and the moment $\tau = T + t$ when the next event in the observed flow occurs. Here we assume that the value of T known exactly.

The aim of this article is to obtain an explicit form of a probability density function $p_T(\tau)$ of interval duration in modulated MAP event flows with unextendable dead time (further observed flow or flow).

3   The Expressions for Probability Density $p_T(\tau)$

A probability density function of interval duration between neighboring events in observed flow is defined by the formula:

$$
\begin{cases}
p_T(\tau) = 0, 0 \leq \tau < T, \\
p_T(\tau - T) = \sum_{i=1}^{2} \pi_i(0|T) \sum_{j=1}^{2} q_{ij}(T) \sum_{k=1}^{2} \widetilde{p}_{jk}(\tau - T), \tau \geq T,
\end{cases} \tag{1}
$$

where $\tilde{p}_{jk}(\tau - T)$ is a probability density function that the process $\lambda(t)$ changes its state from the jth state to the kth state without the event occurrence on the interval $(T, T + t)$ and with the event occurrence at the moment $\tau = T + t, j, k = 1, 2$; $q_{ij}(T)$ is a transition probability that the process $\lambda(t) = \lambda_j$ at the moment of dead time end $\tau = T$ in condition that the process $\lambda(0) = \lambda_i$ at the moment $\tau = 0, i, j = 1, 2$; $\pi_i(0|T)$ is a conditional stationary probability that the process $\lambda(t)$ sojourns in the state i at the time moment $\tau = 0$ in condition that the flow event occurred at the moment $\tau = 0, i = 1, 2$, and a dead time period of duration T occurred at this time moment $(\pi_1(0|T) + \pi_2(0|T) = 1)$. Let us introduce $p_{jk}(t)$ is a transition probability that there are no flow events on the interval $(T, T + t)$ and $\lambda(T + t) = \lambda_k$ at the moment $T + t$ in condition that at the time moment $\tau = T \ \lambda(T) = \lambda_j, j, k = 1, 2$.

As an example, we consider a formula derivation for a probability $p_{11}(t)$. Let us suppose that the moment of a dead time end is $t = 0$. We consider the interval $(0, t + \Delta t)$ and how the process $\lambda(t)$ behaves oneself on this intervals.

The interval $(0, t + \Delta t)$ breaks down into two adjacent intervals: the first interval $(0, t)$, where $t = 0$ is a moment of a dead time end and the second small enough semiinterval $[t, t + \Delta t)$.

Let us consider three possible cases:

1. $p_{11}(t)e^{-\alpha_1 \Delta t}e^{-\lambda_1 \Delta t} = (1 - (\alpha_1 + \lambda_1)\Delta t)p_{11}(t) + o(\Delta t)$ is a probability that there are no events on the interval $(0, t)$, the value of the process $\lambda(t)$ is $\lambda(t) = \lambda_1$ at the moment t in condition that at the time moment $t = 0$ the value of the process $\lambda(t)$ is $\lambda(0) = \lambda_1$ and on the semiinterval $[t, t + \Delta t)$ the first state does not end;

2. $p_{12}(t)(1 - e^{-\alpha_2 \Delta t})e^{-\lambda_2 \Delta t} = \alpha_2 \Delta t p_{12}(t) + o(\Delta t)$ is a probability that there are no events on the interval $(0, t)$, the value of the process $\lambda(t)$ is $\lambda(t) = \lambda_2$ at the moment t in condition that at the time moment $t = 0$ the value of the process $\lambda(t)$ is $\lambda(0) = \lambda_1$ and on the semiinterval $[t, t + \Delta t)$ the second state ends with probability $1 - e^{-\alpha_2 \Delta t}$ and the second state does not end with probability $e^{-\lambda_2 \Delta t}$;

3. $p_{12}(t)e^{-\alpha_2 \Delta t}(1 - e^{-\lambda_2 \Delta t})P_0(\lambda_1|\lambda_2) = \lambda_2 P_0(\lambda_1|\lambda_2)p_{12}(t)\Delta t + o(\Delta t)$ is a probability that there are not events on the interval $(0, t)$, the value of the process $\lambda(t)$ is $\lambda(t) = \lambda_2$ at the moment t in condition that at the time moment $t = 0$ the value of the process $\lambda(t)$ is $\lambda(0) = \lambda_1$ and on the semiinterval $[t, t + \Delta t)$ the second state does not end with probability $e^{-\alpha_2 \Delta t}$ and the second state ends with probability $1 - e^{-\lambda_2 \Delta t}$, the flow event does not occur and the process $\lambda(t)$ transits from the second state to the first state with probability $P_0(\lambda_1|\lambda_2)$.

There are no another possibilities. So the probability $p_{11}(t)$ at the time moment $t + \Delta t$ takes the following form:

$$p_{11}(t + \Delta t) = (1 - (\alpha_1 + \lambda_1)\Delta t)p_{11}(t) + \alpha_2 p_{12}(t)\Delta t + \lambda_2 P_0(\lambda_1|\lambda_2)p_{12}\Delta t + o(\Delta t).$$

Transferring the probability $p_{11}(t)$ from the right-hand side to the left-hand side, dividing the left- and the right-hand side by Δt and passing to the limit for $\Delta t \to 0$, we find the differential equation with the initial condition

$$p'_{11}(t) = -(\alpha_1 + \lambda_1)p_{11}(t) + (\alpha_2 + \lambda_2 P_0(\lambda_1|\lambda_2))p_{12}(t), \ p_{11}(0) = 1.$$

Similarly, we have the following system of differential equations with the initial conditions:

$$p'_{12}(t) = (\alpha_1 + \lambda_1 P_0(\lambda_2|\lambda_1))p_{11}(t) - (\alpha_2 + \lambda_2)p_{12}(t),$$

$$p'_{21}(t) = -(\alpha_1 + \lambda_1)p_{21}(t) + (\alpha_2 + \lambda_2 P_0(\lambda_1|\lambda_2))p_{22}(t),$$

$$p'_{22}(t) = -(\alpha_2 + \lambda_2)p_{22}(t) + (\alpha_1 + \lambda_1 P_0(\lambda_2|\lambda_1))p_{21}(t),$$

$$p_{12}(0) = 0, \ p_{21}(0) = 0, \ p_{22}(0) = 1.$$

Solving obtained system of differential equations, we find the probabilities $p_{jk}(t), j, k = 1, 2$:

$$p_{11}(t) = \frac{1}{z_2 - z_1} \cdot \left[(\lambda_2 + \alpha_2 - z_1)e^{-z_1 t} - (\lambda_2 + \alpha_2 - z_2)e^{-z_2 t}\right],$$

$$p_{12}(t) = \frac{\alpha_1 + \lambda_1 P_0(\lambda_2|\lambda_1)}{z_2 - z_1} \cdot \left[e^{-z_1 t} - e^{-z_2 t}\right],$$

$$p_{21}(t) = \frac{\alpha_2 + \lambda_2 P_0(\lambda_1|\lambda_2)}{z_2 - z_1} \cdot \left[e^{-z_1 t} - e^{-z_2 t}\right],$$

$$p_{22}(t) = \frac{1}{z_2 - z_1} \cdot \left[(\lambda_1 + \alpha_1 - z_1)e^{-z_1 t} - (\lambda_1 + \alpha_1 - z_2)e^{-z_2 t}\right],$$

$$z_1 = \frac{1}{2}\left[(\lambda_1 + \lambda_2 + \alpha_1 + \alpha_2)\right.$$

$$\left.- \sqrt{(\lambda_1 - \lambda_2 + \alpha_1 - \alpha_2)^2 + 4(\alpha_1 + \lambda_1 P_0(\lambda_2|\lambda_1))(\alpha_2 + \lambda_2 P_0(\lambda_1|\lambda_2))}\right], \quad (2)$$

$$z_2 = \frac{1}{2}\left[(\lambda_1 + \lambda_2 + \alpha_1 + \alpha_2)\right.$$

$$\left.+ \sqrt{(\lambda_1 - \lambda_2 + \alpha_1 - \alpha_2)^2 + 4(\alpha_1 + \lambda_1 P_0(\lambda_2|\lambda_1))(\alpha_2 + \lambda_2 P_0(\lambda_1|\lambda_2))}\right],$$

$$0 < z_1 < z_2.$$

To derive a formula for probability density $\tilde{p}_{jk}(t)$ we should consider the interval $(0, t + \Delta t)$ and how the process $\lambda(t)$ behaves oneself on this interval. The interval $(0, t + \Delta t)$ breaks down into two adjacent intervals: the first interval $(0, t)$, where $t = 0$ is a moment of a dead time end and the second small enough semiinterval $[t, t + \Delta t)$.

Then $\tilde{p}_{jk}(t)\Delta t + o(\Delta t), j, k = 1, 2$, is a joint probability that the process $\lambda(t)$ changes its the jth state to the lth state without the event occurrence on the interval $(0, t)$ and on the semiinterval $[t, t + \Delta t)$ the lth state ends, the flow event occurs with intensity λ_l and the process $\lambda(t)$ transits from the lth state to the kth state, $l = 1, 2$.

As an example, we derive a formula for a joint probability $\tilde{p}_{11}(t)\Delta t + o(\Delta t)$. Let us consider two cases:

1. $\tilde{p}_{11}^{(1)}(t)\Delta t + o(\Delta t)$ is a joint probability that the process $\lambda(t)$ remains at the first state without the event occurrence on the interval $(0, t)$ and on the semiinterval $[t, t + \Delta t)$ the first state ends, the flow event occurs with intensity λ_1 and the process $\lambda(t)$ remains at the first state. This joint probability can be written as

$$\tilde{p}_{11}^{(1)}(t)\Delta t + o(\Delta t) = p_{11}(t)(1 - e^{-\lambda_1 \Delta t})P_1(\lambda_1|\lambda_1)$$

$$= \lambda_1 P_1(\lambda_1|\lambda_1)p_{11}(t)\Delta t + o(\Delta t).$$

2. $\tilde{p}_{11}^{(2)}(t)\Delta t + o(\Delta t)$ is a joint probability that the process $\lambda(t)$ changes its state from the first state to the second state without the event occurrence on the interval $(0, t)$ and on the semiinterval $[t, t + \Delta t)$ the second state ends, the flow event occurs with intensity λ_2 and the process $\lambda(t)$ transits from the second state to the first state. This joint probability can be written as

$$\tilde{p}_{11}^{(2)}(t)\Delta t + o(\Delta t) = p_{12}(t)(1 - e^{-\lambda_2 \Delta t})P_1(\lambda_1|\lambda_2)$$

$$= \lambda_2 P_1(\lambda_1|\lambda_2)p_{12}(t)\Delta t + o(\Delta t).$$

Then the joint probability $\tilde{p}_{11}(t)\Delta t + o(\Delta t)$ take the following form

$$\tilde{p}_{11}(t)\Delta t + o(\Delta t) = \lambda_1 P_1(\lambda_1|\lambda_1)p_{11}(t)\Delta t + \lambda_2 P_1(\lambda_1|\lambda_2)p_{12}(t)\Delta t + o(\Delta t).$$

Diving the left- and the right-hand side by Δt and passing to the limit for $\Delta t \to 0$, we find the formula for probability density $\tilde{p}_{11}(t)$:

$$\tilde{p}_{11}(t) = \lambda_1 P_1(\lambda_1|\lambda_1) p_{11}(t) + \lambda_2 P_1(\lambda_1|\lambda_2) p_{12}(t).$$

Similarly, another probability densities $\tilde{p}_{12}(t)$, $\tilde{p}_{21}(t)$, $\tilde{p}_{22}(t)$ are obtained. So the probability densities $\tilde{p}_{jk}(t), j, k = 1, 2$, are of the form:

$$
\begin{aligned}
\tilde{p}_{11}(t) &= \lambda_1 P_1(\lambda_1|\lambda_1) p_{11}(t) + \lambda_2 P_1(\lambda_1|\lambda_2) p_{12}(t), \\
\tilde{p}_{12}(t) &= \lambda_1 P_1(\lambda_2|\lambda_1) p_{11}(t) + \lambda_2 P_1(\lambda_2|\lambda_2) p_{12}(t), \\
\tilde{p}_{21}(t) &= \lambda_1 P_1(\lambda_1|\lambda_1) p_{21}(t) + \lambda_2 P_1(\lambda_1|\lambda_2) p_{22}(t), \\
\tilde{p}_{22}(t) &= \lambda_1 P_1(\lambda_2|\lambda_1) p_{21}(t) + \lambda_2 P_1(\lambda_2|\lambda_2) p_{22}(t),
\end{aligned}
\tag{3}
$$

where the probability $\tilde{p}_{jk}(t), j, k = 1, 2$, are defined in (2).

Since the sequence $\{\lambda(t_k)\}$ at the time moments $t_1, t_2, ..., t_k...$ of the flow events occurence is an embedded Markov chain, the following equations take place:

$$
\begin{aligned}
\pi_1(0|T) &= \pi_1(0|T)\pi_{11} + \pi_2(0|T)\pi_{21}, \\
\pi_2(0|T) &= \pi_1(0|T)\pi_{12} + \pi_2(0|T)\pi_{22},
\end{aligned}
\tag{4}
$$

where π_{ij} is a transition probability that the process $\lambda(t)$ transits from the ith state to the jth $(i, j = 1, 2)$ during the time from the event occurrence at the time moment $\tau = 0$ till the moment of the next flow event occurrence.

To derive π_{ij} let us introduce transition probability $q_{ij}(\tau)$ that the process $\lambda(\tau) = \lambda_j$ at the time moment τ in condition that the process $\lambda(0) = \lambda_i$ at the time moment $\tau = 0, i, j = 1, 2$. Using a Δ-method described above for introduced probabilities $q_{ij}(\tau)$ we obtain the following system of the differential equations with initial conditions:

$$
\begin{aligned}
q'_{11}(\tau) &= \big[-(\alpha_1 + \lambda_1) + \lambda_1 P_1(\lambda_1|\lambda_1) \big] q_{11}(\tau) \\
&\quad + (\alpha_2 + \lambda_2[1 - P_1(\lambda_2|\lambda_2)]) q_{12}(\tau), \\
q'_{12}(\tau) &= \big[-(\alpha_2 + \lambda_2) + \lambda_2 P_1(\lambda_2|\lambda_2) \big] q_{12}(\tau) \\
&\quad + (\alpha_1 + \lambda_1[1 - P_1(\lambda_1|\lambda_1)]) q_{11}(\tau), \\
q'_{21}(\tau) &= \big[-(\alpha_1 + \lambda_1) + \lambda_1 P_1(\lambda_1|\lambda_1) \big] q_{21}(\tau) \\
&\quad + (\alpha_2 + \lambda_2[1 - P_1(\lambda_2|\lambda_2)]) q_{22}(\tau), \\
q'_{22}(\tau) &= \big[-(\alpha_2 + \lambda_2) + \lambda_2 P_1(\lambda_2|\lambda_2) \big] q_{22}(\tau) \\
&\quad + (\alpha_1 + \lambda_1[1 - P_1(\lambda_1|\lambda_1)]) q_{21}(\tau), \\
q_{11}(0) &= 1, \quad q_{12}(0) = 0, \quad q_{21}(0) = 0, \quad q_{22}(0) = 1.
\end{aligned}
\tag{5}
$$

Solving the system (5), we find the probabilities $q_{ij}(\tau), i, j = 1, 2$:

$$
\begin{aligned}
q_{11}(\tau) &= \pi_1 + \pi_2 e^{-a\tau}, \quad q_{12}(\tau) = \pi_2 - \pi_2 e^{-a\tau}, \\
q_{21}(\tau) &= \pi_1 - \pi_1 e^{-a\tau}, \quad q_{22}(\tau) = \pi_2 + \pi_1 e^{-a\tau}, \\
a &= \alpha_1 + \alpha_2 + \lambda_1[1 - P_1(\lambda_1|\lambda_1)] + \lambda_2[1 - P_1(\lambda_2|\lambda_2)], \\
\pi_1 &= \frac{\alpha_2 + \lambda_2[1 - P_1(\lambda_2|\lambda_2)]}{\alpha_1 + \alpha_2 + \lambda_1[1 - P_1(\lambda_1|\lambda_1)] + \lambda_2[1 - P_1(\lambda_2|\lambda_2)]}, \\
\pi_2 &= \frac{\alpha_1 + \lambda_1[1 - P_1(\lambda_1|\lambda_1)]}{\alpha_1 + \alpha_2 + \lambda_1[1 - P_1(\lambda_1|\lambda_1)] + \lambda_2[1 - P_1(\lambda_2|\lambda_2)]},
\end{aligned}
\tag{6}
$$

where the prior final probabilities π_i, $i = 1, 2$, of the ith state of the process $\lambda(t)$ are obtained in [7].

Letting in (6) the time moment $\tau = T$, we can rewrite (6) as

$$q_{11}(T) = \pi_1 + \pi_2 e^{-aT}, \quad q_{12}(T) = \pi_2 - \pi_2 e^{-aT},$$
$$q_{21}(T) = \pi_1 - \pi_1 e^{-aT}, \quad q_{22}(T) = \pi_2 + \pi_1 e^{-aT}. \tag{7}$$

Since the process $\lambda(t)$ is a Markov process, the transition probabilities $q_{ij}(\tau)$ and p_{ij} let us write the expressions for the transition probabilities $\pi_{ij}(\tau), i, j = 1, 2$, by the Chapman-Kolmogorov formula as

$$\pi_{11} = q_{11}(T)p_{11} + q_{12}(T)p_{21}, \quad \pi_{12} = q_{12}(T)p_{22} + q_{11}(T)p_{12},$$
$$\pi_{21} = q_{22}(T)p_{21} + q_{21}(T)p_{11}, \quad \pi_{22} = q_{21}(T)p_{12} + q_{22}(T)p_{22}, \tag{8}$$
$$\pi_{11} + \pi_{12} = 1, \quad \pi_{21} + \pi_{22} = 1,$$

where p_{ij} is a transition probability that the process $\lambda(t)$ transits from the ith state to the jth during the time interval between the moment $\tau = 0$ (the moment of event occurrence) and the moment of the next event occurrence. Since τ is a undefined time moment, the transition probabilities $p_{ij}, i, j = 1, 2$, are determined as

$$p_{ij} = \int_0^\infty \tilde{p}_{ij}(\tau) d\tau. \tag{9}$$

Substituting at first (3) into (9) and then (2) into (9) and calculating the corresponding integrals, we find

$$p_{11} = \frac{\lambda_1 P_1(\lambda_1|\lambda_1)(\lambda_2 + \alpha_2) + \lambda_2 P_1(\lambda_1|\lambda_2)(\alpha_1 + \lambda_1 P_0(\lambda_2|\lambda_1))}{(\lambda_1 + \alpha_1)(\lambda_2 + \alpha_2) - (\alpha_1 + \lambda_1 P_0(\lambda_2|\lambda_1))(\alpha_2 + \lambda_2 P_0(\lambda_1|\lambda_2))},$$
$$p_{12} = \frac{\lambda_1 P_1(\lambda_2|\lambda_1)(\lambda_2 + \alpha_2) + \lambda_2 P_1(\lambda_2|\lambda_2)(\alpha_1 + \lambda_1 P_0(\lambda_2|\lambda_1))}{(\lambda_1 + \alpha_1)(\lambda_2 + \alpha_2) - (\alpha_1 + \lambda_1 P_0(\lambda_2|\lambda_1))(\alpha_2 + \lambda_2 P_0(\lambda_1|\lambda_2))},$$
$$p_{21} = \frac{\lambda_2 P_1(\lambda_1|\lambda_2)(\lambda_1 + \alpha_1) + \lambda_1 P_1(\lambda_1|\lambda_1)(\alpha_2 + \lambda_2 P_0(\lambda_1|\lambda_2))}{(\lambda_1 + \alpha_1)(\lambda_2 + \alpha_2) - (\alpha_1 + \lambda_1 P_0(\lambda_2|\lambda_1))(\alpha_2 + \lambda_2 P_0(\lambda_1|\lambda_2))},$$
$$p_{22} = \frac{\lambda_2 P_1(\lambda_2|\lambda_2)(\lambda_1 + \alpha_1) + \lambda_1 P_1(\lambda_2|\lambda_1)(\alpha_2 + \lambda_2 P_0(\lambda_1|\lambda_2))}{(\lambda_1 + \alpha_1)(\lambda_2 + \alpha_2) - (\alpha_1 + \lambda_1 P_0(\lambda_2|\lambda_1))(\alpha_2 + \lambda_2 P_0(\lambda_1|\lambda_2))}. \tag{10}$$

Taking into account the explicit formulas for $q_{ij}(T)$ into (7) the transition probabilities (8) take the following form:

$$\pi_{11} = p_{11} - \pi_2(p_{11} - p_{21})[1 - e^{-aT}],$$
$$\pi_{12} = p_{12} + \pi_2(p_{22} - p_{12})[1 - e^{-aT}],$$
$$\pi_{21} = p_{21} + \pi_1(p_{11} - p_{21})[1 - e^{-aT}],$$
$$\pi_{22} = p_{22} - \pi_1(p_{22} - p_{12})[1 - e^{-aT}]. \tag{11}$$

Substituting (11) into (4) and taking into account that $\pi_1(0|T) + \pi_2(0|T) = 1$, we obtain the explicit formulas for $\pi_i(0|T), i, j = 1, 2$:

$$\pi_1(0|T) = \frac{p_{21} + \pi_1(p_{11} - p_{21})[1 - e^{-aT}]}{p_{12} + p_{21} - (1 - p_{11} - p_{22})[1 - e^{-aT}]},$$
$$\pi_2(0|T) = \frac{p_{12} + \pi_2(p_{22} - p_{12})[1 - e^{-aT}]}{p_{12} + p_{21} - (1 - p_{11} - p_{22})[1 - e^{-aT}]}, \tag{12}$$

where a and π_i are defined by (6) and p_{ij} are defined by (10). We rewrite the formula (1) as:

$$p_T(\tau - T) = [\pi_1(0|T)q_{11}(T) + \pi_2(0|T)q_{21}(T)] \sum_{k=1}^{2} \tilde{p}_{1k}(\tau - T)$$

$$+[\pi_1(0|T)q_{12}(T) + \pi_2(0|T)q_{22}(T)] \sum_{k=1}^{2} \tilde{p}_{2k}(\tau - T), \tau \geq T. \tag{13}$$

We consider the expressions $\pi_1(0|T)q_{1i}(T) + \pi_2(0|T)q_{2i}(T), i = 1, 2$, in the formula (13). According to formulas (7), these expressions become

$$\pi_1(0|T)q_{11}(T) + \pi_2(0|T)q_{21}(T) = \pi_1 + [\pi_2 - \pi_2(0|T)]e^{-aT},$$
$$\pi_1(0|T)q_{12}(T) + \pi_2(0|T)q_{22}(T) = \pi_2 - [\pi_2 - \pi_2(0|T)]e^{-aT}. \tag{14}$$

We denote by $\pi_1(t) = \pi_1 + [\pi_2 - \pi_2(0|T)]e^{-aT}$. We can show that $\pi_1(t)$ is the conditional probability that the process $\lambda(t)$ sojourns in the state i at the time moment $\tau = T$ in condition that the flow event occurred at the time moment $\tau = 0$, $i = 1, 2$, and a dead time period of a duration T occurred at this time moment. Similarly, $\pi_2(t) = \pi_2 - [\pi_2 - \pi_2(0|T)]e^{-aT}$. Then we can rewrite (14) as

$$\pi_1(T) = \pi_1 + [\pi_2 - \pi_2(0|T)]e^{-aT},$$
$$\pi_2(T) = \pi_2 - [\pi_2 - \pi_2(0|T)]e^{-aT}. \tag{15}$$

Substituting (15) into (13), we find

$$p_T(\tau - T) = \pi_1(T) \sum_{k=1}^{2} \tilde{p}_{1k}(\tau - T) + \pi_2(T) \sum_{k=1}^{2} \tilde{p}_{2k}(\tau - T), \tau \geq T.$$

Let $t = \tau - T, t \geq 0$. Then we have

$$p_T(t) = \pi_1(T) \sum_{k=1}^{2} \tilde{p}_{1k}(t) + \pi_2(T) \sum_{k=1}^{2} \tilde{p}_{2k}(t), t \geq 0. \tag{16}$$

Substituting at first (3) then (2) into (16) and making sufficiently difficult transformations, we obtain the explicit form of a probability density function $p_T(\tau)$:

$$p_T(\tau) = \begin{cases} 0, 0 \leq \tau < T, \\ \gamma(T)z_1 e^{-z_1(\tau - T)} + (1 - \gamma(T))z_2 e^{-z_2(\tau - T)}, \tau \geq T, \end{cases}$$

where $\gamma(T) = \dfrac{1}{z_2 - z_1}\{z_2 - \lambda_1\pi_1(T)[1 - P_0(\lambda_2|\lambda_1)] - \lambda_2\pi_2(T)[1 - P_0(\lambda_1|\lambda_2)]\}$, $\pi_1(T), \pi_2(T)$ are defined by (15); z_1, z_2 are defined by (2).

4   Conclusion

The obtained results provide the possibility to estimate the unknown parameters of a modulated MAP event flows and also the duration of unextendable dead time. The duration estimation of unextendable dead time is essential to determine the number of the lost events which have the useful information. The obtained formula for probability density $p_T(\tau)$ allows us to carry out an estimation of unknown parameters and duration of unextendable dead time in the modulated MAP event flows with unextendable dead time by the maximum likelihood method or method of moments. In the first case we have to derive the system of moment's equations and in the second case the likelihood function. These estimation problem are the subject of the further research.

Acknowledgments. The work is supported by Tomsk State University Competitiveness Improvement Program.

References

1. Kingman, J.F.C.: On doubly stochastic poisson process. Proc. Camb. Philos. Soc. **60**(4), 923–930 (1964)
2. Basharin, G.P., Kokotushkin, V.A., Naumov, V.A.: On the Equivalent Substitutions Method for Computing Fragments of Communication Networks. Izv. Akad. Nauk USSR. Tekhn. Kibern. **6**, 92–99 (1979)
3. Basharin, G.P., Kokotushkin, V.A., Naumov, V.A.: On the equivalent substitutions method for computing fragments of communication networks. Izv. Akad. Nauk USSR. Tekhn. Kibern. **1**, 55–61 (1980)
4. Neuts, M.E.: A versatile markov point process. J. Appl. Probab. **16**, 764–779 (1979)
5. Lucantoni, D.M.: New results on the single server queue with a batch marcovian arrival process. Commun. Stat. Stoch. Models **7**, 1–46 (1991)
6. Dudin, A.N., Klimenok, V.I.: Queueing Systems with Correlated Flows. Univ, Belarus Gos (2000)
7. Nezhelskaya, L.: Optimal state estimation in modulated MAP event flows with unextendable dead time. In: Dudin, A., Nazarov, A., Yakupov, R., Gortsev, A. (eds.) ITMM 2014. CCIS, vol. 487, pp. 342–350. Springer, Heidelberg (2014)
8. Gortsev, A.M., Nezhel'skaya, L.A., Solov'ev, A.A.: Optimal state estimation in MAP event flows with unextendable died time. Autom. Remote Control **73**(8), 1316–1326 (2012)
9. Gortsev, A.M., Nezhel'skaya, L.A., Shevchenko, T.I.: Estimation of the states of an MC-stream of events in the presence of measurement errors. Russ. Phys. J. **36**(12), 1153–1167 (1993)
10. Gortsev, A.M., Nezhel'skaya, L.A.: An asynchronous double stochastic flow with initiation of superfluous events. Discrete Math. Appl. **21**(3), 283–290 (2011)
11. Bushlanov, I.V., Gortsev, A.M., Nezhel'skaya, L.A.: Estimating parameters of the synchronous twofold-stochastic flow of events. Autom. Remote Control **69**(9), 1517–1533 (2008)
12. Gortsev, A.M., Nezhel'skaya, L.A.: Estimation of the dead time period and intensities of the synchronous double stochastic event flow. Radiotekhnika **10**, 8–16 (2004)

13. Gortsev, A.M., Nezhel'skaya, L.A.: Estimation of the dead-time period and parameters of a semi-synchronous double stochastic stream of events. Meas. Tech. **46**(6), 536–545 (2003)
14. Gortsev, A.M., Nezhel'skaya, L.A.: Estimate of parameters of synchronously alternating poisson stream of events by the moment method. Telecommun. Radio Eng. (English translation of Elektrosvyaz and Radiotekhnika) **50**(1), 56–63 (1996)
15. Gortsev, A.M., Nezhel'skaya, L.A.: Estimation of the parameters of a synchro-alternating poisson event flow by the moment method. Radiotekhnika **40**(7–8), 6–10 (1995)
16. Apanasovich, V.V., Kolyada, A.A., Chernyavskii, A.F.: Statistical Analysis of Random Flows in a Physical Experiment. Universitetskoe, Minsk (1988)

Statistical Modeling of Air-Sea Turbulent Heat Fluxes by Finite Mixtures of Gaussian Distributions

Victor Korolev[1,2], Andrey Gorshenin[2,3]([✉]), Sergey Gulev[1,4],
and Konstantin Belyaev[1,4]

[1] Lomonosov Moscow State University, Moscow, Russia
victoryukorolev@yandex.ru, gul@sail.msk.ru, kosbel55@gmail.com
[2] Institute of Informatics Problems, Federal Research Center "Computer Science
and Control" of the Russian Academy of Sciences, Moscow, Russia
a.k.gorshenin@gmail.com
[3] Radioengineering and Electronics, Moscow State University of Information
Technologies, Moscow, Russia
[4] P.P. Shirshov Institute of Oceanology, Moscow, Russia

Abstract. The approach originally developed for the investigation of the traffic, that is, the intensities of information flows in financial markets, is applied for the statistical analysis of climatic data. The statistical regularities in the behavior of sensible and latent turbulent heat fluxes recomputed from 6-hourly NCEP-NCAR for the period 1948 − 2008 in Atlantic are analyzed. It is proposed to represent these regularities by probability distributions that are mixtures of several normal (Gaussian) laws with parameters varying in time. The method of moving separation of mixtures is used to obtain the values of the parameters of the mixtures. This approach allows to analyze the regularities in the variation of the parameters and, hence, to capture the low-term variability which can be considered as a trend and high-term dynamics associated with diffusion or irregular variability.

Keywords: Finite mixtures of normal distributions · Moving separation of mixtures · Data mining · Probabilistic models

1 Introduction

Surface turbulent air-sea sensible and latent heat fluxes, the language of air-sea communication, are critically important in many areas of geosciences. Surface flux data are available from several data sources which each has its strengths and weaknesses. The most long-term global surface flux time series (for the period of a century and longer) are available from Voluntary Observing Ship (VOS) data [4], while for the last several decades satellite observations, reanalyses and blended products (e.g., OA-FLUX [5,6]) provide global datasets with reasonably high space and time resolution.

© Springer International Publishing Switzerland 2015
A. Dudin et al. (Eds.): ITMM 2015, CCIS 564, pp. 152–162, 2015.
DOI: 10.1007/978-3-319-25861-4_13

Our knowledge about variability of surface turbulent heat fluxes is limited in most cases by the first (in few cases the second) moment of the distribution of fluxes. These are typically conventionally computed from surface flux time series and constitute the basis for climatological analyses and validation activities [7,8]. However, the detailed assessment of surface heat flux characteristics, including estimation and evaluation of extreme surface flux values requires accurate knowledge of the entire distribution of turbulent heat fluxes and the analysis of variability in the parameters of this distribution in time and in space. The situation when one of key ocean-atmosphere variables is extensively used and studied without explicit knowledge of its statistical distribution makes it difficult to evaluate surface fluxes in the ocean and climate models, and, thus, damps the models predictive potential. One more reason why the accurate distributions of surface turbulent heat fluxes is required is the necessity to quantify and potentially minimize sampling errors in VOS-based surface flux products [9,10]. Sampling uncertainties being large in magnitude, affect both mean estimates of surface fluxes and characteristics of extreme fluxes.

An attempt to justify probability distribution for turbulent surface fluxes has been done in [11] where such a distribution was found to be reasonably well approximated by so-called Fisher-Tippet (FT) distribution controled by two parameters, namely locale and scale parameters. In paper [11] these parameters were evaluated along with performing goodness-of-fit-tests. The authors of paper [11] presented global climatologies of the FT distribution parameters for surface fluxes along with estimation of extreme surface fluxes. Further some results of evaluation of FT distribution were used for the analysis of centennial long time series of surface turbulent fluxes reconstructed from VOS observations for the period from 1880 onward [12]. However, many questions associated with probability distribution of surface turbulent fluxes still remain open. FT distribution tends frequently underestimate extremely high turbulent heat fluxes and does not allow to a full extent for the accurate representation of the cases of the so-called "heavy tails" in flux distributions.

In the present work we further exploit the original idea about generalization of synoptic time-series of surface fluxes. The scheme is based on the representation of the probability distribution of the heat flux increments (first order differences) in form of a mixture of a number of normal (Gaussian) distributions with different time dependent parameters whose weights may also vary in time. The method of moving separation of mixtures is used to obtain the values of the parameters of the mixtures. This approach allows to analyze the regularities in variation of the distribution parameters and, hence, to capture the low-frequency and short-term variability in surface fluxes which can be respectively attributed to longer term changes on the seasonal and interannual timescales and to irregular variability.

The approach is essentially based on the mathematical models and methods which were originally proposed for the analysis of the information traffic, that is, the intensities of information flows in financial markets (see, for example, [1–3]).

2   Data Homogenization

Traditionally, statistical analysis of the stochastic regularities in the observed time series analyses all available data without any pre-processing aiming to homogenize the data. For example, in [11] FT distribution was applied to the raw surface flux time series. Such an approach, however, can be hardly used for the analysis of very long time series and the evolution of the distribution parameters in time. However, the sample used for the statistical analysis is not homogeneous with individual samples being likely interdependent. To explain this, we can consider the following schematic example.

Everywhere in what follows the notation $\varphi(x)$ and $\Phi(x)$ will be used for the standard normal (Gaussian) probability distribution density and the standard normal distribution function, respectively:

$$\varphi(x) = \frac{1}{\sqrt{2\pi}}e^{-x^2/2}, \quad \Phi(x) = \int_{-\infty}^{x} \varphi(z)\,dz, \quad x \in \mathbb{R}.$$

Let n be a natural number and $\xi_1, \xi_2, \ldots, \xi_n$ be independent identically distributed random variables with the common distribution function $F(x) = \Phi(x - a)$ (that is, each ξ_j has the normal distribution with the mean a and the unit variance). Let us construct a new set of random variables $\zeta_1, \zeta_2, \ldots, \zeta_n$ by setting

$$\zeta_k = \xi_1 + \ldots + \xi_k, \quad k = 1, \ldots, n.$$

Obviously, for each $k \in \{1, \ldots, n\}$ the element ζ_k has the normal distribution with mean ka and variance k. Therefore, the sample $\zeta_1, \zeta_2, \ldots, \zeta_n$ is by no means homogeneous and independent.

Figure 1 illustrates this effect presenting the histogram constructed from the artificially simulated sample $\xi_1, \ldots, \xi_n$ with $n = 1000$ and $a = 2$ (above) and the corresponding sample $\zeta_1, \ldots, \zeta_n$ (below). The lower histogram is essentially skewed to the right with very few negative values. This is exactly the shape of the distribution proposed in [11].

Stochastic features of ζ_k are to a great extent determined by those of the sum $\xi_1 + \ldots + \xi_{k-1}$ and only slightly depend on those of ξ_k. The greater k, the less the contribution of ξ_k into ζ_k is. Therefore, any analysis of the statistical regularities of ξ_i, $i = 1, \ldots, n$ directly from the sample $\zeta_1, \zeta_2, \ldots, \zeta_n$ should be performed with a serious caution. Moreover, from a athematical viewpoint it is incorrect to apply standard statistical procedures to the sample $\zeta_1, \zeta_2, \ldots, \zeta_n$. Practical applicability of the results based upon such analysis is always questionable.

To avoid the impact of the above mentioned problems (resulting from a long-time maintained conventional practices) onto analysis of statistical regularities in the behavior of time series we consider the transformed time series of increments of the initially observed surface heat fluxes.

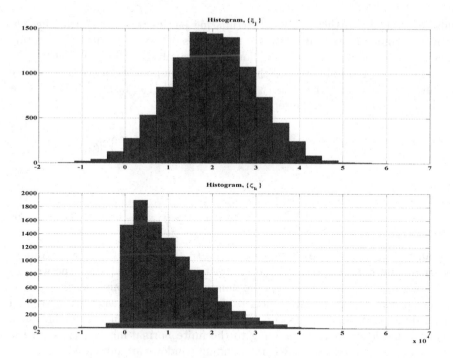

Fig. 1. The histograms constructed from the samples $\xi_1, \ldots, \xi_n$ (above) and $\zeta_1, \ldots, \zeta_n$ (below) with $n = 1000$ and $a = 2$.

3   The Outline of the Method of Moving Separation of Finite Normal Mixtures for the Analysis of the Observed Time Series

To reveal the structural changes of the observed stochastic processes in time, the so-called method of moving separation of mixtures (MSM method) is successfully used. This method was proposed in [1]. Papers [3,13,14] give examples of the efficient performance of this method. They are related to the analysis of financial markets, traffic in information systems and plasma turbulence. The key point of this method is that the variability of the time series can be decomposed into the dynamical and the diffusive components.

In this method, one-dimensional distributions of the increments of the basic process are approximated by finite location-scale mixtures of normal distributions. Theoretical background for these models is based on a statement that finite normal mixtures can effectively approximate general location-scale normal mixtures or normal variance-mixtures. These are the limit laws for the distributions of sums of a random number of independent random variables or non-homogeneous and non-stationary random walks, see details in [1,15,16].

To analyze the changes in the character of stochastic process, the problem of statistical estimation of unknown parameters of distributions should be successively solved for a running sample segment (of a fixed length) forming the sub-sample to be further analysed. Estimating parameters for the running segment (or window), one can derive the time series of these parameters. Resulting time series of the parameters will allow for the analysis of temporal changes in the behavior of the diffusive and the dynamical components in the process. We assume that the cumulative density function for a given segment of data centered at the time moment t can be represented as

$$F_t(x) = \sum_{i=1}^{k} \frac{p_i(t)}{\sigma_i(t)\sqrt{2\pi}} \int_{-\infty}^{x} \exp\left\{ - \frac{(t - a_i(t))^2}{2\sigma_i^2(t)} \right\} dt, \tag{1}$$

where

$$\sum_{i=1}^{k} p_i(t) = 1, \quad p_i(t) \geqslant 0. \tag{2}$$

(for all $x \in \mathbb{R}$, $a_i(t) \in \mathbb{R}$, $\sigma_i(t) > 0$, $i = 1, \ldots, k$). The model (1) is called a finite location-scale normal mixture. The parameters $p_1(t), \ldots, p_k(t)$ are the weights satisfying (2). The parameter k is the number of mixture components.

The parameters of model (1) noticeably depend on time, as it can be seen on Fig. 2 showing the histograms constructed from different windows of the width 200 and the densities corresponding to the finite normal mixture (1) with the parameters estimated from the corresponding windows are presented.

The parameter k may be also treated as depending on time. However, for both purposes of the effective settings of the method and the interpretation of the results it is preferable to fix the maximum possible value of k in advance. As a rule, the number of components does not exceed 6 or 7. Typically, at least six or seven components provide an excellent approximation of any model. When the parameters of the model (1) are estimated for the moving segments, some weights may be very close to zero or to be evaluated as zeroes. This implies the corresponding component to vanish and the number of components to decrease.

The parameters $a_1(t), \ldots, a_k(t)$ are associated with the dynamic component of the internal variability of the process, and the parameters $\sigma_1(t), \ldots, \sigma_k(t)$ are associated with the diffusive one, see [1]. If Z_t is a random variable with a distribution function (1), then its variance can be represented as the sum of the two components:

$$\mathrm{D}Z_t = \sum_{i=1}^{k} p_i(t) \left[a_i(t) - \overline{a}(t)\right]^2 + \sum_{i=1}^{k} p_i(t)\sigma_i^2(t), \tag{3}$$

where

$$\overline{a}(t) = \sum_{i=1}^{k} p_i(t)a_i(t).$$

The first term in the right-hand side of (3) depends only on the weights $p_i(t)$ and the expected values $a_i(t)$ of the components of mixture (1). Since Z_t is an increment of the basic process, then $a_i(t)$ is the expected value of the increment,

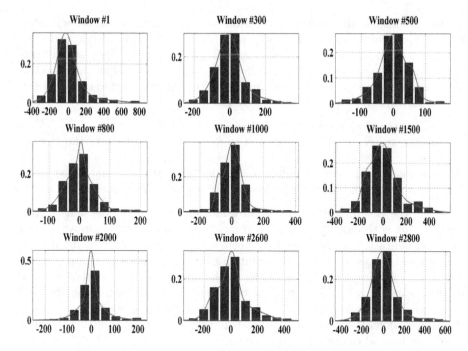

Fig. 2. The histograms constructed from different windows and the fitted finite normal mixture densities.

i.e., the "trend" component. Hence, the first component quantifies the part of the total variance (changeability) which is due to local elementary trends. It is called the *dynamic* component of the variance. Moreover, $a_i(t)$ is the expected value of the random variable whose distribution is just the i-th component of mixture (1). By construction, this random variable is the increment of the initial process at the unit time interval, that is, $a_i(t)$ is the mean velocity of the variation of the i-th component. Thus, the set of pairs $(a_1(t), p_1(t)), \ldots, (a_k(t), p_k(t))$ determines the distribution of the velocities over local trends at time t.

The second term in the right-hand side of (3) depends only on the weights $p_i(t)$ and the variances $\sigma_i^2(t)$ of the components and represents the purely stochastic *diffusive* component of the total variance.

4   Estimation of the Parameters by EM Algorithm

To estimate the parameters of model (1), at each window the classical EM algorithm was used. The EM algorithm is an iterative numerical procedure for the maximization of the multi-parameter likelihood functions. It was suggested in [17] and comprehensively described in [1]. Although very many modifications of this algorithm have already been proposed, the classical EM algorithm remains to be the most reliable tool for the estimation of the parameters of the finite normal mixture model (1).

Nevertheless, in this case this algorithm has some important drawbacks. Perhaps, the main of them is its instability with respect to the initial approximation. The finite normal mixture model likelihood function has a very non-smooth surface with peaks, ravines or rills. Therefore, being a "greedy" algorithm, the EM algorithm converges to the local maximum which is closest to the starting point. The usual ways to treat this obstacle are the following:

- to choose the starting point at random;
- to choose several starting points and to average the results over the number of runs;
- to choose several starting points and to use the estimate that delivers the maximum value of the target likelihood function among the runs as the result.

When the EM algorithm is used in a moving mode as it was done in this work dealing with moving separation of mixtures, much attention must be paid to the visualization of the results. For convenient interpretation of the results, the obtained curves depicting the evolution of the parameters $p_i(t)$, $a_i(t)$, $\sigma_i(t)$, $i = 1, \ldots, k$ in time should be smooth. At the first sight the smoothness of the resulting curves can be achieved, if the final result of the EM algorithm obtained at the previous window is used as the starting point for the EM algorithm at the next window. However, this rule leads to that the danger of hitting the local extremum instead of the global one increases.

The results obtained for the three ways of the choice of starting points mentioned above with some modifications are presented on Fig. 3. On the upmost graph, the temporal variation of the local trend parameters $a_i(t)$ estimated by the EM algorithm with the random choice of the starting points at each window is presented. At each window the EM algorithm is run five times, the starting points are chosen randomly for each run. The results are averaged over runs at each window. The weights of the components are visualized by colors according to the color scale at the right. The second graph presents the temporal variation of the local diffusion parameters $\sigma_i^2(t)$ estimated by the same version of the EM algorithm with the random choice of the starting points. On the third graph, the temporal variation of the local trend parameters $a_i(t)$ estimated by the "normal" EM algorithm with the random choice of the starting points for the weights is presented. At each window the starting point for the rest (location and scale) parameters is one and the same for all components and is equal to the sample mean and sample variance, respectively, calculated from the window. The fourth graph presents the temporal variation of the parameters local diffusion $\sigma_i^2(t)$ estimated by the same version of the EM algorithm. The fifth graph presents the temporal variation of the local trend parameters $a_i(t)$ estimated by the EM algorithm with the random choice of the starting points at each window. At each window the EM algorithm is run five times, the starting points are chosen randomly for each run. At each window, as the result the estimate that delivers the maximum value of the target likelihood function among the five runs is taken. The sixth graph presents the temporal variation of the local diffusion parameters $\sigma_i^2(t)$ estimated by the same version of the EM algorithm aimed at the maximization of the likelihood function over runs.

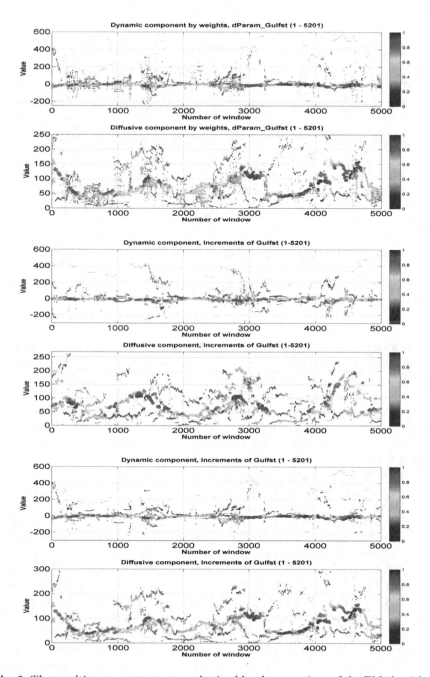

Fig. 3. The resulting parameter curves obtained by three versions of the EM algorithm.

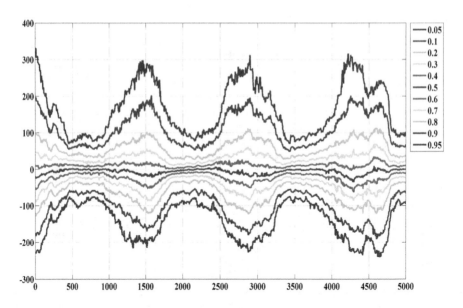

Fig. 4. The quantiles of the probability distribution of the increments of the heat-flux process as the isotopes of the probability density surface.

We can conclude that the third version of the EM algorithm with repeated random choice of starting points and maximization of the results over runs at each window gives most clear results.

This algorithm was applied to the analysis of temporal variation of the parameters of the distribution of increments of the heat-flux process. The obtained quantiles of the distribution are presented on Fig. 4 as isotopes of the density surface (the horizontal axis is time covering the period of about 3.5 years). The seasonal periodicity is clearly seen.

Figure 5 presents the evolution of the moment characteristics of the probability distribution of increments of the heat-flux process. It is clearly seen that the expected value of the increment noticeably oscillates in time with periodically changing amplitudes. Furthermore, at each period, the amplitudes are smaller for the period of seasonal increase of general mean than the amplitudes of oscillations for the period of the seasonal decrease of the general mean. The seasonally periodical variation of the variance is clearly seen. It is very interesting (if not surprising) that the purely stochastic diffusive component of the variance (see the second term on the right-hand side of relation (3)) depicted by the green curve on the upper right graph makes greater contribution to the total variance than the dynamic component (see the first term on the right-hand side of relation (3)) due to systematic trends depicted by the blue curve. It is also interesting that the distribution of the increments is slightly asymmetric with right slope heavier than the left one. Another interesting observation is that the kurtosis of this distribution is maximum during the "calm" period.

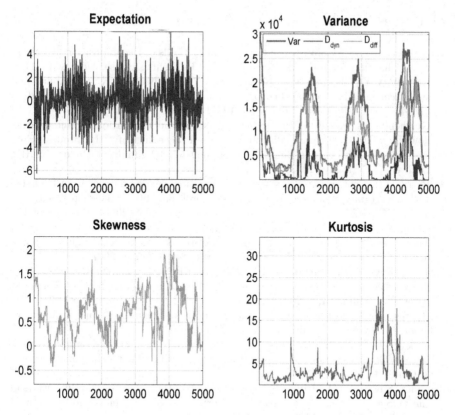

Fig. 5. The moment characteristics of the probability distribution of the increments of the heat-flux process.

5   Conclusions

In the paper, the method of moving separation of mixtures was applied to the analysis of statistical regularities in the temporal evolution of heat-fluxes. This method was realized by a special version of the EM algorithm aimed at the maximization of the likelihood function within the class of finite normal mixture models. It was demonstrated that in the stochastic character of the evolution of heat-fluxes, one basic component with low variance can be identified accompanying by stochastically emerging and disappearing components with large variance. Some regularities in the temporal variation of the moment characteristics of the heat-flux process increments were observed.

Acknowledgments. The research is supported by the Russian Foundation for Basic Research (project 15-37-20851) and by the President Grant for Government Support of Young Russian Scientists (project MK-4103.2014.9).

References

1. Korolev, V.Y.: Probabilistic and Statistical Methods of Decomposition of Volatility of Chaotic Processes. Moscow University Publishing House, Moscow (2011)
2. Korolev, V.Y., Chertok, A.V., Korchagin, A.Y., Gorshenin, A.K.: Probability and statistical modeling of information flows in complex financial systems based on high-frequency data. Inf. Appl. **7**(1), 12–21 (2013)
3. Gorshenin, A., Korolev, V.: Modeling of statistical fluctuations of information flows by mixtures of gamma distributions. In: Proceedings of 27th European Conference on Modelling and Simulation, pp. 569–572. Digitaldruck Pirrot GmbHP, Dudweiler, Germany (2013)
4. Berry, D.I., Kent, E.C.: A new air-sea interaction gridded dataset from ICOADS with uncertainty estimates. Bull. Am. Meteorol. Soc. **90**(5), 645–656 (2009)
5. Yu, L.: Global variations in oceanic evaporation (1958–2005): the role of the changing wind speed. J. Clim. **20**, 5376–5390 (2007)
6. Yu, L., Weller, R.A.: Objectively analyzed air-sea heat fluxes for the global ice-free oceans (1981–2005). Bull. Amer. Meteor. Soc. **88**, 527–539 (2007)
7. Josey, S.A.: A comparison of ECMWF, NCEP-NCAR and SOC surface heat fluxes with moored buoy measurements in the subduction region of the Northeast Atlantic. J. Clim. **14**, 1780–1789 (2001)
8. Bouras, D.: Comparison of five satellite-derived latent heat flux products to moored buoy data. J. Clim. **19**, 6291–6313 (2006)
9. Gulev, S.K., Jung, T., Ruprecht, E.: Estimation of the impact of sampling errors in the VOS observations on air-sea fluxes. Part I. Uncertainties in climate means. J. Clim. **20**, 279–301 (2007)
10. Gulev, S.K., Jung, T., Ruprecht, E.: Estimation of the impact of sampling errors in the VOS observations on air-sea fluxes. Part II. Impact on trends and interannual variability. J. Clim. **20**, 302–315 (2007)
11. Gulev, S.K., Belyaev, K.P.: Probability distribution characteristics for surface air-sea turbulent heat fluxes over the global ocean. J. Clim. **25**(1(1)), 184–206 (2012)
12. Gulev, S.K., Latif, M., Keenlyside, N., Park, W., Koltermann, K.P.: North Atlantic Ocean control on surface heat flux on multidecadal timescales. Nature **499**, 464–467 (2013)
13. Gorshenin, A.K., Korolev, V.Y., Tursunbayev, A.M.: Median modification of EM- and SEM-algorithms for separation of mixtures of probability distributions and their application to the decomposition of volatility of financial time series. Inf. Appl. **2**(4), 12–47 (2008)
14. Gorshenin, A.K.: The information technology to research the fine structure of chaotic processes in plasma by the analysis of the spectra. Syst. Means Inf. **24**(1), 116–125 (2014)
15. Korolev, V.Y.: Generalized hyperbolic laws as limit distributions for random sums. Theor. Probab. Appl. **58**(1), 63–75 (2014)
16. Korolev, V.Yu., Zaks, L.M.: Generalized variance gamma distributions as limit laws for random sums. Inf. Appl. **7**(1), 105–115 (2013)
17. Dempster, A., Laird, N., Rubin, D.: Maximum likelihood estimation from incompleted data. J. Roy. Stat. Soc. Ser. B **39**(1), 1–38 (1977)

Research of Mathematical Model of Insurance Company in the Form of Queueing System with Unlimited Number of Servers Considering "Implicit Advertising"

Diana Dammer[✉]

Tomsk State University, Tomsk, Russia
di.dammer@yandex.ru

Abstract. This paper is devoted to the research of the model of insurance company with an unlimited insurance field and the parameter of arrival process of insurance risks, which depends on the risks that are already insured in the company. Using method of characteristic functions we got joint probability distribution of a two-dimensional stochastic process of a number of risks that are insured in the company and a number of benefit payments. We also got expressions for the expected values and variances of components of a two-dimensional process. Total benefit payments is reviewed and its distribution and numerical characteristic are found.

Keywords: Mathematical model · Insurance company · Benefit payments · Queueing system · Characteristic function

1 Introduction

In modern economics mathematical methods are widely used, both for solving practical tasks and for theoretical modeling of sociology-economic process. These models and their researches are getting pretty much of attention nowadays. Models of actuarial mathematics, which studies insurance, are not left aside either. Generally, all the papers devoted to the research of insurance company's mathematical models have such characteristics of a company's work as: expected values of risk's number, capital, bankruptcy possibility and so on. Thus, paper [1] is about model of insurance company takes into account advertising expenses, paper [2] is about model with possibility of reassurance of some company's risks. In [3] we got the distribution of number of benefit payments with random variable of the duration of the contract and the stationary Poisson arrival process of insurance risks. In [4] by using method of asymptotic analysis we have found probability distribution of two-dimensional process of a number benefit payments and a number of insurance risks, given that the arrival process of insurance risks is stationary Poisson. In this paper we research two-dimensional process of a

A. Dudin et al. (Eds.): ITMM 2015, CCIS 564, pp. 163–174, 2015.
DOI: 10.1007/978-3-319-25861-4_14

number of benefit payments and a number of risks, that are insured in the company, in case when the parameter of the arrival process of insurance risks depends on the risks that are already insured in the company, which considers possibility of an implicit advertising, which is no doubt present in real life. Models with this arrival process of insurance risks are reviewed in [5], but methods of model's research are of different nature and the process of a number of benefit payments is ignored.

2   Mathematical Model and Formulation of the Problem

Let's review the model of insurance company with an unlimited insurance field [6] in the form of queuing system with an unlimited number of servers (Fig. 1). The validity of the insurance contract matches the server's duration of request handling. We will assume that risks are flowing into company, forming the arrival process with intensity that depends on a number of insured risks. Intensity of that arrival process will be determined by two components: parameter λ, which determines the arrival process of risks that come independently from insured ones, and parameter α, which determines the arrival process of risks that are under the influence of "implicit advertising". Every risk that has been in the company for the period of insurance police validity regadless of other risks generate a demand for insurance payment with γ intensity. And these requests also form the stationary Poisson process of events. Its natural to assume that benefit payment is determined by insured accident. We will assume that duration of the insurance contract for each risk located in the company will be random variable that is distributed by exponential law with parameter μ.

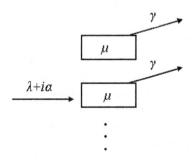

Fig. 1. Model of the insurance company in form of queuing system with an unlimited number of servers.

Designations: $n(t)$ — number of benefit payments during the time interval $[0,t]$, $i(t)$ — number of insurance risks located in the company at instant of time t, $P(i,n,t) = \mathbf{P}\{i(t) = i, n(t) = n\}$ — probability distribution of a two-dimensional process of a number of benefit payments and a number of insurance risks at instant of time t. The task is to find this distribution.

3 Joint Probability Distribution of Two-Dimensional Stochastic Process of a Number of Insurance Risks and a Number of Benefit Payments

Let's set up a system of Kolmogorov differential equations [7] for probability distribution $P(i, n, t)$ using the Δt method. First, the prelimit equalities:

$$
\begin{aligned}
P(i, n, t + \Delta t) = {} & P(i, n, t)(1 - (\lambda + i\alpha)\Delta t)(1 - i\gamma\Delta t)(1 - i\mu\Delta t) \\
& + (\lambda + (i - 1)\alpha)\Delta t P(i - 1, n, t) \\
& + i\gamma\Delta t P(i, n - 1, t) + (i + 1)\mu\Delta t P(i + 1, n, t) + o(\Delta t).
\end{aligned} \tag{1}
$$

System of differential equations will have this form:

$$
\begin{aligned}
\frac{\partial P(i, n, t)}{\partial t} = {} & -[\lambda + i(\alpha + \mu + \gamma)]P(i, n, t) + (\lambda + (i - 1)\alpha)P(i - 1, n, t) \\
& + (i + 1)\mu P(i + 1, n, t) + i\gamma P(i, n - 1, t).
\end{aligned} \tag{2}
$$

To solve system (2) let's introduce function:

$$
\sum_{i=0}^{\infty} \sum_{n=0}^{\infty} e^{jui} z^n P(i, n, t) = H(u, z, t), \tag{3}
$$

that is characteristic by u and generating by z, where j is the imaginary unit. We will continue solving the task of determining the form of this function. Then, form system (2), considering properties of characteristic functions, we will get partial differential equation of first order for the function $H(u, z, t)$:

$$
\begin{aligned}
\frac{\partial H(u, z, t)}{\partial t} = {} & -\lambda H(u, z, t)(1 - e^{ju}) \\
& + j\frac{\partial H(u, z, t)}{\partial u}(\alpha + \mu + \gamma - \alpha e^{ju} - \mu e^{-ju} - \gamma z).
\end{aligned} \tag{4}
$$

Solution for this differential equation is determined by solving of the following system of ordinary differential equations for characteristic curves [8]:

$$
\frac{dt}{1} = \frac{du}{-j(\alpha + \mu + \gamma - \alpha e^{ju} - \mu e^{-ju} - \gamma z)} = \frac{dH(u, z, t)}{H(u, z, t)\lambda(e^{ju} - 1)}. \tag{5}
$$

We will start by finding the two first integrals of this system. First, let's take a look at this equation:

$$
dt = \frac{du}{j(\alpha(e^{ju} - 1) + \mu(e^{-ju} - 1) - \gamma(1 - z))}. \tag{6}
$$

We will change variables $e^{ju} - 1 = v$, and, considering

$$
u = \frac{\ln(v + 1)}{j}, \quad du = \frac{dv}{j(v + 1)}, \quad e^{-ju} = \frac{1}{v + 1}, \quad j^2 = -1, \tag{7}
$$

the Eq. (6) will have this form:

$$dt = \frac{dv}{-(\alpha v^2 + (\alpha - \mu - \gamma(1 - z))v - \gamma(1 - z))}. \tag{8}$$

Let's take a look at right part of the last equation. We can write down

$$\alpha v^2 + (\alpha - \mu - \gamma(1 - z))v - \gamma(1 - z) = \alpha(v - v_1)(v - v_2), \tag{9}$$

where v_1 and v_2 are the roots of said quadratic equation. Let's write down expressions for v_1 and v_2:

$$v_1 = \frac{1}{2}\left[\left(1 - \frac{\mu}{\alpha} - \frac{\gamma}{\alpha}(1 - z)\right) + \sqrt{D}\right],$$

$$v_2 = \frac{1}{2}\left[\left(1 - \frac{\mu}{\alpha} - \frac{\gamma}{\alpha}(1 - z)\right) - \sqrt{D}\right], \tag{10}$$

where discriminant is

$$D = \left(1 - \frac{\mu}{\alpha} - \frac{\gamma}{\alpha}(1 - z)\right)^2 + 4\frac{\gamma}{\alpha}(1 - z) > 0. \tag{11}$$

Therefore, roots v_1 and v_2 are real and different. Besides, given that natural condition $\alpha < \mu$, roots $v_1 > 0$ and $v_2 \leq 0$.

Thus, based on the foregoing, Eq. (8) could be written in this form:

$$dt = \frac{dv}{-\alpha(v - v_1)(v - v_2)}. \tag{12}$$

Solution for Eq. (12) will have this form:

$$t = \frac{1}{\alpha(v_1 - v_2)} \ln\left(\frac{v - v_2}{v - v_1}\right) - \ln(\tilde{C}_1), \tag{13}$$

which will be determining our first integral. Lets write down expression for constant $\tilde{C}_1$, we have:

$$\tilde{C}_1 = e^{-t}\left(\frac{v - v_2}{v - v_1}\right)^{\frac{1}{\alpha(v_1 - v_2)}}. \tag{14}$$

We denote $C_1 = \tilde{C}_1^{\alpha(v_1 - v_2)}$, then

$$C_1 = e^{-t\alpha(v_1 - v_2)}\left(\frac{v - v_2}{v - v_1}\right). \tag{15}$$

Other first integral will be found from equation:

$$\frac{dH(u, z, t)}{H(u, z, t)\lambda(e^{ju} - 1)} = \frac{du}{-j(\alpha + \mu + \gamma - \alpha e^{ju} - \mu e^{-ju} - \gamma z)}. \tag{16}$$

Let's make similar change of variables $e^{ju} - 1 = v$. We will introduce function $H_1(v, z, t) = H(u, z, t)$. Lets write down equation (16) for the function $H_1(v, z, t)$ while splitting variables:

$$\frac{dH_1(v, z, t)}{H_1(v, z, t)} = \frac{\lambda v \, dv}{-(\alpha v^2 + (\alpha - \mu - \gamma(1 - z))v - \gamma(1 - z))}, \tag{17}$$

or considering (9)

$$\frac{dH_1(v, z, t)}{H_1(v, z, t)} = \frac{\lambda v \, dv}{-\alpha(v - v_1(z))(v - v_2(z))}, \tag{18}$$

where v_1 and v_2 are determined by expressions (10). Let's write down the solution for Eq. (18), assuming that $v_1 = v_1(z)$ and $v_2 = v_2(z)$:

$$H_1(v, z, t) = C_2 \left[\frac{(v - v_2)^{v_2}}{(v - v_1)^{v_1}} \right]^{\frac{\lambda}{\alpha(v_1 - v_2)}}. \tag{19}$$

We will introduce arbitrary differentiable function $\phi(C_1) = C_2$. Then the general solution of Eq. (18) considering (15) will have this form:

$$H_1(v, z, t) = \phi \left[e^{-\alpha(v_1 - v_2)t} \left(\frac{v - v_2}{v - v_1} \right) \right] \left[\frac{(v - v_2)^{v_2}}{(v - v_1)^{v_1}} \right]^{\frac{\lambda}{\alpha(v_1 - v_2)}}. \tag{20}$$

We define particular solution with the help of initial conditions. To do this, we will write down value of function $H(u, z, t)$ at $t = 0$. Then

$$H(u, z, 0) = \sum_{i=0}^{\infty} \sum_{n=0}^{\infty} e^{jui} z^n P(i, n, 0) = \sum_{i=0}^{\infty} e^{jui} P(i), \tag{21}$$

because at the initial time (i.e. at the moment when insurance company starts their work) there were no benefit payment, which means $P(i, n, 0) = P(i)$, if $n = 0$, and $P(i, n, 0) = 0$, if $n > 0$.

Let's denote $H(u, z, 0) = G(u)$, then by using equation (4) we can write down the equation for function $G(u)$:

$$j(\mu - \alpha e^{ju}) \frac{dG(u)}{du} + \lambda e^{ju} G(u) = 0. \tag{22}$$

Solution will have this form:

$$G(u) = C_3 \left(e^{ju} - \frac{\mu}{\alpha} \right)^{-\frac{\lambda}{\alpha}}. \tag{23}$$

We will find constant C_3 from condition $G(0) = 1$. We have:

$$C_3 = \left(1 - \frac{\mu}{\alpha} \right)^{\frac{\lambda}{\alpha}}, \tag{24}$$

then

$$G(u) = \left(\frac{1 - \frac{\alpha}{\mu} e^{ju}}{1 - \frac{\alpha}{\mu}} \right)^{-\frac{\lambda}{\alpha}} . \tag{25}$$

Considering (20) we can write down

$$H_1(v, z, 0) = \phi \left(\frac{v - v_2}{v - v_1} \right) \left[\frac{(v - v_2)^{v_2}}{(v - v_1)^{v_1}} \right]^{\frac{\lambda}{\alpha(v_1 - v_2)}} , \tag{26}$$

or

$$\left(\frac{1 - \frac{\alpha}{\mu}(v + 1)}{1 - \frac{\alpha}{\mu}} \right)^{-\frac{\lambda}{\alpha}} = \phi \left(\frac{v - v_2}{v - v_1} \right) \left(\frac{(v - v_2)^{v_2}}{(v - v_1)^{v_1}} \right)^{\frac{\lambda}{\alpha(v_1 - v_2)}} . \tag{27}$$

Now the task is to define the form of function $\phi(.)$. Let's denote

$$x = \frac{v - v_2}{v - v_1} . \tag{28}$$

Then

$$\phi(x) = \left[\frac{\left(1 - \frac{\alpha}{\mu} \right)(v_2 - v_1)}{(1 - x) - \frac{\alpha}{\mu}(1 + v_2 - x(1 + v_1))} \right]^{\frac{\lambda}{\alpha}} x^{\frac{\lambda v_2}{\alpha(v_2 - v_1)}} , \tag{29}$$

where it is considered that

$$v = \frac{v_2 - x v_1}{1 - x} . \tag{30}$$

Now we can write down the expression for function $\phi(.)$:

$$\phi \left[e^{-\alpha(v_1 - v_2)t} \left(\frac{v - v_2}{v - v_1} \right) \right] = e^{\lambda v_2 t} \left[\left(1 - \frac{\alpha}{\mu} \right)(v_2 - v_1)(v - v_1) \right]^{\frac{\lambda}{\alpha}}$$
$$\times \left[(v - v_1) - (v - v_2)e^{\alpha(v_2 - v_1)t} - \right.$$
$$\left. - \frac{\alpha}{\mu} \left((v - v_1)(1 + v_2) - (v - v_2)(1 + v_1)e^{\alpha(v_2 - v_1)t} \right) \right]^{-\frac{\lambda}{\alpha}} \tag{31}$$
$$\times \left(\frac{v - v_2}{v - v_1} \right)^{\frac{\lambda v_2}{\alpha(v_2 - v_1)}} .$$

Accordingly, we will write down the expression for function $H_1(v, z, t)$, taking into account that v_1 and v_2 are functions of z and have the form (10). We have:

$$H_1(v, z, t) = e^{\lambda v_2(z) t} \left[\left(1 - \frac{\alpha}{\mu} \right)(v_1(z) - v_2(z)) \right]^{\frac{\lambda}{\alpha}}$$
$$\times \left\{ (v_1(z) - v) \left[1 - \frac{\alpha}{\mu}(1 + v_2(z)) \right] \right.$$
$$\left. - (v_2(z) - v)e^{\alpha(v_2(z) - v_1(z))t} \left[1 - \frac{\alpha}{\mu}(1 + v_1(z)) \right] \right\}^{-\frac{\lambda}{\alpha}} . \tag{32}$$

By passing from variable v to variable u, let's write down the expression for function $H(u, z, t)$:

$$
H(u, z, t) = e^{\lambda v_2(z)t} \left[\left(1 - \frac{\alpha}{\mu} \right) (v_1(z) - v_2(z)) \right]^{\frac{\lambda}{\alpha}}
$$
$$
\times \left\{ (v_1(z) - e^{ju} + 1) \left[1 - \frac{\alpha}{\mu}(1 + v_2(z)) \right] \right.
$$
$$
\left. - (v_2(z) - e^{ju} + 1)e^{\alpha(v_2(z) - v_1(z))t} \left[1 - \frac{\alpha}{\mu}(1 + v_1(z)) \right] \right\}^{-\frac{\lambda}{\alpha}}.
$$

(33)

Thus, resulting function (33) is characteristic function of two-dimensional stochastic process of a number of risks that are insured in the company and a number of benefit payments. Knowing this function, we can find one-dimensional marginal distributions of processes $i(t)$ and $n(t)$.

4 Probability Distributions of a Number of Insurance Risks and a Number of Benefit Payments

Let's suppose that in (33) $u = 0$, now we can get generating function of process $n(t)$:

$$
H(0, z, t) = F(z, t) = e^{\lambda v_2(z)t} \left[\left(1 - \frac{\alpha}{\mu} \right) (v_1(z) - v_2(z)) \right]^{\frac{\lambda}{\alpha}}
$$
$$
\times \left\{ v_1(z) \left[1 - \frac{\alpha}{\mu}(1 + v_2(z)) \right] \right.
$$
$$
\left. - v_2(z)e^{\alpha(v_2(z) - v_1(z))t} \left[1 - \frac{\alpha}{\mu}(1 + v_1(z)) \right] \right\}^{-\frac{\lambda}{\alpha}}.
$$

(34)

We write down characteristic function of process $i(t)$ by assuming that in (33) $z = 1$. Because of

$$
v_1(1) = \frac{\mu}{\alpha} - 1, \ v_2(1) = 0,
$$

we have

$$
H(u, 1, t) = G(u) = \left(\frac{1 - \dfrac{\alpha}{\mu}}{1 - \dfrac{\alpha}{\mu} e^{ju}} \right)^{\frac{\lambda}{\alpha}}.
$$

(35)

Since the resulting characteristic function (35) does not depend of time, we can say that process of a number of insurance risks is stationary.

Let's find probability distributions for a number of insurance risks $P_1(i)$ and a number of benefit payments $P_2(n, t)$, by looking at numerical example. Figures 2 and 3 show distributions $P_1(i)$ and $P_2(n, t)$ for the following parameters: $\lambda = 0.6$, $\mu = 1$, $\alpha = 0.9$, $\gamma = 0.1$, $t = 1$.

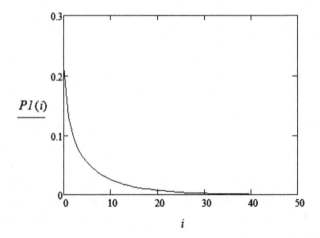

Fig. 2. Probability distribution of a number of insurance risks

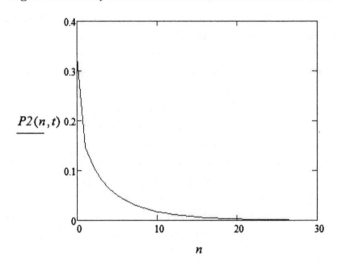

Fig. 3. Probability distribution of a number of benefit payments

5   Numerical Characteristics of a Number of Insured Risks and a Number of Benefit Payments

Now we can write down expected values for a number of risks and a number of benefit payments:

$$\mathsf{E}\{i(t)\} = \frac{1}{j}\frac{dG(u)}{du}\bigg|_{u=0} = \frac{\lambda}{\mu - \alpha}, \tag{36}$$

and

$$\mathsf{E}\{n(t)\} = \frac{\partial F(z,t)}{\partial z}\bigg|_{z=1} = \frac{\lambda\gamma}{\mu - \alpha}t. \tag{37}$$

Following expressions are for variances:

$$D\{i(t)\} = \frac{\lambda\mu}{(\mu - \alpha)^2},$$

(38)

and

$$D\{n(t)\} = 2\frac{\lambda\mu\gamma^2}{(\mu - \alpha)^3}t + \frac{\lambda\gamma}{\mu - \alpha}t - 2\frac{\lambda\mu\gamma^2}{(\mu - \alpha)^4}\left(1 - e^{-(\mu - \alpha)t}\right).$$

(39)

Formulas (36) and (38) match with the result we got in [5], where one-dimensional process of a number of insured risks considering "implicit advertising" is researched.

Let's review correlation coefficient of processes $i(t)$ and $n(t)$. Knowing function $H(u, z, t)$, we can find joint moment of studied processes. We have:

$$\frac{1}{j}\frac{\partial^2 H(u, z, t)}{\partial u \partial z}\bigg|_{u = 0, z = 1} = E\{i(t)n(t)\},$$

(40)

then, considering characteristics we got earlier, let's write down the expression for correlation coefficient:

$$r_{in}(t) = \frac{\lambda\gamma\mu(1 - e^{(\alpha - \mu)t})}{\sqrt{\lambda\mu[2\lambda\mu\gamma^2(\mu - \alpha)t + \lambda\gamma(\mu - \alpha)^3 t - 2\lambda\mu\gamma^2\left(1 - e^{-(\mu - \alpha)t}\right)]}}.$$

(41)

Nonzero correlation coefficient shows the presence of dependence between processes $i(t)$ and $n(t)$.

6 Numerical Characteristics of Value of the Total Benefit Payments

We will denote $S(t)$ as a value of the total benefit payments for all insured accidents during the time interval $[0, t]$, ξ — the value of the payment for one insured accident. Let's introduce characteristic function of the value $S(t)$:

$$\Psi(\eta, t) = E\{e^{-\eta S(t)}\}.$$

(42)

Let's take a closer look at this function. We have:

$$\Psi(\eta, t) = E\left\{e^{-\eta S(t)}\right\} = E\left\{e^{-\eta \sum_{i=0}^{n(t)} \xi_i}\right\}$$

$$= \sum_{n=0}^{\infty} E\left\{e^{-\eta \sum_{i=0}^{n(t)} \xi_i}\bigg| n(t) = n\right\}P(n, t)$$

(43)

$$= \sum_{n=0}^{\infty} E\left\{\prod_{i=0}^{n} e^{-\eta \xi_i}\bigg| n(t) = n\right\}P(n, t) = \sum_{n=0}^{\infty} \theta^n(\eta)P(n, t),$$

where $\theta(\eta) = \mathsf{E}\{e^{-\eta\xi}\}$ is the characteristic function of the value ξ. With this in mind we can write down:

$$\Psi(\eta, t) = \sum_{n=0}^{\infty} \theta^n(\xi)P(n, t) = F(\theta(\eta), t). \tag{44}$$

Let's introduce functions

$$w_1(\eta) = v_1(\phi(\eta)) = \frac{1}{2}\left[\left(1 - \frac{\mu}{\alpha} - \frac{\gamma}{\alpha}(1 - \phi(\eta))\right) + \sqrt{D(\theta(\eta))}\right],$$

$$w_2(\eta) = v_2(\phi(\eta)) = \frac{1}{2}\left[\left(1 - \frac{\mu}{\alpha} - \frac{\gamma}{\alpha}(1 - \phi(\eta))\right) - \sqrt{D(\theta(\eta))}\right], \tag{45}$$

where

$$D(\theta(\eta)) = \left[1 - \frac{\mu}{\alpha} - \frac{\gamma}{\alpha}(1 - \phi(\eta))\right]^2 + 4\frac{\gamma}{\alpha}(1 - \phi(\eta)). \tag{46}$$

Expressions (45), (46) are written considering (10) and (11). Then function $\Psi(\eta, t)$ will have this form

$$\Psi(\eta, t) = F(\theta(\eta), t) = e^{\lambda w_2(\eta)t}\left[\left(1 - \frac{\alpha}{\mu}\right)(w_2(\eta) - w_1(\eta))\right]^{\frac{\lambda}{\alpha}}$$

$$\times \left(-w_1(\eta) + w_2(\eta)e^{\alpha(w_2(\eta) - w_1(\eta)t}\right. \tag{47}$$

$$\left. -\frac{\alpha}{\mu}\left[(-w_1(\eta)(1 + w_2(\eta)) + w_2(\eta)(1 + w_1(\eta))e^{\alpha(w_2(\eta) - w_1(\eta)t}\right]\right)^{-\frac{\lambda}{\alpha}}.$$

Now, that we know the form of the characteristic function of a value of the total benefit payments, we can obtain the expected value and the variance of value $S(t)$. Let's denote $\mathsf{E}\{\xi\} = a_1$, $\mathsf{E}\{\xi^2\} = a_2$. Because of

$$\left.\frac{\partial\Psi(\eta, t)}{\partial\eta}\right|_{\eta=0} = -\mathsf{E}\{S(t)\}, \tag{48}$$

after transformations we will get

$$\mathsf{E}\{S(t)\} = \frac{\lambda\gamma a_1}{\mu - \alpha}t. \tag{49}$$

For the second initial moment $S(t)$ we can write down

$$\left.\frac{\partial^2\Psi(\eta, t)}{\partial\eta^2}\right|_{\eta=0} = \mathsf{E}\{S^2(t)\}. \tag{50}$$

Then the variance of the total benefit payments will have the following form:

$$\mathsf{D}\{S(t)\} = \frac{\lambda\gamma a_2}{\mu - \alpha}t + 2\frac{\lambda\mu\gamma^2 a_1^2}{(\mu - \alpha)^3}t - 2\frac{\lambda\mu\gamma^2 a_1^2}{(\mu - \alpha)^4}\left(1 - e^{-(\mu-\alpha)t}\right). \tag{51}$$

Let's take a look at another characteristic of $S(t)$ - coefficient of variation $V\{S(t)\}$. It is defined as the ratio of the standart devation to the expected value:

$$V\{S(t)\} = \frac{\sqrt{D\{S(t)\}}}{E\{S(t)\}}.$$

Behavior of coefficient of variation $V(t)$ is shown at Fig. 3 with the following parameters: $\lambda = 5$, $\mu = 1$, $\alpha = 0.8$, $\gamma = 0.1$, $a_1 = 10$, $a_2 = 60$. Numerical calculations show that $V\{S(t)\}$ is significantly decreasing with the passage of time, reaching value of 0.01 at $t = 540$, which allows us to find pretty accurate prognosed value of the capital of insurance company (Fig. 4).

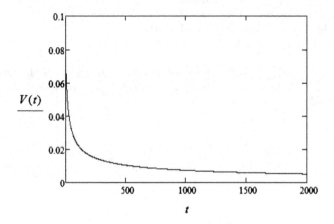

Fig. 4. Coefficient of variation of the total benefit payments

7   Conclusions

Thereby, in this paper we have researched mathematical model of the insurance company in the form of queueing system with an unlimited number of servers. We have found the expression for characteristic function of a two-dimensinal process of a number of benefit payments and a number of insurance risk. Also we have found expressions for numerical characteristics of said processes. It is shown that the results are the generalization of particular cases. Characteristic function, expected value and variance of a value of the total benefit payments have also been found. These results may be used for analysis of indicators of economic activity of insurance companies.

References

1. Akhmedova, D.D., Terpugov, A.F.: Mathevatical model of the insurance company considering advertising expenses. Univerisities News Phys. **1**, 25–29 (2001)
2. Glukhova, E.V., Kapustin, E.V.: Calculation of probability of insurance company's bankruptcy considering reassurance. Universities News Phys. **4**, 3–9 (2000)
3. Nazarov, A.A., Dammer, D.D.: Research of a number of requests for insurance payments in the company with arbitrary length of duration of the contract. Omsk State Univ. J. **15**(2), 24–32 (2011)
4. Dammer, D.D., Nazarov, A.A.: Research of the mathematical model of the insurance company in form of the infinite queuing system by using method of asymptotic analysis. In: Materials of VII Ferghan Conference Limit Theorems and Its Applications, Namangan pp. 191–196 (2015)
5. Glukhova, E.V., Zmeev, O.A., Livshits, K.I.: Mathematical models of insurance. Published by Tomsk University, Tomsk (2004)
6. Gafurov, S.R., Gugnin, V.I., Amanov, S.N.: Business language. Shark, Tashkent (1995)
7. Nazarov, A.A., Terpugov, A.F.: Theory of queuing. Publishment NTL, Tomsk (2005)
8. Elsgolts, L.E.: Differential equations and calculus of variations.Science. Nauka, Moscow (1969)

Study of the Queuing Systems $M|GI|N|\infty$

Ekaterina Lisovskaya and Svetlana Moiseeva$^{(\boxtimes)}$

Tomsk State University, Lenina, 36, Tomsk, Russia
{ekaterina_lisovs,smoiseeva}@mail.ru
http://www.tsu.ru

Abstract. In this paper, we study the queuing system with unlimited queue and with N servers. We obtain the approximation of probability distribution of the number of customers in the system. We obtain the formula of the probability of immediate service and the characteristic function of a positive waiting time. The optimal number of servers can be determined by the obtained characteristics.

Keywords: Queuing systems · Waiting time · Approximation of the probability distribution · Queue

1 Introduction

Mathematical models of queuing systems (QS) is widely used in the solution of important practical problems arising in connection with the rapid development of communication systems, the emergence of information systems, the emergence of a variety and complexity of technological systems, the creation of automated control systems.

Multiserver QS are mathematical models of real systems and processes in the area of telecommunications, communication networks, etc. There are papers by modeling call-centers [1,2].

In this paper we consider queuing system $M|GI|N|\infty$. The system arrival process is distributed by Poisson law with rate λ. The system has N servers. Service times on each servers are i.i.d. with distribution function $A(x)$. The arriving customer occupies any free server or goes to the queue in case of all servers are busy.

Its known that for the system $M|M|N|K$ Erlang formulas have been obtained [3]. However, for general service time and infinity queue the obtained problem theoretically didnot solve, and an analytical solution is not possible, therefore, the task of obtaining the approximation of stationary probability distribution $P(i)$, $0 \leq i < \infty$ of the number of customers in the system $M|GI|N|\infty$. We obtain the formulas for the probability distribution of positive waiting time using the approximation.

A. Dudin et al. (Eds.): ITMM 2015, CCIS 564, pp. 175–184, 2015.
DOI: 10.1007/978-3-319-25861-4_15

2   Approximation of Probability Distribution of the Number of Customers in the System $M|GI|N|\infty$

Denote the number of customers in the system at time t by $i(t)$. Then $P(i) = P\{i(t) = i\}$ is the probability distribution of the number of customers in the system at time t.

Let π_i be an approximation of the probability distribution which is defined as a composite distribution [4]

$$\pi_i = \begin{cases} C_1 P_1(i), 0 \leq i \leq N, \\ C_2 P_2(i - N + 1), i \geq N. \end{cases} \tag{1}$$

The probabilities $P_1(i)$, where $0 \leq i < N$, are the probabilities of the number of occupied servers in N-server QS with customers losses $(M|GI|N|0)$, when all servers are busy. Erlang formula defines the probability $P_1(i)$ [5]

$$P_1(i) = \frac{\frac{(\lambda a)^i}{i!}}{\sum\limits_{k=0}^{N} \frac{(\lambda a)^k}{k!}}, \tag{2}$$

where $a = \int\limits_0^\infty (1 - A(x))dx$ is the average service time.

The probabilities $P_2(i)$ are defined when all servers are busy. In this case, the block of occupied servers is considered as a single and its service has distribution function $B(x)$. Therefore, the probabilities $P_2(i)$, where $i = 0, 1, ...$ are defined as the probabilities of the number of customers in the single-server system $M|GI|1|\infty$ with waiting.

In this case, the Pollaczek-Khinchin formula for the generating function can be use:

$$G(x) = \sum_{n=0}^{\infty} x^n P_2(n) = (1 - \lambda b) \frac{(1 - x)B^*(\lambda - \lambda x)}{B^*(\lambda - \lambda x) - x}. \tag{3}$$

To determine the distribution function $B(x)$ we consider the output process of serviced customers when all N servers are occupied.

3   Distribution of Sum of Independent Recurrent Process

Consider the sum of N independent recurrent process, with the same distribution functions $A(x)$.

Let τ be the value of the jump [5] for the total process. Then it is obvious that $\tau = \min(\tau_1, \tau_2, ..., \tau_N)$, where $\tau_1, \tau_2, ..., \tau_N$ are independent value of jump for total process.

Therefore

$$P\{\tau > x\} = 1 - \frac{1}{b} \int\limits_0^x (1 - B(z))dz = P\{\min\{\tau_1, \tau_2, ..., \tau_N\} > x\}$$

$$= P\{\tau_1 > x\}P\{\tau_2 > x\}...P\{\tau_N > x\} = \left(1 - \frac{1}{a} \int\limits_0^x (1 - A(z))dz\right)^N.$$

Hence, for the total process we have the following equation:

$$1 - \frac{1}{a} \int\limits_0^x (1 - B(z))dz = \left(1 - \frac{1}{a} \int\limits_0^x (1 - A(z))dz\right)^N,$$

then we differentiate equation by x and obtain the following formula:

$$B(x) = 1 - Nb \left(1 - \frac{1}{a} \int\limits_0^x (1 - A(z))dz\right)^{N-1} \frac{1}{a}(1 - A(x)).$$

Knowing $\frac{1}{b} = N\frac{1}{a}$ [5], then in the system $M|GI|N|\infty$, the distribution function of the lengths of the intervals of the total process has the form:

$$B(x) = 1 - (1 - A(x)) \left(1 - \frac{1}{a} \int\limits_0^x (1 - A(z))dz\right)^{N-1}. \tag{4}$$

and the density distribution has the following form:

$$b(x) = \left\{ A'(x) \left(1 - \frac{1}{a} \int\limits_0^x (1 - A(z))dz\right) + \frac{N-1}{a}(1 - A(x))^2 \right\}$$

$$\times \left(1 - \frac{1}{a} \int\limits_0^x (1 - A(z))dz\right)^{N-2}.$$

4   Expansion of the Function Pollaczek-Khinchin

Probabilities $P_2(i)$ can be found using the inverse Fourier transform, or expending function (3) to a power series in x.

To determine the probability by the second method, we write the following expansion

$$B^*(\lambda - \lambda x) = \int\limits_0^\infty e^{-(\lambda - \lambda x)z} dB(z) = \int\limits_0^\infty e^{-\lambda z} e^{-\lambda x z} dB(z)$$

$$= \int\limits_0^\infty e^{-\lambda z} \sum_{n=0}^\infty \frac{(\lambda z)^n}{n!} dB(z) = \sum_{n=0}^\infty x^n \int\limits_0^\infty e^{-\lambda z} \frac{(\lambda z)^n}{n!} dB(z).$$

Denote

$$\beta_n = \int_0^{\infty} e^{-(\lambda z)} \frac{(\lambda z)^n}{n!} dB(z),$$

we obtain the expansion

$$B^*(\lambda - \lambda x) = \sum_{n=0}^{\infty} x^n \beta_n.$$

Hence

$$(1-x)B^*(\lambda - \lambda x) = \sum_{n=0}^{\infty} x^n \beta_n - \sum_{n=0}^{\infty} x^{n+1} \beta_n = \sum_{n=0}^{\infty} x^n \beta_n - \sum_{n=1}^{\infty} x^n \beta_{n-1}$$

$$= \beta_0 + \sum_{n=1}^{\infty} x^n (\beta_n - \beta_{n-1}) = \sum_{n=1}^{\infty} x^n b_n.$$

where $b_0 = \beta_0$, $b_n = \beta_n - \beta_{n-1}$.

The denominator of the expression (3) is writen in the form

$$(B^*(\lambda - \lambda x) - x)^{-1} = \sum_{n=0}^{\infty} x^n \alpha_n. \tag{5}$$

To determine α_n we rewrite the formula (5) as:

$$1 = (B^*(\lambda - \lambda x) - x) \sum_{n=0}^{\infty} x^n \alpha_n = \sum_{n=0}^{\infty} x^n \alpha_n \sum_{n=0}^{\infty} x^n \beta_n - \sum_{n=0}^{\infty} x^{n+1} \alpha_n$$

$$= \sum_{n=0}^{\infty} x^n \sum_{k=0}^{n} \alpha_k \beta_{n-k} - \sum_{n=1}^{\infty} x^n \alpha_{n-1} = \alpha_0 \beta_0 + \sum_{n=1}^{\infty} x^n \sum_{k=0}^{n} \alpha_k \beta_{n-k} - \alpha_{n-1}.$$

Equating coefficients of same powers of x in this expression, we obtain recurrence formulas:

$$\alpha_0 = \frac{1}{\beta_0}, \quad \alpha_n = \frac{1}{\beta_0} \left[\alpha_{n-1} - \sum_{k=0}^{n-1} \alpha_k \beta_{n-k} \right].$$

Considering the expansion of the function $G(x)$, can be written expression

$$G(x) = \sum_{n=0}^{\infty} x^n P_2(n) = (1 - \lambda b) \frac{(1-x)B^*(\lambda - \lambda x)}{B^*(\lambda - \lambda x) - x} = (1 - \lambda b) \sum_{i=0}^{\infty} x^i b_i \sum_{i=0}^{\infty} x^i \alpha_i$$

$$= (1 - \lambda b) \sum_{i=0}^{\infty} x^i \sum_{k=0}^{i} \alpha_k b_{i-k}.$$

Thus

$$P_2(i) = (1 - \lambda b) \sum_{k=0}^{i} \alpha_k b_{i-k}.$$

5   Finding of Constants

Constants C_1 and C_2 can be found from the normalization condition and the conditions of "stitching":

$$\begin{cases} \sum_{i=0}^{\infty} \pi_i = 1, \\ C_1 P_1(N) = C_2 P_2(1), \end{cases}$$

we obtain

$$1 = \sum_{i=0}^{\infty} \pi_i = C_1 \sum_{i=0}^{N} P_1(i) + C_2 \sum_{i=N+1}^{\infty} P_2(i - N + 1) = C_1 + C_2 \sum_{n=2}^{\infty} P_2(n)$$

$$= C_1 + C_2(1 - (P_2(0) + P_2(1))),$$

thus

$$C_1 = \frac{P_2(1)}{P_2(1) + P_1(N)(1 - (P_2(0) + P_2(N)))},$$

$$C_2 = \frac{P_1(N)}{P_2(1) + P_1(N)(1 - (P_2(0) + P_2(N)))}. \tag{6}$$

So expression (1) has the form:

$$\pi_i = \begin{cases} \frac{P_2(1)}{P_2(1)+P_1(N)(1-(P_2(0)+P_2(N)))} P_1(i), 0 \le i \le N, \\ \frac{P_1(N)}{P_2(1)+P_1(N)(1-(P_2(0)+P_2(N)))} P_2(i - N + 1), i > N. \end{cases}$$

6   Probability of Immediate Service

Let τ be the waiting time of customer service start. Using (1) the probability of immediate service can be written as

$$P_0 = \sum_{i=0}^{N-1} \pi_i = C_1 \sum_{i=0}^{N-1} P_1(i) = C_1(1 - P_1(i))$$

$$= \frac{P_2(1)(1 - P_1(N))}{P_2(1) + P_1(N)[1 - (P_2(0) + P_2(1))]}, \tag{7}$$

where considering expressions (2) and (3) the following equalities

$$P_2(0) = G(0) = 1 - \lambda b,$$

$$P_2(1) = G'(0) = (1 - \lambda b)\frac{1 - B^*(\lambda)}{B^*(\lambda)},$$

$$P_1(N) = \frac{\frac{(\lambda a)^N}{N!}}{\sum_{i=0}^{N} \frac{(\lambda a)^i}{i!}}.$$

7 Probability Distribution of a Positive Waiting Time

If the customer arrives in the system at time when all servers are busy, then its waiting time $\tau > 0$ and this value we call as a positive waiting time τ^+.

We find the conditional probability distribution $P_q(m)$, where $m > 0$ that there are m customers in the queue considering that all servers are busy.

Using expression (1), we written:

$$
\begin{aligned}
P_q(m) &= \frac{\pi_{N+m}}{\sum\limits_{i=0}^{\infty} \pi_{N+i}} = \frac{C_2 P_2(1+m)}{C_2 \sum\limits_{i=0}^{\infty} P_2(1+m)} \\
&= \frac{P_2(1+m)}{(1-P_2(0))} = \frac{1}{\lambda b} P_2(1+m).
\end{aligned}
\tag{8}
$$

Expression (9) is the conditional probability distribution that there are m customers in the queue considering that all servers in the system are busy

$$
P_q(m) = \frac{1}{\lambda b} P_2(m+1).
\tag{9}
$$

We find the generating function $G_q(x)$ of this distribution

$$
\begin{aligned}
G_q(x) &= \sum_{m=0}^{\infty} x^m P_q(m) = \sum_{m=0}^{\infty} x^m \frac{1}{\lambda b} P_2(m+1) \\
&= \frac{1}{\lambda b x} \sum_{\nu=1}^{\infty} x^\nu P_q(\nu) = \frac{1}{\lambda b x} [G(x) - P_0] \\
&= \frac{1}{\lambda b x} [G(x) - (1 - \lambda b)] \\
&= \frac{1}{\lambda b x} \left[(1 - \lambda b) \frac{(1-x)B^*(\lambda - \lambda x)}{B^*(\lambda - \lambda x) - x} - (1 - \lambda b) \right] \\
&= \frac{1 - \lambda b}{\lambda b x} \frac{(1-x)B^*(\lambda - \lambda x) - B^*(\lambda - \lambda x) + x}{B^*(\lambda - \lambda x) - x} \\
&= \frac{1 - \lambda b}{\lambda b} \frac{1 - B^*(\lambda - \lambda x)}{B^*(\lambda - \lambda x) - x},
\end{aligned}
$$

then

$$
G_q(x) = \frac{1 - \lambda b}{\lambda b} \frac{1 - B^*(\lambda - \lambda x)}{B^*(\lambda - \lambda x) - x}.
\tag{10}
$$

This generating function is obtained on the period when all servers are busy. In this condition, the N-server block of servers is defined by the distribution function $A(x)$ is permissible to replace the single-server with the distribution function $B(x)$ from formula (4).

Customer arriving in the system when all servers are busy finds m customers in the queue with probability $P_q(m)$. So, the waiting time consists of the total time service of customers, each having the distribution function $B(x)$ from formula (4) and residual service time of one customer with the distribution function

$$B_0(x) = \frac{1}{b} \int_0^x (1 - B(z))\, dz.$$

We denote residual service time by ξ_0 and service times of the first, the second and the m-th customers in the queue by $\xi_1, \xi_2, ..., \xi_m$ respectively. Then the waiting time can be determined by

$$\tau^+ = \xi_0 + \xi_1 + \xi_2 + ... + \xi_m.$$

We find the characteristic function $h(u)$ of the positive waiting time customers in the system $M|GI|N|\infty$. Using total probability law for mean, we can write

$$h(u) = M\left\{e^{jut}\right\} = \sum_{m=0}^{\infty} M\left\{exp\left\{ju(\xi_0 + ... + \xi_m)\right\}\,|m(t) = m\right\} P_q(m)$$

$$= \sum_{m=0}^{\infty} M\left\{e^{ju\xi_0}\right\} \left(M\left\{e^{ju\xi_1}\right\}\right)^m P_q(m) = \phi(u)^m P_q(m),$$

where $\phi_0(u)$ and $\phi(u)$ are characteristic functions of the residual and total times service of one customer, here

$$\phi(u) = \int_0^{\infty} e^{jux}\, dB(x). \tag{11}$$

The last equation for $h(u)$ we rewrite as

$$h(u) = \phi_0(u) G_q(\phi(u)) = \phi_0(u) \frac{1 - \lambda b}{\lambda b} \frac{1 - B^*(\lambda - \lambda\phi(u))}{B^*(\lambda - \lambda\phi(u)) - \phi(u)}. \tag{12}$$

So

$$B_0(x) = \frac{1}{b} \int_0^x (1 - B(z))\, dz,$$

then

$$\phi_0(u) = \int_0^{\infty} e^{jux}\, dB_0(x) = \frac{1}{b} \int_0^x e^{jux}(1 - B(x)) dx = \frac{1}{jub}(\phi(u) - 1),$$

therefore, the characteristic function $h(u)$ of formula (12) is written as

$$h(u) = \frac{1}{jub}(1 - \lambda b) \frac{\phi(u) - 1}{\lambda b} \frac{1 - B^*(\lambda - \lambda\phi(u))}{B^*(\lambda - \lambda\phi(u)) - \phi(u)}. \tag{13}$$

Here $\phi(u)$ has the form (11).

Formulas (5) and (13) completely characterize the waiting time of customer in the queue N-server system $M|GI|N|\infty$.

8   Mean of the Positive Waiting Time

Applying the characteristic function $h(u)$ from formula (13), the mean of positive waiting time customer in the queue is written in the form

$$\bar{\tau}^+ = \frac{1}{j} h'(u)|_{u=0} = \frac{b_2}{2b(1-\lambda b)}, \tag{14}$$

where b is the mean, and b_2 is the second initial moment, defined by distribution function $B(x)$ from (4).

9   Optimal Number of Servers in the Multiserver System

In order for steady-state regime to exists in N-server queuing system with the waiting, it is necessary that the system load $\rho = \lambda b = \lambda \frac{a}{N}$ is less than one.

Therefore, N has to satisfy inequality

$$N > \lambda a. \tag{15}$$

The optimal value N_{opt} of the number of servers is defined by criteria:

$$N_{opt} \left[\min_N \{ N : P(\tau > \tau_{\max}) \le \delta \} \right].$$

Let τ be waiting time of customer service start, $\tau_{\max}$ is the upper limit waiting time customer service start, δ is allowable share of customers who will wait for the start of service longer than $\tau_{\max}$. The condition $P(\tau > \tau_{\max}) \le \delta$ can be replaced by the following equivalent condition

$$(1 - P_0)P(\tau^+ > \tau_{\max}) \le \delta, \tag{16}$$

where the probability of immediate service is given by (5).

Applying inverse Fourier transform to the function $h(u)$, the probability $P(\tau^+ > \tau_{\max})$ can be written as the following integration formula:

$$P(\tau^+ > \tau_{\max}) = 1 - \frac{1}{2\pi} \int_{-\infty}^{\infty} \frac{1 - e^{-ju\tau_{\max}}}{ju} h(u) du. \tag{17}$$

Here $h(u)$ has the form (13), and

$$\phi(u) = \int_0^{\infty} e^{jux} dB(x) = 1 + ju \int_0^{\infty} e^{jux} [1 - B(x)] dx,$$

$$B^*(\alpha) = \int_0^{\infty} e^{-\alpha x} dx = 1 - \alpha \int_0^{\infty} e^{-\alpha x} [1 - B(x)] dx = \phi(j\alpha), \tag{18}$$

and hence

$$\phi(u) = B^*(-ju). \tag{19}$$

The most consuming is the finding of probability $P(\tau^+ > \tau_{\max})$ in integral formula (17), because it requires numerical calculation of three-dimensional integrals: firstly, finding the function $B(x)$ by the formula (4), secondly, finding the Fourier and Laplace transform by the formulas (18)–(19), thirdly, the calculations probability by the formula (17).

This problem is solved by considering for this task under heavy load.

10   Asymptotic Analysis of a Positive Waiting Time Under Heavy Load

The characteristic function of a positive waiting time $h(u)$ has the form (13). We find its limit value under heavy load, when $1 - \lambda b = \varepsilon$ and $\varepsilon \to 0$. For this, in expression (13) performs substitutions

$$u = \varepsilon\omega, \quad h(u) = H(\omega, \varepsilon),$$

we obtain the following function

$$H(\omega, \varepsilon) = \frac{(\phi(\varepsilon\omega) - 1)(1 - \lambda b)}{\lambda b^2 j\varepsilon\omega} \frac{1 - B^*(\lambda - \lambda\phi(\varepsilon\omega))}{B^*(\lambda - \lambda\phi(\varepsilon\omega)) - \phi(\varepsilon\omega)}. \tag{20}$$

Let us find its limit for $\varepsilon \to 0$. We can write

$$\phi(\varepsilon\omega) = Me^{j\varepsilon\omega\xi} = 1 + j\varepsilon\omega\xi b + \frac{(j\varepsilon\omega\xi)^2}{2}b_2 + o(\varepsilon^3),$$

$$\lambda - \lambda\phi(\varepsilon\omega) = -j\varepsilon\omega\xi b - \frac{(j\varepsilon\omega\xi)^2}{2}\lambda b_2 + o(\varepsilon^3),$$

$$B^*(\alpha) = Me^{-\alpha\xi} = 1 - \alpha b + \frac{\alpha^2}{2}b_2 + o(\alpha^3),$$

$$B^*(\lambda - \lambda\phi(\varepsilon\omega)) = B^*\left(j\varepsilon\omega b - \frac{(j\varepsilon\omega)^2}{2}\lambda b_2\right) = B^*\left(-j\varepsilon\omega(1 - \varepsilon) - \frac{(j\varepsilon\omega)^2}{2}\frac{b_2}{b}\right)$$

$$= 1 + b\left(j\varepsilon\omega(1 - \varepsilon) + \frac{(j\varepsilon\omega)^2}{2}\frac{b_2}{b}\right) + \frac{(j\varepsilon\omega)^2}{2}b_2 + o(\varepsilon^3).$$

Substituting these expressions in expression (20), we obtain:

$$H(\omega, \varepsilon) = \frac{1 + j\varepsilon\omega b - 1}{\lambda b^2 j\varepsilon\omega}\varepsilon \frac{1 - j\varepsilon\omega b}{1 + j\varepsilon\omega b - j\varepsilon\omega b\varepsilon + (j\varepsilon\omega)^2 b_2 - \left[1 + j\varepsilon\omega b + \frac{(j\varepsilon\omega)^2}{2}b_2\right]}$$

$$= \frac{1}{\lambda b}\varepsilon \frac{-j\varepsilon\omega b}{-j\varepsilon\omega b\varepsilon + \frac{(j\varepsilon\omega)^2}{2}b_2} = \frac{1}{1 - j\omega\frac{b_2}{2b}}.$$

Performing here the following inverse transformation

$$\omega = \frac{u}{\varepsilon} = \frac{u}{1 - \lambda b},$$

we obtain the approximate expression for $\lambda b \uparrow 1$

$$h(u) \approx \frac{1}{1 - ju\frac{b_2}{2b(1-\lambda b)}}.$$

for the characteristic function, which has the form of the characteristic function of the exponential distribution with rate

$$\gamma = \frac{2b(1 - \lambda b)}{b_2}. \tag{21}$$

Note that the mean $1/\gamma$ of the asymptotic distribution is equal to the mean $\bar{\tau}^+$ of positive waiting time (14).

Therefore, the probability $P(\tau^+ > \tau_{\max})$ under heavy load can be defined from the formula:

$$P(\tau^+ > \tau_{\max}) = e^{-\gamma\tau_{\max}} = \exp\left\{-\frac{2b(1-\lambda b)}{b_2}\tau_{\max}\right\}. \tag{22}$$

11   Conclusion

In this paper, we study the queuing system $M|GI|N|\infty$. We obtain the approximation of probability distribution of the number of customers in the system. We derive the formula of the probability of immediate service and the characteristic function of a positive waiting time. The optimal number of servers can be determined by the obtained characteristics.

Acknowledgments. The work is performed under the state order of the Ministry of Education and Science of the Russian Federation (No. 1.511.2014/K).

References

1. Brown, L., Gans, N., Mandelbaum, A., et al.: Statistical analysis of a telephone call center: a queueing-science perspective. J. Am. Stat. Assoc. **100**, 36–50 (2005)
2. Jouini, O., Aksin, Z., Dallery, Y.: Call centers with delay information: models and insights. Manuf. Serv. Operat. Manag. **13**, 534–548 (2011)
3. Kleinrock, L., Grushko, I.I., Neiman, V.I.: Queueing Theory. Translated from English. Mechanical Engineering (1979)
4. Lisovskaya, E.Yu., Moseeva, S.P.: Study of the process the number of customers in the system $M|GI|N|\infty$. Probability theory, stochastic processes, mathematical statistics and applications. In: Proceedings of the International Scientific Conference Devoted to the 80th Anniversary of Professor, Doctor of Physical and Mathematical Sciences Gennady Alekseevich Medvedev, pp. 123–127 (2015). (in Russian)
5. Nazarov, A., Terpugov, A.F.: Queueing Theory: Textbook. Publishing House of the NTL, Tomsk (2010). (in Russian)

Methods for Analysis of Queueing Models with Instantaneous and Delayed Feedbacks

Agassi Melikov[1]([✉]), Leonid Ponomarenko[2], and Anar Rustamov[3]

[1] Institute of Control Systems, ANAS, F. Agayev 9, AZ1141 Baku, Azerbaijan
agassi.melikov@gmail.com
http://www.cyber.az
[2] Institute of Information Technologies and Systems, NASU, Qlushkov Ave. 40,
Kiev 03680, Ukraine
laponomarenko@ukr.net
http://www.ukr.net
[3] Qafqaz University, H. Aliyev 120, AZ0101 Khirdalan, Baku, Azerbaijan
anrustemov@qu.edu.az
http://www.qu.edu.az

Abstract. The new Markov models of multi-channel queueing systems with instantaneous and delayed feedback are proposed. In these models part of already serviced calls instantaneously feeds back to channel while the rest part either leaves the system or feeds back to channel after some delay in orbit. Behavior of already serviced calls is handled by randomized parameters. Both exact and asymptotic methods to calculate the quality of service (QoS) metrics of the proposed models are developed. Exact method is based on the system of balance equations (SBE) for steady-state probabilities of appropriate three dimensional Markov chain (3-D MC) while asymptotic method uses the new hierarchical space merging algorithm for 3-D MC. Results of numerical experiments are demonstrated.

Keywords: Queueing model · Instantaneous and delayed feedback · Three-dimensional markov chain · Exact analysis · Asymptotic analysis

1 Introduction

In examining the literature on the queueing systems, it is evident that major interest has been focused on the models without feedback phenomena. However, queueing systems with feedback are adequate mathematical models of many real situations in which part of already serviced calls return to the system to get additional service.

Among models of queueing systems with feedback two kinds of models should be distinguished: (1) models with instantaneous feedback (i.e. models without orbit) and (2) models with delayed feedback (i.e. models with orbit).

In the available literature both kinds of models have been investigated separately. But models of queueing systems with simultaneously instantaneous and delayed feedback have not been investigated.

© Springer International Publishing Switzerland 2015
A. Dudin et al. (Eds.): ITMM 2015, CCIS 564, pp. 185–199, 2015.
DOI: 10.1007/978-3-319-25861-4_16

In this paper, the Markov models of multichannel queueing systems with instantaneous and delayed feedback are examined. Let's note that taking into account of both types of feedback mechanisms leads to increase of vector dimension which describe the state of the system. As a result an approach to investigate the models based on the system of balance equations (SBE) for steady-state probabilities becomes inefficient especially for the large scale models. So, developing efficient methods for asymptotic analysis of models with a large number of channels and large size of orbit is highly desired. Below we propose both exact and asymptotic methods to calculate the quality of service (QoS) metrics of the queueing models with instantaneous and delayed feedback.

The paper is organized as follows. In Sect. 2, brief review of related works devoted to queueing models with feedback is given. The description of the model with state-dependent feedback probabilities is presented in Sect. 3. Exact and asymptotic methods to calculate the QoS metrics are developed in Sect. 4 and Sect. 5 respectively. The results of numerical experiments performed by using the developed methods are demonstrated in Sect. 6. Conclusion remarks are given in Sect. 7.

2   Related Works

Let us first consider the works in which models without orbits (instantaneous feedback) are investigated. Pioneer work was the paper Takacs [1]. In this paper model $M/G/1/$ with instantaneous Bernoulli feedback was examined. The main result of the paper is the recurrent relations to obtain joint distribution of sojourn time and number of calls in the system.

Model $M/GI/1/\infty$ with vacations and instantaneous Bernoulli feedback has been investigated in [2]. The queue length probability generating functions embedded at the moments of (i) call departures, (ii) call feedbacks and (iii) server vacation completions are developed.

In paper [3] the model $M_2/G_2/1/\infty$ with two independent Poisson traffics and instantaneous Bernoulli feedback were examined. Each type of traffic has its own service time distribution and the decision to feedback or not is based on the type of call completing service. Type-1 calls have a non-preemptive priority over type-2 calls and if a call feeds back it always becomes a type-2 call in the priority scale. Conditions for the existence of stationary mode are found; both joint and marginal distributions of the queue lengths are found as well. It is shown that several earlier known results (for instance, results of the paper [1]) can be obtained simply from the results given here.

In paper [4] the following model $M/M/1$ with general feedback mechanism was considered. At the completion of the i-th feeds back cycle the call feedback with probability $p(i)$ or departs from the system with probability $1-p(i)$. It is shown that under some conditions the given model approaches to the $M/G/1$ with processor sharing (PS) queue discipline. An expression for the Laplace-Stieltjes transforms (LST) of the sojourn time distribution in the given queueing system is found.

In paper [5] LST of the sojourn time distribution in the $M/G/1/N$ with state-dependent feedback is developed and the author notes that apart from the special cases $N=1$ and $N=2$ general algebraic expressions for inverse of LST will be difficult to obtain.

Model $MAP/PH/1/N$ in which both departure and feedback probabilities depend on the states of some random environment considered in paper [6]. To calculate the steady-state probabilities of the appropriate 4-D Markov chain (MC) the authors apply the Neuts matrix-geometric method [7].

Models of queuing systems with instantaneous feedback and infinite number of channels were investigated in papers [8,9].

Let's note that in all above mentioned works it is assumed that primary and feedback calls have the same channel holding times. Moreover in these works (except the paper [6]) it is assumed that both departure and feedback probabilities are governed by Bernoulli schema with constant parameters, i.e. they are state-independent. These assumptions essentially restricted applicability of their results in real systems.

Two-dimensional models of multi-channel queuing systems in which primary and feedback calls are not identical in terms of channel occupancy time were investigated in paper [10]. Moreover in this paper it is assumed that departure and feedback probabilities are state-dependent. In order calculate their QoS metrics the space merging approach for two-dimensional Markov chains is used [11,12].

Now let us consider the works in which models with orbit (delayed feedback) are examined. First work in this direction was done by Takacs also [13]. In this paper model $M/M/1$ with Bernoulli feedback and infinite orbit has been considered. At each feedback time to channel, besides entire calls return from the orbit, additional number of calls are also arriving. To find mean length of the queue, waiting time in the queue and sojourn time in the system the method of two dimensional generating functions is used.

Model $\cdot/G/k/b, b < \infty$ with stationary ergodic flow process and infinite orbit was examined in [14]. It is assumed that number of feedback for each call is random variable. It is shown that under some mild conditions feedback flow converges to a Poisson process as the feedback delay distribution is scaled up.

In the paper [15] model $M/G/1$ with Bernoulli feedback and finite orbit is considered. Here it is assumed that when orbit becomes full then all calls from the orbit instantaneously feed back to the buffer of the queuing system. Joint distribution of number of calls in system and in orbit is carried out. Similar model was examined in paper [16]. The only difference is the following: required switching time for feeds back of call from orbit to buffer is not zero and it is random quantity with known distribution function.

Model $M/G/1$ with infinite buffer and exponential sojourn time in orbit was investigated in [17]. Here feedback probabilities depend on the both number of calls in the system and in the orbit as well as holding time of calls. Distribution functions of length of queue, busy period of channel and departure flow are found.

Model M/M/n, $n > 1$, with finite buffer and exponential sojourn time in orbit was investigated in [18]. Here feedback probabilities depend on the number of busy channels in the system during the call departure. Both loss probabilities of primary and feedback calls, average number of busy channels and average number of feedback calls in the orbit are found.

It is important to note that in all above mentioned works models with instantaneous and delayed feedbacks are investigated separately. To our best knowledge, models with both instantaneous and delayed feedbacks are not investigated.

3   Formulation of the Model

This system contains $N > 1$ identical channels which are used by Poisson flow of primary calls (p-calls) with intensity λ_p. The channel occupancy times of p-calls are assumed to be independent and have identical exponential distribution with mean $1/\mu_p$.

After completion of the service of the p-calls the following decisions might be accepted: (i) it leaves the system with probability $\sigma_1(x)$; (ii) it feeds back instantaneously with probability $\sigma_2(x)$; (iii) it enters the orbit with probability $\sigma_3(x) = 1 - \sigma_1(x) - \sigma_2(x)$. These probabilities depend on the parameter x which denotes the state of some external random environment.

The orbit size for repeated calls (r-calls) is R, $0 < R < \infty$. It means that an arrived to orbit call will be accepted if upon its arrival the number of r-calls in orbit is less than R; otherwise an arrived call will be lost. Sojourn times of r-calls in the orbit are independent and identically distributed random variables and they have common exponential distribution with mean $1/\lambda_r$. It is assumed that r-calls from the orbit are not persistent, i.e. if upon arrival of r-call all channels of the system are busy then it is lost eventually.

Distribution functions of channel occupancy time of heterogeneous r-calls (instantaneous and delayed) are assumed to be independent and exponential with common mean $1/\mu_r$ generally speaking For simplicity the model as well as in order to find analytically tractable results it is assumed that both kinds of r-calls (instantaneous and delayed) do not return again in future (although as it seen below the developed approach allows to take into account multiple repetition as well).

The main performance metrics of the given queueing systems are the following parameters: (i) loss probability of p-calls (P_p); (ii) loss probability of r-calls from the orbit (P_r); (iii) mean number of p-calls in channels (L_p); (iv) mean number of r-calls in channels (L_r); (v) mean number of r-calls in the orbit (L_o); (vi) coefficient of utilization of channels (C_u).

4   Exact Method for Calculation of the Performance Metrics

By taking into account the form of distribution functions of random variables involved in the formation of the model, we conclude that operating of the investigated queuing system might be described by the three-dimensional Markov chain

(3-D MC). So, states of the system at equilibrium at any time are described by three-dimensional vectors $\mathbf{n} = (n_p, n_r, n_o)$ where the first (n_p) and second components (n_r), respectively, indicate the number of initial and repeated calls in the channels, and the third component (n_o) indicates the number of calls in orbit. The state space (i.e. set of all possible states) S is defined as

$$S = (\boldsymbol{n} : n_p = 0, 1, ..., N; n_r = 0, 1, ..., N; n_p + n_r \leq N; n_o = 0, 1, ..., R) \quad (1)$$

Let us suppose that the behavior of call which already received initial service in the system is determined by the number of repeated calls in orbit, i.e. states of the external random environment are determined by scalar x, which defines the number of repeated calls in orbit, $x=0,1,...,$ R Since the call is received in orbit only when at the moment of its receipt total number of repeat calls is smaller than R, then we have $\sigma_3(R)=0$.

The intensity of transition from state $\boldsymbol{n}$ to state $\boldsymbol{n}'$ is denoted as $q=(\boldsymbol{n}, \boldsymbol{n}')$, $\boldsymbol{n}, \boldsymbol{n}' \in S$ The combination of these values involve the Q-matrix of the given MC. They are d below:

$$q(\boldsymbol{n}, \boldsymbol{n}') = \begin{cases} \lambda_p, & \text{if } \boldsymbol{n}' = \boldsymbol{n} + \boldsymbol{e_1} \\ n_p \mu_p \sigma_1(n_0), & \text{if } \boldsymbol{n}' = \boldsymbol{n} - \boldsymbol{e_1} \\ n_p \mu_p \sigma_2(n_0), & \text{if } \boldsymbol{n}' = \boldsymbol{n} - \boldsymbol{e_1} + \boldsymbol{e_2} \\ n_p \mu_p \sigma_3(n_0), & \text{if } n_o \leq R, \boldsymbol{n}' = \boldsymbol{n} - \boldsymbol{e_1} + \boldsymbol{e_3} \\ n_r \mu_r, & \text{if } \boldsymbol{n}' = \boldsymbol{n} - \boldsymbol{e_2} \\ n_o \lambda_r, & \text{if } n_p + n_r \leq N, \boldsymbol{n}' = \boldsymbol{n} + \boldsymbol{e_2} - \boldsymbol{e_3}, \text{ or} \\ & \quad n_p + n_r = N, \boldsymbol{n}' = \boldsymbol{n} - \boldsymbol{e_3} \\ 0, & \text{in other cases.} \end{cases} \quad (2)$$

Here e_i is the i-th unit vector of the three-dimensional Euclidean space, $i = 1$, 2, 3. The given three-dimensional MC with finite number of states is irreducible, which determines the existence of a stationary regime. Let $p(\boldsymbol{n})$ means a steady-state probability of $\boldsymbol{n} \in S$. These values comply with the relevant equilibrium equations system (EES), which is constructed on the basis of (2) and has the following form:

$$(\lambda_p I(n_p + n_r < N) + n_p \mu_p + n_r \mu_r + n_o \lambda_r) p(\boldsymbol{n})$$
$$= \lambda_p p(\boldsymbol{n} - \boldsymbol{e_1}) I(n_p > 0) + (n_p + 1) \mu_p (p(\boldsymbol{n} + \boldsymbol{e_1}) \sigma_1(n_o)$$
$$+ p(\boldsymbol{n} + \boldsymbol{e_1} - \boldsymbol{e_2}) \sigma_2(n_o) + p(\boldsymbol{n} + \boldsymbol{e_1} - \boldsymbol{e_3}) \sigma_3(n_o - 1) I(n_o > 0)) \quad (3)$$
$$+ (n_r + 1) \mu_r p(\boldsymbol{n} + \boldsymbol{e_2}) + (n_o + 1) \lambda_r (p(\boldsymbol{n} + \boldsymbol{e_3}) \delta(n_p + n_r, N)$$
$$+ p(\boldsymbol{n} - \boldsymbol{e_2} + \boldsymbol{e_3}) I(n_o < R, n_r > 0))$$

Herein after $I(A)$ denotes the indicator function of event A and $\delta(i, j)$ represents Kroneckers symbols. Normalizing condition of the EES can be shown as below:

$$\sum_{n \in S} p(\boldsymbol{n}) = 1 \quad (4)$$

After finding a solution of the EES (3), (4) the performance metrics of the investigated system are determined by the marginal distributions of the 3-D MC. Thus, since the flow of the primary call is Poisson, their loss probability can be determined by PASTA theorem [19]:

$$P_p = \sum_{n \in S} p(n)\delta(n_p + n_r, N) \tag{5}$$

Based on the results of the work [14] we conclude that traffic of retrial calls is considered as Poisson one, and therefore in order to calculate the loss probability of retrial calls can also be used PASTA theorem. Since the retrial calls are generated only when the orbit is not empty, then the desired characteristic is defined as follows:

$$P_r = \sum_{n \in S} p(n)\delta(n_p + n_r, N)(1 - \delta(n_o, 0)) \tag{6}$$

The average number of primary and retrial calls in channels, and retrial calls in the orbit is defined as the expectation of the appropriate discrete random variables:

$$L_x = \sum_{j=1}^{N} j\Phi_x(j) \tag{7}$$

where $\Phi_x(j) = \sum_{n \in S} p(n)\delta(n_x, j), x \in \{p, r\}$;

$$L_o = \sum_{j=1}^{R} j\Psi(j) \tag{8}$$

where $\Psi(j) = \sum_{n \in S} p(n)\delta(n_o, j)$.
Channel utilization coefficient (C_u) is determined based on the formula (7):

$$C_u = (L_p + L_r)/N. \tag{9}$$

Unfortunately, due to the complex structure of the Q-matrix of the given 3-D MC it is too complicate (most probably it is impossible) to find analytical solution to the EES (3), (4). Therefore the only way to solve them is to use numerical methods of linear algebra (the effective and well-known method for solving EES is Gauss-Seidel one).

5   Asymptotic Method for Calculation of the Performance Metrics

The dimension of the EES (3), (4) is determined based on the dimension of the state space (1), which consists of $(N + 1)(N + 2)(R + 1)/2$ states, i.e. it is estimated to be $O(N^2 R)$. Therefore, the above-given exact method makes it possible to calculate the performance metrics (5)–(9) only for models with moderate dimensions of the state space (1), but for large scale models it encounters great computational difficulties. To eliminate them, we can use the method of the state space merging theory of stochastic systems [11,12]. Here is provided a hierarchical space merging algorithm (SMA) for calculating the steady-state probabilities of the investigated 3-D MC according to the certain asymptotic conditions.

For the correct application of this method the following condition is considered: $\lambda >> max\{\mu_p, \mu_r\}$, i.e. it is assumed that the system is operating under high load.

It should be noted that in the proposed models, as a rule, the arrival intensity of the p-call significantly exceeds the arrival intensity of r-call from orbit, i.e. condition $\lambda_p >> \lambda_r$ is natural. Then, having this assumption we can say that transition intensity between states inside the planes that are parallel to the base of the prism is much greater than the transitions intensity between states of different planes. In that case we can consider the following splitting of the state space (1):

$$S = \bigcup_{k=0}^{R} S_k, S_k \cup S_{k'} = \emptyset, \text{ if } k \neq k' \tag{10}$$

where $S_k = \{\boldsymbol{n} \in S : n_o = k\}, k = 0, 1, ..., R$. In other words, it is considered that entire state space (1) is sliced into different planes that are the parallel to the base of the prism.

The merging function is determined based on the splitting (10) as follows:

$$U_1(\boldsymbol{n}) = < k >, \text{ if } \boldsymbol{n} \in S_k, \tag{11}$$

where $< k >$ is a merged state, which includes all states of class S_k. Let $\Omega_1 = \{< k >: k = 0, 1, ..., R\}$.

According to SMA [11] (see the Appendix of this work) state probabilities of the initial model are defined as follows

$$p_{\boldsymbol{n}} \approx \rho_k(n_p, n_r)\pi_1(< k >) \tag{12}$$

where $\rho_k(n_p, n_r)$ denotes the probability of the state (n_p, n_r) within the splitting model with state space S_k, and $\pi_1(< k >)$ is the probability of the merged state $< k > \in \Omega_1$.

Therefore, for the calculation of the stationary distribution of the initial 3-D MC we need to find stationary distributions of 2-D MC (their number is equal $R+1$) and one 1-D MC. For large number of channels computational difficulties arise when calculating the stationary distributions of these 2-D MC with state space $S_k, k = 0, 1, ..., R$. Therefore in order to calculate stationary distributions within the classes $S_k, k = 0, 1, ..., R$ it is required to apply SMA to each class. In other words, we consider the hierarchy of the merged models.

From (10) it is clear that all the splitting models with state spaces $S_k, k = 0, 1, ..., R$ involve identical 2-D MC. Therefore, in the future the value of k is fixed and splitting model with state space S_k is analyzed.

In the state space S_k the following splitting is considered:

$$S_k = \bigcup_{i=0}^{N} S_k^i, \, S_k^i \cup S_k^j = \emptyset, \, \text{if } i \neq j \tag{13}$$

where $S_k^i = \{\boldsymbol{n} \in S_k : n_r = i\}, i = 0, 1, ..., N$ In other words, we consider a partitioning of the state space of the splitting model by row.

Further, based on the splitting (13) in the state space S_k the following merged function is determined:

$$U_2(\boldsymbol{n}) = < i >, \text{ if } \boldsymbol{n} \in S_k^i, \tag{14}$$

where $< i >$ is a merged state, which includes all states of class S_k^i. Let $\Omega_2 = \{< i >: i = 0, 1, ..., N\}$.

According to the SMA we have:

$$\rho_k(n_p, n_r) \approx \rho_{n_r}^k(n_p)\pi_2^k(< n_r >) \tag{15}$$

where $\rho_{n_r}^k(n_p)$ denotes the probability of the state (n_p, n_r) within the splitting model with state space $S_k^{n_r}$, and $\pi_2^k(< n_r >)$ is the probability of the merged state $< n_r > \in \Omega_2$.

Lets consider the problem of calculating the state probabilities within the classes S_k^i. First of all, lets mention that since the class of states S_k^N has only one state $(0, N)$, then below it is assumed that $\rho_N^k(0) = 1$.

In the class of states $S_k^i, i \neq N$ the second component is constant. Therefore, in splitting models with state space S_k^i microstate $(n_p, i) \in S_k^i$ can be described only with the first component. Further, for the sake of convenience in the splitting model with state space S_k^i the state (n_p, i) is just referred as $n_p, n_p = 0, 1, ..., N-i$.

The transition intensity between the states n_p and n_p' of the splitting model with state space S_k^i is denoted as $q_k(n_p, n_p')$. From (2) we get that these parameters are defined as follows:

$$q_k(n_p, n_p') = \begin{cases} \lambda_p, & \text{if } n_p' = n_p + 1 \\ n_p \mu_p \sigma_1(k), & \text{if } n_p' = n_p - 1 \\ 0, & \text{in other cases.} \end{cases} \tag{16}$$

Hence, from (16) we get that state probabilities within the splitting model with state space $S_k^i, i = 0, 1, ..., N-1$ are calculated as state probabilities of 1-D birth and death process (BDP), i.e.

$$\rho_i^k(j) = \frac{(\nu_p/\sigma_1(k))^j}{j!}\rho_i^k(0), j = 1, 2, ..., N - i \tag{17}$$

where $\nu_p = \lambda_p/\mu_p$ and $\rho_i^k(0)$ is determined from the normalizing condition, i.e. $\sum_{j=0}^{N-i} \rho_i^k(j) = 1$.

The transition intensity from merged state $< i >$ to other merged state $< j >$ is denoted as $q_k(< i >, < j >), < i >, < j >\in \Omega_2$ After certain mathematical transformations using SMA we get:

$$q_k(< i >, < j >) = \begin{cases} \mu_p \alpha_k(i), & \text{if } j = i + 1 \\ i\mu_r & \text{if } j = i - 1 \\ 0, & \text{in other cases.} \end{cases} \qquad (18)$$

where $\alpha_k(i) = \sigma_2(k) \sum_{j=1}^{N-i} j\rho_i^k(j), i = 0, 1, ..., N - 1$.

Thus, from (18) we get the following relations to calculate the probabilities of merged states $\pi_2^k(< n_r >), < n_r >\in \Omega_2$:

$$\pi_2^k(< n_r >) = \frac{(\mu_p/\mu r)^{n_r}}{n_r!} \prod_{i=0}^{n_r-1} \alpha_k(i)\pi_2^k(< 0 >), n_r = 1, 2, ..., N \qquad (19)$$

where $\pi_2^k(< 0 >)$ is derived from the normalizing condition, i.e., $\sum_{j=0}^{N} \pi_2^k(< j >) = 1$.

Now in order to calculate the stationary distribution of the initial 3-D MC it is required to find the probabilities of the merged states $\pi_1(< k >), < k >\in \Omega_1$ (see Eq. (12)). For this purpose we need to determine transition intensities between classes (layers) $S_k, k = 0, 1, ..., R$ (see splitting (10)).

Lets denote the transition intensity between classes S_k and S_k' as $q(S_k, S_k')$ These intensities are determined by the relations (2), (17) and (19). Then, using the SMA after certain mathematical transformations we get:

$$q(S_k, S_k') = \begin{cases} \Lambda(k), & \text{if } k' = k + 1 \\ k\lambda_r, & \text{if } k' = k - 1 \\ 0, & \text{in other cases.} \end{cases} \qquad (20)$$

where $\Lambda(k) = \mu_p \sigma_3(k) \sum_{j=1}^{N-i} j\rho_i^k(j), k = 0, 1, ..., R - 1$.

Hence, from (20) we get that the required probabilities of the merged state $\pi_1(< k >), < k >\in \Omega_1$ are defined as a stationary distribution of 1-D BDP with variable parameters, i.e.

$$\pi_1(< k >) = \frac{1}{k!\lambda_r^k} \prod_{i=0}^{k-1} \Lambda(i)\pi_1(< 0 >), k = 1, 2, ..., R \qquad (21)$$

where $\pi_1(< 0 >)$ is derived from the normalizing condition, i.e., $\sum_{k=0}^{R} \pi_1(< k >) = 1$.

Finally, the stationary distribution of the initial 3-D MC is derived from (12) by (17), (19) and (21):

$$p(n_p, n_r, n_o) \approx \rho_{n_r}^{n_o}(n_p)\pi_1(< n_o >)\pi_2^{n_o}(< n_r >) \qquad (22)$$

In summary, we get the following expressions for the approximate calculation of the required performance metrics of the investigated system:

$$P_p \approx \sum_{k=0}^{R} \pi_1(<k>) \sum_{i=0}^{N} \rho_i^k (N-i)\pi_2^k(<i>) \tag{23}$$

$$P_r \approx \sum_{k=1}^{R} \pi_1(<k>) \sum_{i=0}^{N} \rho_i^k (N-i)\pi_2^k(<i>) \tag{24}$$

$$L_p \approx \sum_{k=0}^{R} \pi_1(<k>) \sum_{j=1}^{N} j \sum_{i=0}^{N-j} \rho_i^k(j)\pi_2^k(<i>) \tag{25}$$

$$L_r \approx \sum_{k=0}^{R} \pi_1(<k>) \sum_{i=0}^{N} i\pi_2^k(<i>) \tag{26}$$

$$L_o \approx \sum_{k=1}^{R} k\pi_1(<k>) \tag{27}$$

Channel utilization coefficient is approximately determined according to (9), (25) and (26).

6   Numerical Results

The proposed methods (exact and approximate) allow to study the performance metrics of the investigated systems with instantaneous and delayed feedback versus their structural (the number of channels and the size of the orbit) and load of parameters (arrival and services intensities of primary and retrial calls) as well as the parameters that determine the behavior of the call after receiving the primary service (i.e. the probabilities $\sigma_i(k), i = 1, 2, 3$. However, as noted above, precise analysis is possible only for the models with moderate size.

In order to be short, here are only studied the dependence of performance metrics on the number of channels for fixed values of other parameters.

The results of the numerical experiments for a hypothetical model are shown in Figs. 1–5 (for the convenience in the graphical representation, in some cases, the ordinate axis is illustrated in logarithmic scale). Here are given figures of the performance metrics for the following values of the load parameters of the model: $\lambda_p = 4, \mu_p = 0.5, \lambda_r = 2, \mu_r = 0.2$. For simplicity, it is assumed that the probabilities $\sigma_i(k), i = 1, 2, 3$ are constants, in other words, $\sigma_1(k) = 0.4, \sigma_2(k) = 0.3, \sigma_3(k) = 0.3$ for any $k = 0, 1, ..., R - 1$, and $\sigma_1(k) = 0.5, \sigma_2(k) = 0.5$.

The increase in the number of channels leads to a systematic decrease in the loss probability of the primary challenges (P_p); however, the loss probability of retrial calls (P_r) increases at small number of channels (N), and after reaching a certain maximum value, the indicated performance metrics again systematically decreases (see Figs. 1 and 2). This phenomenon is explained by the effect of the

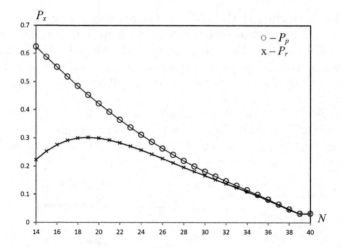

Fig. 1. Dependence of the loss probability of the primary and retrial calls on the number of channels, $R = 3$

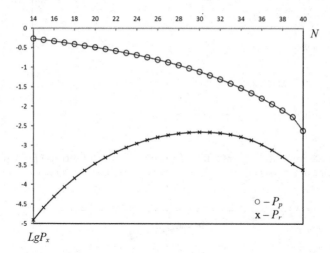

Fig. 2. Dependence of the loss probability of the primary and retrial calls on the number of channels, $R = 10$

feedback at small number of channels. As the number of channels is getting bigger and bigger, the effect of repeated calls almost disappears. In large values of N the values of the function P_r asymptotically approaches to the values of P_p but $P_r < P_p$.

It is interesting to note that the increase in the size of orbit for retrial calls leads to reduce in the values of both functions P_p and P_r. In other words, the values of these functions at $R = 3$ is greater than the values for $R = 10$ (see. Figs. 1 and 2).

In the case of function P_r this property has quite logical explanation, but for the function P_p it has negative relation, so that increase in the size of orbit has to result in increase of the loss probability of primary calls. This is due to the fact that the selected source data for the hypothetical model for any number of channels (N) and the average number of retrial calls in the orbit (L_o) at $R = 3$ are substantially greater than at $R = 10$.

Dependence of average number of primary (L_p) and retrial (L_r) calls, and the average number of retrial calls in the orbit (L_o) on the number of channels (N) of the system given in Fig. 3. It is apparent that for the selected data functions

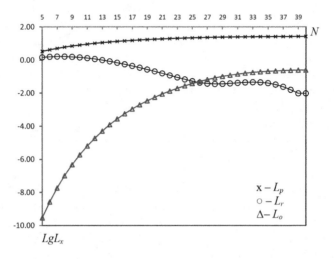

Fig. 3. Dependence of average number of primary (L_p) and retrial (L_r) calls, and the average number of retrial calls in the orbit (L_o) on the number of channels (N), $R = 10$

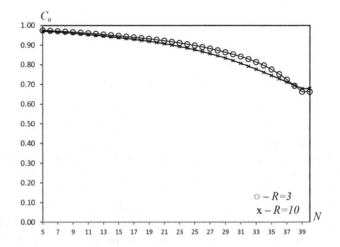

Fig. 4. Dependence of the channel utilization on the number of channel

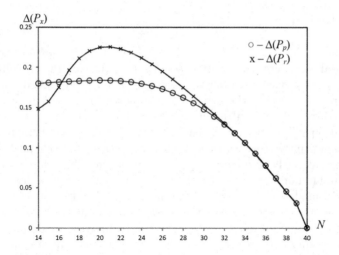

Fig. 5. Dependence of the behavior of the absolute error of loss probabilities of primary ($\triangle P_p$) and retrial ($\triangle P_r$) call on the number of channels of the system, $\lambda_r = 2; \lambda_p = 4; \mu_r = 0.2; \mu_p = 0.5; R = 3$

L_o and L_p systematically increase, while function L_r is not increasing; wherein the number of repeated calls in the channels is significantly smaller than the number of the primary challenges in the channels.

Dependence of the channel utilization on the number of channel is shown in Fig. 4. As expected, the function of C_u is decreasing function, and moreover threefold increase in the size of orbit doesn't affect its value.

Another purpose of the numerical experiments is to determine the accuracy of the proposed approximate method, and comparison of execution time exact and approximate algorithms. Regarding the accuracy of the approximate formulas, we can say that they are strongly dependent on the dimension of the model, on the absolute values of the load parameters, as well as their relationship. For the above-mentioned initial data of proposed model the steady-state probabilities in exact and approximate approaches are very close to each other. However, the required characteristics is determined as marginal distributions of initial model so in some cases it is expected that the values of these characteristics might strongly differentiate from each other. The behavior of these characteristics with respect to the structural parameters in the exact and approximate approaches are identical. Furthermore, numerical experiments have shown that an increase in the dimension of the model, their absolute values are asymptotically approaching to zero. This is very significant, because approximate approach is proposed only for very large dimension of the model.

Due to the limited size of work here in Fig. 5 is shown only dependence of the behavior of the absolute error of loss probabilities of primary ($\triangle P_p$) and retrial ($\triangle P_r$) call on the number of channels of the system.

7   Conclusion

In the paper model of multi-channel queuing system with instantaneous and delayed retrial calls were studied. Here some part of the calls may require instantaneous repeated service just after the having primary service. And remain part can either return through the orbit within random time, or leave the system at all. We consider a general model where it is assumed that service time of primary and repeated calls vary from each other. Exit probabilities of the calls after receiving primary service, their immediate return to the channels or entering orbit depend on the number of repeated calls in the orbit. More ever systems with finite and infinite dimensions of the orbit were analyzed. It is shown that the mathematical models of the investigated queuing system is a three-dimensional Markov chain. The methods of exact and asymptotic analysis of the characteristics of mentioned system were proposed. Accurate analysis is based on a system of equilibrium equations for the state probabilities, but asymptotic analysis uses the principles of merging of the state space of stochastic systems. The hierarchical state space merging algorithm were proposed in order to calculate the stationary distribution of the three-dimensional Markov chains.

The proposed approach can be used for studying similar models in which more complex access schemas with priority for different call types are implemented. In addition this approach can be applied in models in which the exit probability from the system, return probability to the channels or entering probability to the orbit depend on the number of different types of calls in the channels or depends on the sum of them. Solution of the optimizing problems for investigated queueing system is also important ones. This problems are included to our future research question.

References

1. Takacs, L.: A single-server queue with feedback. Bell Syst. Tech. J. **42**, 505–519 (1963)
2. Wortman, M.A., Disney, R.L., Kiessler, P.C.: The M/GI/1 Bernoulli feedback queue with vacations. Queueing Syst. **9**(4), 353–363 (1991)
3. D'Avignon, G.R., Disney, R.L.: Queues with instantaneous feedback. Manag. Sci. **24**(2), 168–180 (1977)
4. Berg, J.L., Boxma, O.J.: The M/G/1 queue with processor sharing and its relation to feedback queue. Queueing Syst. **9**(4), 365–402 (1991)
5. Hunter, J.J.: Sojourn time problems in feedback queue. Queueing Syst. **5**(1–3), 55–76 (1989)
6. Dudin, A.N., Kazimirsky, A.V., Klimenok, V.I., Breuer, L., Krieger, U.: The queueing model MAP/PH/1/N with feedback operating in a Markovian random environment. Austrian J. Stat. **34**(2), 101–110 (2005)
7. Neuts, F.: Matrix-geometric solutions in stochastic models: an algorithmic approach. John Hopkins University Press, Baltimore (1981)
8. Nazarov, A.A., Moiseeva, S.P., Morozova, A.S.: Investigation of queuing system with repeated servicing and infinite number of channels by the method of limit decomposition. Comput. Technol. **13**(5), 88–92 (2008). (in Russian)

9. Moiseeva, S.P., Zacharolnaya, I.A.: Mathematical model of parallel servicing of multiple calls with repeated servicing. Auto Metric. **47**(6), 51–58 (2011). (in Russian)

10. Melikov, A.Z., Ponomarenko, L.A., Kuliyeva, K.H.N: Calculation of the characteristics of multichannel queuing system with pure losses and feedback. J. Autom. Inf. Sci. **47**(6) (2015)

11. Ponomarenko, L., Kim, C.S., Melikov, A.: Performance Analysis and Optimization of Multi-traffic on Communication Networks. Springer, Heidelberg (2010)

12. Korolyuk, V.S., Korolyuk, V.V.: Stochastic Models of Systems. Kluwer, Boston (1999)

13. Takacs, L.: A queueing model with feedback. Oper. Res. **11**(4), 345–354 (1977)

14. Pekoz, E.A., Joglekar, N.: Poisson traffic flow in a general feedback. J. Appl. Probab. **39**(3), 630–636 (2002)

15. Lee, H.W., Seo, D.W.: Design of a production system with feedback buffer. Queueing Syst. **26**(1), 187 (1997)

16. Lee, H.W., Ahn, B.Y.: Analysis of a production system with feedback buffer and general dispatching time. Math. Prob. Eng. **5**, 421–439 (2000)

17. Foley, R.D., Disney, R.L.: Queues with delayed feedback. Adv. Appl. Probab. **15**(1), 162–182 (1983)

18. Melikov, A.Z., Ponomarenko, L.A., Kuliyeva, K.N.: Numerical analysis of the queuing system with feedback. Cybern. Syst. Anal. **51**, 4 (2015)

19. Wolff, R.W.: Poisson arrivals see time averages. Oper. Res. **30**(2), 223–231 (1992)

Gaussian Approximation of Distribution of States of the Retrial Queueing System with r-Persistent Exclusion of Alternative Customers

Anatoly Nazarov and Yana Chernikova[✉]

Tomsk State University, Lenina, 36, Tomsk, Russia
nazarov.tsu@gmail.com, evgenevna.92@mail.ru
http://www.tsu.ru

Abstract. In this paper, we study the retrial queueing system with two arrival processes and two orbits with r-persistent exclusion of alternative customers by method of asymptotic analysis under condition of long delay. Stationary probability distribution of server states and values of asymptotic means of the number of customers in the orbits are obtained.

Keywords: Retrial queuing system · r-persistent exclusion of alternative customers · Orbit · Asymptotic analysis

1 Introduction

Queueing systems, in which arriving customers who find all servers and waiting positions (if any) occupied may retry for service after a period of time, are called Retrial queues [1–3]. A review of the main results on this topic can be found in [4]. Retrial queues have been widely used as mathematical models of different communication systems: shared bus local area networks operating under transmission protocols like CSMA/CD (Carrier Sense Multiple Access with Collision Detection), cellular mobile networks, computer and communications networks, IP networks. Priority control is also wildely used in production practice, transportation management, etc. Several authors including Choi, B.D. [6–10], Rengnanathan, N. [11], Krishna Reedy, G.V. [12], Zhu, Y.J. [13] have studied priority queues. These authors and several others have studied single server or multi-server queues with two or more priority classes under preemptive or non-preemptive priority rules. Choi, B.D. We analyzed a M/G/1 retrial queueing systems with two types of calls and finite capacity, Moreno, P. considered an M/G/1 retrial queue with recurrent customers and general retrial times [14]. In [15] retrial queue system M/G/1 with queue length r and the priority of the primary customers is studied. In [16], generalization of [15] is implemented.

In this paper, we study the retrial queueing system $M^{(2)}/M^{(2)}/1$ with r-persistent exclusion of alternative customers.

A. Dudin et al. (Eds.): ITMM 2015, CCIS 564, pp. 200–208, 2015.
DOI: 10.1007/978-3-319-25861-4_17

2   Problem Statement

We consider retrial queueing system with two arrival processes and two orbits with r-persistent collision of alternative customers (Fig. 1).

We assume that two arrival processes to the system are described by the stationary Poisson process with intensity λ_1 and λ_2, respectively. Customer, which finds the free server, occupies it during a random time which is exponentially distributed with intensity μ_1 and μ_2, respectively. If, at the moment of arrival, customer of the first type finds the server busy with a customer of the first type, then it goes to the orbit 1 (the orbit for customer of the first type), where it performs a random delay with duration determined by exponential distribution with intensity σ_1. From the orbit 1, after the random delay, the customer tries to occupy the server again. If at the time of arrival, customer of the first type finds the server busy with a customer of the second type, then the arrived customer with probability r_1 replaces the customer, which was in service, and occupies the server, and with probability $1 - r_1$ it goes to the orbit 1.

The same goes for the second type customer. If at the moment of arrival, customer of the second type finds the server busy with a customer of the second type, then it goes to the orbit 2 (the orbit for customer of the second type), where it performs a random delay with duration determined by exponential distribution with intensity σ_2. From the orbit 2, after the random delay, the customer tries to occupy the server again. If, at the time of arrival, customer of the second type finds the server busy with a customer of the first type, then an arrived customer

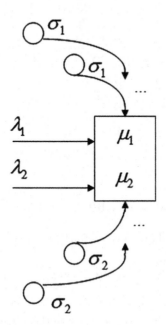

Fig. 1. Retrial queueing system $M^{(2)}/M^{(2)}/1$.

with probability r_2 replaces the customer, which was in service, and occupies the server, and with probability $1 - r_2$ it goes to the orbit 2.

Let $i_1(t)$ be the number of customers in the orbit 1 and $i_2(t)$ be the number of customers in the orbit 2, and the process $k(t)$ defines the server state at the moment t in the following way:

$$
k(t) = \begin{cases} 0, & \text{if server is free,} \\ 1, & \text{if server is busy with a customer of the first type,} \\ 2, & \text{if server is busy with a customer of the second type.} \end{cases}
$$

We would like to solve a problem of computation of stationary probability distribution of the number of customers in the orbits 1 and 2 and server state.

3 System of Kolmogorov Differential Equations

We consider Markovian process $\{k(t), i_1(t), i_2(t)\}$, $t \geq 0$.

Let us denote by $P\{k(t) = k, i_1(t) = i_1, i_2(t) = i_2\} = P_k(i_1, i_2, t)$ a probability that, at the moment t, the server in the state k and i_1 customers are in the orbit 1, i_2 customers are in the orbit 2.

We write system of differential Kolmogorovs equations for the probability distribution $\{P_0(i_1, i_2, t), P_1(i_1, i_2, t), P_2(i_1, i_2, t)\}$:

$$
\frac{\partial P_0(i_1, i_2, t)}{\partial t} = -(\lambda_1 + \lambda_2 + i_1\sigma_1 + i_2\sigma_2)P_0(i_1, i_2, t) + \mu_1 P_1(i_1, i_2, t) + \mu_2 P_2(i_1, i_2, t),
$$

$$
\begin{aligned}
\frac{\partial P_1(i_1, i_2, t)}{\partial t} = {} & -(\lambda_1 + \lambda_2 + \mu_1 + r_2 i_2\sigma_2)P_1(i_1, i_2, t) + (1 - r_2)\lambda_2 P_1(i_1, i_2 - 1, t) \\
& + \lambda_1 P_0(i_1, i_2, t) + (i_1 + 1)\sigma_1 P_0(i_1 + 1, i_2, t) + \lambda_1 P_1(i_1 - 1, i_2, t) \\
& + r_1\lambda_1 P_2(i_1, i_2 - 1, t) + r_1(i_1 + 1)\sigma_1 P_2(i_1 + 1, i_2 - 1, t), \qquad (1)
\end{aligned}
$$

$$
\begin{aligned}
\frac{\partial P_2(i_1, i_2, t)}{\partial t} = {} & -(\lambda_1 + \lambda_2 + \mu_2 + r_1 i_1\sigma_1)P_2(i_1, i_2, t) + (1 - r_1)\lambda_1 P_2(i_1 - 1, i_2, t) \\
& + \lambda_2 P_0(i_1, i_2, t) + (i_2 + 1)\sigma_2 P_0(i_1, i_2 + 1, t) + \lambda_2 P_2(i_1, i_2 - 1, t) \\
& + r_2\lambda_2 P_1(i_1 - 1, i_2, t) + r_2(i_2 + 1)\sigma_2 P_1(i_1 - 1, i_2 + 1, t).
\end{aligned}
$$

4 Equations for Partial Characteristic Function

We introduce the partial characteristic function in the following form:

$$
H_k(u_1, u_2, t) = \sum_{i_1=0}^{\infty} \sum_{i_2=0}^{\infty} e^{ju_1 i_1} e^{ju_2 i_2} P_k(i_1, i_2, t), \quad k = 0, 1, 2,
$$

where $j = \sqrt{-1}$ is imaginary unit. We rewrite the system (1) for partial characteristic function.

We can rewrite system (1) as:

$$- (\lambda_1 + \lambda_2) H_0(u_1, u_2) + j\sigma_1 \frac{\partial H_0(u_1, u_2)}{\partial u_1} + j\sigma_2 \frac{\partial H_0(u_1, u_2)}{\partial u_2} \qquad (2)$$

$$+ \mu_1 H_1(u_1, u_2) + \mu_2 H_2(u_1, u_2) = 0,$$

$$- (\lambda_1 + \lambda_2 + \mu_1) H_1(u_1, u_2) + j\sigma_2 r_2 \frac{\partial H_1(u_1, u_2)}{\partial u_2} - j\sigma_1 e^{-ju_1} \frac{\partial H_0(u_1, u_2)}{\partial u_1}$$

$$+ (1 - r_2) \lambda_2 e^{ju_2} H_1(u_1, u_2) + \lambda_1 H_0(u_1, u_2) + \lambda_1 e^{ju_1} H_1(u_1, u_2)$$

$$+ r_1 \lambda_1 e^{ju_2} H_2(u_1, u_2) - j r_1 \sigma_1 e^{j(u_2 - u_1)} \frac{\partial H_2(u_1, u_2)}{\partial u_1} = 0,$$

$$- (\lambda_1 + \lambda_2 + \mu_2) H_2(u_1, u_2) + j\sigma_1 r_1 \frac{\partial H_2(u_1, u_2)}{\partial u_1} - j\sigma_2 e^{-ju_2} \frac{\partial H_0(u_1, u_2)}{\partial u_2}$$

$$+ (1 - r_1) \lambda_1 e^{ju_1} H_2(u_1, u_2) + \lambda_2 H_0(u_1, u_2) + \lambda_2 e^{ju_2} H_2(u_1, u_2)$$

$$+ r_2 \lambda_2 e^{ju_1} H_1(u_1, u_2) - j r_2 \sigma_2 e^{j(u_1 - u_2)} \frac{\partial H_1(u_1, u_2)}{\partial u_2} = 0.$$

We will solve system (2) using the method of asymptotic analysis under condition of long delay ($\sigma \to 0$).

5 The First-Order Asymptotic Analysis

In system (2) we make substitutions:

$\sigma_m = \sigma \gamma_m; \ \sigma = \varepsilon; u_m = \varepsilon w_m, m = 1, 2; \ H_k(u_1, u_2) = F_k(w_1, w_2, \varepsilon), \quad k = 0, 1, 2.$

We can rewrite system (2) in the following form:

$$- (\lambda_1 + \lambda_2) F_0(w_1, w_2, \varepsilon) + j\gamma_1 \frac{\partial F_0(w_1, w_2, \varepsilon)}{\partial w_1} + j\gamma_2 \frac{\partial F_0(w_1, w_2, \varepsilon)}{\partial w_2} \qquad (3)$$

$$+ \mu_1 F_1(w_1, w_2, \varepsilon) + \mu_2 F_2(w_1, w_2, \varepsilon) = 0,$$

$$- (\lambda_1 + \lambda_2 + \mu_1) F_1(w_1, w_2, \varepsilon) + j\gamma_2 r_2 \frac{\partial F_1(w_1, w_2, \varepsilon)}{\partial w_2} - j\gamma_1 e^{-j\varepsilon w_1} \frac{\partial F_0(w_1, w_2, \varepsilon)}{\partial w_1}$$

$$+ (1 - r_2) \lambda_2 e^{j\varepsilon w_2} F_1(w_1, w_2, \varepsilon) + \lambda_1 F_0(w_1, w_2, \varepsilon) + \lambda_1 e^{j\varepsilon w_1} F_1(w_1, w_2, \varepsilon)$$

$$+ r_1 \lambda_1 e^{j\varepsilon w_2} F_2(w_1, w_2, \varepsilon) - j r_1 \gamma_1 e^{j\varepsilon(w_2 - w_1)} \frac{\partial F_2(w_1, w_2, \varepsilon)}{\partial w_1} = 0,$$

$$- (\lambda_1 + \lambda_2 + \mu_2) F_2(w_1, w_2, \varepsilon) + j\gamma_1 r_1 \frac{\partial F_2(w_1, w_2, \varepsilon)}{\partial w_1} - j\gamma_2 e^{-j\varepsilon w_2} \frac{\partial F_0(w_1, w_2, \varepsilon)}{\partial w_2}$$

$$+ (1 - r_1) \lambda_1 e^{j\varepsilon w_1} F_2(w_1, w_2, \varepsilon) + \lambda_2 F_0(w_1, w_2, \varepsilon) + \lambda_2 e^{j\varepsilon w_2} F_2(w_1, w_2, \varepsilon)$$

$$+ r_2 \lambda_2 e^{j\varepsilon w_1} F_1(w_1, w_2, \varepsilon) - j r_2 \gamma_2 e^{j\varepsilon(w_1 - w_2)} \frac{\partial F_1(w_1, w_2, \varepsilon)}{\partial w_2} = 0.$$

Theorem 1. *Limiting values* $\{F_k(w_1, w_2)\}$ *of the solution* $\{F_k(w_1, w_2, \varepsilon)\}$ *of the system (3) have the following form:*

$$F_k(w_1, w_2) = R_k e^{jw_1 x_1 + jw_2 x_2},$$

where values R_0, R_1, R_2, x_1, x_2 *is the solution of the following system:*

$$-(\lambda_1 + \lambda_2 + \gamma_1 x_1 + \gamma_2 x_2) R_0 + \mu_1 R_1 + \mu_2 R_2 = 0, \tag{4}$$
$$(\lambda_1 + \gamma_1 x_1) R_0 - (\lambda_2 + \mu_1 + r_2 \gamma_2 x_2 - (1 - r_2)\lambda_2) R_1 + (r_1 \lambda_1 + r_1 \gamma_1 x_1) R_2 = 0,$$
$$(\lambda_2 + \gamma_2 x_2) R_0 + (r_2 \lambda_2 + r_2 \gamma_2 x_2) R_1 - (\lambda_1 + \mu_2 + r_1 \gamma_1 x_1 - (1 - r_1)\lambda_1) R_2 = 0,$$
$$-\gamma_1 x_1 R_0 + (\lambda_1 + r_2 \lambda_2 + r_2 \gamma_2 x_2) R_1 + (r_1 \gamma_1 x_1 + (1 - r_1)\lambda_1) R_2 = 0,$$
$$-\gamma_2 x_2 R_0 + (r_2 \gamma_2 x_2 + (1 - r_2)\lambda_2) R_1 + (\lambda_1 + r_1 \lambda_1 + r_1 \gamma_1 x_1) R_2 = 0.$$

6   The Second-Order Asymptotic Analysis

To find the asymptotic of the second order we must execute following substitute at system (2):

$$H_k(u_1, u_2) = H_k^{(2)}(u_1, u_2) \exp\left\{ j\frac{u_1}{\sigma} x_1 + j\frac{u_2}{\sigma} x_2 \right\}$$

$$\sigma_k = \gamma_k \sigma, \quad \sigma = \varepsilon^2, \quad u_k = \varepsilon w_k, \quad H_k^{(2)}(u_1, u_2) = F_k(w_1, w_2, \varepsilon).$$

We can rewrite system (2) as:

$$-(\lambda_1 + \lambda_2) F_0(w_1, w_2, \varepsilon) + j\gamma_1 \varepsilon \frac{\partial F_0(w_1, w_2, \varepsilon)}{\partial w_1} + j\gamma_2 \varepsilon \frac{\partial F_0(w_1, w_2, \varepsilon)}{\partial w_2} \tag{5}$$
$$+ \mu_1 F_1(w_1, w_2, \varepsilon) + \mu_2 F_2(w_1, w_2, \varepsilon) - \gamma_1 x_1 F_0(w_1, w_2, \varepsilon) - \gamma_2 x_2 F_0(w_1, w_2, \varepsilon) = 0,$$

$$-(\lambda_1 + \lambda_2 + \mu_1) F_1(w_1, w_2, \varepsilon) + j\gamma_2 r_2 \varepsilon \frac{\partial F_1(w_1, w_2, \varepsilon)}{\partial w_2} - r_2 \gamma_2 x_2 F_1(w_1, w_2, \varepsilon)$$

$$- j\gamma_1 e^{-j\varepsilon w_1} \varepsilon \frac{\partial F_0(w_1, w_2, \varepsilon)}{\partial w_1} + \gamma_1 x_1 e^{-j\varepsilon w_1} F_0(w_1, w_2, \varepsilon)$$

$$+ (1 - r_2)\lambda_2 e^{j\varepsilon w_2} F_1(w_1, w_2, \varepsilon) + \lambda_1 F_0(w_1, w_2, \varepsilon)$$

$$+ \lambda_1 e^{j\varepsilon w_1} F_1(w_1, w_2, \varepsilon) + r_1 \lambda_1 e^{j\varepsilon w_2} F_2(w_1, w_2, \varepsilon)$$

$$- j r_1 \gamma_1 \varepsilon e^{j\varepsilon(w_2 - w_1)} \frac{\partial F_2(w_1, w_2, \varepsilon)}{\partial w_1} + r_1 \gamma_1 x_1 e^{j\varepsilon(w_2 - w_1)} F_2(w_1, w_2, \varepsilon) = 0,$$

$$-(\lambda_1 + \lambda_2 + \mu_2) F_2(w_1, w_2, \varepsilon) + j\gamma_1 r_1 \varepsilon \frac{\partial F_2(w_1, w_2, \varepsilon)}{\partial w_1} - r_1 \gamma_1 x_1 F_2(w_1, w_2, \varepsilon)$$

$$- j\gamma_2 e^{-j\varepsilon w_2} \varepsilon \frac{\partial F_0(w_1, w_2, \varepsilon)}{\partial w_2} + \gamma_2 x_2 e^{-j\varepsilon w_2} F_0(w_1, w_2, \varepsilon)$$

$$+ (1 - r_1)\lambda_1 e^{j\varepsilon w_1} F_2(w_1, w_2, \varepsilon) + \lambda_2 F_0(w_1, w_2, \varepsilon)$$

$$+ \lambda_2 e^{j\varepsilon w_2} F_2(w_1, w_2, \varepsilon) + r_2 \lambda_2 e^{j\varepsilon w_1} F_1(w_1, w_2, \varepsilon)$$

$$- j r_2 \gamma_2 \varepsilon e^{j\varepsilon(w_1 - w_2)} \frac{\partial F_1(w_1, w_2, \varepsilon)}{\partial w_2} + r_2 \gamma_2 x_2 e^{j\varepsilon(w_1 - w_2)} F_1(w_1, w_2, \varepsilon) = 0.$$

Theorem 2. *Limiting values $\{F_k(w_1, w_2)\}$ of the solution $\{F_k(w_1, w_2, \varepsilon)\}$ of the system (5) have the following form:*

$$F_k(w_1, w_2) = R_k \Phi(w_1, w_2),$$

where values R_0, R_1, R_2, x_1, x_2 is the solution of the system (4).
We write function $\Phi(w_1, w_2)$ in the following form:

$$\Phi(w_1, w_2) = \exp\left\{ \frac{(jw_1)^2}{2} Q_{11} + \frac{(jw_2)^2}{2} Q_{22} + jw_1 jw_2 Q_{12} \right\},$$

where values Q_{11}, Q_{12}, Q_{22} is the solution of the following system:

$$Q_{11}(\gamma_1 R_0 y_0 - \gamma_1 R_0 y_1 - r_1\gamma_1 R_2 y_1 + r_1\gamma_1 R_2 y_2 - \gamma_1 R_0 - r_1\gamma_1 R_2)$$
$$+ Q_{12}(\gamma_2 R_0 y_0 + r_2\gamma_2 R_1 y_1 - \gamma_2 R_0 y_2 - r_2\gamma_2 R_1 y_2 + r_2\gamma_2 R_1)$$
$$= \lambda_1 R_1 y_1 - x_1\gamma_1 R_0 y_1 - r_1\gamma_1 x_1 R_2 y_1 + (1 - r_1)\lambda_1 R_2 y_2 + r_2\lambda_2 R_1 y_2$$
$$+ r_2\gamma_2 x_2 R_1 y_2 - \frac{1}{2}\gamma_1 x_1 R_0 - \frac{1}{2}\lambda_1 R_1 - \frac{1}{2}r_2\lambda_2 R_1 - \frac{1}{2}r_2\gamma_2 x_2 R_1$$
$$- \frac{1}{2}r_1\gamma_1 x_1 R_2 - \frac{1}{2}(1 - r_1)\lambda_1 R_2,$$

$$Q_{22}(\gamma_2 R_0 d_0 - \gamma_2 R_0 d_2 - r_2\gamma_2 R_1 d_2 + r_2\gamma_2 R_1 d_1 - \gamma_2 R_0 - r_2\gamma_2 R_1)$$
$$+ Q_{12}(\gamma_1 R_0 d_0 + r_1\gamma_1 R_2 d_2 - \gamma_2 R_0 d_2 - r_1\gamma_1 R_2 d_1 + r_1\gamma_1 R_2)$$
$$= \lambda_2 R_2 d_2 - x_2\gamma_2 R_0 d_2 - r_2\gamma_2 x_2 R_1 d_2 + (1 - r_2)\lambda_2 R_1 d_1 + r_1\lambda_1 R_2 d_1$$
$$+ r_1\gamma_1 x_1 R_2 d_1 - \frac{1}{2}\gamma_2 x_2 R_0 - \frac{1}{2}\lambda_2 R_2 - \frac{1}{2}r_1\lambda_2 R_2 - \frac{1}{2}r_1\gamma_1 x_1 R_2$$
$$- \frac{1}{2}r_2\gamma_2 x_2 R_1 - \frac{1}{2}(1 - r_2)\lambda_2 R_1,$$

$$Q_{11}(\gamma_1 R_0 z_0^{(0)} - \gamma_1 R_0 z_1^{(0)} - r_1\gamma_1 R_2 z_1^{(0)} + r_1\gamma_1 R_2 z_2^{(0)} + r_1\gamma_1 R_2)$$
$$+ Q_{12}(\gamma_2 R_0 z_0^{(0)} + r_2\gamma_2 R_1 z_1^{(0)} - \gamma_2 R_0 z_2^{(0)} - r_2\gamma_2 R_1 z_2^{(0)} + r_2\gamma_2 R_1 + \gamma_1 R_0 z_0^{(1)}$$
$$- \gamma_1 R_0 z_1^{(1)} - r_1\gamma_1 R_2 z_1^{(1)} + r_1\gamma_1 R_2 z_2^{(1)} + r_1\gamma_1 R_2 - \gamma_1 R_0 - \gamma_2 R_0)$$
$$+ Q_{22}(\gamma_2 R_0 z_0^{(1)} - \gamma_2 R_0 z_2^{(1)} - r_2\gamma_2 R_1 z_2^{(1)} + r_2\gamma_2 R_1 z_1^{(1)} + r_2\gamma_2 R_1)$$
$$= \lambda_1 R_1 z_1^{(0)} - x_1\gamma_1 R_0 z_1^{(0)} - r_1\gamma_1 x_1 R_2 z_1^{(0)} + (1 - r_1)\lambda_1 R_2 z_2^{(0)} + r_2\lambda_2 R_1 z_2^{(0)}$$
$$+ r_2\gamma_2 x_2 R_1 z_2^{(0)} + \lambda_2 R_2 z_2^{(1)} - x_2\gamma_2 R_0 z_2^{(1)} - r_2\gamma_2 x_2 R_1 z_2^{(1)} + (1 - r_2)\lambda_2 R_1 z_1^{(1)}$$
$$+ r_1\lambda_1 R_2 z_1^{(1)} + r_1\gamma_1 x_1 R_2 z_1^{(1)} + r_1\gamma_1 x_1 R_2 + r_2\gamma_2 x_2 R_1,$$

Values y_0, y_1, y_2; d_0, d_1, d_2; $z_0^{(0)}$, $z_1^{(0)}$, $z_2^{(0)}$; $z_0^{(1)}$, $z_1^{(1)}$, $z_2^{(1)}$ are the solutions of the system (6)-(9), respectively.

$$-(a_1 + a_2)y_0 + a_1 y_1 + a_2 y_2 = \lambda_1 - a_1, \tag{6}$$
$$\mu_1 y_0 - (\mu_1 + a_2 r_2)y_1 + r_2 a_2 y_2 = \lambda_1 + r_2 a_2,$$
$$\mu_2 y_0 + r_1 a_1 y_1 - (\mu_2 + a_1 r_1)y_2 = \lambda_1 - r_1 a_1.$$

$$- (a_1 + a_2)d_0 + a_1 d_1 + a_2 d_2 = \lambda_2 - a_2, \tag{7}$$
$$\mu_1 d_0 - (\mu_1 + a_2 r_2)d_1 + r_2 a_2 d_2 = \lambda_2 - r_2 a_2,$$
$$\mu_2 d_0 + r_1 a_1 d_1 - (\mu_2 + a_1 r_1)d_2 = \lambda_2 + r_1 a_1.$$

$$- (a_1 + a_2)z_0^{(0)} + a_1 z_1^{(0)} + a_2 z_2^{(0)} = \lambda_2 - a_2, \tag{8}$$
$$\mu_1 z_0^{(0)} - (\mu_1 + a_2 r_2)z_1^{(0)} + r_2 a_2 z_2^{(0)} = \lambda_2 - r_2 a_2,$$
$$\mu_2 z_0^{(0)} + r_1 a_1 z_1^{(0)} - (\mu_2 + a_1 r_1)z_2^{(0)} = \lambda_2 + r_1 a_1.$$

$$- (a_1 + a_2)z_0^{(1)} + a_1 z_1^{(1)} + a_2 z_2^{(1)} = \lambda_1 - a_1, \tag{9}$$
$$\mu_1 z_0^{(1)} - (\mu_1 + a_2 r_2)z_1^{(1)} + r_2 a_2 z_2^{(1)} = \lambda_1 + r_2 a_2,$$
$$\mu_2 z_0^{(1)} + r_1 a_1 z_1^{(1)} - (\mu_2 + a_1 r_1)z_2^{(1)} = \lambda_1 - r_1 a_1.$$

7 Numerical Realization

For example, we take the parameters of arrival processes as:

$$\lambda_1 = 3, \lambda_2 = 2.$$

If the parameters of exponential law service are fixed as follow is:

$$\mu_1 = 10, \mu_2 = 20.$$

The parameters of a random delay with duration determined by exponential distribution are fixed in following form:

$$\sigma_1 = 0.02, \sigma_2 = 0.03.$$

So as $\sigma_1 = \gamma_1 \sigma$, $\sigma_2 = \gamma_2 \sigma$, then we will take $\gamma_1 = 2$, $\gamma_2 = 3$. Probability of displacement of the customer from the server by the customer of the first type $r_1 = 1$. Probability of displacement of the customer from the server by the customer of the second type $r_2 = 1$.

We have values of asymptotic means of the number of customers in the orbits with these values of parameters

$$x_1 = 100, x_2 = 44$$

and variance

$$Q_{11} = 1.152, Q_{22} = 0.308$$

and correlation coefficient

$$r = 0.421.$$

8   Conclusion

In the paper we study the retrial queueing system $M^{(2)}$—$M^{(2)}$— 1 with r-persistent exclusion of alternative customers by method of asymptotic analysis under condition of long delay. Stationary probability distribution of server states and values of asymptotic means of the number of customers in the orbits are obtained. Two-demension marginal distribution of the number of customers in the orbit 1, in the orbit 2 is asymptotically Gaussian. We obtain the numerical realization for the condidered parameters.

Acknowledgments. The work is performed under the state order of the Ministry of Education and Science of the Russian Federation (No. 1.511.2014/K).

References

1. Artalejo, J.R.: A classified bibliography of research on retrial queues: progress in 1990–1999. Top **7**(2), 187–211 (1999)
2. Artalejo, J.R.: Accessible bibliography on retrial queues. Math. Comput. Model **30**(1–2), 1–6 (1999)
3. Artalejo, J.R.: Accessible bibliography on retrial queues: progress in 2000–2009. Math. Comput. Model **51**, 1071–1081 (2010)
4. Falin, G.I.: A survey of retrial queues. Queuing Syst. **7**, 127–167 (1990)
5. Falin, G.I., Artalejo, J.R., Martin, M.: On the single retrial queue with priority customers. Queueing Syst. **14**(3–4), 439–455 (1993)
6. Choi, B.D., Chang, Y.: Single server retrial queues with priority calls. Math. Comput. Model. **30**(3–4), 7–32 (1999)
7. Choi, B.D., Choi, K.B., Lee, Y.W.: M/G/1 retrial queueing systems with two types of calls and finite capacity. Queueing Syst. **19**, 215–229 (1995)
8. Choi, B.D., Park, K.K.: The M/G/1 retrial queue with bernoulli schedule. Queueing Syst. **7**(2), 219–227 (1990)
9. Choi, B.D., Shin, Y.W., Ahn, W.C.: Retrial queues with collision arising from unslotted CSMA/CD protocol. Queueing Syst. **11**(4), 335–356 (1992)
10. Choi, B.D., Park, K.K., Pearce, C.E.M.: An M/M/1 retrial queue with control policy and general retrial times. Queueing Syst. **14**(3–4), 275–292 (1993)
11. Rengnanathan, N., Kalayanaraman, R., Srinivasan, B.: A finite capacity single server retrial queue with two types of calls. Int. J. Inf. Manage. Sci. **13**(3), 47–56 (2002)
12. Reedy, G.V.K., Nadarajan, R.: A non-preemptive priority multi-server queueing system with general bulk service and hertergeneous arrivals. Comput. Oper. Res. **4**, 447–453 (1993)
13. Zhu, Y.J., Zhou, Z.H., Feng, Y.G.: M/G/1 retrial queue system with priority and repair. Zidonghua Xuebao/Acta Automatica Sin **34**(2), 195–201 (2008)
14. Moreno, P.: An M/G/1 retrial queue with recurrent customers and general retrial times. Appl. Math. Comput. **159**(3), 651–666 (2004)
15. Bocharov, P.P., Pavlova, O.I., Puzikova, D.A.: M/G/1r retrial queueing systems with priority of primary customers. Math. Comput. Model. **30**(3–4), 89–98 (1999)
16. DApice, C., De Simone, T., Manzo, R., Rizelian, G.: Priority service of primary customers in the M/G/1/r retrial queueing system with server searching for customers. Informacionny Processy **4**(1), 13–23 (2004)

17. Nazarov, A.A., Chernikova, J.E.: Study of RQ system M/ GI/ 1 with replacement in condition of long delay. Bull. Tomsk Polytech. Univ. Manage. Comput. Eng. Inf. **323**(5), 16–20 (2013)
18. Nazarov, A.A., Chernikova, J.E.: Research of the RQ- system M/ GI/ 1 with the priority of arriving customers by method of asymptotic cumulants. In: Proceedings of the 2nd Russian National Youth Scientific Conference with International Participation: Mathematics and Program Supporting of Information, Technical and Economical Systems, Tomsk (2014)

Determination of Loss Characteristics in Queueing Systems with Demands of Random Space Requirement

Oleg Tikhonenko$^{(\boxtimes)}$

Institute of Computer Sciences, Cardinal Stefan Wyszynski University in Warsaw,
Ul. Woycickiego 1/3, 01-938 Warsaw, Poland
oleg.tikhonenko@gmail.com

Abstract. We investigate queueing systems with demands having some random space requirements (capacities) and service times generally depending on their capacities. For such systems, we discus the problem of steady-state loss characteristics determination, calculate these characteristics for some special cases and present a way for their estimation.

Keywords: Queueing system · Demand space requirement (capacity) · Total demands capacity

1 Introduction

We consider queueing systems of $M/G/n$-type with identical servers, unbounded queue and demands of random space requirement. It means that each demand is characterized by some non-negative random indication named the demand space requirement or demand capacity ζ. We also assume in general that demand service time ξ and its space requirement ζ are dependent.

The joint distribution of ζ and ξ random variables we characterize by the joint distribution function

$$F(x,t) = \mathsf{P}\{\zeta < x, \xi < t\}.$$

Note that in this case the marginal distribution function of the random variable ζ and ξ takes the form $L(x) = \mathsf{P}\{\zeta < x\} = F(x, \infty)$ and $B(t) = \mathsf{P}\{\xi < t\} = F(\infty, t)$ consequently. Denote by $\eta(t)$ the number of demands present in the system at the time instant t. We also denote by $\sigma(t)$ the total sum of space requirements of these demands. The random process $\sigma(t)$ is called the total (demands) capacity. We assume that the values of this process are bounded by positive number V that will be named the capacity or the buffer space of the system (in our models we sometimes assume that $V = \infty$).

The buffer space is occupied by the demand at the epoch it arrives and is released entirely at the epoch it completes service. If the value V is finite, some of demands can be lost. A demand having space requirement (capacity) x, which

© Springer International Publishing Switzerland 2015
A. Dudin et al. (Eds.): ITMM 2015, CCIS 564, pp. 209–215, 2015.
DOI: 10.1007/978-3-319-25861-4_18

arrives at the epoch τ, will be admitted to the system, if $\sigma(\tau-0)+x \leq V$; then we have $\eta(\tau) = \eta(\tau-0)+1$, $\sigma(\tau) = \sigma(\tau-0)+x$. Otherwise (if $\sigma(\tau-0)+x > V$), the demand will be lost and we have $\eta(\tau) = \eta(\tau-0)$, $\sigma(\tau) = \sigma(\tau-0)$. If t is an epoch of service termination of the demand having space requirement y, we have $\eta(t) = \eta(t-0)-1$, $\sigma(t) = \sigma(t-0)-y$.

The system under consideration we shall denote by $M/G/n/(\infty, V)$. If $V = \infty$, we obtain the classical system $M/G/n/\infty$ with demands of random space requirement without losing. Let a be the rate of the demands arrival process, β_1 be the first moment of service time, $\rho = a\beta_1$. We assume that, for all systems under consideration, the following limits exist in the sense of a weak convergence: $\eta(t) \Rightarrow \eta$, $\sigma(t) \Rightarrow \sigma$, or, in other words, we can analyse a steady-state behavior of these systems.

Such systems have been used to model and solve the various practical problems occurring in the design of computer and communicating systems [1,2].

In the paper, we discus the problem of calculation and estimation of loss characteristics for the systems of $M/G/n/(\infty, V)$ type.

2   Loss Characteristics for Systems with Bounded Capacity

It is clear that the most familiar characteristic of losing in the system under consideration is the loss probability P_{loss}. Intuitively, it is a part of losing demands. This characteristic can be obtain from the following equilibrium condition [1]: the mean number of demands admitting to the system during unit of time must be equal (in steady state) to the mean number of service terminations during this time. If there are no other limitation in the system, except of the system capacity one, we can calculate P_{loss} using the relation

$$P_{loss} = 1 - \int_0^V D_V(V-x)\,\mathrm{d}L(x), \tag{1}$$

where $D_V(x)$ is the steady-state distribution function of the total demands capacity σ. If the random variables ζ and ξ are independent, we can also calculate P_{loss} for one-server system using the relation [3]

$$P_{loss} = 1 - \frac{1-p_0}{\rho}, \tag{2}$$

where $p_0 = \mathsf{P}\{\eta = 0\} = \mathsf{P}\{\sigma = 0\}$ is the steady-state probability that the system is empty.

For some systems, we can obtain explicit formulas for P_{loss}. Consider, for example, the one-server system $M/M/1/(\infty, V)$ with exponentially distributed demand space requirement. Let random variables ζ and ξ be independent. Denote by f and μ the parameters of random variables ζ and ξ distributions, respectively.

Then, we have [1]:

$$p_0 = \begin{cases} \dfrac{1-\rho}{1 - \rho e^{-(1-\rho)fV}}, & \text{if } \rho \neq 1, \\[2ex] (1 + fV)^{-1}, & \text{if } \rho = 1; \end{cases}$$

$$D_V(x) = \begin{cases} p_0[1 - \rho e^{-(1-\rho)fx}], & \text{if } x \leq V, \\ 1, & \text{if } x > V. \end{cases} \tag{3}$$

It follows from (1) and (3), or from (2) that

$$P_{loss} = \begin{cases} \dfrac{1-\rho}{e^{(1-\rho)fV} - \rho}, & \text{if } \rho \neq 1, \\[2ex] (1 + fV)^{-1}, & \text{if } \rho = 1. \end{cases}$$

The same result we obtain for the processor sharing system with from the previous example distributions of demand space requirement and demand length (which are assumed be independent).

Another example is processor sharing system $M/M/1/(\infty, V) - EPS$ with demand length proportional to its capacity ($\xi = c\zeta$, $c > 0$, where ζ has an exponential distribution with parameter f). In this case, we obtain [4]:

$$p_0 = \begin{cases} \dfrac{1-\rho}{1 - \sqrt{\rho} e^{-fV} \left[\sinh(\sqrt{\rho}fV) + \sqrt{\rho}\cosh(\sqrt{\rho}fV)\right]}, & \text{if } \rho \neq 1, \\[2ex] \dfrac{4}{3 + 2fV + e^{-2fV}}, & \text{if } \rho = 1; \end{cases}$$

$$D_V(x) = \begin{cases} \dfrac{p_0}{1-\rho} \left\{1 - \sqrt{\rho} e^{-fx} \left[\sinh(\sqrt{\rho}fx) + \sqrt{\rho}\cosh(\sqrt{\rho}fx)\right]\right\}, & \text{if } x \leq V, \\[2ex] 1, & \text{if } x > V \end{cases} \tag{4}$$

for $\rho \neq 1$, and

$$D_V(x) = \begin{cases} \dfrac{p_0}{4} \left(3 + 2fx + e^{-2fx}\right), & \text{if } x \leq V, \\[2ex] 1, & \text{if } x > V \end{cases} \tag{5}$$

for $\rho = 1$.

From (1), (4) and (5) the next relation follows:

$$P_{loss} = p_0 e^{-fV} \cosh(\sqrt{\rho}fV),$$

where $\rho = a/\mu = ac/f$.

It is clear that P_{loss} is less informative loss characteristic of the system under consideration, because it is a part of losing demands, not a part of losing information. Therefore, more informative loss characteristic is the loss probability of a unit of demand capacity, Q_{loss}, that can be determined as

$$Q_{loss} = 1 - \frac{1}{\varphi_1} \int_0^V x D_V(V - x) dL(x), \qquad (6)$$

where $\varphi_1 = \mathsf{E}\zeta = \int_0^\infty x\, dL(x)$.

Note that, for systems under consideration, the inequality $Q_{loss} \geq P_{loss}$ holds. It follows from the obvious inequality

$$\varphi_1 = \int_0^\infty \varphi_1 dL(x) = \int_0^\infty x\, dL(x) \geq \int_0^V x\, dL(x),$$

if we write out the relation for P_{loss} in the form

$$P_{loss} = 1 - \frac{1}{\varphi_1} \int_0^V \varphi_1 D_V(V - x)\, dL(x).$$

In some cases, we can also obtain the explicit formulas for Q_{loss}. Consider, for example, the above systems $M/M/1/(\infty, V)$ and $M/M/1/(\infty, V) - EPS$. For the first of them, from (3) and (6), we have:

$$Q_{loss} = \frac{p_0 e^{-fV}}{\rho} \left[(1 + \rho)e^{\rho fV} - 1 \right].$$

For the second system we obtain from (4), (5) and (6) that

$$Q_{loss} = \frac{p_0 e^{-fV}}{\sqrt{\rho}} \left[\sinh\left(\sqrt{\rho}fV\right) + \sqrt{\rho}\cosh\left(\sqrt{\rho}fV\right) \right].$$

Note that, for the second system, the relation

$$Q_{loss} = 1 - \frac{1 - p_0}{\rho}$$

holds (see [3]).

It is clear that these explicit formulas exist thanks to existence of the explicit formulas for $D_V(x)$.

3   Estimation of Loss Characteristics

Consider a steady-state system QS_∞ with Poisson entry and with no losses ($V = \infty$). Consider also the system QS_V with the same parameters, which differs from the first one in limitation of system capacity ($V < \infty$) only. It is clear that we can analyze the steady state behavior of the second system as well.

Let $D_\infty(x)$ and $D_V(x)$ be the distribution function of total demands capacity of the first and second system, respectively. We shall mark all characteristics of the system QS_∞ by lower index ∞, and all ones of the system QS_V – by lower index V. The following statement holds.

Theorem. *For arbitrary real x, the inequality $D_\infty(x) \le D_\infty(x)$ holds.*
Proof. Introduce the following notation: $\chi(t)$ is the total capacity of demands arriving to the system on the time interval $[0; t)$; $\nu(t)$ is the total capacity of demands being lost on this interval; $\mu(t)$ is the total capacity of demands being served on this interval. It is obvious that, for arbitrary system, the equality $\chi(t) = \mu(t) + \nu(t) + \sigma(t)$ holds, and, obviously, $\nu_\infty(t) = 0$. Hence, we have

$$\mu_V(t) + \nu_V(t) + \sigma_V(t) = \mu_\infty(t) + \sigma_\infty(t),$$

whereas

$$\sigma_\infty(t) - \sigma_V(t) = \mu_V(t) + \nu_V(t) - \mu_\infty(t). \tag{7}$$

Now we prove the inequality

$$\mu_V(t) + \nu_V(t) - \mu_\infty(t) \ge 0.$$

Introduce the system QS_* with only difference from the system QS_∞ that demands being lost in the system QS_V, in this system are served immediately. Its characteristics we mark by lower index $*$. It is clear that

$$\nu_*(t) = 0, \ \mu_*(t) = \mu_V(t) + \nu_V(t).$$

All demands in QS_* begin its service not later than in the system QS_∞, whereas we obtain $\mu_*(t) \ge \mu_\infty(t)$, and it follows from the relation (7) that

$$\sigma_\infty(t) - \sigma_V(t) = \mu_*(t) - \mu_\infty(t) \ge 0.$$

Thus for arbitrary $t \ge 0$ we have $\sigma_\infty(t) \ge \sigma_V(t)$.

Then, for distribution functions $D_\infty(x, t)$ and $D_V(x, t)$ of the random variables $\sigma_\infty(t)$ and $\sigma_V(t)$ consequently, we obtain $D_\infty(x, t) \le D_V(x, t)$, implying in particular that $D_\infty(x) \le D_V(x)$, as $t \to \infty$.

The theorem is proved.

From this theorem we obtain the following corollaries.

Corollary 1. *For the loss probability of a demand P_{loss}, the following inequality holds:*

$$P_{loss} = 1 - \int_0^V D_V(V - x)\, \mathrm{d}L(x) \le 1 - \int_0^V D_\infty(V - x)\, \mathrm{d}L(x) = P_{loss}^*. \tag{8}$$

Corollary 2. *For the loss probability of a unit of demand capacity Q_{loss}, the following inequality holds:*

$$Q_{loss} = 1 - \frac{1}{\varphi_1} \int_0^V x D_V(V - x)\, \mathrm{d}L(x) \le 1 - \frac{1}{\varphi_1} \int_0^V x D_\infty(V - x)\, \mathrm{d}L(x) = Q_{loss}^*. \tag{9}$$

The values P^*_{loss} and Q^*_{loss} can be interpreted as an upper boundaries for P_{loss} and Q_{loss} consequently. They can be used for estimating the system capacity that guarantees not exceeding of the values P_{loss} and Q_{loss} consequently, when the distribution function $D_\infty(x)$ is known.

Consider, for example, the system $M/M/1/(\infty, V)$ with exponentially distributed demand capacity $(L(x) = 1 - e^{-fx}, x > 0, f > 0)$ and demand service time proportional to its capacity $(\xi = c\zeta, c > 0)$, for the case of $\rho = ac/f < 1$. We cannot calculate the exact values P_{loss} and Q_{loss} in this case, but we can estimate this characteristics by calculating P^*_{loss} and Q^*_{loss}. Indeed, for similar system $M/G/1/\infty$ with unbounded system capacity, we have [1]:

$$D_\infty(x) = \begin{cases} 1 + \dfrac{\rho^2 e^{-(1-\rho)fx}}{1-2\rho} - \\ \quad - \dfrac{\rho(1-\rho)}{\sqrt{\rho(4+\rho)}}\left(\dfrac{1-b_1}{1-b_1-\rho}e^{-b_1 fx} - \dfrac{1-b_2}{1-b_2-\rho}e^{-b_2 fx}\right), & \text{if } \rho \neq \tfrac{1}{2}, \\[2ex] 1 + \tfrac{1}{9}e^{-2fx} - \tfrac{1}{3}\left(\tfrac{11}{6} + \tfrac{fx}{4}e^{-fx/2}\right), & \text{if } \rho = \tfrac{1}{2}, \end{cases}$$

where $b_1 = \dfrac{2+\rho-\sqrt{\rho(4+\rho)}}{2}$, $b_2 = \dfrac{2+\rho+\sqrt{\rho(4+\rho)}}{2}$, and, as it follows from this relation and inequalities (8), (9),

$$P^*_{loss} = \begin{cases} \dfrac{\rho(1-\rho)}{\sqrt{\rho(4+\rho)}}\left(\dfrac{e^{-b_1 fV}}{1-b_1-\rho} - \dfrac{e^{-b_2 fV}}{1-b_2-\rho}\right) - \dfrac{\rho e^{-(1-\rho)fV}}{1-2\rho}, & \text{if } \rho \neq \tfrac{1}{2}, \\[2ex] 1 + \tfrac{1}{9}e^{-2fx} - \tfrac{1}{3}\left(\tfrac{11}{6} + \tfrac{fx}{4}e^{-fx/2}\right), & \text{if } \rho = \tfrac{1}{2}; \end{cases}$$

$$Q^*_{loss} = \begin{cases} (1 + fV)e^{-fV} + \dfrac{(1+\rho fV - e^{\rho fV})e^{-fV}}{1-2\rho} + \dfrac{\rho(1-\rho)}{\sqrt{\rho(4+\rho)}} \\ \quad \times \left\{\dfrac{e^{-b_1 fV}-[1+(1-b_1)fV]e^{-fV}}{(1-b_1)(1-b_1-\rho)} - \dfrac{e^{-b_2 fV}-[1+(1-b_2)fV]e^{-fV}}{(1-b_2)(1-b_1-\rho)}\right\}, & \text{if } \rho \neq \tfrac{1}{2}, \\[2ex] \tfrac{1}{9}\left(10e^{-fV/2} - e^{-2fV}\right) + \tfrac{1}{4}fVe^{-fV/2}, & \text{if } \rho = \tfrac{1}{2}. \end{cases}$$

It is clear that we often can't calculate the explicit relation for $D_\infty(x)$. But, to estimate P_{loss}, we can calculate the steady-state first and second moments of total demands capacity in the system $M/G/1/\infty$. Denote by $\alpha_{ij} = \mathsf{E}(\zeta^i \xi^j)$ the mixed $(i+j)$th moment of the random variables ζ and ξ, $i,j = 1, 2, \dots$. Let φ_i and β_i be the ith moment of the random variable ζ and ξ consequently. Then, for the first moment $\delta_1 = \mathsf{E}\sigma$ we obtain [1]:

$$\delta_1 = a\alpha_{11} + \frac{a^2\beta_2\varphi_1}{2(1-\rho)};$$

and, for the second moment $\delta_2 = \mathsf{E}\sigma^2$ we have [1]:

$$\delta_2 = a(\alpha_{21} + a\varphi_1\alpha_{12}) + \frac{a^3\beta_2\varphi_1\alpha_{11}}{1-\rho} + \frac{a^2\beta_2\varphi_2}{2(1-\rho)} + \frac{a^3\beta_3\varphi_1^2}{3(1-\rho)} + \frac{a^4\beta_2^2\varphi_1^2}{2(1-\rho)^2}.$$

Introduce the notation $\Phi(x) = \int_0^x D_\infty(x - u)\,\mathrm{d}L(u)$. Then, we have from the inequality (8) that $P_{loss}^* = 1 - \Phi(V)$. It is obvious that $\Phi(x)$ is the distribution function of the random variable $\kappa = \sigma + \zeta$, where the random variables σ and ζ are independent. The first and second moments of the random variable κ can be calculated by the following formulas:

$$f_1 = \mathsf{E}\kappa = \delta_1 + \varphi_1, \; f_2 = \mathsf{E}\kappa^2 = \delta_2 + \varphi_2 + 2\delta_1\varphi_1.$$

We propose to approximate $\Phi(x)$ by the function $\Phi^*(x) = \frac{\gamma(p,gx)}{\Gamma(p)}$, where $\gamma(p, gx) = \int_0^{gx} t^{p-1}e^{-t}\mathrm{d}t$ is incomplete Gamma function and $\Gamma(p) = \gamma(p, \infty)$ is Gamma function. The values of the parameters p and g we choose so that the first and second moments $f_1^* = p/g$ and $f_2^* = p(p+1)/g^2$ of the approximate distribution are equal to f_1 and f_2 consequently. Hence, we obtain

$$p = \frac{f_1^2}{f_2 - f_1^2}, \; g = \frac{f_1}{f_2 - f_1^2}. \tag{10}$$

A good quality of this approximation has been confirmed by simulation. So, we can use the approximate relation $P_{loss}^* \approx 1 - \Phi^*(V)$, where the parameters p and g are determined by formulas (10).

4   Conclusions

In the paper, we investigate queueing systems with demands of random space requirements (capacity) and demand service time depending on its capacity. For this systems, we analyze loss characteristics: loss probability and probability of unit of demand capacity losing. We calculate these characteristics for some special cases and consider a possibility of their estimation. It is shown that such estimation is possible, if the distribution function of total demands capacity is known. For loss probability, we propose to estimate this characteristic by approximation of a distribution function for the sum of total demands capacity and demand space requirement with Gamma distribution function.

References

1. Tikhonenko, O.: Computer Systems Probability Analysis. Akademicka Oficyna Wydawnicza EXIT, Warsaw (2006) (in Polish)
2. Tikhonenko, O.M.: Destricted capacity queueing systems: determination of their characteristics. Autom. Remote Control. **58**(6), 969–972 (1997)
3. Morozov, E., Nekrasova, R., Potakhina, L., Tikhonenko, O.: Asymptotic analysis of queueing systems with finite buffer space. In: Kwiecień, A., Gaj, P., Stera, P. (eds.) CN 2014. CCIS, vol. 431, pp. 223–232. Springer, Heidelberg (2014)
4. Tikhonenko, O.M.: Queuing systems with processor sharing and limited resources. Autom. Remote Control. **71**(5), 803–815 (2010)

Queueing System $GI|GI|\infty$ with n Types of Customers

Ekaterina Pankratova and Svetlana Moiseeva[✉]

National Research Tomsk State University, Lenin Avenue. 36,
634050 Tomsk, Russia
pankate@sibmail.com, smoiseeva@mail.ru

Abstract. The research of the queuing system with renewal arrival process, infinite number of n different types servers and arbitrary service time distribution is proposed. Expressions for the characteristic function of the number of busy servers for different types of customers in the system under the asymptotic condition that service time infinitely grows equivalently to each type of customers are derived.

Keywords: Queuin system · Renewal arrival process · Different types servers · Arbitrary service time · Characteristic function · Asymptotic analysis

1 Introduction

The results of research of the queuing system with infinite number of servers can be found in articles of A.V. Pechinkin [1–3], A.A. Nazarov, P. Abaev, R. Razumchik [4], B. D'Auria [5], D. Baum and L. Breuer [6,7], J. Bojarovich and L. Marchenko [8], E.A. van Doorn and A.A. Jagers [9], N.G. Duffield [10], C. Fricker and M. R. Jaïbi [11], E. Girlich [12], A. K. Jayawardene and O. Kella [13], M. Parulekar and A. M. Makowski [14] and others.

Numerous studies of real flows in various subject areas, in particular, telecommunication flows and flows in economic systems led to the conclusion about the inadequacy of the classic models of flows of random events to real data. There is an interest in investigation of flows, in which the customers are not identical and therefore require fundamentally different services [23,24]. The queuing systems with heterogeneous devices include systems of parallel service, which can be found in articles of G.P. Basharin, K.E. Samuylov [15], A. Movaghar [16], M. Kargahi [17], J.A. Morrisson, C. Knessl [18], D.G. Down [19], N. Bambos, G. Michalidis [20] and others. In these works, all systems have a Poisson input and exponential service time. In the papers [21,22], systems with parallel service of MMPP and renewal arrivals with paired customers are investigated.

In this paper, we study a queueing system with renewal arrival process and heterogeneous service. The main difference between the system in the paper from the previously considered ones is that when the customer comes in the system it is marked by i-th$(i = 1, \ldots, n)$ type in order to given probabilities. Service times for customers of different types has different arbitrary distribution function.

© Springer International Publishing Switzerland 2015
A. Dudin et al. (Eds.): ITMM 2015, CCIS 564, pp. 216–225, 2015.
DOI: 10.1007/978-3-319-25861-4_19

2   Statement of the Problem

Consider the queuing system with infinite number of servers of n different types and arbitrary service time. Incoming flow is a renewal arrival process with n types of customers. Recurrent incoming flow is determined by the distribution function $A(x)$ of the lengths of the intervals between the time of occurrence of renewal arrival process. At the time of occurrence of the event in this stream only one customer comes in the system. The type of incoming customer is defined as i-type with probability p_i $(i = 1, \ldots, n)$. It is servicing during a random time having an arbitrary distribution function B_i corresponding to the type of the customer.

Set the problem of analysis of n-dimensional stochastic process $\{l_1(t), l_2(t), \ldots, l_n(t)\}$ of the number of busy servers of each type at the moment t. Incoming stream is not Poisson, therefore the n-dimensional process $\{l_1(t), l_2(t), \ldots, l_n(t)\}$ is non-Markov. Consider a $(n + 1)$-dimensional Markov process $\{z(t), l_1(t), l_2(t), \ldots, l_n(t)\}$, where $z(t)$ —the remaining time from t until the occurrence of the following event of renewal arrival process.

Denote: $\{r_1(T), \ldots, r_n(T)\}$ —the number of customers who have not completed service at time T and enrolled in at the time t, $t < T$;

$S_i(t) = P\{\tau_k^{(i)} > T - t\} = 1 - B_i(T - t)$ —the probability of non-completion of the service application type i, $(i = 1, \ldots, n)$;

$1 - S_i(t)$ —the probability of completion of the service application type i, $(i = 1, \ldots, n)$.

Let at the initial moment of time $t_0 < T$ the system is empty, i.e. $l_1(t_0) = \ldots = l_n(t_0) = 0$. Then $l_1(T) = r_1(T), \ldots, l_n(T) = r_n(T)$. Thus to study the process $\{l_1(t), \ldots, l_n(t)\}$ it is necessary to investigate the n-dimensional process $\{r_1(t), \ldots, r_n(t)\}$ at any point of time $t_0 \leq t \leq T$ and put $t = T$.

A random $(n + 1)$-dimensional process $\{z(t), r_1(t), \ldots, r_n(t)\}$ is a $(n + 1)$-dimensional non-stationary Markov chain. Write the system of Kolmogorov differential equations for the joint probability distribution $P\{z, r_1, \ldots, r_n, t\}$

$$\frac{\partial P(z, r_1, \ldots, r_n, t)}{\partial t} = \frac{\partial P(z, r_1, \ldots, r_n, t)}{\partial z} + \frac{\partial P(0, r_1, \ldots, r_n, t)}{\partial z}(A(z) - 1)$$

$$+ \frac{\partial P(0, r_1 - 1, \ldots, r_n, t)}{\partial z} p_1 S_1(t) A(z) + \ldots + \frac{\partial P(0, r_1, \ldots, r_n - 1, t)}{\partial z} p_n S_n(t) A(z)$$

$$- \frac{\partial P(0, r_1, \ldots, r_n, t)}{\partial z} A(z) \sum_{i=1}^{n} p_i S_i(t). \tag{1}$$

Introduce the characteristic function of the form:

$$H(z, u_1, \ldots, u_n, t) = \sum_{r_1=0}^{\infty} \cdots \sum_{r_n=0}^{\infty} e^{ju_1 r_1} \times \cdots \times e^{ju_n r_n} P(z, r_1, \ldots, r_n, t),$$

where $j = \sqrt{-1}$ – imaginary unit.

Using (1) write the system of differential equations for the characteristic function $H(z, u_1, \ldots, u_n, t)$

$$\frac{\partial H(z, u_1, \ldots, u_n, t)}{\partial t} = \frac{\partial H(z, u_1, \ldots, u_n, t)}{\partial z} + \frac{\partial H(0, u_1, \ldots, u_n, t)}{\partial z}(A(z) - 1)$$

$$+ \frac{\partial H(0, u_1, \ldots, u_n, t)}{\partial z} A(z) \sum_{i=1}^{n} p_i S_i(t)(e^{ju_i} - 1), \tag{2}$$

$$H(z, u_1, \ldots, u_n, t_0) = R(z),$$

where $R(z)$ - stationary probability distribution of the stochastic process $z(t)$.

3   Method of the Asymptotic Analysis

3.1   Asymptotics of the First Order

We will solve the basis equation for the characteristic function (2) in the asymptotic condition that service time on appliances growths equivalently to each other, viz. $b_i \to \infty$, where $b_i = \int_0^\infty (1 - B_i(x))dx$, $i = 1, \ldots, n$ —the average value of the service time customer such as the i-th.

Denote

$$t\varepsilon = \tau, \ t_0\varepsilon = \tau_0, \ b_i = \frac{1}{q_i\varepsilon}, \ u_i = \varepsilon x_i, \tag{3}$$

$$S_i(t) = \tilde{S}_i(\tau), \ i = 1, \ldots, n, \ H(z, u_1, \ldots, u_n, t) = F_1(z, x_1, \ldots, x_n, \tau, \varepsilon).$$

Taking into account (3) we can write (2) as

$$\varepsilon \frac{\partial F_1(z, x_1, \ldots, x_n, \tau, \varepsilon)}{\partial \tau} = \frac{\partial F_1(z, x_1, \ldots, x_n, \tau, \varepsilon)}{\partial z} \tag{4}$$

$$+ \frac{\partial F_1(0, x_1, \ldots, x_n, \tau, \varepsilon)}{\partial z}(A(z) - 1) + \frac{\partial F_1(0, x_1, \ldots, x_n, \tau, \varepsilon)}{\partial z} A(z) \sum_{i=1}^{n} p_i \tilde{S}_i(\tau)(e^{j\varepsilon x_i} - 1).$$

Lemma 1. *Limit value function $F_1(z, x_1, \ldots, x_n, \tau, \varepsilon)$ at $\varepsilon \to 0$ has the form*

$$\lim_{\varepsilon \to 0} F_1(z, x_1, \ldots, x_n, \tau, \varepsilon) = F_1(z, x_1, \ldots, x_n, \tau)$$

$$= R(z) \exp\left\{ j\lambda \sum_{i=1}^{n} p_i x_i \int_{\tau_o}^{\tau} \tilde{S}_i(w)dw \right\}, \tag{5}$$

where $\lambda = \frac{\partial R(0)}{\partial z}$.

Proof. If $\varepsilon \to 0$ in (4), then obtain:

$$\frac{\partial F_1(z, x_1, \ldots, x_n, \tau)}{\partial z} + \frac{\partial F_1(0, x_1, \ldots, x_n, \tau)}{\partial z}(A(z) - 1) = 0. \tag{6}$$

Then we look for $F_1(z, x_1, \ldots, x_n, \tau)$ as

$$F_1(z, x_1, \ldots, x_n, \tau) = R(z)\Phi_1(x_1, \ldots, x_n, \tau), \tag{7}$$

where $\Phi_1(x_1, \ldots, x_n, \tau)$ - the desired function.

If $z \to \infty$ in (4), then obtain:

$$\varepsilon \frac{\partial F_1(\infty, x_1, \ldots, x_n, \tau, \varepsilon)}{\partial \tau} = \frac{\partial F_1(0, x_1, \ldots, x_n, \tau, \varepsilon)}{\partial z} \sum_{i=1}^{n} p_i \tilde{S}_i(\tau)(e^{j\varepsilon x_i} - 1). \tag{8}$$

Expand exponents in the Eq. (8) into a Taylor series, divide the left and right side of it by ε, substitute into the received expression the function $F_1(z, x_1, \ldots, x_n, \tau)$ in the form (7) and let $\varepsilon \to 0$:

$$\frac{\partial \Phi_1(x_1, \ldots, x_n, \tau)}{\partial \tau} = j \frac{\partial R(0)}{\partial z} \Phi(x_1, \ldots, x_n, \tau) \sum_{i=1}^{n} p_i \tilde{S}_i(\tau) x_i. \tag{9}$$

Taking into account the initial condition $\Phi_1(x_1, \ldots, x_n, \tau_0) = 1$ we obtain the following expression

$$\Phi_1(x_1, \ldots, x_n, \tau) = \exp\left\{ j\lambda \sum_{i=1}^{n} p_i x_i \int_{\tau_0}^{\tau} \tilde{S}_i(w) dw \right\}. \tag{10}$$

Thus,

$$F_1(z, x_1, \ldots, x_n, \tau) = R(z) \exp\left\{ j\lambda \sum_{i=1}^{n} p_i x_i \int_{\tau_0}^{\tau} \tilde{S}_i(w) dw \right\}.$$

$\square$

Taking into account Lemma 1 and substitutions (3) we can write the asymptotic approximate equality $(\varepsilon \to 0)$:

$$H(z, u_1, \ldots, u_n, t) = F_1(z, x_1, \ldots, x_n, \tau, \varepsilon) \approx F_1(z, x_1, \ldots, x_n, \tau)$$

$$= R(z) \exp\left\{ j\lambda \sum_{i=1}^{n} p_i u_i \int_{t_0}^{t} S_i(w) dw \right\}. \tag{11}$$

For the characteristic function of process $\{l_1(t), \ldots, l_n(t)\}$ at $t = T = 0$ denote

$$h_1(u_1, \ldots, u_n) = \exp\left\{ j\lambda \sum_{i=1}^{n} p_i u_i \int_{-\infty}^{0} (1 - B_i(-w)) dw \right\}$$

$$= \exp\left\{ j\lambda \sum_{i=1}^{n} p_i u_i b_i \right\}. \tag{12}$$

The function $h_1(u_1, \ldots, u_n)$ will be called the asymptotics of the first order for the system $GI|GI|\infty$ with heterogeneous service.

Defenition 1. *The functions*

$$h_1{}^{(i)}(u_i) = Me^{ju_i l_i(t)} = h_1(0, \ldots, u_i, \ldots, 0) = \exp\{j\lambda p_i u_i b_i\}, \; i = 1, \ldots, n,$$

will be called the asymptotics of the first order for the characteristic function of the busy servers of any type in system $GI|GI|\infty$ with heterogeneous service.

Consider the asymtotics of the second order for more accurate approximation.

3.2　Asymptotics of the Second Order

Consider the function $H(z, u_1, \ldots, u_n, t)$ in the form of

$$H(z, u_1, \ldots, u_n, t) = H_2(z, u_1, \ldots, u_n, t) \exp\left\{ j\lambda \sum_{i=1}^{n} p_i u_i \int_{t_0}^{t} S_i(w)dw \right\}. \quad (13)$$

Using (13) in (2) obtain the expression for $H_2(z, u_1, \ldots, u_n, t)$:

$$\frac{\partial H_2(z, u_1, \ldots, u_n, t)}{\partial t} + H_2(z, u_1, \ldots, u_n, t) j\lambda \sum_{i=1}^{n} p_i S_i(t) u_i$$

$$= \frac{\partial H_2(z, u_1, \ldots, u_n, t)}{\partial z} + \frac{\partial H_2(0, u_1, \ldots, u_n, t)}{\partial z}(A(z) - 1) \quad (14)$$

$$+ \frac{\partial H_2(0, u_1, \ldots, u_n, t)}{\partial z} A(z) \sum_{i=1}^{n} p_i S_i(t)(e^{ju_i} - 1),$$

where $\lambda = \frac{\partial R(0)}{\partial z}$.

Substitute the following in (14):

$$t\varepsilon^2 = \tau, \; t_0\varepsilon^2 = \tau_0, \; b_i = \frac{1}{q_i\varepsilon^2}, \; u_i = \varepsilon x_i, \quad (15)$$

$$S_i(t) = \tilde{S}_i(\tau), \; i = 1, \ldots, n, \; H_2(z, u_1, \ldots, u_n, t) = F_2(z, x_1, \ldots, x_n, \tau, \varepsilon)$$

and obtain:

$$\varepsilon^2 \frac{\partial F_2(z, x_1, \ldots, x_n, \tau, \varepsilon)}{\partial \tau} + F_2(z, x_1, \ldots, x_n, \tau, \varepsilon) j\lambda\varepsilon \sum_{i=1}^{n} p_i x_i \tilde{S}_i(\tau)$$

$$= \frac{\partial F_2(z, x_1, \ldots, x_n, \tau, \varepsilon)}{\partial z} + \frac{\partial F_2(0, x_1, \ldots, x_n, \tau, \varepsilon)}{\partial z}(A(z) - 1) \quad (16)$$

$$+ \frac{\partial F_2(0, x_1, \ldots, x_n, \tau, \varepsilon)}{\partial z} A(z) \sum_{i=1}^{n} p_i \tilde{S}_i(\tau)(e^{j\varepsilon x_i} - 1).$$

Theorem 1. *Limit value function* $F_2(z, x_1, \ldots, x_n, \tau, \varepsilon)$ *at* $\varepsilon \to 0$ *has the form*

$$\lim_{\varepsilon \to 0} F_2(z, x_1, \ldots, x_n, \tau, \varepsilon) = F_2(z, x_1, \ldots, x_n, \tau)$$

$$= R(z) \exp \left\{ j^2 \left[\lambda \sum_{i=1}^{n} p_i \frac{x_i^2}{2} \int_{\tau_0}^{\tau} \tilde{S}_i(w) dw \right. \right. \tag{17}$$

$$\left. \left. + \sum_{i=1}^{n} p_i^2 x_i^2 \frac{\partial f_i(0)}{\partial z} \int_{\tau_0}^{\tau} \tilde{S}_i^2(w) dw + \sum_{i=1}^{n} \sum_{g=1, g \neq i}^{n} p_i p_g x_i x_g \int_{\tau_0}^{\tau} \tilde{S}_i(w) \tilde{S}_g(w) dw \right] \right\},$$

where $\lambda = \frac{\partial R(0)}{\partial z}$ *and functions* $f_i(z)$ *are defined by the following system of equations*

$$\frac{\partial f_i(z)}{\partial z} + \frac{\partial f_i(0)}{\partial z} (A(z) - 1) + \lambda A(z) = \lambda R(z), \ i = 1, \ldots, n. \tag{18}$$

Proof. Desirable solution of the Eq. (16) should be like the following:

$$F_2(z, x_1, \ldots, x_n, \tau, \varepsilon) = \Phi_2(x_1, \ldots, x_n, \tau)$$

$$\times \left\{ R(z) + j\varepsilon \sum_{i=1}^{n} p_i x_i f_i(z) \tilde{S}_i(\tau) \right\} + O(\varepsilon^2). \tag{19}$$

Using (19) in (16), obtain:

$$R(z) j\varepsilon \lambda \sum_{i=1}^{n} p_i x_i \tilde{S}_i(\tau) = \frac{\partial R(z)}{\partial z} + \frac{\partial R(0)}{\partial z} (A(z) - 1) \tag{20}$$

$$+ j\varepsilon \sum_{i=1}^{n} p_i x_i \tilde{S}_i(\tau) \left\{ \frac{\partial f_i(z)}{\partial z} + (A(z) - 1) \frac{\partial f_i(0)}{\partial z} + \lambda A(z) \right\} + O(\varepsilon^2).$$

Hence taking into account $\frac{\partial R(z)}{\partial z} + \frac{\partial R(0)}{\partial z} (A(z) - 1) = 0$ may earn the following system of equations for the functions $f_i(z)$, $i = 1, \ldots, n$ when $\varepsilon \to 0$:

$$\frac{\partial f_i(z)}{\partial z} + \frac{\partial f_i(0)}{\partial z} (A(z) - 1) + \lambda A(z) = \lambda R(z),$$

which coincides with (18).

Expand exponents in the Eq. (16) into a Taylor series:

$$\varepsilon^2 \frac{\partial F_2(z, x_1, \ldots, x_n, \tau, \varepsilon)}{\partial \tau} = (j\varepsilon)^2 A(z) \sum_{i=1}^{n} p_i \frac{x_i^2}{2} \tilde{S}_i(\tau) \frac{\partial F_2(0, x_1, \ldots, x_n, \tau, \varepsilon)}{\partial z}$$

$$+ (j\varepsilon) \left[A(z) \sum_{i=1}^{n} p_i x_i \tilde{S}_i(\tau) \frac{\partial F_2(0, x_1, \ldots, x_n, \tau, \varepsilon)}{\partial z} - \lambda \sum_{i=1}^{n} p_i x_i \tilde{S}_i(\tau) F_2(z, x_1, \ldots, x_n, \tau, \varepsilon) \right]$$

$$+ \frac{\partial F_2(z, x_1, \ldots, x_n, \tau, \varepsilon)}{\partial z} + (A(z) - 1) \frac{\partial F_2(0, x_1, \ldots, x_n, \tau, \varepsilon)}{\partial z} + O(\varepsilon^3).$$

Substitute into received expression (19). Since $\frac{\partial R(z)}{\partial z} + \frac{\partial R(0)}{\partial z}(A(z) - 1) = 0$ we can write

$$\varepsilon^2 \frac{\partial \Phi_2(x_1, \ldots, x_n, \tau)}{\partial \tau} R(z) = (j\varepsilon)^2 \Phi_2(x_1, \ldots, x_n, \tau)$$

$$\times \left[A(z)\lambda \sum_{i=1}^{n} p_i \frac{x_i^2}{2} \tilde{S}_i(\tau) + A(z) \sum_{i=1}^{n} p_i x_i \tilde{S}_i(\tau) \sum_{g=1}^{n} p_g x_g \tilde{S}_g(\tau) \frac{\partial f_g(0)}{\partial z} \right.$$

$$\left. - \lambda \sum_{i=1}^{n} p_i x_i \tilde{S}_i(\tau) \sum_{g=1}^{n} p_g x_g \tilde{S}_g(\tau) f_g(z) \right] + j\varepsilon \Phi(x_1, \ldots, x_n, \tau) \sum_{i=1}^{n} p_i x_i \tilde{S}_i(\tau)$$

$$\times \left[\lambda A(z) - \lambda R(z) + \frac{\partial f_i(z)}{\partial z} + (A(z) - 1)\frac{\partial f_i(0)}{\partial z} \right] + O(\varepsilon^3).$$

Using (18) we obtain the following expression:

$$\varepsilon^2 \frac{\partial \Phi_2(x_1, \ldots, x_n, \tau)}{\partial \tau} R(z) = (j\varepsilon)^2 \Phi_2(x_1, \ldots, x_n, \tau)$$

$$\times \left[A(z)\lambda \sum_{i=1}^{n} p_i \frac{x_i^2}{2} \tilde{S}_i(\tau) + A(z) \sum_{i=1}^{n} p_i x_i \tilde{S}_i(\tau) \sum_{g=1}^{n} p_g x_g \tilde{S}_g(\tau) \frac{\partial f_g(0)}{\partial z} \right. \tag{21}$$

$$\left. - \lambda \sum_{i=1}^{n} p_i x_i \tilde{S}_i(\tau) \sum_{g=1}^{n} p_g x_g \tilde{S}_g(\tau) f_g(z) \right] + O(\varepsilon^3).$$

Divide both sides of the expression (21) by ε^2 and pass to the limit provided $\varepsilon \to 0$ and $z \to \infty$:

$$\frac{\partial \Phi_2(x_1, \ldots, x_n, \tau)}{\partial \tau} = j^2 \Phi_2(x_1, \ldots, x_n, \tau)$$

$$\times \left[\lambda \sum_{i=1}^{n} p_i \frac{x_i^2}{2} \tilde{S}_i(\tau) + \sum_{i=1}^{n} p_i x_i \tilde{S}_i(\tau) \sum_{g=1}^{n} p_g x_g \tilde{S}_g(\tau) \frac{\partial f_g(0)}{\partial z} \right]. \tag{22}$$

Solution of the differential Eq. (22) corresponding to the initial condition $\Phi_2(x_1, \ldots, x_n, \tau_0) = 1$ is the function $\Phi_2(x_1, \ldots, x_n, \tau)$ of the form:

$$\Phi_2(x_1, \ldots, x_n, \tau) = \exp\left\{ j^2 \left[\lambda \sum_{i=1}^{n} p_i \frac{x_i^2}{2} \int_{\tau_0}^{\tau} \tilde{S}_i(w)dw + \sum_{i=1}^{n} p_i^2 x_i^2 \frac{\partial f_i(0)}{\partial z} \int_{\tau_0}^{\tau} \tilde{S}_i^2(w)dw \right. \right.$$

$$\left. \left. + \sum_{i=1}^{n} \sum_{g=1, g \neq i}^{n} p_i p_g x_i x_g \frac{\partial f_i(0)}{\partial z} \int_{\tau_0}^{\tau} \tilde{S}_i(w)\tilde{S}_g(w)dw \right] \right\}. \tag{23}$$

$\square$

Taking into account the approximate equations of the form

$$H_2(z, u_1, \ldots, u_n, t) = F_2(z, x_1, \ldots, x_n, \tau, \varepsilon)$$

$$\approx F_2(z, x_1, \ldots, x_n, \tau) = R(z)\Phi_2(x_1, \ldots, x_n, \tau).$$

Using (15) write expression for the function $H_2(z, u_1, \ldots, u_n, t)$:

$$H_2(z, u_1, \ldots, u_n, t) = R(z) \exp \left\{ j^2 \left[\lambda \sum_{i=1}^{n} p_i \frac{u_i^2}{2} \int_{t_0}^{t} S_i(w) dw \right. \right.$$

$$+ \sum_{i=1}^{n} p_i^2 u_i^2 \frac{\partial f_i(0)}{\partial z} \int_{t_0}^{t} S_i^2(w) dw + \sum_{i=1}^{n} \sum_{g=1, g \neq i}^{n} p_i p_g u_i u_g \frac{\partial f_i(0)}{\partial z} \int_{t_0}^{t} S_i(w) S_g(w) dw \left. \right] \right\}.$$

Then using (13) we obtain:

$$H(z, u_1, \ldots, u_n, t) = R(z) \exp \left\{ j\lambda \sum_{i=1}^{n} p_i u_i \int_{t_0}^{t} S_i(w) dw \right.$$

$$+ j^2 \left[\lambda \sum_{i=1}^{n} p_i \frac{u_i^2}{2} \int_{t_0}^{t} S_i(w) dw + \sum_{i=1}^{n} p_i^2 u_i^2 \frac{\partial f_i(0)}{\partial z} \int_{t_0}^{t} S_i^2(w) dw \right.$$

$$+ \sum_{i=1}^{n} \sum_{g=1, g \neq i}^{n} p_i p_g u_i u_g \frac{\partial f_i(0)}{\partial z} \int_{t_0}^{t} S_i(w) S_g(w) dw \left. \right] \right\}.$$

Denote

$$\int_{-\infty}^{0} S_i^2(w) dw = \int_{-\infty}^{0} (1 - B_i(-w))^2 dw = \int_{0}^{\infty} (1 - B_i(w))^2 dw = \beta_i,$$

$$\int_{-\infty}^{0} S_i(w) S_g(w) dw = \int_{-\infty}^{0} (1 - B_i(-w))(1 - B_g(-w)) dw$$

$$= \int_{0}^{\infty} (1 - B_i(w))(1 - B_g(w)) dw = \beta_{ig},$$

$$i = 1, \ldots, n, \ g = 1, \ldots, n.$$

Then for the characteristic function of the random process $\{l_1(t), l_2(t), \ldots,$
$l_n(t)\}$ $h_2(u_1, \ldots, u_n) = M e^{j \sum_{i=1}^{n} u_l l_i(T)} = H(\infty, u_1, \ldots, u_n, T)$ at $t = T = 0$ and
$t_0 \to -\infty$ we obtain

$$h_2(u_1, \ldots, u_n) = \exp \left\{ j\lambda \sum_{i=1}^{n} p_i u_i b_i + j^2 \left[\lambda \sum_{i=1}^{n} p_i \frac{u_i^2}{2} b_i \right. \right.$$

$$+ \sum_{i=1}^{n} p_i^2 u_i^2 \frac{\partial f_i(0)}{\partial z} \beta_i + \sum_{i=1}^{n} \sum_{g=1, g \neq i}^{n} p_i p_g u_i u_g \frac{\partial f_i(0)}{\partial z} \beta_{ig} \left. \right] \right\}. \tag{24}$$

The expression (24) will be called the asymptotics of the second order for the system $GI|GI|\infty$ with heterogeneous service.

4　Conclusion

In this paper, we construct and investigate the mathematical model of the queuing system with the renewal arrival process and heterogeneous service. The system under consideration is studied using asymptotic analysis. Namely, the expression for the asymptotic of the first and the second order are obtained for the characteristic function of the busy servers of each type.

Acknowledgments. The work is performed under the state order of the Ministry of Education and Science of the Russian Federation (No. 1.511.2014/K).

References

1. Pechinkin, A.V.: Boundary of change of stationary queue in queuing systems with various service disciplines. In: Proceedings of the Seminar "Problems of Stability of Stochastic Models", 109. All-Union Scientific Research Institute for System Studies, Moscow, pp. 118–121 (1985) (in Russian)
2. Pechinkin, A.V.: The inversion procedure with probabilistic priority in queuing system with extraordinary incoming flow. Stochastic processes and their applications. Mathematical research, Shtiintsa, Kishinev (1989) (in Russian)
3. Pechinkin, A.V., Sokolov, I.A., Chaplygin, V.V.: Stationary characteristics ofmultiline queuing system with simultaneous failures of devices. Comput. Sci. Appl. **1**(2), 28–38 (2007). (in Russian)
4. Abaev, P.: On mean return time in queueing system with constant service time and bi-level hysteric policy. In: Modern Probabilistic Methods for Analysis and Optimization of Information and Telecommunication Networks. Proceedings of the International Conference, Minsk, pp. 11–19 (2013)
5. Auria, B.D.: $M|M|\infty$ queues in semi-Markovian random environment. Queueing Syst. **58**(3), 221–237 (2008)
6. Baum, D.: The infinite server queue with Markov additive arrivals in space. In: Probabilistic Analysis of Rare Events. Proceedings of the International Conference, Riga, pp. 136–142 (1999)
7. Baum, D., Breuer, L.: The Inhomogeneous $BMAP|G|\infty$ queue. In: Proceedings of the 11th GI/ITG Conference on Measuring, Modelling and Evaluation of Computer and Communication Systems (MMB 2001), Aachen, pp. 209–223 (2001)
8. Bojarovich, J., Marchenko, L.: An open queueing network with temporarily nonactive customers and rounds. In: Proceedings of the International Conference "Modern Probabilistic Methods for Analysis and Optimization of Information and Telecommunication Networks", Minsk, pp. 33–36 (2013)
9. Doorn, E.A., Jagers, A.A.: Note on the $GI|GI|\infty$ system with identical service and interarrival-time distributions. J. Queueing Syst. **47**, 45–52 (2004)
10. Duffield, N.G.: Queueing at large resources driven by long-tailed $M|G|\infty$-modulated processes. Queueing Syst. **28**(1–3), 245–266 (1998)
11. Fricker, C., Jaïbi, M.R.: On the fluid limit of the $M|G|\infty$ queue. Queueing Syst. **56**(3–4), 255–265 (2007)
12. Girlich, E., Kovalev, M., Listopad, N.: Optimal choice of the capacities of telecommunication networks to provide QoS-Routing. In: Proceedings of the International Conference "Modern Probabilistic Methods for Analysis and Optimization of Information and Telecommunication Networks", Minsk, pp. 93–104 (2013)

13. Jayawardene, A.K., Kella, O.: $M|G|\infty$ with alternating renewal breakdowns. Queueing Syst. **22**(1–2), 79–95 (1996)
14. Parulekar, M., Makowski, A.M.: Tail probabilities for $M|G|\infty$ input processes: I. Preliminary asymptotics. Queueing Syst. **27**(3–4), 271–296 (1997)
15. Basharin, G.P., Samouylov, K.E., Yarkina, N.V., Gudkova, I.A.: A new stage in mathematical teletraffic theory. Autom. Remote Contr. **70**(12), 1954–1964 (2009)
16. Movaghar, A.: Analysis of a dynamic assignment of impatient customers to parallel queues. Queueing Syst. **67**(3), 251–273 (2011)
17. Kargahi, M., Movaghar, A.: Utility accrual dynamic routing in real-time parallel systems. In: Transactions on Parallel and Distributed Systems (TDPS), vol. 21(12), pp. 1822–1835. IEEE (2010)
18. Knessl, C.A., Morrison, J.: Heavy traffic analysis of two coupled processors. Queueing Syst. **43**(3), 173–220 (2003)
19. Down, D.G., Wu, R.: Multi-layered round robin routing for parallel servers. Queueing Syst. **53**(4), 177–188 (2006)
20. Bambos, N., Michailidis, G.: Queueing networks of random link topology: stationary dynamics of maximal throughput schedules. Queueing Syst. **50**(1), 5–52 (2005)
21. Ivanovskaya (Sinyakova), I., Moiseeva, S.: Investigation of the queuing system $MMP^{(2)}|M_2|\infty$ by method of the moments. In: Proceedings of the Third International Conference "Problems of Cybernetics and Informatics", Baku, vol. 2, pp. 196–199 (2010)
22. Sinyakova, I., Moiseeva, S.: Investigation of queuing system $GI^{(2)}|M_2|\infty$. In: Proceedings of the International Conference "Modern Probabilistic Methods for Analysis and Optimization of Information and Telecommunication Networks", Minsk, pp. 219–225 (2011)
23. Pankratova, E., Moiseeva, S.: Queueing system $MAP|M|\infty$ with n types of customers. In: Proceedings of the 13th International Science Conference, ITMM 2014 named after A.F.Terpugov, Anzhero-Sudzhensk, pp. 356–366 (2014)
24. Pankratova, E., Moiseeva, S.: Queueing system with renewal arrival process and two types of customers. In: Ultra Modern Telecommunications and Control Systems and Workshops (ICUMT), pp. 514–517. IEEE (2014)

Performance Analysis of Unreliable Queue with Back-Up Server

Valentina Klimenok[1], Alexander Dudin[1]([✉]), and Vladimir Vishnevsky[2]

[1] Department of Applied Mathematics and Computer Science,
Belarusian State University, 220030 Minsk, Belarus
{klimenok,dudin}@bsu.by
[2] Institute of Control Sciences of Russian Academy of Sciences and Closed
Corporation "Information and Networking Technologies", Moscow, Russia
vishn@inbox.ru

Abstract. We consider an unreliable single-server queueing system with a reserve (back-up) server which is appropriate for modeling, e.g., the hybrid communication systems having the atmospheric optic channel (FSO-Free Space Optics) and standby radio channel IEEE 802.11n. We assume that the main server of the system (optic channel, server 1) is unreliable while the reserve server (radio channel, server 2) is absolutely reliable. It is assumed that the service time on the server 1 is essentially smaller than the service time on the server 2. If the server 1 fails during the service of a customer, the customer immediately occupies the reserve server. After failure occurence, the server 1 immediately goes to repair. When the repair period of the server 1 ends and the customer is served by the server 2, the customer immediately moves to the server 1. We consider exponential distributions of all random variables describing the operation of the system. This assumption allows to minimize the use of matrix-geometric technique and provide more or less simple analytical analysis of the system. We derive ergodicity condition and determine the stationary distribution of the Markov chain describing the process of the system states using the generating functions and roots methods, calculate some key performance measures and derive an expression for the Laplace-Stieltjes transform of the sojourn time distribution of an arbitrary customer. Illustrative numerical results are presented.

Keywords: Unreliable queueing system · Stationary state distribution · Sojourn time distribution

1 Introduction

Queueing theory is the well recognized mathematical tool for solving the important problems in logical and technical design, capacity planning, performance evaluation, statical and dynamic optimization of many real world objects and processes, especially in telecommunications, manufacturing, computer engineering, etc. Essential feature of the majority of real life systems and devices is their

A. Dudin et al. (Eds.): ITMM 2015, CCIS 564, pp. 226–239, 2015.
DOI: 10.1007/978-3-319-25861-4_20

unreliability, i.e., their ability to fail at a random moment and require repair or replacement. So, the existing literature in theory of unreliable queueing systems is huge. For recent references see, e.g., paper [1]. Because the failure of a server may lead to the break in providing the service to customers, reservation of the server is highly desirable. So, complementary to the main server, some back-up server should be ready to resume the service when the main server breaks down. Such situation occurs, e.g., in modelling hybrid communication systems where the ultra-high speed atmospheric optic channel (FSO-Free Space Optics) has a millimeter-wave (71–7 GHz, 81–86 GHz) radio channel as a backup communication channel for situations of the unfavorable weather conditions, e.g., fog or mist, see [2]. This model suits also for description of processing of some low priority flow of customers with occasional possibility of use additional bandwidth of a channel at periods when the high priority customers are absent in the system.

   In this short paper, we analyse an unreliable single-server queueing system with a reliable reserve (back-up) server which suits for modelling operation of the hybrid communication systems. The rest of the paper consists of the following. Mathematical model is described in Sect. 2. The process of the system states is described in Sect. 3. Necessary and sufficient condition for ergodicity of this process is presented. Stationary distribution of the process states is derived in the form of generating functions. Problem of computation of stationary distribution of sojourn time of an arbitrary customer in the system is solved in terms of Laplace-Stieltjes transform in Sect. 4. Illustrative numerical examples are presented in Sect. 5. Section 6 concludes the paper.

2   Mathematical Model

We consider a queueing system consisting of an infinite buffer and two servers: the main working server (server 1) and the back-up (reserve) server (server 2). We assume that the server 1 is unreliable while the server 2 is absolutely reliable. The input flow is defined by the stationary Poisson arrival process with parameter λ. The service times on the server 1 and on the server 2 are exponentially distributed with parameters μ_1 and μ_2, respectively. It is assumed that the server 1 is much high-speed than the server 2. A customer, which arrives into the system and sees idle and fault- free server 1, immediately occupies this server. If the server is busy, the customer is placed at the end of the queue in the buffer and is picked-up for a service later on, according the FIFO discipline. If at the arrival epoch the server 1 is under repair and the server 2 is idle, the latter server begins providing the service to the customer. If the server 2 is busy, the customer joins the buffer.

   The server 1 is subject to breakdowns and repairs. The flow of breakdowns is defined as the stationary Poisson process with parameter h. When the server fails, the repair period starts immediately. This period has exponential distribution with parameter τ. Breakdowns arriving during the repair time are ignored by the system. A customer whose service was interrupted moves to server 2 and this server starts its service. If the service of a customer is not finished until the

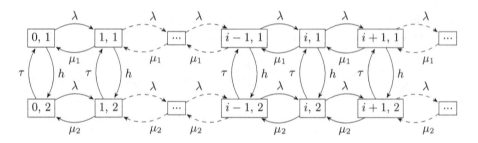

Fig. 1. State space and transitions of the Markov chain ξ_t

end of repair period, the customer returns on server 1 and this server starts its service.

Note, that, due the memoryless property of exponential distribution, the above scenario of customer service after the transition from server 1 to server 2 and back describes both so called "hot" and "cold" kinds of redundancy.

3   Process of the System States: Stationary Distribution

Let

- i_t be the number of calls in the system at the moment t, $i_t \geq 0$;
- $n_t = \begin{cases} 1, & \text{if the server 1 is fault-free (it is idle or working) at the moment } t; \\ 2, & \text{if the server 1 is under repair and the server 2 serves a customer} \end{cases}$

at the moment t.

It is easy to see that the process of the system states

$$\xi_t = \{i_t, n_t\},\ t \geq 0,$$

is a regular irreducible continuous time Markov chain.

The state space and transitions of the chain are shown in Fig. 1.

Let $-Q_{(i,n),(i,n)}$ be the parameter of exponentially distributed time of a stay of the process ξ_t in the state $(i,n), i \geq 0, n = 1, 2$, and $Q_{(i,n),(i',n')}$ be the intensity of transition of the Markov chain ξ_t from the state (i, n) to the state (i', n'). Let

$$Q_{i,j} = \begin{pmatrix} Q_{(i,1),(j,1)} & Q_{(i,1),(j,2)} \\ Q_{(i,2),(j,1)} & Q_{(i,2),(j,2)} \end{pmatrix}.$$

Lemma 1. Infinitesimal generator $Q = (Q_{i,j})$ of the Markov chain ξ_t can be represented in the block matrix form as

$$Q = \begin{pmatrix} Q_{0,0} & Q_{0,1} & O & O & \cdots \\ Q_{-1} & Q_0 & Q_1 & O & \cdots \\ O & Q_{-1} & Q_0 & Q_1 & \cdots \\ O & O & Q_{-1} & Q_0 & \cdots \\ \vdots & \vdots & \vdots & \vdots & \ddots \end{pmatrix}$$

where non-zero blocks are defined by formulas

$$Q_{0,0} = \begin{pmatrix} -(\lambda+h) & h \\ \tau & -(\lambda+\tau) \end{pmatrix}, \ Q_{0,1} = \begin{pmatrix} \lambda & 0 \\ 0 & \lambda \end{pmatrix},$$

$$Q_{-1} = \begin{pmatrix} \mu_1 & 0 \\ 0 & \mu_2 \end{pmatrix}, \ Q_0 = \begin{pmatrix} -(\lambda+\mu_1+h) & h \\ \tau & -(\lambda+\mu_2+\tau) \end{pmatrix}, \ Q_1 = \begin{pmatrix} \lambda & 0 \\ 0 & \lambda \end{pmatrix}.$$

It follows from the form of the generator that the chain under consideration belongs to the class of Quasi Birth-and- Death process, see, e.g., [3].

Theorem 1. The necessary and sufficient condition for existence of the stationary distribution of the Markov chain ξ_t is the fulfillment of the inequality

$$\lambda < \frac{\tau}{h+\tau}\mu_1 + \frac{h}{h+\tau}\mu_2. \tag{1}$$

Proof. It follows from [3], that a necessary and sufficient condition for existence of the stationary distribution of the chain ξ_t, $t \geq 0$, can be formulated in terms of the blocks of the generator Q and has the form of the inequality

$$\mathbf{x}Q_{-1}\mathbf{e} > \mathbf{x}Q_1\mathbf{e} \tag{2}$$

where the row vector $\mathbf{x}$ is the unique solution of the system of linear algebraic equations

$$\mathbf{x}(Q_{-1} + Q_0 + Q_1) = \mathbf{0}, \ \mathbf{x}\mathbf{e} = 1$$

and $\mathbf{e}$ denotes a unit column vector.

It is easily verified by the direct substitution that the solution of this system is as follows:

$$\mathbf{x} = (\frac{\tau}{h+\tau}\mu_1, \ \frac{h}{h+\tau}\mu_2). \tag{3}$$

Using (3) in (2), we easily reduce inequality (2) to the form (1). Theorem is proved.

In what follows we assume inequality (1) be fulfilled.

Denote the stationary state probabilities of the chain ξ_t, $t \geq 0$, as

$$\alpha_i = \lim_{t \to \infty} P\{i_t = i, n_t = 1\}, \ \beta_i = \lim_{t \to \infty} P\{i_t = i, n_t = 2\}, \ i \geq 0.$$

The system of balance (Chapman-Kolmogorov) equations for these probabilities is evidently written as follows:

$$\alpha_0(\lambda+h) = \beta_0\tau + \alpha_1\mu_1, \tag{4}$$

$$\beta_0(\lambda+\tau) = \beta_1\mu_2 + \alpha_0h, \tag{5}$$

$$\alpha_i(\lambda+h+\mu_1) = \alpha_{i+1}\mu_1 + \beta_i\tau + \alpha_{i-1}\lambda, \ i > 0, \tag{6}$$

$$\beta_i(\lambda+\tau+\mu_2) = \beta_{i+1}\mu_2 + \alpha_ih + \beta_{i-1}\lambda, \ i > 0. \tag{7}$$

Complementing these equations by the normalization equation

$$\sum_{i=0}^{\infty}(\alpha_i + \beta_i) = 1 \tag{8}$$

we get the system of linear algebraic equations that has the unique solution.

To calculate this solution, we will use the generating functions method. Introduce the generating functions

$$\alpha(z) = \sum_{i=0}^{\infty}\alpha_i z^i, \ \beta(z) = \sum_{i=0}^{\infty}\beta_i z^i, \ |z| \le 1.$$

Theorem 2. The generating functions $\alpha(z)$ and $\beta(z)$ have the following form:

$$\alpha(z) = \frac{\mu_1 \alpha_0 (1 - z) - \tau\beta(z) z}{\lambda z^2 - (\lambda + \mu_1 + h) z + \mu_1}, \tag{9}$$

$$\beta(z) = [\alpha_0 h\mu_1 z - \beta_0\mu_2 (\lambda z^2 - z (h + \lambda + \mu_1) + \mu_1)]/[\lambda^2 z^3 - (h\lambda + \lambda^2$$
$$+ \lambda\mu_1 + \lambda\mu_2 + \lambda\tau)z^2 + (h\mu_2 + \lambda\mu_1 + \lambda\mu_2 + \mu_1\mu_2 + \mu_1\tau) z - \mu_1\mu_2] \tag{10}$$

where the probabilities α_0 and β_0 are expressed as follows:

$$\alpha_0 = \frac{(\lambda\sigma^2 - (\lambda + \mu_1 + h)\sigma + \mu_1)(h\mu_2 - \lambda(h + \tau) + \mu_1\tau)}{\mu_1 (1 - \sigma)(h + \tau)(\mu_1 - \lambda\sigma)}, \tag{11}$$

$$\beta_0 = \frac{\alpha_0 h\mu_1\sigma}{\lambda\sigma^2 - (\lambda + \mu_1 + h)\sigma + \mu_1}, \tag{12}$$

and the quantity σ is the unique solution of the cubic equation

$$\lambda^2 z^3 - (h\lambda + \lambda^2 + \lambda\mu_1 + \lambda\mu_2 + \lambda\tau) z^2 + (h\mu_2 + \lambda\mu_1 + \lambda\mu_2$$

$$+ \mu_1\mu_2 + \mu_1\tau)z - \mu_1\mu_2 = 0 \tag{13}$$

in the region $|z| < 1$.

Proof. Multiplying Eqs. (4)–(6) by the corresponding degrees of z and summing over i, we obtain the equation

$$\alpha_0 (\lambda + h) + (\lambda + \mu_1 + h) \sum_{i=1}^{\infty}\alpha_i z^i = \alpha_1\mu_1 + \beta_0\tau + \lambda \sum_{i=1}^{\infty}\alpha_{i-1}z^i + \mu_1 \sum_{i=1}^{\infty}\alpha_{i+1}z^i + \tau \sum_{i=1}^{\infty}\beta_i z^i.$$

Similarly, we use Eqs. (5)–(7) to get the following equation:

$$\beta_0 (\lambda + \tau) + (\lambda + \mu_2 + \tau) \sum_{i=1}^{\infty}\beta_i z^i = \alpha_0 h + \beta_1\mu_2 + \lambda \sum_{i=1}^{\infty}\beta_{i-1}z^i + \mu_2 \sum_{i=1}^{\infty}\beta_{i+1}z^i + h \sum_{i=1}^{\infty}\alpha_i z^i.$$

After some algebra these two equations take the following form:

$$\left((\lambda + \mu_1 + h) z - \mu_1 - \lambda z^2\right) \alpha(z) = \tau \beta(z) z + \mu_1 \alpha_0 (z-1), \tag{14}$$

$$\left((\lambda + \mu_2 + \tau) z - \mu_2 - \lambda z^2\right) \beta(z) = h \alpha(z) z + \mu_2 \beta_0 (z-1). \tag{15}$$

Excluding the function $\alpha(z)$ from Eqs. (14)–(15), we get $\beta(z)$ as a rational function whose numerator and denominator have multiplier $(z-1)$. After division, we obtain expression (10) for $\beta(z)$. Equation (9) immediately follows from relation (14).

Now we focus on the problem of calculating the unknown probabilities α_0, β_0. We will solve this problem using the reasonings of analyticity of a generating function in the unit disk $|z| < 1$ of the complex plane.

Let us consider the function $\beta(z)$ given by formula (10). The denominator of this function is a third order polynomial. It can be shown that, under fulfillment of ergodicity condition (1), this polynomial has the unique and positive real root in the unit disk of the complex plane. Denote this root as σ, $0 < \sigma < 1$. As $\beta(z)$ is an analytical function, the numerator of (10) must vanish in the point $z = \sigma$. Equating the numerator to zero, we derive expression (12) for β_0.

It follows from (9) that

$$\alpha(1) = \frac{\tau}{h} \beta(1). \tag{16}$$

Using (16) in normalizing Eq. (8), we obtain that

$$\alpha(1) = \frac{\tau}{\tau + h}, \quad \beta(1) = \frac{h}{\tau + h}. \tag{17}$$

Using (12) and (17) in (10) under $z = 1$, we derive that probability α_0 is of the form (11). Theorem is proved.

Remark 1. Balance Eqs. (4)–(7) can be derived using the conservation principle of flows in the graph depicted in Fig. 1. Namely, they can be derived by cutting the graph via isolation of the nodes of the graph and equating the flow arriving to any node (this flow is assumed to be equal to the sum of products of probabilities of the states corresponding to the nodes, from which transition to this node is possible, by the intensity of the corresponding transition) to the flow departing from this node (this flow is assumed to be equal to the product of the probability of this state by the sum of intensities of possible transition from this node). By means of another cuts of the graph, namely, via cutting the graph by the vertical lines we can derive another set of equations, namely,

$$(\alpha_i + \beta_i)\lambda = \alpha_{i+1}\mu_1 + \beta_{i+1}\mu_2, \quad i \geq 0.$$

From this system, one more equation for generating functions $\alpha(z)$ and $\beta(z)$ is easily derived in the form

$$\alpha(z)(\mu_1 - \lambda z) + \beta(z)(\mu_2 - \lambda z) = \alpha_0 \mu_1 + \beta_0 \mu_2.$$

From this relation, alternative to formula (9) expression of the generating function $\alpha(z)$ via the generating function $\beta(z)$ can be written as

$$\alpha(z) = (\mu_1 - \lambda z)^{-1}(\alpha_0\mu_1 + \beta_0\mu_2 - \beta(z)(\mu_2 - \lambda z)).$$

It is worth to note that formulas (17) are easy obtained using the cut of the graph depicted in Fig. 1 by the horizontal line.

Corollary 1. The key performance measures of the system are as follows:

- Probability that the system is empty is given by

$$P^{(0)} = \alpha(0) + \beta(0).$$

- Probability $P^{(1)}$ that, at an arbitrary time moment, the server 1 is fault-free and probability $P^{(2)}$ that the server 1 is under repair are given by

$$P^{(1)} = \frac{\tau}{\tau + h}, \quad P^{(2)} = \frac{h}{\tau + h}.$$

- Probability $P_{serv}^{(1)}$ that, at an arbitrary time moment, the server 1 serves a customer and probability $P_{serv}^{(2)}$ that the server 1 is under repair and the server 2 serves a customer are given by

$$P_{serv}^{(1)} = \alpha(1) - \alpha_0, \quad P_{serv}^{(2)} = \beta(1) - \beta_0.$$

- Mean number of customers in the system

$$EL = \alpha'(1) + \beta'(1).$$

Remark 2. Presented here expressions for probabilities $P^{(1)}$ that the server 1 is fault-free and probability $P^{(2)}$ that the server 1 is under repair make condition (1) intuitively clear. The left hand side of inequality (1) is the mean arrival rate while the right hand side is the mean departure rate from the system when it is overloaded because the value of the service rate is equal to μ_1 when the server 1 is not broken and is equal to μ_2 when this server is broken.

4   Stationary Distribution of Sojourn Time

In the system under consideration, the sojourn time essentially depends on the breakdowns arrival process and on duration of the repair period. During the sojourn time of a customer, it may repeatedly move from server 1 to server 2 and vice versa what complicates the analysis. In this section, we derive the Laplace-Stieltjes transform of the sojourn time distribution of an arbitrary customer in the system.

Let $V(t)$ be the stationary distribution function of the sojourn time and $v(s) = \int_0^\infty e^{-st}dV(t), Res \geq 0$, be the Laplace-Stieltjes transform of this distribution.

Theorem 3. The Laplace-Stieltjes transform of the sojourn time of an arbitrary customer in the system is calculated by

$$v\left(s\right) = \sum_{i=0}^{\infty} \mathbf{p}_i \mathbf{v}_i^T\left(s\right), \tag{18}$$

where

$$\mathbf{p}_i = \left(\alpha_i, \beta_i\right), \ i \geq 0,$$

and the row vectors $\mathbf{v}_i(s)$ are calculated as

$$\mathbf{v}_i\left(s\right) = \mathbf{v}_0\left(s\right)\left[A\left(s\right)\left[I - B\left(s\right)\right]^{-1}\right]^i, \ i \geq 1, \tag{19}$$

where the row vector $\mathbf{v}_0(s)$ is of the form

$$\mathbf{v}_0(s) = \left(\frac{\mu_1\left(\mu_2 + \tau + s\right) + h\mu_2}{\left(\mu_1 + h + s\right)\left(\mu_2 + \tau + s\right) - h\tau}, \right.$$
$$\left. \frac{1}{\mu_2 + \tau + s}\left[\mu_2 + \tau \frac{\mu_1\left(\mu_2 + \tau + s\right) + h\mu_2}{\left(\mu_1 + h + s\right)\left(\mu_2 + \tau + s\right) - h\tau}\right]\right), \tag{20}$$

and the matrices $A(s)$ and $B(s)$ are given by

$$A(s) = \begin{pmatrix} \frac{\mu_1}{\mu_1 + h + s} & 0 \\ 0 & \frac{\mu_2}{\mu_2 + \tau + s} \end{pmatrix}, \quad B(s) = \begin{pmatrix} 0 & \frac{\tau}{\mu_2 + \tau + s} \\ \frac{h}{\mu_1 + h + s} & 0 \end{pmatrix}. \tag{21}$$

Proof. Let $\mathbf{v}_i(s) = \left(v_i(s, 1), v_i(s, 2)\right)$ where $v_i(s, n)$, $Re \ s \geq 0$, is the Laplace-Stieltjes transform (LST) of the stationary distribution of the sojourn time of a customer which finds, upon arrival, i customers in the system and the server 1 in the state n, $i \geq 0$, $n = 1, 2$.

To derive formula (18) for $\mathbf{v}_i(s)$, we use the probabilistic interpretation of the LST, see, e.g. [4,5]. To this end, we assume that, independently of the considered system operation, the stationary Poisson flow of so called catastrophes arrives. Let s, $s > 0$, be the intensity of this flow. Then $v_i(s, n)$ can be interpreted as the probability of no catastrophe arrival during the sojourn time of an arriving customer which finds, upon arrival, i customers in the system and the server 1 in the state n, $i \geq 0$, $n = 1, 2$. This allows us to derive the expression for $v_i(s, n)$ by means of probabilistic reasonings.

Using the probabilistic interpretation of the LST, we are able to write the following system of linear algebraic equation for $v_0(s, 1)$ and $v_0(s, 2)$:

$$\begin{cases} v_0\left(s, 1\right) = \int_0^{\infty} e^{-st}e^{-ht}e^{-\mu_1 t}\mu_1 \, dt + \int_0^{\infty} e^{-st}e^{-ht}e^{-\mu_1 t} h \, dt \, v_0\left(s, 2\right), \\ \\ v_0\left(s, 2\right) = \int_0^{\infty} e^{-st}e^{-\tau t}e^{-\mu_2 t}\mu_2 \, dt + \int_0^{\infty} e^{-\tau t}e^{-st}e^{-\mu_2 t}\tau \, dt \, v_0\left(s, 1\right). \end{cases} \tag{22}$$

Integrating in (22), we obtain the system

$$\begin{cases} v_0(s,1) = \dfrac{\mu_1}{\mu_1 + h + s} + \dfrac{h}{\mu_1 + h + s} v_0(s,2), \\[3mm] v_0(s,2) = \dfrac{\mu_2}{\mu_2 + \tau + s} + \dfrac{\tau}{\mu_2 + \tau + s} v_0(s,1). \end{cases}$$

The solution of this system is as follows:

$$v_0(s,1) = \frac{\mu_1(\mu_2 + \tau + s) + h\mu_2}{(\mu_1 + h + s)(\mu_2 + \tau + s) - h\tau}, \tag{23}$$

$$v_0(s,2) = \frac{1}{\mu_2 + \tau + s} \left[\mu_2 + \tau \frac{\mu_1(\mu_2 + \tau + s) + h\mu_2}{(\mu_1 + h + s)(\mu_2 + \tau + s) - h\tau} \right]. \tag{24}$$

Formulas (23) and (24) prove validity of formula (20) in the statement of the theorem.

Now, using the probabilistic interpretation of the LST, we write the following system of linear algebraic equation for the functions $v_i(s,1)$ and $v_i(s,2)$, $i \geq 1$:

$$\begin{cases} v_i(s,1) = \dfrac{\mu_1}{\mu_1 + h + s} v_{i-1}(s,1) + \dfrac{h}{\mu_1 + h + s} v_i(s,2), \\[3mm] v_i(s,2) = \dfrac{\mu_2}{\mu_2 + \tau + s} v_{i-1}(s,2) + \dfrac{\tau}{\mu_2 + \tau + s} v_i(s,1), \ i \geq 1. \end{cases} \tag{25}$$

It is evidently easily seen that system (25) can be rewritten in the vector form as

$$\mathbf{v}_i(s) = \mathbf{v}_{i-1}(s) A(s) + \mathbf{v}_i(s) B(s), \ i \geq 1, \tag{26}$$

where the matrices $A(s)$ and $B(s)$ have form (21).

From (26), we get the following recursive formula for sequential calculation of the vectors $\mathbf{v}_i(s)$:

$$\mathbf{v}_i(s) = \mathbf{v}_{i-1}(s) A(s) [I - B(s)]^{-1}, \ i \geq 1.$$

Using this formula, we express all vectors $\mathbf{v}_i(s)$, $i \geq 1$, in terms of the vector $\mathbf{v}_0(s)$ defined by formulas (20) and (21). The resulting expression for the vectors $\mathbf{v}_i(s)$, $i \geq 1$, is given by (19).

Now, using the formula of total probability, we write the desired Laplace-Stieltjes transform of the sojourn time distribution in the form (18).

Theorem is proved.

Corollary 2. The mean sojourn time of an arbitrary customer is calculated using the formula

$$Ev = -v'(0) = -\sum_{i=0}^{\infty} \mathbf{p}_i \frac{d\mathbf{v}_i^T(s)}{ds} \Big|_{s=0}.$$

Corollary 3. The LST $v^{(n)}(s)$ of the sojourn time distribution of a customer, which finds, upon arrival, the server 1 in the state n, $n = 1, 2$, is defined by

$$(v^{(1)}(s), v^{(2)}(s)) = \sum_{i=0}^{\infty} \mathbf{v}_i(s) \begin{pmatrix} \alpha_i & 0 \\ 0 & \beta_i \end{pmatrix}.$$

Corollary 4. The mean sojourn times $Ev^{(n)}$ of customers, that find, upon arrival, the server 1 in the state n, $n = 1, 2$, are calculated using the formula

$$(Ev^{(1)}, Ev^{(2)}) = -\sum_{i=0}^{\infty} \frac{d\mathbf{v}_i(s)}{ds} \big|_{s=0} \begin{pmatrix} \alpha_i & 0 \\ 0 & \beta_i \end{pmatrix}.$$

5   Numerical Examples

In this section, we present the illustrative numerical examples which demonstrate the feasibility of the methods developed in the paper and show the behavior of the mean number of customers in the system depending on parameters of the system.

Figures 2 illustrates the behavior of the mean number of customers in the system, EL, as a function of the arrival rate λ for different values of the rate of breakdowns. In this example, the service rates and the repair rate are taken as: $\mu_1 = 4$, $\mu_2 = 2$, $\tau = 4$.

As it is anticipated, the value EL increases when the arrival rate increases and the system becomes more congested. Also, the value of EL increases when

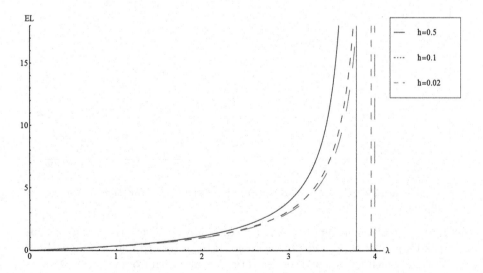

Fig. 2. The mean number of customers in the system as a function of the arrival rate for different values of the rate of breakdowns

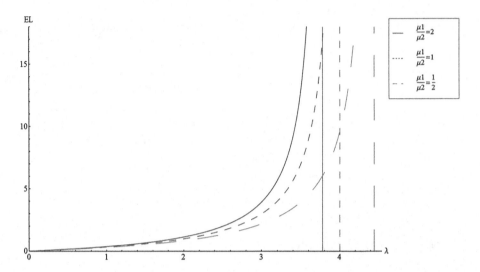

Fig. 3. The mean number of customers in the system as a function of the arrival rate for different values of the ratio $\frac{\mu_1}{\mu_2}$

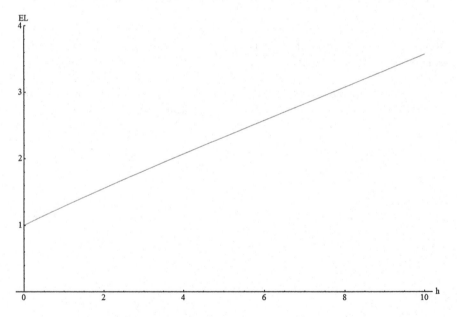

Fig. 4. The mean number of customers in the system as a function of the rate of breakdowns

the rate of breakdowns increases. This is also expected because, when the rate of breakdowns increases, a customer is forced to be serviced more frequently by the server 2 which has a lower service rate. It is also seen from Fig. 2, that every curve has an asymptote that corresponds to the values of λ and h for which the condition (1) for existence of the stationary regime in the system is violated.

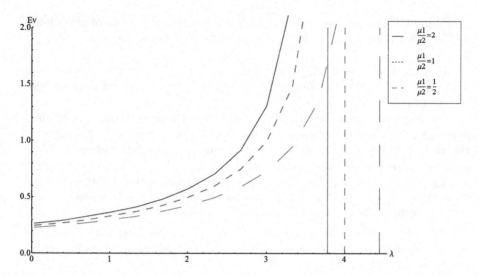

Fig. 5. The mean sojourn time as a function of the arrival rate for different values of the ratio $\frac{\mu_1}{\mu_2}$

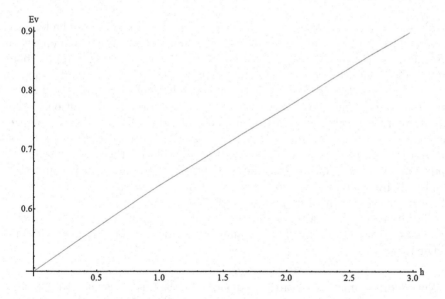

Fig. 6. The mean sojourn time as a function of the rate of breakdowns

Figure 3 shows the behavior of the value EL as a function of arrival intensity λ under different ratios of the service rates μ_1 and μ_2. In this example we assume that $\tau = 4$, $h = 0.5$. The value of service rate at the server 1 is assumed to be fixed as $\mu_1 = 4$ and the value of μ_2 is varied to get the predetermined ratio $\frac{\mu_1}{\mu_2}$. In such a situation, it is clear that decreasing of the ratio under the fixed values of μ_1 and λ implies the decreasing of the mean EL.

Figure 4 illustrates the behavior of EL as a function of the rate of breakdowns h. The other parameters of the system are fixed as follows: $\lambda = 2$, $\mu_1 = 4$, $\mu_2 = 2$, $\tau = 4$.

As it was noted above, the mean number of customers in the system increases with increasing h. More interesting conclusion is that the corresponding function is close to the linear one.

The next two figures show the behavior of the mean sojourn time Ev. Figure 5 depicts the mean sojourn times Ev as a function of λ under different values of the ratio $\frac{\mu_1}{\mu_2}$. The procedures of changing this ratio and the values of others parameters are the same as in the case of Fig. 3.

It is seen that the curves in Fig. 5 behave as the corresponding curves in Fig. 3 if the latter ones are shifted down along the y-axis of the plot. Moreover, under the fixed values of λ and the relation $\frac{\mu_1}{\mu_2}$, the values of EL and Ev are related as $Ev = \lambda^{-1}EL$. The same effect we observe comparing the graphs in Figs. 6 and 4. This reinforce our belief that Little's formula is valid for the queueing system under consideration.

6   Conclusion

Single-server queueing system with main server subject to failures and a back-up server, which is appropriate for modelling operation of the hybrid communication systems, is considered. We assume that the main server of the system (optic channel) is unreliable while the reserve server (radio channel) is absolutely reliable. All distributions describing customers service and server breakdowns and repair are exponential. This allows to get relatively simple formulas for stationary distribution of the Markov chain describing behavior of the system states using the generating functions. Formulas for computation of the Laplace-Stieltjes transform of the sojourn time distribution are presented with the partial use of the matrix analytic technique and the method of catastrophes. Illustrative numerical results giving some insight into the system behavior are presented. Correctness of Little's formula is numerically established.

Results can be extended to the cases of more general, e.g., Markov arrival processes of arrivals and breakdowns and phase type distributions of service and repair times.

Acknowledgments. The research is supported by the Russian Foundation for Basic Research (grant No. 14-07-90015) and the Belarusian Republican Foundation for Fundamental Research (grant No. F14R-126).

References

1. Krishnamoorthy, A., Pramod, P.K., Chakravarthy, S.R.: Queues with interruptions: a survey. Top **22**, 290–320 (2014)
2. Vishnevsky, V., Kozyrev, D., Semenova, O.V.: Redundant queueing system with unreliable servers. In: Proceedings of the 6th International Congress on Ultra Modern Telecommunications and Control Systems and Workshops (ICUMT), pp. 383–386 (2014)
3. Neuts, M.: Matrix-geometric Solutions in Stochastic Models - An Algorithmic Approach. Johns Hopkins University Press, Baltimore (1981)
4. Kesten, H., Runnenburg, J.T.: Priority in waiting line problems. Mathematisch Centrum, Amsterdam (1956)
5. van Danzig, D.: Chaines de Markof dans les ensembles abstraits et applications aux processus avec regions absorbantes et au probleme des boucles. Ann. de l'Inst. H. Pioncare **14**(fasc. 3), 145–199 (1955)

Compound Poisson Demand Inventory Models with Exponential Batch Size's Distribution

Anna V. Kitaeva[1]([⊠]), Valentina I. Subbotina[1],
and Oleg I. Zhukovskiy[2]

[1] Tomsk State University, Lenina Avenue 36, Tomsk, Russia
kit1157@yandex.ru, valsubbotina@mail.ru
[2] Tomsk State University of Control Systems and Radioelectronics,
Lenina Avenue 40, 634050 Tomsk, Russia

Abstract. We consider exact probability distributions of demand and selling time for compound Poisson demand process with exponential batch size's distribution in the newsvendor problem's framework and compare the results with the normal and diffusion approximation respectively. Also we receive the inventory level's distribution for on-off control of inventory level when the process of delivering the product to outlets is also a compound Poisson independent from the first one.

Keywords: Inventory management · Compound poisson demand · Exponential distribution · Distribution of inventory level · Distribution of selling time

1 The Problem Statement

Nowadays a set of stochastic models are available to solve the inventory control problem under various conditions encountered in practice, for example, see Chopra and Meindl [1], and Beyer, Cheng, Sethi, and Taksar [2]. Due to complexity of arising tasks numerical or approximate analytical methods are widely used. In the latter case the statistical simulation is needed to confirm the theoretical results.

Here we compare the exact results for an exponential batch size's distribution with the approximate ones which were used to solve the optimization tasks in Kitaeva, Subbotina, and Zmeev [3], Kitaeva, Subbotina, and Stepanova [4].

Firstly we consider the model described in Kitaeva, Subbotina, and Zmeev [3]. This model is connected with the newsvendor problem. The newsvendor problem is one of the classical problems of inventory management, see, for example, Arrow, Harris, and Marshak [5], Silver, Pyke, and Peterson [6]. It has been studied since the eighteenth century and widely used to analyse systems with perishable products in such different fields as, for example, health insurances,

V.I. Subbotina—This work is performed under the state order No. 1.511.2014/K of the Ministry of Education and Science of the Russian Federation.

airlines, sports and fashion industries. And nowadays a lot of papers related to this problem are still being published, see reviews by Khouja [7], Qin et al. [8], a handbook editing by Tsan-Ming Choi [9].

Let the demand be generated by customers arriving according a Poisson process and requiring amounts of varying size independent of the arrival process. The amounts required at each arrival (batch sizes) are independent identically distributed continuous random variables with exponential distribution

$$p_\xi(x) = \frac{1}{a} \exp\left(-\frac{x}{a}\right), x \geq 0, a > 0. \tag{1}$$

The duration of the product's lifetime is T. At the beginning of a time period T the buyer has a lot size Q.

In general case, i.e., only the first and second moments of batch sizes distribution, a_1 and a_2 respectively, are known, the normal approximation of a random customer demand at $[0, T]$ is considered in Kitaeva, Subbotina, and Zmeev [3] for fast moving items, i.e., $\lambda T = d >> 1$, and a task of selling (retail) price optimization is solved. As also shown in Kitaeva, Subbotina, and Zmeev [3] the asymptotic, i.e., $Q/a_1 = q >> 1$, distribution of the length of time τ it takes to sell the lot in general case is given by the density function

$$\tilde{u}(t) = \frac{Q}{\sqrt{2\pi a_2 \lambda} t^{3/2}} \exp\left(-\frac{a_1^2 \lambda}{2 a_2 t}\left(t - \frac{Q}{a_1 \lambda}\right)^2\right).$$

In this paper we get the exact distributions for an exponential distributed batch sizes and compare the results with the approximate ones. For exponential distribution $a_1 = a$ and $a_2 = 2a^2$.

Secondly we consider the following inventory model. Let the product flow be continuous with fixed rate ν_0, the demand be a Poisson process with constant intensity λ, the values of purchases be independent identically distributed random variables having an exponential distribution with the mean a.

Let $Q(t)$ denotes the level of inventory at time t. The storage capacity is bounded. Consider the following control of the inventory level (on-off control): if $Q(\cdot)$ is above a base-stock level Q_0 we begin to deliver the product to outlets.

Let the process of delivering be a compound Poisson with intensity λ^* and exponential batch sizes distribution with mean a^*. The process is independent from the process of customers' consumption. In this case we get the exact distribution of $Q(\cdot)$.

The general on-off control model has been investigated in Kitaeva [10], Kitaeva, Subbotina, and Zmeev [11], where the values of purchases have any distribution with known the first and second moments, and the process of delivering is deterministic with a fixed rate. The diffusion approximation of Marcovian process $Q(\cdot)$ has been considered.

2   The Newsvendor Model

Let $X(t)$ be a random customer demand at $[0, t]$, $p(\cdot)$ be a probability density function of $X(T) = X$.

Let n be a fixed number of customers at $[0, T]$, then a demand at $[0, T]$

$$X_n = \xi_1 + \xi_2 + \xi_3 + \ldots + \xi_n,$$

where ξ_i are independent identically distributed random variables with distribution (1).

Taking into account the properties of exponential distribution we have that probability density function of X_n

$$p_n(x) = \begin{cases} \delta(x), & \text{if } n = 0, \\ \dfrac{x^{n-1}}{a^n(n-1)!} e^{-x/a}, & \text{if } n \geq 1. \end{cases}$$

Then the probability density function of X

$$p(x) = \delta(x)e^{-\lambda T} + \sum_{n=1}^{\infty} \frac{x^{n-1}}{a^n(n-1)!} \cdot \frac{(\lambda T)^n}{n!} e^{-x/a - \lambda T}$$

$$= \delta(x)e^{-\lambda T} + e^{-x/a - \lambda T} \sqrt{\frac{\lambda T}{ax}} \sum_{s=0}^{\infty} \frac{1}{s!(s+1)!} \left(\frac{1}{2} \cdot 2\sqrt{\frac{\lambda T x}{a}} \right)^{2s+1}$$

$$= \delta(x)e^{-\lambda T} + e^{-x/a - \lambda T} \sqrt{\frac{\lambda T}{ax}} I_1 \left(2\sqrt{\frac{\lambda T x}{a}} \right),$$

where $I_1(\cdot)$ is the modified Bessel function of the first kind and first order, $\delta(\cdot)$ is the Dirac delta function.

Denote dimensionless quantity $\lambda T = d$ and random value $\dfrac{X}{a} = Y$. Then probability density function of Y

$$p_Y(x) = \delta(x)e^{-d} + e^{-x-d} \sqrt{\frac{d}{x}} I_1 \left(2\sqrt{xd} \right). \tag{2}$$

Compare the exact result (2) with the normal approximation used in Kitaeva, Subbotina, and Zmeev [3]: $\dfrac{X - a\lambda T}{a\sqrt{2\lambda T}}$ converges in distribution to a standard normal random variable $N(0, 1)$ as λT tends to infinity, or Y converges to $N(d, \sqrt{2d})$.

In Fig. 1 the graphics of the exact (solid line) and approximate normal (dashed line) probability density functions of the demand divided by the purchase's mean value are shown for different values of parameter $d = \lambda T$. As d increases the approximation becomes more accurate.

Let us find exact distribution $u(\cdot)$ of the length of time τ it takes to sell the lot for the model under consideration.

1. If the lot is bought completely by the first customer conditional probability density function of τ $u(t|1) = \lambda e^{-\lambda t}$, and probability of this event is $\int_{Q}^{\infty} \frac{1}{a} e^{-x/a} \, dx = e^{-Q/a}$.

2. If the lot is over when the n-th customer, $(n > 1)$, make a purchase, conditional probability density function of τ

$$u(t|n) = \frac{\lambda^n t^{n-1}}{(n-1)!}e^{-\lambda t},$$

and probability of this event is

$$\int_0^Q p_{n-1}(x)dx \int_{Q-x}^\infty p(y)dy = e^{-Q/a} \int_0^Q e^{x/a} p_{n-1}(x)dx$$

$$= e^{-Q/a} \int_0^Q e^{x/a} \frac{x^{n-2}}{a^{n-1}(n-2)!}e^{-x/a}\,dx = \frac{Q^{n-1}}{a^{n-1}(n-1)!}e^{-Q/a}.$$

So we receive

$$u(t) = \lambda e^{-\lambda t - Q/a} + \sum_{n=2}^\infty \frac{\lambda^n t^{n-1}}{(n-1)!}e^{-\lambda t} \cdot \frac{Q^{n-1}}{a^{n-1}(n-1)!}e^{-Q/a}$$

$$= \lambda e^{-\lambda t - Q/a}\left[1 + \sum_{s=1}^\infty \frac{1}{(s!)^2}\left(\frac{\lambda t Q}{a}\right)^s\right].$$

Taking into account following equation from Abramowitz and Stegun [12]

$$1 + \sum_{s=1}^\infty \frac{1}{(s!)^2}\left(\frac{\lambda t Q}{a}\right)^s = I_0\left(2\sqrt{\frac{\lambda t Q}{a}}\right),$$

we finally get

$$u(t) = \lambda e^{-\lambda t - Q/a} I_0\left(2\sqrt{\frac{\lambda t Q}{a}}\right),$$

where $I_0(\cdot)$ is the modified Bessel function of the first kind and zeroth order.

Denote dimensionless quantity $\dfrac{Q}{a} = q$ and random value $\lambda\tau = \gamma$. Probability density function of γ

$$u_\gamma(t) = e^{-t-q} I_0\left(\sqrt{2tq}\right).$$

In Kitaeva, Subbotina, and Zmeev [3] the approximate $(q \gg 1)$ probability density function of τ has been considered

$$\tilde{u}(t) = \frac{q}{2\sqrt{\pi\lambda}t^{3/2}} \exp\left(-\frac{\lambda}{4t}\left(t - \frac{q}{\lambda}\right)^2\right).$$

It follows

$$\tilde{u}_\gamma(t) = \frac{q}{2\sqrt{\pi}t^{3/2}} \exp\left(-\frac{(t-q)^2}{4t}\right).$$

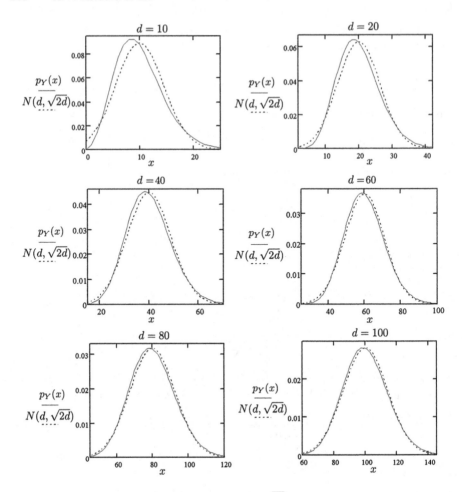

Fig. 1. The exact $p_Y(x)$ and approximate $N(d, \sqrt{2d})$ probability density functions of Y for different d.

In Fig. 2 the graphics of the exact (solid line) and approximate (dashed line) probability density functions of selling time multiplied by λ are shown for different values of parameter $q = \dfrac{Q}{a}$. As q increases the approximation becomes more accurate.

3   Random Product's Delivering

Denote stationary probability density function of $Q(\cdot)$

$$p(s) = \begin{cases} p_1(s), & s < Q_0, \\ p_2(s), & s > Q_0. \end{cases}$$

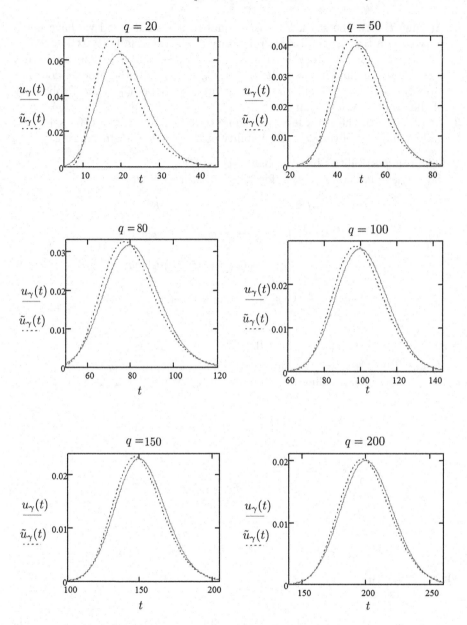

Fig. 2. The exact $\tilde{u}(t)$ and approximate $u(t)$ probability density functions of γ for different q.

We begin with the case $s < Q_0$. There are three possibilities of how stock level process $Q(\cdot)$ enters the state s during Δt:

1. At time $t - \Delta t$ the stock level was equal to $s - \nu_0 \Delta t$, and we have not a purchase on time interval Δt. Probability of the event is $1 - \lambda \Delta t + o(\Delta t)$.
2. At time $t - \Delta t$ the stock level was equal to $s - \nu_0 \Delta t + x$, and we have a purchase x on time interval Δt. If $x < Q_0 - s + \nu_0 \Delta t$ the process does not cross the base-stock level, otherwise the process crosses Q_0 at Δt. Probability of the event is $\lambda \Delta t + o(\Delta t)$.
3. At time $t - \Delta t$ the stock level was equal to $s - \nu_0 \Delta t + x$, $x > Q_0 - s + c_0 \Delta t$, and we have a delivering x. Probability of the event is $\lambda^* \Delta t + o(\Delta t)$.

The rest possibilities have probabilities $o(\Delta t)$.

So we have integral equation for Marcovian process $Q(\cdot)$

$$
\begin{aligned}
p_1(s) = {} & p_1(s - \nu_0 \Delta t)(1 - \lambda \Delta t) \\
& + \lambda \Delta t \int_0^{Q_0 - s + \nu_0 \Delta t} p_1(s - \nu_0 \Delta t + x) p_\xi(x) dx \\
& + \lambda \Delta t \int_{Q_0 - s + \nu_0 \Delta t}^{\infty} p_2(s - \nu_0 \Delta t + x) p_\xi(x) dx \\
& + \lambda^* \Delta t \int_{Q_0 - s + \nu_0 \Delta t}^{\infty} p_2(s - \nu_0 \Delta t + x) p_\eta(x) dx + o(\Delta t),
\end{aligned}
\tag{3}
$$

where $p_\eta(x) = \dfrac{1}{a^*} \exp\left(-\dfrac{x}{a^*}\right), x \geq 0$.

Using Taylor's expansion for $p_1(\cdot)$ and $p_2(\cdot)$: $p_i(s - \nu_0 \Delta t) = p_i(s) - \nu_0 \Delta t p'_i(s) + o(\Delta t)$, dividing by Δt and tending Δt to zero in (3) we receive

$$
\lambda p_1(s) + \nu_0 p'_1(s) = \lambda \int_s^{Q_0} p_1(x) p_\xi(x - s) dx
$$

$$
+ \lambda \int_{Q_0}^{\infty} p_2(x) p_\xi(x - s) dx + \lambda^* \int_{Q_0}^{\infty} p_2(x) p_\eta(x - s) dx.
$$

Taking derivative twice with respect to s we have

$$
p'''_1(s) + (b_1 - a^{-1} - a^{*-1}) p''_1(s) + a^{*-1}(a^{-1} - b_1) p'_1(s) = 0,
$$

where $b_1 = \lambda / \nu_0$.

The solution

$$
p_1(s) = C_1 \exp\left(\frac{s}{a^*}\right) + C_2 \exp((a^{-1} - b_1)s) + C_3.
\tag{4}
$$

Due to normalization condition $C_3 = 0$. Let $a^{-1} - b_1 > 0$, i.e., $\nu_0 > \lambda a$. In essence we need this condition to ensure the existence of the stationary distribution of inventory level.

Rewrite (4) in following form

$$
p_1(s) = C_1 \exp\left(\frac{s - Q_0}{a^*}\right) + C_2 \exp[(a^{-1} - b_1)(s - Q_0)], \quad s < Q_0.
$$

Let us consider the case $Q > Q_0$.

States and probabilities of the corresponding changes of process $Q(\cdot)$ during Δt are as follows:

1. At time $t - \Delta t$ the stock level was equal to $s - \nu_0 \Delta t$, and we have neither a delivering nor a purchase on time interval Δt. The probability of the event is $1 - \lambda \Delta t - \lambda^* \Delta t + o(\Delta t)$.
2. At time $t - \Delta t$ the stock level was equal to $s - \nu_0 \Delta t + x$, and we have a purchase x on time interval Δt, $x > \nu_0 \Delta t$. Probability of the event is $\lambda \Delta t + o(\Delta t)$.
3. At time $t - \Delta t$ the stock level was equal to $s - \nu_0 \Delta t + x$, and we have a delivering x, $x > \nu_0 \Delta t$. Probability of the event is $\lambda^* \Delta t + o(\Delta t)$.
   The rest possibilities have probabilities $o(\Delta t)$.

So we receive equation

$$p_2(s) = p_2(s - \nu_0 \Delta t)(1 - \lambda \Delta t - \lambda^* \Delta t)$$

$$+\lambda \Delta t \int_{\nu_0 \Delta t}^{\infty} p_2(s - \nu_0 \Delta t + x) p_\xi(x) dx$$

$$+\lambda^* \Delta t \int_{\nu_0 \Delta t}^{\infty} p_2(s - \nu_0 \Delta t + x) p_\eta(x) dx + o(\Delta t).$$

Analogously the previous case we get

$$\nu_0 p'''_2(s) + (b_1 + b_2 - a^{-1} - a^{*-1}) p''_2(s) + (a^{-1} a^{*-1} - b_1 a^{*-1} - b_2 a^{-1}) p'_2 = 0,$$

where $b_2 = \lambda^*/\nu_0$.

Solution of the above equation

$$p_2(s) = C_1 \exp(k_1 s) + C_2 \exp(k_2 s) + C_3,$$

where $k_{1,2} = (a^{-1} + a^{*-1} - b_1 - b_2 \pm \sqrt{D})/2$, $D = (a^{-1} - a^{*-1} + b_2 - b_1)^2 + 4b_1 b_2 > 0$.

Let the condition holds $\nu_0 < \lambda a + \lambda^* a^*$. It is the second condition of the stationary distribution's existence. Then $k_2 < 0$, $k_1 > 0$.

From normalization condition it follows $C_1 = C_3 = 0$. So we get

$$p_2(s) = C \exp[k_2(s - Q_0)], \quad s > Q_0.$$

Let us define constants C, C_1, C_2.

It is easy to see that

$$p_2(Q_0) = O(\Delta t) + p_1(Q_0 - \nu_0 \Delta t)(1 - \lambda \Delta t) + o(\Delta t). \tag{5}$$

From (5) the boundary condition follows: $p_1(Q_0) = p_2(Q_0)$, and we get

$$C = C_1 + C_2. \tag{6}$$

On the other hand

$$
\begin{aligned}
p_1(Q_0) = {} & p_1(Q_0 - \nu_0 \Delta t)(1 - \lambda \Delta t) \\
& + \lambda \Delta t \int_{\nu_0 \Delta t}^{\infty} p_2(Q_0 - \nu_0 \Delta t + x) p_\xi(x) dx \\
& + \lambda^* \Delta t \int_{\nu_0 \Delta t}^{\infty} p_2(Q_0 - \nu_0 \Delta t + x) p_\eta(x) dx + o(\Delta t).
\end{aligned} \tag{7}
$$

From (7), taking (6) into account, we have equation

$$a^{*-1}C_1 + b_1 C_1 + a^{-1}C_2 = \left(\frac{b_2}{1 - a^* k_2} + \frac{b_1}{1 - a k_2} \right)(C_1 + C_2). \tag{8}$$

The third equation is normalization condition: $\int_{-\infty}^{Q_0} p_1(s)ds + \int_{Q_0}^{\infty} p_2(s)ds = 1$,

or, taking (6) into account,

$$C_1 \left(a^* - \frac{1}{k_2} \right) + C_2 \left(\frac{a}{1 - ab_1} - \frac{1}{k_2} \right) = 1. \tag{9}$$

Solving system of Eqs. (6), (8), and (9) we get

$$C_2 = \frac{k_2(1 - ab_1)(1 + a^* b_1 - a^* h)}{(1 + a^* b_1 - a^* h)(ak_2 - 1 + ab_1) - a^{*2}(h - a^{-1})(1 - ab_1)(a^{-1} - k_2)},$$

$$C = \frac{k_2(1 - ab_1)(1 + a^* b_1 - a^* a^{-1})}{(1 + a^* b_1 - a^* h)(ak_2 - 1 + ab_1) - a^{*2}(ah - 1)(a^{-1} - b_1)(a^{-1} - k_2)},$$

where $h = \left(\dfrac{b_2}{1 - a^* k_2} + \dfrac{b_1}{1 - a k_2} \right)$.

So we obtain stationary probability density function of the inventory level

$$
p(s) = \begin{cases}
C_1 \exp\left(\dfrac{s - Q_0}{a^*} \right) + C_2 \exp[(a^{-1} - b_1)(s - Q_0)], & s < Q_0, \\
C \exp[k_2(s - Q_0)], & s > Q_0.
\end{cases}
$$

4   Conclusion

So we see that the approximate selling time's distribution derived from diffusion approximation of the stock level process gives good results, at least for exponentially distributed jump sizes of a compound Poisson demand.

Note that a sum of independent compound Poisson processes is also a compound Poisson process with the intensity equals the sum of the intensities and jump sizes' distribution equals the weighted sum of the corresponding jump sizes' distributions. It means that for the model described in 3-rd section of the paper the diffusion approximation considered in [10,11] can be used. The compare of the approximate and exact results for the on-off control model is the theme of the next research.

References

1. Chopra, S., Meindl, P.: Supply Chain Management. Prentice Hall, London (2001)
2. Beyer, D., Cheng, F., Sethi, S.P., Taksar, M.: Markovian Demand Inventory Models. Springer, New York (2010)
3. Kitaeva, A., Subbotina, V., Zmeev, O.: The Newsvendor problem with fast moving items and a compound poisson price dependent demand. In: 15th IFAC Symposium on Information Control Problems in Manufacturing, INCOM 2015, (IFAC-PapersOnLine), vol. 48, pp. 1375–1379. Elsevier (2015)
4. Kitaeva, A., Subbotina, V., Stepanova, N.: Estimating the compound poisson demand's parameters for single period problem for large lot size. In: 15th IFAC Symposium on Information Control Problems in Manufacturing, INCOM 2015, (IFAC-PapersOnLine), vol. 48, pp. 1357–1361. Elsevier (2015)
5. Arrow, K.J., Harris, T.E., Marschak, J.: Optimal inventory policy. Econometrica 19(3), 205–272 (1951)
6. Silver, E.A., Pyke, D.F., Peterson, R.: Inventory management and production planning and scheduling. Wiley, New York (1998)
7. Khouja, M.: The single-period (news-vendor) problem: literature review and suggestionsfor future research. OMEGA-INT J. 27(5), 537–553 (1999)
8. Qin, Y., Wang, R., Vakharia, A., Chen, Y., Hanna-Seref, M.: The newsvendor problem: review and directions for future research. Eur. J. Oper. Res. 213, 361–374 (2011)
9. Tsan-Ming, C. (ed.) Handbook of Newsvendor Problems: Models, Extensions and Applications. Springer, New York (2012)
10. Kitaeva, A.V.: Stabilization of inventory system performance: on/off control. In: the 19th World Congress of the International Federation of Automatic Control, IFAC Proceedings Volumes (IFAC-PapersOnline), vol. 19, pp. 10748–10753 (2014)
11. Kitaeva, A., Subbotina, V., Zmeev, O.: Diffusion appoximation in inventory management with examples of application. In: Dudin, A., Nazarov, A., Yakupov, R., Gortsev, A. (eds.) ITMM 2014. CCIS, vol. 487, pp. 189–196. Springer, Heidelberg (2014)
12. Abramowitz, M., Stegun, I. (eds.) Handbook of Mathematical Functions with Formulas, Graphs, and Mathematical Tables. Dover Publications, New York (1972)

On an M/G/1 Queue with Vacation in Random Environment

Achyutha Krishnamoorthy[1](✉), Jaya Sivadasan[2],
and Balakrishnan Lakshmy[1]

[1] Cochin University of Science and Technology, Kochi, India
{achyuthacusat,jayasreelakam}@gmail.com, luck@cusat.ac.in
[2] Maharaja's College, Kochi, India

Abstract. We study an $M/G/1$ queue with multiple vacation and vacation interruption. Both normal vacation (type I) and working vacation (type II) are considered. The exhaustive service discipline is assumed in this paper. At the end of a busy period, depending on the environment, the server either opts for normal vacation or working vacation. On completion of type I vacation if the server finds the system empty he goes for type II vacation. On completion of type II vacation if the server finds the system empty goes for another type II vacation. On completion of service in type II vacation, if the server finds one or more customers in queue he returns to normal service, interrupting the vacation. An arriving customer, during type I vacation, joins the queue with probability q or leaves the system with probability $1 - q$ and during type II vacation all the arriving customers join the queue. Using supplementary variable technique we derive the distributions for the queue length and service status under steady state condition. Laplace-Stieltjes transform of the stationary waiting time is also developed. Some numerical illustrations are also given.

Keywords: Multiple vacation · Random environment · Working vacation · Vacation interruption

1 Introduction

In every queueing system there is a facility providing service. If this facility is continuously available in the queueing system, the queueing system will work more efficiently. But this is an ideal condition. The service facility may break down or sometimes it may require maintenance. In such situations the server will not be available for a short duration. That duration is considered as vacation. i.e. the unavailability of server to primary customers is called vacation.

A. Krishnamoorthy—Research supported by Kerala State Council for Science, Technology & Environment (No. 001/KESS/2013/CSTE))
Jaya S—Research supported by FDP of UGC, No.F.FIP/12th Plan/KLMG009 TF-12.

If the queue is empty the server remains idle. The idle time of the server can be utilized for some other work. If the customers in the queue is less, the functioning of the server in a slow rate will reduce the operating cost, energy consumption, and the start up cost. These advantages are pointing towards working vacation. Working vacation is an extension of regular vacation. In working vacation, instead of completely stopping the service the server provides service at a slow rate. In working vacation the minor maintenance of the server can be performed successfully without the dissatisfaction of the customers. Working vacation reduces the chance of reneging of the customers compared to normal vacation. In this era of high demand for commodities and services which are available in a short spell, the concept of working vacation is very useful. This may be the main reason of the extensive research work going on in working vacation queueing models.

Vacation queueing system is one of the important research fields due to applicability in communication, network and so on. The wide application of this concept in all walks of life is the reason for the staggering development of working vacation models. For more details of vacation models, readers are referred to the survey paper by Doshi [3]. The monograph by Takagi [5] is an excellent and detailed study of vacation models. For basic concepts and detailed discussions on vacation models we refer to [8]. Different types of vacation models are discussed in the survey paper by Ke et al. [16].

Working vacation was first studied by Servi and finn [6]. They applied $M/M/1$ queue with multiple working vacation to model a wave length division multiplexing optical access network. They derived the transform formula for the distribution of the number of customers in the system and sojourn time distributions in the steady state. Li et al. [11] proved that the stochastic decomposition property of vacation queues holds for the $M/M/1$ queueing system with working vacation. [18] discuss about a queueing system with two heterogeneous servers, one of which is always available but the other goes on vacation in the absence of customers waiting for service. The vacationing server, returns to serve at a lower rate as an arrival finds the other server busy. In [19] Sreenivasan et al. discuss about a MAP/PH/1 queueing model in which the server is subject to taking vacations and offering services at a lower rate during vacation. The service is returned to normal rate whenever the vacation gets over or when the queue length hits a specific threshold value. [7,9] discuss $M/G/1$ queueing system with working vacation. In this, the normal service period, service period during vacation and vacation duration follows general distribution. Li et al. [15] extend the result for $M/M/1$ working vacation system to $M/G/1$ case. [7,9,15] are generalizations of [6]. Kim et al. [12] investigated $M/G/1$ queue with working vacation. Tian et al. [14] discuss about discrete time $Geo/Geo/1$ queue with multiple working vacation using matrix analytic method. Fuhrmann and Cooper [2] and Shanthikumar [4] establish stochastic decomposition results for classical $M/G/1$ queue with general vacation.

Li and Tian [10,20] introduced and developed the concept of vacation interruption. Vacation interruption is the coming back of server to service from vacation without completing vacation due to the arrival of customers in the queue.

In [13] performance analysis of GI/M/1 queue with working vacation and vacation interruption is dicussed by Li et al. Zhang and Hou [17] analyze an $M/G/1$ queue with working vacation and vacation interruption.

The important features of the model discussed in the present work are

- The server goes for vacation only if the queue is empty. i.e. the exhaustive discipline has been applied.
- Both normal vacation (type I vacation) and multiple working vacation (type II vacation) are considered here.
- During normal vacation, if a customer arrives, service is not provided until completion of vacation whereas while in working vacation service is provided in case customer arrives, however, at a slower rate.
- At the end of a busy period, depending on the environment, the server opts for normal vacation or working vacation.
- On completion of type I vacation, if the server finds the system empty he goes for type II vacation.
- On completion of type II vacation if the server finds the system empty he goes for another type II vacation.
- On completion of service in type II vacation if the server finds one or more customers in queue he returns to normal service interrupting the vacation.
- A customer arriving during type I vacation, joins the queue with probability q or leaves the system with probability $1 - q$.
- A customer arriving during type II vacation, joins the queue with probability 1.

The rest of the paper is arranged as follows. In Sect. 2 the model and notations are described. Stability of the system is discussed in Sect. 3. Steady state distributions are calculated in Sect. 4. Waiting time analysis is done in Sect. 5. Some numerical illustrations are given in Sect. 6.

2   Model Description

Consider an $M/G/1$ queue with Poisson arrival of rate λ. Vacation to server starts whenever the system turns empty at a service completion epoch. There are two types of vacations. Depending on the environment the server goes either for type I vacation with probability p_1 or for type II vacation with probability p_2 such that $p_1 + p_2 = 1$. During type I vacation the arriving customer joins the queue with probability q or leaves the system with probability $1 - q$. On completion of type I vacation if the server finds the system empty, he goes for type II vacation. Type II vacation is a working vacation in which a customer is served on arrival at a lower rate if the server is idle during vacation. On completion of type II vacation if the server finds the system empty again goes for type II vacation. On completion of service in working vacation if the server finds one or more customers in the system it shifts to normal service, interrupting the vacation. Otherwise the server continues the vacation. If the vacation is completed before service completion the service is restarted at normal rate.

Table 1. Distribution functions

Operation	Distribution function	PDF	LST	Mean
Normal service	$S(t)$	$s(t)$	$S^*(s)$	$1/\mu$
Vacation service	$S_v(t)$	$s_v(t)$	$S_v^*(s)$	$1/\mu_v$
type I vacation	$V_1(t)$	$v_1(t)$	$V_1^*(s)$	$1/\gamma_1$
type II vacation	$V_2(t)$	$v_2(t)$	$V_2^*(s)$	$1/\gamma_2$

The duration of vacations and services follow mutually independent general distributions. The distribution functions and the corresponding density functions are as defined above:(Table 1).

- $N(t)$ - Number of customers in the system;
- $C(t)$ - Status of server;
- $S'(t)$ - Elapsed service time of the customer in normal service mode at time t;
- $S_v'(t)$ - Elapsed service time in working vacation;
- $V_1'(t)$ - Elapsed vacation time duration of type I vacation;
- $V_2'(t)$ - Elapsed vacation time duration of type II vacation.

$$C(t) = \begin{cases} 0, \text{ if server is busy with normal service;} \\ 1, \text{ if the server is in type I vacation;} \\ 2, \text{ if the server is in type II vacation;} \end{cases}$$

The states of the system at time t can be represented by the continuous time Markov process $\{N(t), C(t), t \geq 0\}$.

$$P_{n,0}(x,t)dx = P\{N(t) = n, C(t) = 0, x \leq S'(t) < x + dx\},$$
for $t \geq 0, x \geq 0, n \geq 1$

$$P_{n,1}(x,t)dx = P\{N(t) = n, C(t) = 1, x \leq V_1'(t) < x + dx\},$$
for $t \geq 0, x \geq 0, n \geq 0$

$$P_{n,2}(x,y,t)dxdy = P\{N(t) = n, C(t) = 2, x \leq V_2'(t) < x + dx,$$
$y \leq S_v'(t) < y + dy\}$, for $t \geq 0, x > 0, y \geq 0, n \geq 0$

Let $\mu(x), \mu_v(x), \gamma_1(x)$, and $\gamma_2(x)$ be the conditional completion rates of normal service, vacation service, type I vacation and type II vacation respectively.
Then $\mu(x)dx = \dfrac{dS(x)}{1 - S(x)}, \mu_v(x)dx = \dfrac{dS_v(x)}{1 - S_v(x)},$
$\gamma_1(x)dx = \dfrac{dV_1(x)}{1 - V_1(x)}, \gamma_2(x)dx = \dfrac{dV_2(x)}{1 - V_2(x)}.$
$P_{n,0}(x) = \lim\limits_{t\to\infty} P_{n,0}(x,t), P_{n,1}(x) = \lim\limits_{t\to\infty} P_{n,1}(x,t),$

$$P_{n,2}(x,y) = \lim\limits_{t\to\infty} P_{n,2}(x,y,t)$$

Define $a_k = \int_0^\infty \dfrac{(\lambda x)^k}{k!} e^{-\lambda x} dS(x),\ b_k = \int_0^\infty \dfrac{(q\lambda x)^k}{k!} e^{-q\lambda x} dV_1(x),$

$c_k = \int_0^\infty \int_0^x \dfrac{(\lambda x)^k}{k!} e^{-\lambda x} dV_2(x) dS_v(y)$ and $d_k = \int_0^\infty \int_0^y \dfrac{(\lambda y)^k}{k!} e^{-\lambda y} dS_v(y) dV_2(x)$

where a_k, b_k, c_k and d_k are the probability for k arrivals during normal service, type I vacation, type II vacation and vacation service respectively. The corresponding probability generating functions are

$$A(z) = S^*(\lambda(1-z)),\ B(z) = V_1^*(q\lambda(1-z)),$$

$$C(z) = \int_0^\infty e^{-\lambda(1-z)x} S_v(x) dV_2(x) \text{ and } D(z) = \int_0^\infty e^{-\lambda(1-z)y} V_2(x) dS_v(y).$$

The diagram (Fig. 1) below provides a pictorial representation of the system evolution.

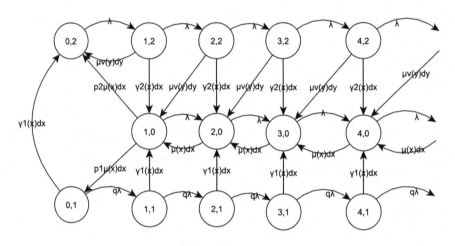

Fig. 1. Model description

3   Stability of the System

Theorem: *The inequality* $\rho = \dfrac{\lambda}{\mu} < 1$ *is necessary and sufficient condition for the system to be stable.*

Let t_n be the departure time of nth customer from the system after service completion or the time at the end of a vacation. X_n be the number of customers in the system just after the nth departure, or just at the end of a vacation.

$$X_{n+1} = \begin{cases} X_n - 1 + M_{n+1}, & \text{for } X_n \geq 1 \\ M_n, & \text{for } X_n = 0 \end{cases}$$

where M_{n+1} is the number of arrivals during the service of a customer or during vacation. The arrivals are independent. Then $\{X_n, n \geq 1\}$ is a Markov chain with state space $Z^+ \cup \{0\}$. This Markov chain is irreducible and aperiodic. Now we have to prove the positive recurrence. For that we use the Foster's Criterion.

Foster's Criterion(see Pakes [1]): An irreducible and aperiodic Markov chain is positive recurrent if there exists a non negative function $f(i)$, $i \in Z^+ \cup \{0\}$ and $\epsilon > 0$ such that the mean drift $\psi(i) = E[f(X_{n+1}) - f(X_n)/X_n = i]$ is finite for all $i \in Z^+ \cup \{0\}$ and $\psi(i) \leq -\epsilon \ \forall i$ except for a finite number.

Here let us consider $f(s) = s$, $s \in Z^+ \cup \{0\}$

Then the mean drift when $i > 0$ is given by

$$\psi(i) = E[f(X_{n+1}) - f(X_n)/X_n = i]$$

$$= \sum_{j=0}^{\infty}(i + j - 1 - i)a_j = \sum_{j=0}^{\infty}(j - 1)a_j = (\rho - 1).$$

When $i = 0$

$$\psi(i) = p_1\rho_1 + p_2(\rho_2 + \rho')$$

where $\rho = \dfrac{\lambda}{\mu}, \rho_1 = \dfrac{q\lambda}{\gamma_1}, \rho_2 = \dfrac{\lambda}{\gamma_2}, \rho' = \dfrac{\lambda}{\mu_b}$.

Obviously $\psi(i) \leq -\epsilon$ except for $i = 0$ which is the sufficient condition for ergodicity. The necessary condition follows from Kaplan's condition which states that $\psi(i) < \infty \ \forall i \in Z^+ \cup \{0\}$ and there exists $j \in Z^+ \cup \{0\}$ such that $\psi(i) \geq 0$ for $i \geq j$.

4 Steady State Distribution

Using supplementary variable technique we get the following system of equations.

$$\frac{dP_{n,0}(x)}{dx} = -(\mu(x) + \lambda)P_{n,0}(x) + \lambda P_{n-1,0}(x)(1 - \delta_{1n}), n \geq 1. \tag{1}$$

$$\frac{dP_{n,1}(x)}{dx} = -(\gamma_1(x) + q\lambda)P_{n,1}(x) + q\lambda P_{n-1,1}(x), n \geq 1. \tag{2}$$

$$\frac{\partial P_{n,2}(x,y)}{\partial x} = -(\gamma_2(x) + \lambda)P_{n,2}(x,y) + \lambda P_{n-1,2}(x,y)(1 - \delta_{1n}), n \geq 1. \tag{3}$$

$$\frac{\partial P_{n,2}(x,y)}{\partial y} = -(\mu_v(y) + \lambda)P_{n,2}(x,y) + \lambda P_{n-1,2}(x,y)(1 - \delta_{1n}), n \geq 1. \tag{4}$$

The steady state boundary conditions at $x = 0$ and $y = 0$ are

$$P_{n,0}(0) = \int_0^{\infty} P_{n,1}(x)\gamma_1(x)dx + \int_0^{\infty} P_{n+1,0}(x)\mu(x)dx$$

$$+ \int_0^\infty P_{n+1,2}(x,y)\mu_v(y)dy + \int_0^\infty P_{n,2}(x,y)\gamma_2(x)dx, n \geq 1. \tag{5}$$

$$P_{0,1}(0) = p_1 \int_0^\infty P_{10}(x)\mu(x)dx. \tag{6}$$

$$P_{0,2}(0) = p_2 \int_0^\infty P_{1,0}(x)\mu(x)dx + \int_0^\infty P_{0,1}(x)\gamma_1(x)dx. \tag{7}$$

$$P_{1,2}(x,0) = \lambda P_{0,2}(x) \tag{8}$$

To solve the system of Eqs. (1)–(4), let us define the following probability generating functions for $|z| < 1$:

$$P_0(x,z) = \sum_{n=1}^\infty P_{n,0}(x)z^n, \quad P_1(x,z) = \sum_{n=0}^\infty P_{n,1}(x)z^n, \quad P_2(x,y,z) = \sum_{n=1}^\infty P_{n,2}(x,y)z^n$$

Multiplying Eqs. (1)–(4) by z^n and summing over n we get

$$\frac{\partial P_0(x,z)}{\partial x} = -[\lambda(1-z) + \mu(x)]P_0(x,z). \tag{9}$$

$$\frac{\partial P_1(x,z)}{\partial x} = -[q\lambda(1-z) + \gamma_1(x)]P_1(x,z). \tag{10}$$

$$\frac{\partial P_2(x,y,z)}{\partial x} = -[\lambda(1-z) + \gamma_2(x)]P_2(x,y,z). \tag{11}$$

$$\frac{\partial P_2(x,y,z)}{\partial y} = -[\lambda(1-z) + \mu_v(y)]P_2(x,y,z). \tag{12}$$

Solving (9) and (10) we get

$$P_0(x,z) = P_0(0,z)(1-S(x))e^{-\lambda(1-z)x}. \tag{13}$$

$$P_1(x,z) = P_1(0,z)(1-V_1(x))e^{-q\lambda(1-z)x} = P_{0,1}(0)(1-V_1(x))e^{-q\lambda(1-z)x}. \tag{14}$$

Solving (11) and (12) we obtain

$$P_2(x,y,z) = P_{0,2}(0)(1-V_2(x))(1-S_v(y))e^{-\lambda(1-z)(x+y)}. \tag{15}$$

Now

$$P_{n,0}(x) = \sum_{i=1}^n P_{i,0}(0)\frac{(\lambda x)^{n-i}e^{-\lambda x}}{(n-i)!}[1-S(x)]. \tag{16}$$

$$P_{n,1}(x) = P_{0,1}(0)\frac{(q\lambda x)^n e^{-q\lambda x}}{n!}[1-V_1(x)]. \tag{17}$$

$$P_{n,2}(x,y) = P_{0,2}(0)\frac{(\lambda y)^{n-1}e^{-\lambda y}}{(n-1)!}[1-S_v(y)][1-V_2(x)]. \tag{18}$$

Solving (6) and (7) using (16) we get

$$P_{0,1}(0) = p_1 P_{1,0}(0)S^*(\lambda) \tag{19}$$

$$P_{0,2}(0) = [p_2 + p_1 V_1^*(\lambda)]P_{1,0}(0)S^*(\lambda) \tag{20}$$

Using the boundary condition we can write $P\Delta = P$

where $P = (P_{0,1}(0), P_{0,2}(0), P_{1,0}(0), P_{2,0}(0), \ldots)$ and

$$\Delta = \begin{bmatrix} 0 & b_0 & b_1 & b_2 & b_3 & \cdots \\ 0 & c_0 + d_0 & c_1 + d_1 & c_2 + d_2 & c_3 + d_3 & \cdots \\ p_1 a_0 & p_2 a_0 & a_1 & a_2 & a_3 & \cdots \\ & & a_0 & a_1 & a_2 & \ddots \\ & & & a_0 & a_1 & \ddots \\ & & & & a_0 & \ddots \\ & & & & & \ddots \end{bmatrix}.$$

It is clear that the matrix Δ is irreducible. It is stochastic since $\sum_{k=0}^{\infty} b_k = B(1) = 1$,

$\sum_{k=0}^{\infty} c_k + d_k = C(1) + D(1) = 1$ and $\sum_{k=1}^{\infty} a_k + p_1 a_0 + p_2 a_0 = A(1) = 1$.

Now we have to prove that Δ is positive recurrent when $\rho < 1$. Δ is positive recurrent when $\sum_{k=1}^{\infty} k a_k < 1$, and this condition is satisfied when $\rho < 1$.

From the matrix Δ,

$$P_{0,1}(0) = P_{1,0}(0)p_1 a_0. \tag{21}$$

$$P_{0,2}(0) = P_{0,1}(0)b_0 + P_{0,2}(0)(c_0 + d_0) + P_{1,0}(0)p_2 a_0. \tag{22}$$

$$P_{j,0}(0) = P_{0,1}(0)b_j + P_{0,2}(0)(c_j + d_j) + \sum_{i=0}^{j} P_{i+1,0}(0)a_{j-i}. \tag{23}$$

From (23),

$$P_0(0, z) = z\frac{[P_{0,1}(0)(B(z) - 1) + P_{0,2}(0)(C(z) + D(z) - 1)]}{z - A(z)}. \tag{24}$$

From (13),

$$P_0(z) = P_0(0, z)\frac{1 - S^*(\lambda(1 - z))}{(\lambda(1 - z))}. \tag{25}$$

From (14),

$$P_1(z) = p_1 P_{1,0}(0)S^*(\lambda)\frac{1 - V_1^*(q\lambda(1 - z))}{(q\lambda(1 - z))}. \tag{26}$$

From (15),

$$P_2(z) = [p_2 + p_1 V_1^*(\lambda)]P_{1,0}(0)S^*(\lambda)\Omega(z), \tag{27}$$

where $\Omega(z) = \int_0^\infty \int_0^x (1 - V_2(x))(1 - S_V(y))e^{-\lambda(1-z)(x+y)}dxdy.$

Let, $P(z) = P_0(z) + P_1(z) + P_2(z)$ be the PGF of the stationary queue size distribution irrespective of the server's state.

Then,

$$P(z) = P_0(0, z)\frac{1 - S^*(\lambda(1-z))}{(\lambda(1-z))} + p_1 P_{1,0}(0)S^*(\lambda)\frac{1 - V_1^*(q\lambda(1-z))}{(q\lambda(1-z))} +$$

$$[p_2 + p_1 V_1^*(\lambda)]P_{1,0}(0)S^*(\lambda)\Omega(z). \qquad (28)$$

Using the condition P(1)=1, we get

$$P_{1,0}(0) = \left[\left(\frac{p_1\rho_1 + (p_2 + p_1 V_1^*(q\lambda))(\rho_2 + \rho_3)}{\mu - \lambda} + \frac{p_1}{\gamma_1} + (p_2 + p_1 V_1^*(q\lambda))\Omega(1)\right)S^*(\lambda)\right]^{-1}$$

$$(29)$$

The expected queue length $E(L) = P'(z)_{z=1}.$

$$= \frac{1}{2}(2P_0'(0, 1)S^{*'}(0) + \lambda P_0(0, 1)S^{*''}(0) - P_0''(0, 1)S^{*'}(0)) + \frac{q}{2}\lambda P_{0,1}(0)V_1^{*''}(0) + P_{0,2}(0)\Omega'(1)$$

$$(30)$$

5   Waiting Time Analysis

To find the waiting time of a customer who joins for service at time t, we have to consider different possibilities depending on the status of server at that time. The server may be in general busy period, vacation I or in vacation II. Let $W(t)$ be the waiting time of a customer who arrives at time t and $W^*(s)$ be the corresponding LST.

Case 1. The customer arrives to the system when the number of customers is 0 and the server is in vacation. It may be either in vacation I or vacation II. If it is in vacation II the customer starts getting service immediately and the waiting time is zero. Let $W_{0,2}^*(s)$ be the corresponding LST. Then

$$W_{0,2}^*(s) = 1.$$

If the server is in vacation I the customer has to wait till the completion of vacation. Let x be the elapsed vacation time until the arrival of the customer and $W_{01}^*(s)$ be the LST of the waiting time of the customer who arrives when the system is empty and the server in vacation I. Then

$$W_{0,1}^*(s) = \int_0^\infty e^{-st}\frac{dV_1(x+t)}{1 - V_1(x)}.$$

Case 2. The waiting time of the customer who arrives to the system when there are n customers in the system and the server is providing normal service to customer is the sum of the remaining service time of the customer in service and the service time of the remaining $n - 1$ customers. Let x be the elapsed service time of the customer in service and $W_{n,0}^*(s)$ be the LST of the waiting time of the customer who arrives to the system when there are n customers and the server is busy. Then

$$W_{n,0}^*(s) = S^{*(n-1)}(s) \int_0^\infty e^{-st} \frac{dS(x+t)}{1 - S(x)}.$$

Case 3. The waiting time of the customer who arrives to the system when there are n customers in the system and the server is is in vacation I is the sum of the remaining vacation time and the service time of the remaining n customers. Let x be the elapsed vacation time and $W_{n,1}^*(s)$ be the LST of the waiting time of the customer who arrives to the system when there are n customers and the server is in vacation I. Then

$$W_{n,1}^*(s) = S^{*(n)}(s) \int_0^\infty e^{-st} \frac{dV_1(x+t)}{1 - V_1(x)}.$$

Case 4. The waiting time of the customer who arrives to the system when there are n customers in the system and the server is is in vacation II is the sum of the remaining vacation time and the service time of the remaining n customers if the vacation completes before vacation service. If the service is completed before vacation completion then the waiting time is the sum of the remaining vacation service time and the service time of the remaining $n - 1$ customers. Let x be the elapsed vacation time, y be the elapsed vacation service time and $W_{n,2}^*(s)$ be the LST of the waiting time when vacation is completed before service and $W_{n,2}^{*'}(s)$ be the LST of the waiting time of the customer when service is completed before vacation of the customer who arrives to the system when there are n customers and the server is in vacation II. Then

$$W_{n,2}^*(s) = S^{*(n)}(s) \int_0^\infty e^{-st} \frac{dV_2(x+t)}{1 - V_2(x)}.$$

$$W_{n,2}^{*'}(s) = S^{*(n-1)}(s) \int_0^\infty e^{-st} \frac{dS_v(y+t)}{1 - S_v(y)}.$$

$$W^*(s) = p_1 \int_0^\infty P_{0,1}(x)dx \int_0^\infty e^{-st} \frac{dV_1(x+t)}{1 - V_1(x)} + p_2 \int_0^\infty P_{0,2}(x,0)dx$$

$$+ \sum_{n=1}^\infty S^{*(n-1)}(s) \int_0^\infty P_{n,0}(x)dx \int_0^\infty e^{-st} \frac{dS(x+t)}{1 - S(x)}$$

$$+ p_1 \sum_{n=1}^\infty S^{*(n)}(s) \int_0^\infty P_{n,1}(x)dx \int_0^\infty e^{-st} \frac{dV_1(x+t)}{1 - V_1(x)}$$

$$+p_2 \sum_{n=1}^{\infty} S^{*(n)}(s) \int_0^{\infty} \int_0^x P_{n,2}(x,y)dxdy \int_0^{\infty} e^{-st} \frac{dV_2(x+t)}{1-V_2(x)}$$

$$+p_2 \sum_{n=1}^{\infty} S^{*(n-1)}(s) \int_0^{\infty} \int_0^y P_{n,2}(x,y)dxdy \int_0^{\infty} e^{-st} \frac{dS_v(y+t)}{1-S_v(y)}.$$

6   Numerical Results

In this section we provides some numerical examples for this model. Assume normal service time is exponentially distributed with parameter μ, vacation service time is exponentially distributed with parameter μ_v, type I vacation duration is exponentially distributed with parameter γ_1 and type II vacation duration is exponentially distributed with parameter γ_2.

6.1   The Variation in Queue Length Due to the Variation in Vacation Service Rate and Arrival Rate

Let $\mu = 5$, $\gamma_1 = 0.4$, $\gamma_2 = 0.3$, $q = 0.5$.

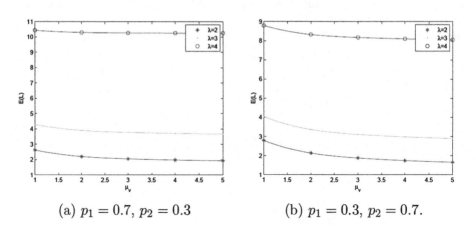

(a) $p_1 = 0.7$, $p_2 = 0.3$ (b) $p_1 = 0.3$, $p_2 = 0.7$.

Fig. 2. Expected queue length $E(L)$ against vacation service rate μ_v

Figure 2 represents the variation in queue length due to the variation in vacation service rate and arrival rate when $p_1 = 0.7$, $p_2 = 0.3$ and $p_1 = 0.3$, $p_2 = 0.7$ respectively. As the value of vacation service rate increases the expected queue length decreases and as the arrival rate increases the queue length also increases which are on expected lines. From Fig. 2 When the probability of opting for type I vacation decreases the expected queue length decreases.

6.2   The Variation in Queue Length Due to the Variation in Vacation Service Rate and Duration of Vacation

Let $\lambda = 4$, $\gamma_2 = 0.05$, $p_1 = 0.3$, $p_2 = 0.7$.

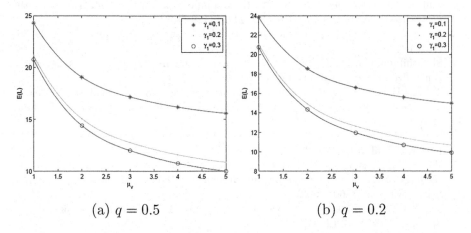

(a) $q = 0.5$ (b) $q = 0.2$

Fig. 3. Expected queue length $E(L)$ against vacation service rate μ_v

Figure 3 represents the variation in queue length due to the variation in vacation service rate and vacation duration when $q = 0.5$ and $q = 0.2$ respectively. As the duration of vacation decreases the queue length decreases. This is due to the early return of server from vacation. When the server returns early from vacation the customer starts getting service earlier and the length of the queue reduces. When the probability of customer joining the queue during vacation I reduces the queue length also reduces which are on expected lines.

Conclusion

In this paper we analyzed an $M/G/1$ queueing model with two types of vacations and vacation interruption. The server goes for vacation only if the queue is empty. Both normal vacation(type I vacation) and multiple working vacation (type II vacation) are considered. At the end of a busy period, depending on the environment, the server opts either normal vacation or working vacation. On completion of type I vacation if the server finds the system empty he goes for type II vacation. On completion of type II vacation if the server finds the system empty he goes for another type II vacation. On completion of service in type II vacation if the server finds one or more customers in queue he returns to normal service interrupting the vacation. If the server is in type I vacation the customer has to wait until the completion of vacation for getting service. An arriving customer during type I vacation, joins the queue with probability q or leaves the system with probability $1-q$. We derived the steady state distribution function. Waiting time analysis is performed. Some numerical illustrations are also provided.

References

1. Pakes, A.G.: Some conditions for ergodicity and recurrence of Markov chains. Oper. Res. **17**, 1058–1061 (1969)
2. Fuhrmann, S., Cooper, S.R.: Stochastic decompositions in the $M/G/1$ queue with generalized vacations. Oper. Res. **32**, 1117–1129 (1985)
3. Doshi, B.T.: Single-server queues with vacations-a survey. Queueing Syst. **1**, 29–66 (1986)
4. Shanthikumar, J.G.: On stochastic decomposition in M/G/1 type queues with generalized server vacations. Oper. Res. **36**, 566–569 (1988)
5. Takagi, H.: Queueing Analysis: A Foundation of Performance Analysis, Vacation and Priority Systems, Part 1, vol. 1. Elsevier Science Publishers B.V, Amsterdam (1991)
6. Servi, L.D., Finn, S.G.: M/M/1 queues with working vacations (M/M/1/WV). Perform. Eval. **50**, 41–52 (2002)
7. Wu, D.A., Takagi, H.: M/G/1 queues with multiple working vacation. In: Proceedings of the Queueing Symposium Stochastic Models and their Applications, Kakegawa, pp. 51–60 (2003)
8. Tian, N., Zhang, G.: Vacation Queueing Models: Theory and Applications. Springer, New York (2006)
9. Wu, D.A., Takagi, H.: M/G/1 queue with multiple working vacations. Perform. Eval. **63**, 654–681 (2006)
10. Li, J., Tian, N.: Discrete time GI/Geo/1 queue with working vacation and vacation interruption. Appl. Math. Comput. **185**, 1–10 (2007)
11. Li, W., Xu, X., Tian, N.: Stochastic decompositions in The $M/M/1$ queue with working vacations. Oper. Res. Lett. **35**, 595–600 (2007)
12. Kim. J.D., Choi, D.W., Chae, K.C.: Analysis of queue-length distribution of the $M/G/1$ queue with working vacations. In: Hawaii International Conference on Statistics and Related Fields, 5–8 June (2008)
13. Li, J., Tian, N., Ma, Z.: Performance analysis of GI/M/1 queue with working vacations and vacation interruption. Appl. Math. Model. **32**, 2715–2730 (2008)
14. Li, J., Tian, N.: Analysis of the discrete time Geo/Geo/1 queue with single working vacation. Qual. Technol. Quant. Manag. **5**(1), 77–89 (2008)
15. Li, J., Tian, N., Zhang, Z., Luh, H.: Analysis of the M/G/1 queue with exponentially working vacations-a matrix analytic approach. Queueing Syst. **61**, 139–166 (2009)
16. Jau-Chuan, K., Chia-Huang, W., Zhang, Z.G.: Recent developments in vacation queueing models: a short survey. Int. J. Oper. Res. **7**(4), 3–8 (2010)
17. Zhang, Z.H. Performance analysis of M/G/1 queue with working vacations and vacation interruption. J. Comput. Appl. Math. **234**, 2977–2985 (2010)
18. Sreenivasan, C., Krishnamoorthy, A.: An M/M/2 queueing system with heterogeneous servers including one with working vacation. Int. J. Stochast. Anal. 2012, 145867 (2012)
19. Sreenivasan, C., Chakravarthy, S.R., Krishnamoorthy, A.: MAP/PH/1 queue with working vacations, vacation interruptions and N policy. Appl. Math. Model **37**(6), 3879–3893 (2013)
20. Li, J., Tian, N.: The M/M/1 queue with working vacations and vacation interruptions. J. Syst. Sci. Syst. Eng. **16**(1), 121–127 (2014)

Switch-Hysteresis Control of the Selling Times Flow in a Model with Perishable Goods

Klimentii Livshits[(⊠)] and Ekaterina Ulyanova

Tomsk State University, Tomsk, Russia
kim47@mail.ru,
katerina_tomsk@sibmail.com

Abstract. In this paper we obtain the probability density function of stock of perishable goods under constant production and hysteresis control of the selling price.

Keywords: Perishable goods · Hysteresis control · Probability density function · Diffusion approximation

1 Introduction

Mathematical models and methods of queueing theory [1,2] are widely used in various fields and, in particular, can be used to analyze the problems of inventory management with a limited shelf life, which have been intensively studied in recent years. Several review articles on the topic appeared during that time, for example S.K. Goyal, B.C. Giri [3], M. Bakker, J. Riezebos, R.H. Teunter [4]. Also worth noting are papers by V.K. Mishra, V.K. Mishra and L.S. Singh [5,6], R. Begum, S.K. Sahu, R.R. Sahoo [7,8], R.P. Tripathi, D. Singh, T. Mishra [9], where authors consider models of inventory management of continuously deteriorating goods under the assumption of a known demand function. In V. Sharma and R.R. Chaudhary [10] a model is considered where demand is known function of time, while the deterioration process is random and follows Weibull distribution. In K. Tripathy and U. Mishra [11] a model is considered in which demand is a known function of price. To analyse the mathematical models one can employ the methods of asymptotic analysis that are widely used in the queuing theory, for example in the mentioned above works by A.A. Nazarov [1], A.A. Nazarov and S.P. Moiseeva [2].

2 Mathematical Model of the Problem

We consider a single-line queueing system (Fig. 1) in the entrance of which applications (perishable goods) with arrival rate c come in. We assume that arrival process can be approximated in such a way that c units arrives per unit time.

The goods continuously deteriorate as they are stored. Let $S(t)$ be the amount of goods at time t. Then during a small time interval Δt a total of $kS(t)\Delta t$

© Springer International Publishing Switzerland 2015
A. Dudin et al. (Eds.): ITMM 2015, CCIS 564, pp. 263–274, 2015.
DOI: 10.1007/978-3-319-25861-4_23

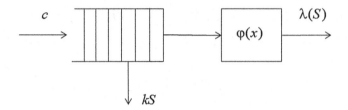

Fig. 1. Mathematical model

is lost. The service, which in this work will be called sales, is provided by parties with random size x, where the values of purchases x are independent random variable with probability density function $\varphi(x)$, mean $M\{x\} = a$ and a second moment $M\{x^2\} = a_2$. Selling times follow a Poisson process with intensity λ that depend on selling price b. We consider the case when the intensity of sales λ monotonically decreases as b grows. For a given price b and, hence, sales process intensity λ the average amount of goods $\bar{S}(t)$ is defined as

$$\bar{S}(t) = S(0)e^{-kt} + \frac{c - \lambda a}{k}(1 - e^{-kt}).$$

Thus if $c - \lambda a > 0$ and $t \gg 1$ we have a constant stock of unsold goods which is undesirable. If $c - \lambda a \leq 0$ we have unsatisfied demand. Hence we need to control either selling price b, or the pace of goods arrival c depending on current stock.

In this paper we assume that sales are controlled in the following way. First, two boundary values for the stock of goods are set, S_1 S_2, such that $S_2 > S_1$. For $S < S_1$ a selling price b_0 is established, for $S > S_2$ a selling price $b_1 < b_0$ is established. For $S_1 \leq S \leq S_2$ the selling price will be either $b = b_0$ or $b = b_1$ depending on the trajectory which the process $S(t)$ followed when it entered this domain. If it crossed the lower bound S_1 upwards then $b = b_0$, while if it crossed the upper bound S_2 downwards, then $b = b_1$. Thus the selling price $b = b_1$ is set as soon as $S(t)$ reaches S_2 and lasts until the stock falls to S_1. The domain $S_1 \leq S \leq S_2$ is in fact what we call the domain of hysteresis stock control. In accordance with this, the intensity of selling times flow at any given moment is given by

$$\lambda(S) = \begin{cases} \lambda_0, S < S_1, \\ \lambda_0 \text{ or } \quad \lambda_1, S_1 \leq S \leq S_2, \\ \lambda_1, S > S_2 \end{cases} \tag{1}$$

It is natural to assume that $C - \lambda_0 a > 0$ and $C - \lambda_1 a < 0$. Finally, there may be a situation when current demand cannot be fully satisfied by the current stock of goods. In such case we assume that $S(t) < 0$. The orders are satisfied in the order of arrival.

The main goal of this paper is to determine the probability density function of the stock of goods in this model and several additional assumptions.

Denote

$$P_i(S,t) = \frac{\Pr\{S \leq S(t) < S + dS, \ \lambda(t) = \lambda_i\}}{dS}, \quad i = 0, 1. \tag{2}$$

Theorem 1. *If $P_i(S,t)$ is differentiable in t and $SP_i(S,t)$ is differentiable in S then functions $P_i(S,t)$ satisfy the following system of equations Kolmogorov*

$$\frac{\partial P_1(S,t)}{\partial t} = -\lambda_1 P_1(S,t) - \frac{\partial}{\partial S}((c - kS)P_1(S,t)) + \lambda_1 \int_0^\infty P_1(S+x)\varphi(x)dx, \quad S \geq S_1,$$

$$(3)$$

$$\frac{\partial P_0(S,t)}{\partial t} = -\lambda_0 P_0(S,t) - \frac{\partial}{\partial S}((c - kS)P_0(S,t))$$
$$+ \lambda_0 \int_S^{S_2} P_0(x,t)\varphi(x - S)dx, \quad S_1 < S < S_2,$$

$$(4)$$

$$\frac{\partial P_0(S,t)}{\partial t} = -\lambda_0 P_0(S,t) - \frac{\partial}{\partial S}((c - kSI(S))P_0(S,t))$$
$$+ \lambda_0 \int_S^{S_2} P_0(x,t)\varphi(x - S)dx + \lambda_1 \int_{S_1}^\infty P_1(x)\varphi(x - S)dx, \quad S \leq S_1,$$

$$(5)$$

where $I(x)$ is a step unit function.

Proof. Consider two close moments of time t and $t + \Delta t$, where $\Delta t \ll 1$. Under given assumptions the conditional probabilities

$$P\{S(t + \Delta t) < z, \lambda(t + \Delta t) = \lambda_1 \,|\, S(t) = S, \lambda(t) = \lambda_1\} =$$
$$(1 - \lambda_1 \Delta t)I(z - S - (c - kS)\Delta t) + \lambda_1 \Delta t \int_0^{S - S_1} I(z - S + x)\varphi(x)dx + o(\Delta t),$$

$$(6)$$

$$P\{S(t + \Delta t) < z, \lambda(t + \Delta t) = \lambda_1 \,|\, S(t) = S, \lambda(t) = \lambda_0\} = 0. \quad (7)$$

Thus for $z \geq S_1$ probability

$$P\{S(t + \Delta t) < z, \lambda(t + \Delta t) = \lambda_1\} =$$
$$(1 - \lambda_1 \Delta t)\int_{S_1}^\infty I(z - S - (c - kS)\Delta t)P_1(S,t)dS$$
$$+ \lambda_1 \Delta t \int_{S_1}^\infty \int_0^{S - S_1} I(z - S + x)\varphi(x)dx P_1(S,t)dS + o(\Delta t).$$

$$(8)$$

For $z \geq S_1$ and a small Δt the integral

$$\int_{S_1}^\infty I(z - S - (c - kS)\Delta t)P_1(S,t)dS = \int_{S_1}^{z - (c - kz)\Delta t + o(\Delta t)} P_1(S,t)dS$$
$$= \int_{S_1}^z P_1(S,t)dS - P_1(z,t)(c - kz)\Delta t + o(\Delta t),$$

and the integral

$$\int_{S_1}^\infty \int_0^{S - S_1} I(z - S + x)\varphi(x)dx P_1(S,t)dt = \int_0^\infty \varphi(x) \int_{S_1 + x}^{z + x} P_1(S,t)dS dx.$$

Substituting the expressions above into (8), differentiating with respect to z and taking the limit $\Delta t \to 0$ we arrive at Eq. (3).

Furthermore, the conditional probabilities

$$P\{S(t+\Delta t) < z, \lambda(t+\Delta t) = \lambda_0 \,|\, S(t) = S, \lambda(t) = \lambda_0\} =$$
$$(1 - \lambda_0 \Delta t)I(z - S - (c - kS)\Delta t) + \lambda_0 \Delta t \int\limits_0^\infty I(z - S + x)\varphi(x)dx + o(\Delta t), \quad (9)$$

$$P\{S(t+\Delta t) < z, \lambda(t+\Delta t) = \lambda_0 \,|\, S(t) = S, \lambda(t) = \lambda_1\} =$$
$$\lambda_1 \Delta t \int\limits_{S-S_1}^\infty I(x - S + x)\varphi(x)dx + o(\Delta t). \quad (10)$$

From where in the domain $z \le S_2$ probability

$$P\{S(t+\Delta t) < z, \lambda(t+\Delta t) = \lambda_0\} =$$
$$(1 - \lambda_0 \Delta t) \int\limits_{-\infty}^{S_2} I(z - S - (c - kS)\Delta t)P_0(S,t)dS$$
$$+\lambda_0 \Delta t \int\limits_{-\infty}^{S_2} \int\limits_0^\infty I(z - S + x)\varphi(x)dxP_0(S,t)dS \quad (11)$$
$$+\lambda_1 \Delta t \int\limits_{S_1}^\infty \int\limits_{S-S_1}^\infty I(z - S + x)\varphi(x)dxP_1(S,t)dS + o(\Delta t).$$

For $z \le S_2$ and a small Δt

$$\int\limits_{-\infty}^{S_2} I(z - S - (c - kS)\Delta t)P_0(S,t)dS = \int\limits_{-\infty}^{z-(c-kz)\Delta t+o(\Delta t)} P_0(S,t)dS =$$
$$\int\limits_{-\infty}^{z} P_0(S,t)dS - (c - kz)P_0(z,t)\Delta t + o(\Delta t),$$

and the integral

$$\int\limits_{-\infty}^{S_2} \int\limits_0^\infty I(z - S + x)\varphi(x)dxP_0(S,t)dS = \int\limits_{-\infty}^z P_0(S,t)ds + \int\limits_z^{S_2} \int\limits_{S-z}^\infty \varphi(x)dxP_0(S,t)dS.$$

Finally, for $S_1 < S < S_2$ the integral

$$\int\limits_{S_1}^\infty \int\limits_{S-S_1}^\infty I(z - S + x)\varphi(x)dxP_1(S,t)dS = \int\limits_{S_1}^\infty \int\limits_{S-S_1}^\infty \varphi(x)dxP_1(S,t)dS,$$

while for $z \le S_1$ the integral

$$\int\limits_{S_1}^\infty \int\limits_{S-S_1}^\infty I(z - S + x)\varphi(x)dxP_1(S,t)dS = \int\limits_{S_1}^\infty \int\limits_{S-z}^\infty \varphi(x)dxP_1(S,t)dS.$$

Substituting the expressions above into (11), differentiating with respect to z and taking the limit $\Delta t \to 0$ we arrive at Eqs. (4) and (5).

The solution of the system (3)–(5) must, apparently, satisfy the following normalising condition

$$\int\limits_{S_1}^{\infty} P_1(S,t)dS + \int\limits_{-\infty}^{S_2} P_0(S,t)dS = 1 \tag{12}$$

while function $P_0(S,t)$ must be continuous at point S_1

$$P_0(S_1+0,t) = P_0(S_1-0,t). \tag{13}$$

The unconditional probability density function $P(S,t)$ of the stock of goods takes the form

$$P(S,t) = \begin{cases} P_1(S,t), & S > S_2, \\ P_1(S,t) + P_0(S,t), & S_1 \le S \le S_2, \\ P_0(S,t), & S < S_1. \end{cases} \tag{14}$$

3   Exponential Distribution of the Sale Amount

Let us consider the simplest case when sales are distributed exponentially

$$\varphi(S) = \frac{1}{a}\exp(-\frac{S}{a}).$$

Denote

$$P_i(S) = \lim_{t\to\infty} P_i(S,t). \tag{15}$$

In the steady state as $t \to \infty$ Eqs. (3)–(5) take the form

$$\lambda_1 P_1(S) + \frac{d}{dS}((c-kS)P_1(S)) - \frac{\lambda_1}{a}e^{\frac{S}{a}}\int\limits_{S}^{\infty} P_1(x)e^{-\frac{x}{a}}dx = 0, \quad S > S_1, \tag{16}$$

$$\lambda_0 P_0(S) + \frac{d}{dS}((c-kS)P_0(S)) - \frac{\lambda_0}{a}e^{\frac{S}{a}}\int\limits_{S}^{S_2} P_0(x)e^{-\frac{x}{a}}dx = 0, \quad S_1 \le S \le S_2, \tag{17}$$

$$\lambda_0 P_0(S) + \frac{d}{dS}((c-kSI(S))P_0(S)) - \frac{\lambda_0}{a}e^{\frac{S}{a}}\int\limits_{S}^{S_2} P_0(x)e^{-\frac{x}{a}}dx$$
$$-\frac{\lambda_1}{a}e^{\frac{S}{a}}\int\limits_{S_1}^{\infty} P_1(x)e^{-\frac{x}{a}}dx = 0, \quad S < S_1. \tag{18}$$

Equation (18) can be differentiated and represented as the following differential equation

$$\frac{d^2}{dS^2}((c-kSI(S))P_0(S)) - \frac{d}{dS}(\frac{c-kSI(S)-\lambda_0 a}{a}P_0(S)) = 0. \tag{19}$$

From here, taking into account boundary condition $P_0(-\infty) = 0$ in the domain $S \leq 0$

$$P_0(S) = De^{\frac{c-\lambda_0 a}{ca}S}. \tag{20}$$

In the domain $0 < S < S_1$ the solution of (19) takes the form

$$P_0(S) = \left[W_1 + W_2 \int_0^S e^{-\frac{x}{a}}(c - kx)^{-\frac{\lambda_0}{k}} dx \right] e^{\frac{S}{a}}(c - kS)^{\frac{\lambda_0}{k}-1}. \tag{21}$$

The condition of continuity of the solution in $S = 0$ yields $D = W_1 c^{\frac{\lambda_0}{k}-1}$. From (18) it follows that in $S = 0$ a condition must holds:

$$cP_0'(0 + 0) - kP_0(0 + 0) = cP_0'(0 - 0).$$

From where $W_2 = 0$. Thus, for $0 < S < S_1$

$$P_0(S) = De^{\frac{S}{a}}(1 - \frac{k}{c}S)^{\frac{\lambda_0}{k}-1}. \tag{22}$$

Equation (17) can be differentiated and represented as the following differential equation

$$\frac{d^2}{dS^2}((c - kS)P_0(S)) - \frac{d}{dS}(\frac{c - kS - \lambda_0 a}{a}P_0(S)) = 0. \tag{23}$$

Its solution takes the form

$$P_0(S) = \left[W_1 + W_2 \int_{S_1}^S e^{-\frac{x}{a}}(1 - \frac{k}{c}x)^{-\frac{\lambda_0}{k}} dx \right] e^{\frac{S}{a}}(1 - \frac{k}{c}S)^{\frac{\lambda_0}{k}-1}. \tag{24}$$

The condition of continuity in the point S_1 of (13) gives

$$W_1 = D. \tag{25}$$

Furthermore, solution (24) must satisfy the initial Eq. (17). Then

$$W_2 = -D[ae^{-\frac{S_2}{a}}(1 - \frac{k}{c}S_2)^{-\frac{\lambda_0}{k}} + \int_{S_1}^{S_2} e^{-\frac{x}{a}}(1 - \frac{k}{c}x)^{-\frac{\lambda_0}{k}} dx]^{-1}. \tag{26}$$

Finally, given that in the model considered the amount of goods is always $S \leq \frac{c}{k}$, the solution of (16) takes the form

$$P_1(S) = Ae^{\frac{S}{a}}(1 - \frac{k}{c}S)^{\frac{\lambda_1}{k}-1}. \tag{27}$$

The relationship between constants A and D is obtained from the condition that the set of found solutions must satisfy (18). Then

$$A = -ae^{-\frac{S_1}{a}}(1 - \frac{k}{c}S_1)^{-\frac{\lambda_1}{k}}W_2, \tag{28}$$

where W_2 is determined by the ratio (26). The last constant D is obtained from the normalising condition (12).

To sum up, the probability density function of the stock of goods is determined by (20), (22), (24), (27), while constants in these expressions are obtained from conditions (25), (26), (28) and (12).

For $S_2 = S_1$ we get the case of switch (threshold) control of the selling price and the probability density function $P(S)$ takes the form

$$P(S) = \begin{cases} De^{\frac{c - \lambda_0 a}{ca} S}, & S < 0, \\ D(1 - \frac{k}{c} S)^{\frac{\lambda_0}{k} - 1} e^{\frac{S}{a}}, & 0 \leq S \leq S_1, \\ D(1 - \frac{k}{c} S_1)^{\frac{\lambda_0 - \lambda_1}{k}} (1 - \frac{k}{c} S)^{\frac{\lambda_1}{k} - 1} e^{\frac{S}{a}}, & S_1 < S \leq \frac{c}{k}, \end{cases} \tag{29}$$

where D is determined by the normalising condition.

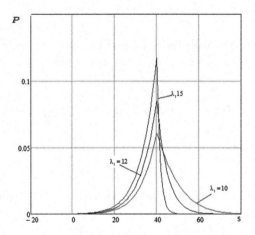

Fig. 2. Relationship between the probability density $P(S)$ and the stock size S.

The relationship between the probability density function $P(S)$ and the stock size S is given on Fig. 1. Threshold $S_1 = 40$, $\lambda_0 = 8$, $c = 10, k = 0.02, a = 1$ (Fig. 2).

4   Diffusion Approximation of the Production/Sales Process Under Switch Control of the Selling Price

In a general case the solution of the system (3)–(5) cannot be obtained even in the stationary mode. Hence in the following we focus on constructing an approximate solution. Consider the case of the switch control of the selling price when thresholds $S_2 = S_1$. The system (3)–(5) can be rewritten to yield

$$\frac{\partial P(S,t)}{\partial t} = \frac{\partial}{\partial S} [(kSI(S) - c)P(S,t)]$$
$$-\lambda(S)P(S,t) + \int_0^\infty \lambda(S + y)P(S + y, t)\varphi(y)dy, \tag{30}$$

where

$$\lambda(S) = \begin{cases} \lambda_0, & S \le S_1, \\ \lambda_1, & S > S_1. \end{cases} \tag{31}$$

Let us assume that the production speed $c = CN$, purchase process' intensities $\lambda_0 = \Lambda_0 N$, $\lambda_1 = \Lambda_1 N$, threshold $S_1 = S_0 N$, where $N \gg 1$. Let us analyse the behaviour of the solution of (30) as $N \to \infty$. Denote $\varepsilon^2 = 1/N$. Let us introduce a function

$$F(S, t, \varepsilon) = P(\frac{S}{\varepsilon}, t). \tag{32}$$

Consider first the domain $S > S_0$. Equation (30) in this domain takes the form

$$\varepsilon^2 \frac{\partial F(y,t,\varepsilon)}{\partial t} + \Lambda_1 F(y, t, \varepsilon) = \\ \varepsilon \frac{\partial}{\partial y} [(k\varepsilon y - C) F(y, t, \varepsilon)] + \Lambda_1 \int_0^\infty F(y + \varepsilon z, t, \varepsilon) \varphi(z) dz. \tag{33}$$

Taking Taylor expansion of $F(y + \varepsilon z, t, \varepsilon)$ with respect to the first argument and focusing our analysis on the first three member of the sum we get

$$\varepsilon^2 \frac{\partial F(y, t, \varepsilon)}{\partial t} = \varepsilon \frac{\partial}{\partial y} [(k\varepsilon y - C + \Lambda_1 a) F(y, t, \varepsilon)] + \Lambda_1 \frac{a_2}{2} \varepsilon 2 \frac{\partial^2 F(y, t, \varepsilon)}{\partial y^2} + o(\varepsilon^2). \tag{34}$$

Introduce new variables

$$t = t, \quad u = y - \frac{1}{\varepsilon} x(t), \tag{35}$$

where the function $x(t)$ will be determined later on, and a function $Q(u, t, \varepsilon)$ such that

$$F(y, t, \varepsilon) = Q(y - \frac{1}{\varepsilon} x(t), t, \varepsilon). \tag{36}$$

We impose an additional condition on $x(t)$ to satisfy equation

$$\dot{x}(t) = -kx(t) + C - \Lambda_1 a. \tag{37}$$

Then for $Q(u, t, \varepsilon)$ we have

$$\frac{\partial Q(u, t, \varepsilon)}{\partial t} = \frac{\partial}{\partial u} [kuQ(u, t, \varepsilon)] + \frac{\Lambda_1 a_2}{2} Q(u, t, \varepsilon) + \frac{o(\varepsilon^2)}{\varepsilon^2}. \tag{38}$$

Let

$$Q(u, t) = \lim_{\varepsilon \to 0} Q(u, t, \varepsilon). \tag{39}$$

Then

$$\frac{\partial Q(u, t)}{\partial t} = \frac{\partial}{\partial u} [kuQ(u, t)] + \frac{\Lambda_1 a_2}{2} \frac{\partial^2 Q(u, t)}{\partial u^2}. \tag{40}$$

The stochastic differential equation that satisfies (40) for the process $u(t)$ is of the form

$$du(t) = -ku(t)dt + \sqrt{\Lambda_1 a_2} dW(t), \tag{41}$$

where $W(t)$ – is a standard Wiener process.

From (37) and (41), accounting for the variable changes been made, we have for the process $\xi(t) = \varepsilon^2 S(t)$ when $\varepsilon \ll 1$ that

$$d\xi(t) = -k\xi(t)dt + (C - \Lambda_1 a)dt + \sqrt{\Lambda_1 a_2}\varepsilon dW(t). \tag{42}$$

Let

$$h(z,t) = \frac{\partial \Pr\{\xi(t) < z\}}{\partial z}. \tag{43}$$

According to (42) probability density function $h(z,t)$ satisfies

$$\frac{\partial h(z,t)}{\partial t} = -\frac{\partial}{\partial z}\left[(C - \Lambda_1 a - kz)h(z,t)\right] + \frac{\Lambda_1 a_2}{2}\varepsilon^2 \frac{\partial^2 h(z,t)}{\partial z^2}. \tag{44}$$

In a steady state we get for probability density function

$$h(z) = \lim_{t \to \infty} h(z,t)$$

$$\frac{\Lambda_1 a_2 \varepsilon^2}{2}\frac{d^2 h(z)}{dz^2} + \frac{d}{dz}\left[(\Lambda_1 a - C + kz)h(z)\right] = 0. \tag{45}$$

From where accounting for the boundary condition $h(\infty) = 0$ we obtain

$$h(z) = Be^{-\frac{(\Lambda_1 a - C + kz)^2}{\Lambda_1 a_2 \varepsilon^2 k}}. \tag{46}$$

Consider now the domain $S < S_0$. Equation (30) with respect to function $F(S,t,\varepsilon)$ (32) now takes the form

$$\varepsilon^2 \frac{\partial F(y,t,\varepsilon)}{\partial t} + \Lambda_0 F(y,t,\varepsilon) =$$
$$\varepsilon \frac{\partial}{\partial y}\left[(k\varepsilon yI(y) - C)F(y,t,\varepsilon)\right] + \Lambda_0 \int_0^\infty F(y + \varepsilon z, t, \varepsilon)\varphi(z)dz + R(y,\varepsilon), \tag{47}$$

where

$$R(y,\varepsilon) = (\Lambda_1 - \Lambda_0) \int_{S_0 - \frac{y}{\varepsilon}}^\infty F(y + \varepsilon z, t, \varepsilon)\varphi(z)dz = o(\varepsilon^2),$$

since the function $F(y,t,\varepsilon)$ is bounded and the second moment a_2 exists. Hence we do not account for the last member of the sum in (47). Taking Taylor expansion of $F(y + \varepsilon z, t, \varepsilon)$ with respect to the first argument we get

$$\varepsilon^2 \frac{\partial F(y,t,\varepsilon)}{\partial t} = \varepsilon \frac{\partial}{\partial y}\left[(k\varepsilon yI(y) - C + \Lambda_0 a)F(y,t,\varepsilon)\right] + \Lambda_0 \frac{a_2}{2}\varepsilon^2 \frac{\partial^2 F(y,t,\varepsilon)}{\partial y^2} + o(\varepsilon^2). \tag{48}$$

Consider $y < 0$. Making substitutions (35) and (36) and assuming

$$\dot{x}(t) = C - \Lambda_0 a, \tag{49}$$

we have for $\varepsilon \to 0$ for the function (13)

$$\frac{\partial Q(u,t)}{\partial t} = \frac{\Lambda_0 a_2}{2} \frac{\partial^2 Q(u,t)}{\partial u^2}.$$

(50)

Let $y > 0$. Making substitutions (36) and (37) and assuming

$$\dot{x}(t) = -kx(t) + C - \Lambda_0 a,$$

(51)

we have for $\varepsilon \to 0$ for the function $Q(u,t)$ (39)

$$\frac{\partial Q(u,t)}{\partial t} = \frac{\partial}{\partial u}[kuQ(u,t)] + \frac{\Lambda_0 a_2}{2}\frac{\partial^2 Q(u,t)}{\partial u^2}.$$

(52)

It follows from (49)–(52) that for $\varepsilon \ll 1$ the process $\xi(t) = \varepsilon^2 S(t)$ satisfies a stochastic differential equation

$$d\xi(t) = -k\xi(t)I(\xi(t))dt + (C - \Lambda_0 a)dt + \sqrt{\Lambda_0 a_2}\varepsilon dW(t).$$

(53)

Thus the probability density function (43) satisfies the following equation

$$\frac{\partial h(z,t)}{\partial t} = -\frac{\partial}{\partial z}[(C - \Lambda_0 a - kzI(z))h(z,t)] + \frac{\Lambda_1 a_2}{2}\varepsilon^2\frac{\partial^2 h(z,t)}{\partial z^2},$$

whereas in a steady state for the probability density function $h(z)$ we have

$$\frac{\Lambda_0 a_2 \varepsilon^2}{2}\frac{d^2 h(z)}{dz^2} + \frac{d}{dz}[(\Lambda_0 a - C + kzI(z))h(z)] = 0.$$

(54)

Taking in account boundary condition $h(-\infty) = 0$ for $z < 0$ we obtain

$$h(z) = De^{\frac{2(C-\Lambda_0 a)}{\Lambda_0 a_2 \varepsilon^2}z}.$$

(55)

For $0 = z = s_0$ the solution of (54) takes the form

$$h(z) = (D_1 + D_2 \int_0^z e^{-\frac{(kx+C-\Lambda_0 a)^2}{k\Lambda_0 a_2 \varepsilon^2}}dx)e^{\frac{(kz+C-\Lambda_0 a)^2}{k\Lambda_0 a_2 \varepsilon^2}}.$$

(56)

In $z = 0$ the continuity conditions $h(0-0) = h(0+0)$, $h'(0-0) = h'(0+0)$ must hold since function $h(z)$ satisfies a second-order differential equation. Hence $D_2 = 0$ and $D = D_1 e^{\frac{(c-\Lambda_0 a)^2}{\Lambda_0 a_2 \varepsilon^2}}$.

Thus the probability density function $h(s)$ is determined by the following expression

$$h(s) = \begin{cases} Ae^{\frac{(C-\Lambda_0 a)^2}{\Lambda_0 a_2 \varepsilon^2 k}}e^{\frac{2(C-\Lambda_0 a)}{\Lambda_0 a_2 \varepsilon^2}s}, & s < 0, \\ Ae^{\frac{(ks+C-\Lambda_0 a)^2}{\Lambda_0 a_2 \varepsilon^2 k}}, & 0 \le s \le s_0, \\ Be^{-\frac{(ks+C-\Lambda_1 a)^2}{\Lambda_1 a_2 \varepsilon^2 k}}, & s > s_0. \end{cases}$$

(57)

The relationship between A and B follows, firstly, from the normalising condition

$$\int\limits_{-\infty}^{0} h(s)ds + \int\limits_{0}^{s_0} h(s)ds + \int\limits_{s_0}^{\infty} h(s)ds = 1 \,,$$

and, secondly, from Eq. (30) when $S = S_0$ in a steady state, which under the above takes the form

$$(kS_0 - c)\frac{\partial P(S_0, \infty)}{\partial S} + (k - \lambda_0)P(S_0, \infty) + \lambda_1 \int\limits_{0}^{\infty} P(S_0 + y, \infty)\varphi(y)dy = 0. \quad (58)$$

Substituting the probability density $P(S, \infty)$ with its approximation (57) we get the second equation that describes the relationship between A and B. To obtain the final expressions one must, evidently, know the explicit form of the probability density function $\varphi(y)$.

5 Conclusion

In this paper we obtain expressions for the probability density function of the stock of perishable goods under constant arrival speed and switch-hysteresis control of the purchase process intensity. We also obtain the explicit solutions for the case of exponentially distributed purchase amounts and a diffusion approximation of the goods production/selling process under switch control of selling intensity. A similar approach can be used when considering other models of control for production and sales of perishable goods, in particular, a model with switch-hysteresis control of the production speed.

References

1. Nazarov, A.A.: Asymptotic Analysis of Markovian Systems, 158 p. Tomsk State University Publishing House, Tomsk (1991) (in Russian)
2. Nazarov, A.A., Moiseeva, S.P.: Method of Asymptotic Analysis in the Queuing Theory, 112 p. NTL, Tomsk (2006) (in Russian)
3. Goyal, S.K., Giri, B.C.: Recent trends in modeling of deteriorating inventory. Eur. J. Oper. Res. **134**(1), 1–16 (2001)
4. Bakker, M., Riezebos, J.J., Teunter, R.H.: Review of inventory systems with deterioration since 2001. Eur. J. Oper. Res. **221**, 275–284 (2012)
5. Mishra, V.K.: An inventory model of instantaneous deteriorating items with controllable deterioration rate for time dependent demand and holding cost. J. Ind. Eng. Manage. **6**(2), 495–506 (2013)
6. Mishra, V.K., Singh, L.S.: Deteriorating inventory model with time dependent demand and partial backlogging. Appl. Math. Sci. **4**(72), 3611–3619 (2010)
7. Begum, R., Sahu, S.K., Sahoo, R.R.: An inventory model for deteriorating items with quadratic demand and partial backlogging. Br. J. Appl. Sci. Technol. **2**(2), 112–131 (2012)

8. Begum, R., Sahu, S.K.: An EOQ model for deteriorating items quadratic demand and shortages. Int. J. Inventory Control Manage. **2**(2), 257–268 (2012)
9. Tripathi, R.P., Singh, D., Mishra, T.: EOQ model for deteriorating items with exponential time dependent demand rate under inflation when supplier credit linked to order quantity. Int. J. Supply Oper. Manage. **1**(I), 20–37 (2014)
10. Sharma, V., Chaudhary, R.: An inventory model for deteriorating items with weibull deterioration with time dependent demand and shortages. Res. J. Manage. Sci. **2**(3), 28–30 (2013)
11. Tripathy, C.K., Mishra, U.: An inventory model for weibull deteriorating items with price dependent demand and time-varying holding cost. Appl. Math. Sci. **44**(4), 2171–2179 (2010)

CUSUM Algorithms for Parameter Estimation in Queueing Systems with Jump Intensity of the Arrival Process

Yulia Burkatovskaya[1]([✉]), Tatiana Kabanova[2],
and Sergey Vorobeychikov[3]

[1] Institute of Cybernetics, Tomsk Polytechnic University,
30 Lenin Prospekt, 634050 Tomsk, Russia
tracey@tpu.ru
http://www.tpu.ru

[2] Department of Applied Mathematics and Cybernetics, Tomsk State University,
36 Lenin Prospekt, 634050 Tomsk, Russia
tvk@bk.ru

[3] International Laboratory of Statistics of Stochastic Processes and Quantitative
Finance, Tomsk State University, 36 Lenin Prospekt, 634050 Tomsk, Russia
sev@mail.tsu.ru
http://www.tsu.ru

Abstract. The problem of Markov-modulated Poisson process intensities estimating is studied. A new approach based on sequential change point detection method is proposed to determine switching points of the flow parameter. Both the intensities of the controlling Markovian chain and the intensities of the flow of events are estimated. The results of simulation are presented.

Keywords: Markov-modulated poisson process · Jump intensity · CUSUM algorithm

1 Introduction

Markovian arrival processes form a powerful class of stochastic processes introduced in [1,2] and thereafter they are widely used now as models for input flows to queueing systems where the rate of the arrival of customers depends on some external factors. MAP is a counting process whose arrival rate is governed by a continuous-time Markov chain.

Queueing systems with jump intensity of customer arrivals is one of the examples of applying MAP. In such models the intensity is supposed to be piecewise constant function depended on the state of random environment. Particulary, this model can be used as a model of a call-center or http-server customers (see [3,4]), healthcare systems (see [5]), etc. Usually the stationary probabilities of system states, sojourn and waiting time distributions, mean length of the queue and other parameters are investigated. To solve such problems there is a need to estimate parameters of customer arrivals.

© Springer International Publishing Switzerland 2015
A. Dudin et al. (Eds.): ITMM 2015, CCIS 564, pp. 275–288, 2015.
DOI: 10.1007/978-3-319-25861-4_24

The typical property of observing time series derived from a MAP is that only the arrivals but not the states of the controlling Markovian chain can be seen. The problem is to estimate both the controlling Markovian chain parameters and parameters of the intensity of the arrival process. A survey of estimation methods is given in [6]. Its emphasis is on maximum likelihood estimation and its implementation via the EM (expectation-maximization) algorithm. The EM iteration alternates between performing an expectation (E) step, which creates a function for the expectation of the log-likelihood evaluated using the current estimator for the parameters, and a maximization (M) step, which computes parameters maximizing the expected log-likelihood found on the E step. These parameter estimators are then used to determine the distribution of the latent variables in the next E step. This approach is developed for different conditions in [7,8], etc. The survey [9] with a huge bibliography is focused on matching moment method which is also widely used for parameter estimation in MAP because of its simplicity. This method is used, for example, in [10]. Bayesian approach based on the a posteriori probability of the controlling chain state is developed in [11]. It provides estimators with the minimum mean square error.

In this paper we propose a different approach. We use the sequential analysis methods for parameter estimation in queueing system with jump intensity of the arrival process. The key idea is to consider time intervals between arrivals as a stochastic process which parameters change in random points. First we are going to detect these points using sequential change point detection methods. Then we are going to estimate the intensity parameters under the assumption that the intensity is constant between detected change points.

The problem of sequential change point detection can be formulated as follows. A stochastic process is observed. Several parameters of the process change in random point. The problem is to detect this change point when the process is observed online. Sequential methods include a special stopping rule that determines a stopping time. At this instant a decision on change point can be made. There are two types of errors typical for sequential change point detection procedures: false alarm, when one makes a decision that change is occurred before a change point (type 1 error), and delay, when the change is not detected (type 2 error).

The CUSUM (or cumulative sum control chart) algorithm was proposed by E.S. Page in [12] and since then it is widely used for online detecting changes in parameters for different time series both with independent and with dependent observations, even for autoregressive type processes. Usually the change in the mean is considered. As far as a change of the state of the controlling Markovian chain causes a jump of the mean length of an interval between arrivals hence the lengths of intervals form a sequence of dependent random variables and it is possible to apply the CUSUM algorithm to this situation. G. Lorden in [14] established that the CUSUM procedure is optimal in a sense that it provides minimum mean time of delay in change detecting when mean time between false alarms is fixed. In this paper we use the CUSUM procedure to determine intervals of the constant intensity of the observed flow of events. After that parameter estimators are constructed.

2   Problem Statement

We consider a Markov-modulated poisson process, i.e. a flow of events, controlled by a Markovian chain with a continuous time. The chain has two states, transition between the states happens at random instants. The time of sojourn of the chain in the i-th state is exponentially distributed with the parameter α_i, where $i = 1, 2$.

The flow of events has the exponential distribution with the intensity parameter λ_1 or λ_2 subject to the state of the Marcovian chain. The parameters of the system λ_1, λ_2 and the instants of switching between the states are supposed to be unknown. We also suppose that $\lambda_i \ll \alpha_i$, i.e., changes of the controlling chain states occur more rarely than observed events. Thus some events occur between switchings of the controlling chain states. This situation is typical for real processes such as call-center or http-server because one of the states can be interpreted as a "usual" state of the system and another state as a "peak-time" state and during each of these states several customers are supposed to arrive. Besides processes having this property are often used for simulation study of algorithms for processes with jump intensity of customer arrivals (for example, see [8,13]).

The sequence of instants of arriving events is observed. The problem is to estimate the parameters λ_1, λ_2, α_1, α_2.

3   Algorithm 1

Let the process $\{t_i\}_{i \geq 0}$ be the sequence of the instants when events of the observed flow occur. Consider the process $\{\tau_i\}_{i \geq 1}$, where $\tau_i = t_i - t_{i-1}$ is the length of the i-th interval between arriving events in the observed flow as it is shown at the diagram (Fig. 1).

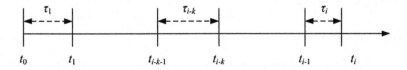

Fig. 1. Construction of the sequence $\{\tau_i\}$.

If the controlling chain is in the l-th state then the mean length between events is equal to $1/\lambda_l$. So at the first stage of our procedure we try to detect the instants of the chain transition from one state to another as the instants of change in the mean of the process $\{\tau_i\}_{i \geq 1}$ using CUSUM procedures.

Let the parameters λ_1, λ_2 satisfy the condition

$$
\begin{aligned}
0 < \lambda_2 < \lambda_1; \\
\frac{1}{\lambda_2} - \frac{1}{\lambda_1} > \Delta,
\end{aligned}
\tag{1}
$$

where Δ is a certain known positive parameter. Choose then an integer parameter $k > 1$ describing the memory depth. The idea is to compare the values τ_i and τ_{i-k}. If there are no changes of the controlling chain state within the interval $[t_{i-k-1}, t_i]$ then the values τ_i and τ_{i-k} have the identical exponential distribution with the mean $1/\lambda_1$ or $1/\lambda_2$. If the chain state changes within the interval $[t_{i-k}, t_{i-1}]$ then the expectations of the values τ_i and τ_{i-k} are different.

On one hand the parameter k should allow us to detect changes with minimal delay, on the other hand it should not be too large to contain more than one chain state change within the interval $[t_{i-k}, t_{i-1}]$. Further we consider the choice of the parameter k in detail.

As the initial state of the chain is unknown, we shall consider two CUSUM procedures simultaneously. The first procedure is set up to detect increase in the mean of the process and hence, decrease of the intensity, and the second procedure is set up to detect decrease in the mean and hence, increase of the intensity. For the first procedure we introduce the sequence of the statistics

$$z_i^{(1)} = \tau_i - \tau_{i-k} - \Delta, \quad i > k. \tag{2}$$

For the second procedure we introduce the sequence of the statistics

$$z_i^{(2)} = \tau_{i-k} - \tau_i - \Delta, \quad i > k. \tag{3}$$

This statistics are calculated at the instant t_i.

Consider then four hypothesis concerning the state of the controlling chain:

- $H_1(t_{i-k-1}, t_i)$ – the intensity of the arrival process on the interval $[t_{i-k-1}, t_i]$ is constant and equal to λ_1;
- $H_2(t_{i-k-1}, t_i)$ – the intensity of the arrival process on the interval $[t_{i-k-1}, t_i]$ is constant and equal to λ_2;
- $H_{1,2}(t_{i-k}, t_{i-1})$ – the intensity of the arrival process on the interval $[t_{i-k}, t_{i-1}]$ changed once from λ_1 to λ_2;
- $H_{2,1}(t_{i-k}, t_{i-1})$ – the intensity of the arrival process on the interval $[t_{i-k}, t_{i-1}]$ changed once from λ_2 to λ_1.

Theorem 1. *If the parameter Δ satisfies condition (1) then the statistics $z_i^{(j)}$, $j \in \{1, 2\}$ (2), (3) have the following properties:*

$$
\begin{aligned}
E\left[z_i^{(1)} \middle| H_l(t_{i-k-1}, t_i) \right] &< 0, \quad l = 1, 2; \\
E\left[z_i^{(1)} \middle| H_{1,2}(t_{i-k}, t_{i-1}) \right] &> 0; \\
E\left[z_i^{(2)} \middle| H_l(t_{i-k-1}, t_i) \right] &< 0, \quad l = 1, 2; \\
E\left[z_i^{(2)} \middle| H_{2,1}(t_{i-k}, t_{i-1}) \right] &> 0.
\end{aligned}
\tag{4}
$$

Proof. Using (1) one obtains

$$E\left[z_i^{(1)}\middle| H_l(t_{i-k-1},t_i)\right] = E\left[\tau_i - \tau_{i-k} - \Delta\middle| H_l(t_{i-k-1},t_i)\right]$$
$$= \frac{1}{\lambda_l} - \frac{1}{\lambda_l} - \Delta < 0;$$
$$E\left[z_i^{(1)}\middle| H_{1,2}(t_{i-k},t_{i-1})\right] = E\left[\tau_i - \tau_{i-k} - \Delta\middle| H_{1,2}(t_{i-k},t_{i-1})\right]$$
$$= \frac{1}{\lambda_2} - \frac{1}{\lambda_1} - \Delta > 0;$$
$$E\left[z_i^{(2)}\middle| H_l(t_{i-k-1},t_i)\right] = E\left[\tau_{i-k} - \tau_i - \Delta\middle| H_l(t_{i-k-1},t_i)\right]$$
$$= \frac{1}{\lambda_l} - \frac{1}{\lambda_l} - \Delta < 0;$$
$$E\left[z_i^{(2)}\middle| H_{2,1}(t_{i-k},t_{i-1})\right] = E\left[\tau_{i-k} - \tau_i - \Delta\middle| H_{2,1}(t_{i-k},t_{i-1})\right]$$
$$= \frac{1}{\lambda_2} - \frac{1}{\lambda_1} - \Delta > 0.$$

So the means of statistics (2), (3) change from negative value to positive when the intensity of the process changes. These properties determine the construction of the procedures. We introduce positive values h_1 and h_2 as the procedures thresholds and construct the cumulative sums $S_i^{(1)}$ and $S_i^{(2)}$ which are recalculated at the instants t_i. For the first procedure it is defined as follows

$$S_0^{(1)} = \Delta;$$
$$S_i^{(1)} = \max\{0, S_{i-1}^{(1)} + z_i^{(1)}\}, \quad i > k; \qquad (5)$$
$$S_i^{(1)} = 0, \quad \text{if} \quad S_i^{(1)} \geq h_1.$$

For the second procedure the cumulative sum is defined as follows

$$S_0^{(2)} = \Delta;$$
$$S_i^{(2)} = \max\{0, S_{i-1}^{(2)} + z_i^{(2)}\}, \quad i > k; \qquad (6)$$
$$S_i^{(2)} = 0, \quad \text{if} \quad S_i^{(2)} \geq h_2.$$

If the cumulative sum $S_i^{(1)}$ reaches the threshold h_1 then the decision is made that the mean time between events increased and hence the intensity of the process decreased, i.e., it changed from λ_1 to λ_2. If the cumulative sum $S_i^{(2)}$ reaches the threshold h_2 then the decision is made that the mean time between events decreased and hence the intensity of the process increased, i.e., it changed from λ_2 to λ_1. Once a sum reaches threshold it is reset to zero and the corresponding procedure is restarted.

Let the sequence $\left\{\sigma_m^{(l)}\right\}_{m\geq 0}$ be the sequence of the instants when the cumulative sum in the l-th procedure reaches the threshold h_l, i.e.

$$\sigma_0^{(l)} = 0;$$
$$\sigma_m^{(l)} = \min\left\{t_j > \sigma_{m-1}^{(l)} : S_j^{(l)} \geq h_l\right\}. \qquad (7)$$

Consider a sequence $\left\{n_i^{(l)}\right\}_{i\geq 0}$ associated with the sequence $\left\{\sigma_m^{(l)}\right\}_{m\geq 0}$ as follows

$$n_0^{(l)} = 0;$$
$$n_m^{(l)} = \max\left\{t_j \leq \sigma_m^{(l)} : S_j^{(l)} > 0, S_{j-1}^{(l)} = 0\right\}. \tag{8}$$

Thus the instant $n_m^{(l)}$ is the first instant when the cumulative sum becomes positive to reach then the threshold. The construction of the sequences are illustrated at Fig. 2. The instants of occurrences t_i are marked by vertical dotted lines. At the diagram above an example of the sum $S_j^{(1)}$ behavior is presented and the instants $\sigma_m^{(1)}$ and $n_m^{(1)}$ are marked out. At the diagram in the middle a similar example for the sum $S_j^{(2)}$ is shown.

We consider the instants $n_i^{(1)}$ as the estimators for the instants when the mean length between the events increases. They are pointed by up arrows at the diagram below. In turn the instants $n_i^{(2)}$ are considered as the estimators for the instants when the mean length between the events increases. They are pointed by down arrows at the diagram below.

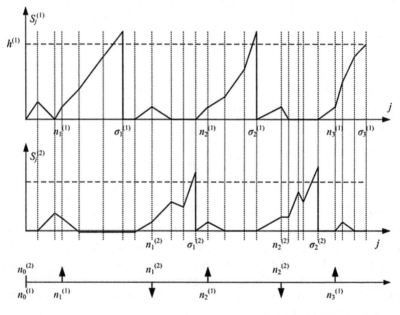

Fig. 2. Construction of the sequences $\left\{\sigma_m^{(l)}\right\}$, $\left\{n_m^{(l)}\right\}$.

In connection with sequential change point detection procedures two type of errors are considered: the false alarm and the skip of the change. A false alarm occurs when one of the cumulative sums reaches the corresponding threshold in the case of the constant intensity of the arrival process. A skip of the change occurs when the change of the parameter occurs but the corresponding cumulative sum does not reach its threshold.

When implementing the procedure it is possible to encounter false alarm situations. We shall record all the exceeding the thresholds by either first or the second cumulative sum. If the same sum reaches threshold several times in a row, we only record the first occurrence.

Thus the procedure for estimation of instants of intensity switching is described as follows. Calculate two cumulative sums given by Eqs. (5), (6). Then construct the sequences $\left\{\sigma_m^{(l)}\right\}$, $\left\{n_m^{(l)}\right\}$ defined by Eqs. (7), (8). Let $n_1^{(1)} < n_1^{(2)}$, then the initial value of the intensity is equal to λ_1. Define the sequence

$$
\begin{aligned}
q_0 &= 0; \\
q_{2l+1} &= \min\left\{n_i^{(1)} : n_i^{(1)} > q_{2l}\right\}, \quad l \geq 0; \\
q_{2l+2} &= \min\left\{n_i^{(2)} : n_i^{(2)} > q_{2l+1}\right\}, \quad l \geq 0.
\end{aligned}
\tag{9}
$$

The values q_1, q_2, ... are calculated using formula (9) while the set

$$
\begin{aligned}
\left\{n_i^{(2)} : n_i^{(1)} > q_{2l}\right\} &\neq \emptyset; \\
\left\{n_i^{(1)} : n_i^{(2)} > q_{2l+1}\right\} &\neq \emptyset.
\end{aligned}
$$

If

$$
\left\{n_i^{(2)} : n_i^{(1)} > q_{2l}\right\} = \emptyset \quad \left(\left\{n_i^{(1)} : n_i^{(2)} > q_{2l+1}\right\} = \emptyset\right)
$$

then we set $q_{2l+1} = N$ ($q_{2l+2} = N$), where N is the instant of the last occurrence. Here the odd instants q_{2l+1} are the estimators of the instants when the intensity changes from λ_1 to λ_2, and the even instants q_{2l+2} are the estimators of the instants when the intensity changes from λ_2 to λ_1.

An example of the sequence construction is illustrated at Fig. 3. The sequences n_i^l are shown at the diagram above. The instants of switching the controlling chain state from 1 to 2 are pointed by up arrows, the instants of switching the controlling chain state from 2 to 1 are pointed by down arrows at the diagram below. The intervals are marked by the numbers of the states of the controlling chain.

Define estimators for parameters λ_1, λ_2

$$
\hat{\lambda}_1 = \frac{N_1}{T_1}, \quad \hat{\lambda}_2 = \frac{N_2}{T_2},
\tag{10}
$$

where N_1 is the total number of events occurred at the intervals $[q_{2l}, q_{2l+1}]$, $q_{2l+1} \leq N$ and T_1 is the total length of these intervals; N_2 is the total number of events occurred at the intervals $[q_{2l+1}, q_{2l+2}]$, $q_{2l+2} \leq N$ and T_2 is the total length of these intervals; $l \geq 0$ (Fig. 3).

Define estimators for parameters α_1, α_2

$$
\hat{\alpha}_1 = \frac{L_1}{T_1}, \quad \hat{\alpha}_2 = \frac{L_2}{T_2},
\tag{11}
$$

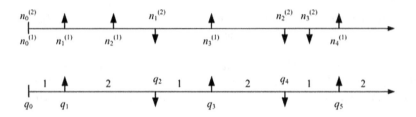

Fig. 3. Construction of the sequences $\{q_m\}$.

where L_1 is the total number of the switching points $q_{2l+1} \leq N$, L_2 is the total number of the switching points $q_{2l+2} \leq N$, $l \geq 0$.

If $n_1^{(2)} < n_1^{(1)}$, then the initial value of the intensity is equal to λ_2 the procedure is similar. Define the sequence

$$q_0 = 0;$$
$$q_{2l+1} = \min \left\{ n_i^{(2)} : n_i^{(1)} > q_{2l} \right\}, \quad l \geq 0; \tag{12}$$
$$q_{2l+2} = \min \left\{ n_i^{(1)} : n_i^{(2)} > q_{2l+1} \right\}, \quad l \geq 0.$$

Here the odd instants q_{2l+1} are the estimators of the instants when the intensity changes from λ_2 to λ_1, and the even instants q_{2l+2} are the estimators of the instants when the intensity changes from λ_1 to λ_2. Estimators for the parameters λ_1, λ_2 are calculated using formula (10), where N_1 is the total number of events occurred at the intervals $[q_{2l+1}, q_{2l+2}]$ and T_1 is the total length of these intervals; N_2 is the total number of events occurred at the intervals $[q_{2l}, q_{2l+1}]$ and T_2 is the total length of these intervals; $l \geq 0$. Estimators for the parameters α_1, α_2 are calculated using formula (11), where L_1 is the total number of the switching points $q_{2l+2} < N$, L_2 is the total number of the switching points $q_{2l+1} < N$, $l \geq 0$.

4   Choice of the Algorithm Parameters

In this section the problem of choice of the parameters k, Δ and h_l is discussed.

We suppose that changes of the controlling chain states occur more rarely than observed events. First, we consider the memory depth parameter k. Let n be a lower bound of the mean number of events between switchings of the controlling chain states. For the model under consideration it means that $n\alpha_i \leq \lambda_i$. It means that it is not effective to choose the memory depth $k \geq n$ or close to n because in this case there can be many situations when more than one chain state change occur within the interval $[t_{i-k}, t_{i-1}]$. On the other hand, the sum $S_i^{(l)}$ should reach the corresponding threshold h_i after switching of the controlling chain state, i.e. some statistics $z_i^{(l)}$ should be positive. It follows from these considerations and numerical calculations that a good choice of the parameter k is

$$k \approx \frac{n}{2}. \tag{13}$$

Then, turn to the parameters Δ and h_l. Condition (1) provides properties (4). The properties make it possible to construct CUSUM procedures. Thus the parameter Δ can be chosen from the interval $(0, 1/\lambda_2 - 1/\lambda_1)$, i.e., it is positive and does not exceed the difference between the mean lengths of the intervals τ_i when the controlling chain is in different states. Let this difference be not less than some $d > 0$:

$$\frac{1}{\lambda_2} - \frac{1}{\lambda_1} \geq d. \tag{14}$$

The parameter Δ and h_l affects the characteristics of the CUSUM procedure, i.e., the mean delay and the mean time between false alarms (see [14]). If the parameter h_l is fixed then increase of the parameter Δ results in decrease of the mean of the statistic $z_i^{(l)}$ and hence the sum $S_i(l)$ reaches the threshold h_l more slowly and hence, a switching of the controlling chain state from the state l can be skipped. Consequently, the number of false detection of the controlling chain state switchings decreases but on the other hand the number of skips of the controlling chain state switchings increases. If the parameter Δ is fixed then increase of the parameter h_l results in the same effect. Vice versa, decrease of the parameter Δ or the parameter h_l while the other parameter is fixed result in increase of the number of false detection of the controlling chain state switchings and decrease of the number of skips of the controlling chain state switchings.

If there are no additional conditions then the procedure is considered to be optimal when the probabilities of the false detection and the skip of the change are equal. It can be guaranteed by the following conditions

$$\begin{aligned}
E\left[z_i^{(1)} \middle| H_l(t_{i-k-1}, t_i)\right] &= -E\left[z_i^{(1)} \middle| H_{1,2}(t_{i-k}, t_{i-1})\right] ; \\
E\left[z_i^{(2)} \middle| H_l(t_{i-k-1}, t_i)\right] &= -E\left[z_i^{(2)} \middle| H_{2,1}(t_{i-k}, t_{i-1})\right].
\end{aligned} \tag{15}$$

It results in the equations (see Theorem 1)

$$-\Delta = -\frac{1}{\lambda_2} + \frac{1}{\lambda_1} + \Delta.$$

Hence the best choice of the parameter Δ is

$$\Delta = \frac{d}{2} \tag{16}$$

where d is defined by the Eq. (14), i.e., Δ is the half of the difference between the mean lengths of the intervals τ_i when the controlling chain is in different states. If the difference is unknown then one has to choose as d a lower bound of the difference. In other words, one has to define the minimal difference that should be detected by the algorithm.

Consider now the parameter h_l. If the memory depth is equal to k then the sum $S_i^{(l)}$ to reach the threshold h_l in not more then k steps (while $E z_i^{(l)} > 0$).

If the parameter Δ satisfies the condition (16) then using (14) and Theorem 1 one obtains

$$E\left[z_i^{(1)}\middle| H_{1,2}(t_{i-k},t_{i-1})\right] = \frac{1}{\lambda_2} - \frac{1}{\lambda_1} - \Delta \geq d - \frac{d}{2} = \frac{d}{2};$$
$$E\left[z_i^{(2)}\middle| H_{2,1}(t_{i-k},t_{i-1})\right] = \frac{1}{\lambda_2} - \frac{1}{\lambda_1} - \Delta \geq d - \frac{d}{2} = \frac{d}{2}.$$

So if the sum $S_i^{(l)}$ starts from zero it reaches the threshold h_l on the average in $2h_l/d$ steps. Hence it is supposed to choose the threshold h_l from the condition $2h_l/d < k$, i.e.

$$h_l < \frac{kd}{2} \approx \frac{nd}{4}. \tag{17}$$

Note that the parameter h_l should not be significantly than its upper bound because it can increase the number of false alarms.

In general the choice of the CUSUM parameters is a rather difficult problem requiring further theoretical investigations. Nevertheless, numerical simulations demonstrated a good quality of the proposed algorithm with the parameters (13), (16), (17).

5   Algorithm 2

The second algorithm is very similar to the first except of the definition of the statistics $z_i^{(l)}$.

Let we have a certain period of observation $[0,T]$ and N is the number of occurrences at the interval. First, we calculate the mean of the length between occurrences using the usual formula

$$\hat{\tau} = \frac{T}{N}. \tag{18}$$

The value $\hat{\tau}$ exceeds the mean length of the interval τ_i when the controlling chain is in the first state, and vice versa, the mean length of the interval τ_i exceeds the value $\hat{\tau}$ when the controlling chain is in the second state, i.e.

$$\frac{1}{\lambda_1} < E\hat{\tau} < \frac{1}{\lambda_2}. \tag{19}$$

Using this property we can construct statistics as follows. For the first procedure we introduce the sequence of the statistics

$$z_i^{(1)} = \tau_i - \hat{\tau}. \tag{20}$$

For the second procedure we introduce the sequence of the statistics

$$z_i^{(2)} = -\tau_i + \hat{\tau}. \tag{21}$$

Consider then two hypothesis concerning the state of the controlling chain:

- $H_1(t_{i-1}, t_i)$ – the intensity of the arrival process on the interval $[t_{i-1}, t_i]$ is constant and equal to λ_1;
- $H_2(t_{i-1}, t_i)$ – the intensity of the arrival process on the interval $[t_{i-1}, t_i]$ is constant and equal to λ_2;

Theorem 2. *The statistics $z_i^{(j)}$, $j \in \{1,2\}$ (20), (21) have the following properties:*

$$
\begin{aligned}
E\left[z_i^{(1)} \middle| H_1(t_{i-1}, t_i)\right] &< 0; \\
E\left[z_i^{(1)} \middle| H_2(t_{i-1}, t_i)\right] &> 0; \\
E\left[z_i^{(2)} \middle| H_1(t_{i-1}, t_i)\right] &> 0; \\
E\left[z_i^{(2)} \middle| H_2(t_{i-1}, t_i)\right] &< 0.
\end{aligned}
\tag{22}
$$

Proof. Using (19) one obtains

$$
\begin{aligned}
E\left[z_i^{(1)} \middle| H_1(t_{i-1}, t_i)\right] &= E\left[\tau_i - \hat{\tau} \middle| H_1(t_{i-1}, t_i)\right] = \frac{1}{\lambda_1} - E\hat{\tau} < 0; \\
E\left[z_i^{(1)} \middle| H_2(t_{i-1}, t_i)\right] &= E\left[\tau_i - \hat{\tau} \middle| H_2(t_{i-1}, t_i)\right] = \frac{1}{\lambda_2} - E\hat{\tau} > 0; \\
E\left[z_i^{(2)} \middle| H_1(t_{i-1}, t_i)\right] &= E\left[-\tau_i + \hat{\tau} \middle| H_1(t_{i-1}, t_i)\right] = -\frac{1}{\lambda_1} + E\hat{\tau} > 0; \\
E\left[z_i^{(2)} \middle| H_2(t_{i-1}, t_i)\right] &= E\left[-\tau_i + \hat{\tau} \middle| H_2(t_{i-1}, t_i)\right] = -\frac{1}{\lambda_2} + E\hat{\tau} < 0.
\end{aligned}
$$

So one can see that the statistics $Z_i^{(l)}$ change their means when the intensity of the arrival process changes. Using in Algorithm 1 statistics (20), (21) instead of (2), (3) we obtain Algorithm 2.

Consider now the choice of the parameters h_l. If n is a lower bound of the mean number of events between switchings of the controlling chain states then the sum $S_i^{(l)}$ should reach the threshold h_l on the average less then in n steps, for example, in $n/2$ steps. For Algorithm 2 we can not estimate the mean of the statistic $E\left[z_i^{(1)} \middle| H_2(t_{i-1}, t_i)\right]$ if the parameters α_i are unknown because we can not calculate $E\hat{\tau}$. Hence, we use a rather crude estimator

$$
\begin{aligned}
E\left[z_i^{(1)} \middle| H_2(t_{i-1}, t_i)\right] &\approx \frac{d}{2}; \\
E\left[z_i^{(2)} \middle| H_1(t_{i-1}, t_i)\right] &\approx \frac{d}{2}.
\end{aligned}
$$

So we come to the inequality

$$
h_l < \frac{nd}{4}
\tag{23}
$$

which is the same as in Algorithm 1.

6   Numerical Simulation

The model for the considered flow and the suggested algorithms was implemented with varying parameters. The results are presented in the tables below (Tables 1 and 2).

Table 1. The results of the simulation for Algorithm 1.

T	λ_1	λ_2	α_1	α_2	h_1	h_2	k	Δ	$\hat{\lambda}_1$	$\hat{\lambda}_2$	$\hat{\alpha}_1$	$\hat{\alpha}_2$
1000	5	1	0,4	0,2	1	1	5	0,2	4,2750	1,0951	0,2458	0,1641
1000	5	1	0,4	0,2	1,8	1,8	5	0,2	3,8424	1,0621	0,1687	0,1226
1000	5	1	0,4	0,2	1,8	1,8	8	0,2	4,3561	1,1852	0,1686	0,1180

Here we use the following notations:

- T is the time of simulation;
- λ_1 and λ_2 are the intensities of the arrival process in the first and the second state, correspondingly;
- α_1 and α_2 are the switching intensities from the first to the second state and vise versa, correspondingly;
- h_1 and h_2 are the CUSUM thresholds;
- k is the parameter of the algorithm, the difference between the numbers of the compared intervals at the statistics (2), (3);
- Δ is the parameter of the algorithm;
- $\hat{\lambda}_1$ and $\hat{\lambda}_2$ are the estimators of the parameters λ_1 and λ_2;
- $\hat{\alpha}_1$ and $\hat{\alpha}_2$ are the estimators of the parameters α_1 and α_2.

Table 2. The results of the simulation for Algorithm 2.

T	λ_1	λ_2	α_1	α_2	h_1	h_2	$\hat{\tau}$	$\hat{\lambda}_1$	$\hat{\lambda}_2$	$\hat{\alpha}_1$	$\hat{\alpha}_2$
1000	5	1	0,3	0,2	0,5	0,5	0,3883	5,2336	1,2276	0,322	0,1732
1000	5	1	0,3	0,2	0,8	0,8	0,3929	5,0591	1,2370	0,2573	0,1322
1000	5	1	0,3	0,2	1	1	0,4355	4,7475	1,1972	0,2587	0,1144
1000	5	1	0,1	0,2	0,5	0,5	0,2668	5,6804	2,0053	0,2912	0,2604
1000	5	1	0,1	0,2	0,8	0,8	0,2924	5,1544	1,6825	0,1180	0,1880
1000	5	1	0,1	0,2	1	1	0,2501	5,2498	2,6283	0,1207	0,1297
1000	5	2	0,1	0,2	0,5	0,5	0,2351	6,1085	2,8854	0,3632	0,2656
1000	5	2	0,1	0,2	0,8	0,8	0,2564	5,1785	2,8092	0,1652	0,1289
1000	5	2	0,1	0,2	1	1	0,2486	5,1949	2,8831	0,1219	0,1162
10000	5	1	0,3	0,2	0,8	0,8	0,3830	4,8379	1,3439	0,2316	0,1318
10000	5	1	0,3	0,2	1	1	0,3766	4,6783	1,4326	0,1917	0,1157

Here we use the same notations as above, $\hat{\tau}$ is the mean length of the interval between occurrences calculated by (18).

First, the quality of the proposed algorithms on the threshold parameters h_i was studied. Increasing of h_i leads to decreasing of probability for the cumulative sums to reach the thresholds and hence an intensity change can be undetected. It causes increasing of error of the estimators $\hat{\lambda}_l$ because of not correct estimation of the controlling chain current state.

On the other hand, increasing of h_i results in decreasing the total number of false alarms. These theoretical conclusions are supported by the simulation results. As the thresholds increase the estimators of the switching parameters $\hat{\alpha}_l$ decrease because less switching points are detected on the first stage of the procedures. In Table 2 for $h_1 = h_2 = 1$ one can see that the estimators $\hat{\alpha}_i$ considerably less the real values of the parameters α_i. The best results are obtained for $h_1 = h_2 = 0,8$ for all intensity parameter values. It supports our considerations concerning the parameter h_l. According to (23) for $\lambda_1 = 5$ and $\lambda_2 = 1$ minimal difference between the mean length of the intervals τ_i is $d = 1/1 - 1/5 = 0.8$ and the recommended choice of h_l is $h_l < (0.8 \times 5)/4 = 1$, but it should not be significantly less.

Increase of the simulation time from 1000 to 10000 does not influence significantly the estimators quality. This result stresses the fact that the proposed algorithms can be used for small sample size.

7   Conclusion

Markovian arrival processes serve as models for real processes, particularly, for call-centers or http-server customers, healthcare systems, etc. Input flow intensity estimation and pertinent model setup is necessary to develop dispatching rule, to calculate optimal number of servers, etc. The suggested algorithms do not use the distribution function of the observing flow and, hence, can be applied to parameter estimation of other types of flows.

Acknowledgements. This paper is supported by The National Research Tomsk State University Academic D.I. Mendeleev Fund Program (NU 8.1.55.2015 L) in 2014–2015.

References

1. Neuts, M.F.: A versatile Markovian point process. J. Appl. Probab. **16**, 764–774 (1979)
2. Lucantoni, D.M., Meier-Hellstern, K.S., Neuts, M.F.: A single server queue with server vacations and a class of non-renewal arrival processes. Adv. Appl. Probab. **22**, 676–705 (1990)
3. Dudina, O., Dudin, S.: Queueing system $MAP/M/N/N + K$ operating in random environment as a model of call center. In: Dudin, A., Klimenok, V., Tsarenkov, G., Dudin, S. (eds.) BWWQT 2013. CCIS, vol. 356, pp. 83–92. Springer, Heidelberg (2013)

4. Dudin, S.: The servicing system MAP(PH) plus MAP/PH/N/R as a model of optimizing an HTTP server with blockings. Autom. Remote Control **1**, 28–38 (2010)
5. Fackrell, M.: Health Care Manage. Sci. Modelling healthcare systems with phase-type distributions **12**, 11–26 (2008). Springer Science + Business Media, LLC
6. Asmussen, S.: Phase-type distributions and related point processes: fitting and recent advances. In: Chakravarthy, S.R., Alfa, A.S. (eds.) Matrix-analytic Methods in Stochastic Models. Lecture Notes in Pure and Applied Mathematics, pp. 137–149. Marcel Dekker, New York (1997)
7. Breuer, L., Kume, A.: An EM algorithm for Markovian arrival processesobserved at discrete times. In: Müller-Clostermann, B., Echtle, K., Rathgeb, E.R. (eds.) MMB&DFT 2010. LNCS, vol. 5987, pp. 242–258. Springer, Berlin (2010)
8. Okamura, H., Dohi, T., Trivedi, K.S.: Markovian arrival process parameter estimation with group data. IEEE/ACM Trans. Netw. **17**(4), 1326–1340 (2009)
9. Gerhardt, I., Nelson, B.L.: On capturing dependence in point processes: matching moments and other techniques. Technical report, Northwestern university (2009)
10. Duffie, D., Glynn, P.: Estimation of continuous-time markov processes sampled at random time intervals. Econometrica **72**, 1773–1808 (2004)
11. Gortsev, A.M., Zuevich, V.L.: Optimal estimation of parameters of an asynchronous doubly stochastic flow of events with arbitrary number of the states. Tomsk State Univ. J. Control Comput. Sci. **4**(17), 25–40 (2011)
12. Page, E.S.: Continuous inspection schemes. Biometrica **42**(1), 100–115 (1956)
13. Gortsev, A.M., Kalyagin, A.A., Nezhelskaya, L.A.: Optimum estimation of states in generalized semi-synchronous flow of events. Tomsk State Univ. J. Control Comput. Sci. **2**(11), 66–81 (2010)
14. Lorden, G.: Procedures for reacting to a change in distribution. Ann. Math. Stat. **42**, 1897–1908 (1971)

On Stochastic Models of Internet Traffic

Tadeusz Czachórski[1], Joanna Domańska[1],
and Michele Pagano[2]([⊠])

[1] Institute of Theoretical and Applied Informatics, Polish Academy of Sciences,
Baltycka 5, 44-100 Gliwice, Poland
{joanna,tadek}@iitis.pl
[2] Department of Information Engineering, University of Pisa,
Via Caruso 16, 56122 Pisa, Italy
m.pagano@iet.unipi.it

Abstract. The paper discusses various models of self-similar Internet traffic and techniques for estimating the intensity of Long-Range Dependence (LRD). In the experimental part real data sets collected in IITiS PAN are used together with synthetic LRD flows generated using Fractional Gaussian noise and Markov modulated Poisson processes. We are especially interested in Markov models since they can be incorporated in Markov queueing models, for which powerful analytical and numerical techniques are available.

1 Introduction

In the last decades a growing interest stimulated by experimental measurements was paid to processes characterized by a slowly decaying correlation structure. In this paper we review some of the most relevant results in the framework of traffic modeling and illustrate them with our measurements and their analysis. In more detail we present different classes of stochastic models which have been used in the literature, highlighting their main features and the motivation behind their use. Then we concentrate on Markov models as the most interesting from our point of view because they may be incorporated in Markov queueing models.

Traffic models can be employed in two different ways: as part of an analytical model (e.g., as input process in a queue model) or to drive a discrete-event simulation. In the first case analytical tractability is essential, while in the second one the key factor is the availability of an efficient generation algorithm; this problem is particularly relevant for Long-Range Dependence (LRD) processes because of their infinite memory: for that reason only approximate models can be used.

In general, traffic models can be used to evaluate networks and protocols (for instance, TCP-Friendly Rate Control in DCCP [1] is based on the throughput achieved by TCP connections in similar network conditions) or to estimate the suitable size of network components (for instance, buffer sizes, server capacities and link rates) to cope with QoS (Quality of Service) users requirements, usually measured in terms of (end–to–end) delay, delay jitter and packet loss rate.

© Springer International Publishing Switzerland 2015
A. Dudin et al. (Eds.): ITMM 2015, CCIS 564, pp. 289–303, 2015.
DOI: 10.1007/978-3-319-25861-4_25

In order to get accurate performance indexes, it is essential to use realistic traffic models, i.e. to employ stochastic processes which are able to capture the essential features of real traffic flows while keeping a reasonable level of tractability.

One of the key points in the evolution of traffic modeling is represented by the availability of high-quality traffic data. The first (and most famous) data set was collected between August 1989 and February 1992 on several Ethernet LANs at the Bellcore Morristown Research and Engineering Center by Leland and Wilson [2,3]. They were able to record hundreds of millions of Ethernet packets without loss (irrespective of the traffic load) and with recorded time-stamps accuracy within 100 μs. The statistical analysis of this dataset highlighted the presence of "burstiness" (the arrival points appear to form visual clusters, i.e., runs of several relatively short interarrival times are followed by a relatively long one) across an extremely wide range of time scales, corresponding to a self-similar (it is not surprising that the Hurst parameter H can be used as a measure of burstiness via the concept of self-similarity) or fractal-like behavior of the aggregate Ethernet LAN traffic.

This feature makes packet traffic very different from conventional telephone traffic (for which queuing theory has been developed starting from the pioneering works by Erlang [4] more than one century ago – see, for instance [5] for a general overview) and basically derives from a strong correlation among arrivals over very long time intervals; formally, the data exhibit Long Range Dependence and this behavior can be easily highlighted by plotting, for instance, the index of dispersion for counts (IDC), defined as [6]

$$I_c(\tau) \;=\; \frac{\text{Var}\,[N(\tau)]}{\mathbb{E}\,[N(\tau)]} \tag{1}$$

where $N(\tau)$ represents the number of arrivals during the interval $[0,\tau)$ and its variance takes account of the temporal dependence in the analyzed traffic sequence. The IDC of a LRD process is an increasing function of τ while for a Poisson process its value is 1 for all τ.

2 Long Range Dependence and Self-similarity

In the literature the terms long-range dependence and self-similarity are often used without distinction, although they are not equivalent concepts, [7].

A continuous time process $Y(t)$ is exactly self-similar with the Hurst parameter H if it satisfies the following condition [8]:

$$Y(t) \stackrel{d}{=} a^{-H} Y(at)$$

for $t \geq 0$, $a \geq 0$ and $0 < H < 1$. Above equality is in the sense of finite dimensional distributions and the Hurst parameter expresses the degree of the self-similarity [9]. The process $Y(t)$ may be non-stationary [10].

In the case of network traffic, one usually has to deal with time series rather than a continuous process. In that context the above definition can be summarized as follows. Let $X(t)$ be a stationary sequence representing incremental

process (e.g. in bytes/second). The corresponding aggregated sequence with a
level of aggregation m:

$$X^{(m)}(k) = \frac{1}{m} \sum_{i=1}^{m} X((k-1)m+i), \quad k = 1, 2, \ldots$$

is obtained by averaging $X(t)$ over nonoverlapping blocks of length m. The
following condition is satisfied for a self-similar process:

$$X \overset{d}{=} m^{1-H} X^{(m)}$$

for all integers m. A stationary sequence X is second-order self-similar if
$m^{1-H} X^{(m)}$ has the same variance and auto-correlation as X for all m. A stationary sequence X is asymptotically second-order self-similar if $m^{1-H} X^{(m)}$ has
the same variance and auto-correlation as X as $m \to \infty$.

Asymptotically second-order self-similar processes are also called long-range
dependent processes and this property is exhibited by network traffic [8]. Long-
range dependence of data refers to temporal similarity present in the data. Let
$X(n)$ be a second order stationary process (representing for instance the amount
of traffic arriving in consecutive time intervals) with covariance function

$$r(k) = \mathrm{Cov}\,(X(n), X(n-k)) \overset{\Delta}{=} \mathbb{E}\left[(X(n) - \mu)(X(n-k) - \mu)\right],$$

autocorrelation function

$$\rho(k) \overset{\Delta}{=} \frac{r(k)}{r(0)}$$

and power density spectrum $S(\omega)$, defined as the Discrete Fourier Transform of
$r(k)$, i.e.

$$S(\omega) = \frac{1}{2\pi} \sum_{k=-\infty}^{\infty} r(k)\, e^{ik\omega}\,.$$

The process $X(n)$ exhibits long range dependence if and only if it exhibits the
following properties which are all equivalent [11]):

- *slowly decaying autocorrelation function:* $\rho(k)$ decreases as a non summable
  power law when k tends to infinity

$$\rho(k) \sim k^{-\alpha} \qquad \text{as } k \to \infty \quad \text{where } 0 < \alpha < 1 \tag{2}$$

- *divergence of the power density spectrum at null frequency:* $S(\omega)$ diverges as
  an integrable power law near the origin

$$S(\omega) \sim \omega^{-\beta} \qquad \text{as } \omega \to 0 \tag{3}$$

where $0 < \beta < 1$ and $\beta = 1 - \alpha$ (due to the duality property of the Discrete
Fourier Transform)

– *bad averaging properties:* the variance of the aggregated process decays more
  slowly than the sample size

$$\mathrm{Var}\left(\frac{1}{n}\sum_{i=0}^{n-1}X(i)\right) \sim n^{-\alpha} \qquad \text{as } n \to \infty \tag{4}$$

The latter property is often used as a graphical test for LRD, known as
Variance–time plot, which can be easily implemented according to the following
steps:

– Average the signal over nonoverlapping windows of size n

$$X^{(n)}(k) = \frac{1}{n}\sum_{i=(k-1)n+1}^{kn}X(i)$$

– Compute the variance σ_n^2 of the averaged signal for many values of n
– Plot σ_n^2 against n in a log-log scale
– The slope of the interpolating straight line gives an estimation of the para-
  meter α

The aggregate variance method was described e.g. in [7,12,13]. The estimated
value of Hurst parameter is obtained by fitting a simple least squares line through
the resulting points in the plane. The asymptotic slope between -1 and 0 sug-
gests LRD and estimated Hurst parameter is given by $H = 1 - \frac{\text{slope}}{2}$. Figure 1
shows as an example the variance-time plot for IITiS trace.

Instead, if the autocorrelation function is summable (it is enough to require
that $\rho(k)$ decays geometrically fast), the process exhibits short range dependence
(SRD).

Roughly speaking, the non-summability of the autocorrelations captures the
essence of LRD: even though the high-lag autocorrelations are individually small,
their cumulative effect is of importance, and gives rise to a behavior of the underlying stochastic process that is markedly different from that of the conventionally considered SRD processes; as highlighted before, this long-term *memory*
captures the persistence phenomenon (burstiness or, in Mandelbrot's terminol-
ogy, *Joseph Effect*) observed in many naturally occurring empirical time series
(not only in internet traffic modeling, but also in econometrics, hydrology and
linguistics, just to cite the most famous examples).

3   Estimation of the Hurst Parameter

Starting from traffic data, one of the key problems is to verify the presence of
some kind of self-similarity (at least asymptotically), which is typically done
through the estimation of the Hurst parameter. Unfortunately, only finite data
sets are available and real traffic data exhibits self-similarity only starting from
some time–scale (indeed, unlike self-similar processes, traffic has a discrete

nature!) and, in case of long measurement campaigns, it is necessary to take into account the presence of non stationarity and its effect over the estimation.

Different estimation techniques have been proposed in the literature [11, 14, 15], working directly on the aggregated time series or in a transformed domain. The above-mentioned *variance time plot* is the most known method of the first kind, this class also includes other worth-mentioning approaches:

- **Rescaled Adjusted Range (R/S) statistics**, one of the better known methods, introduced by the British hydrologist Harold Edwin Hurst. Let $X(t)$ be the increments of $Y(t)$, i.e. $Y(t) = \sum_{i=1}^{t} X(i)$. Then, the R/S statistics or rescaled adjusted range is the ratio $Q(t, k) = R(t, k)/S(t, k)$ where the range $R(t, k)$ takes into account the maximum and minimum deviations

$$R(t, k) = \max_{0 \le i \le k} [Y(t+i) - Y(t) - \frac{i}{k}(Y(t+k) - Y(t))]$$

$$- \min_{0 \le i \le k} [Y(t+i) - Y(t) - \frac{i}{k}(Y(t+k) - Y(t))]$$

and $S(t, k)$ denotes the sample standard deviation, given by

$$S(t, k) = \sqrt{k^{-1} \sum_{i=1,k} (X(t+i) - \bar{X}(t,k))^2}$$

with $\bar{X}(t, k) = k^{-1} \sum_{i=1,k} X(t+i)$.

If $Y(t)$ is self-similar with parameter H then $\mathbb{E}(Q(t, k)) \sim k^H$.

- **Higuchi's method** involves calculating the length of a path and, in principle, finding its fractal dimension D. Denoting by N the number of samples of the time series, the normalized length of the corresponding curve is estimated as follows:

$$L(m) = \frac{N-1}{m^3} \sum_{i=1}^{m} \left\lfloor \frac{N-i}{m} \right\rfloor^{-1} \sum_{k=0}^{\lfloor (N-i)/m \rfloor - 1} |Y(i+km) - Y(i+(k-1)m)|$$

where m is essentially a block size and $\lfloor \cdot \rfloor$ denotes the greatest integer function. Then $\mathbb{E}L(m) \sim C_H m^{-D}$ where $D = 2 - H$. Thus a log-log plot of $L(m)$ vs. m should produce a straight line with the slope of $D = 2 - H$

- **Moment method** investigates self-similarity through the behavior of absolute moments of the aggregated processes $X^{(m)}$

$$\mu^{(m)}(q) \triangleq \mathbb{E} \left| X^{(m)} \right|^q$$

According to the definition of self-similarity, $\mu^{(m)}(q)$ is proportional to $m^{\beta(q)}$, where $\beta(q) = q(H - 1)$.

As far as the analysis in a transformed domain is concerned, the main approaches are the following:

- **Periodogram method** is based on the estimation of the spectral density through the periodogram

$$I(\omega) = \frac{1}{2\pi N} \left| \sum_{j=0}^{N-1} X(j) \, e^{ij\omega} \right|^2$$

By definition, a series with LRD should have a periodogram which is proportional to ω^{1-2H} close to the origin. Therefore, a regression of the logarithm of the periodogram on the logarithm of ω should give a coefficient of $1 - 2H$, hence providing an estimation of the parameter H,
- **Whittle estimator**, also based on the periodogram $I(\omega)$, involves the function

$$\mathcal{Q} \triangleq \int_{-\pi}^{\pi} \frac{I(\omega)}{f(\omega; \boldsymbol{\eta})} \, d\omega$$

where $f(\omega; \boldsymbol{\eta})$ is the spectral density and $\boldsymbol{\eta}$ denotes the vector of unknown parameters. The Whittle estimator is the value of $\boldsymbol{\eta}$ which minimizes the function $\mathcal{Q}$. Unlike the other estimators discussed here, the Whittle estimator is obtained through a non-graphical method. It also assumes that the parametric form of the spectral density, i.e. the function $f(\omega; \boldsymbol{\eta})$, is known,
- **Wavelet estimator** is based on the analysis of the numerical series in the wavelet domain [16], taking advantages of the fact that the wavelet coefficients $d_{j,\cdot}$ at a given resolution level j are quasi–decorrelated; indeed, the mother wavelet $\psi(t)$ is a band-pass signal and the power–law behavior of its Fourier transform at frequencies near 0 cancels the power–law divergence of the spectrum at the origin. Roughly speaking, the coefficient $|d_{j,k}|^2$ measures the amount of energy in the analyzed signal about the time instant $2^j k$ and frequency $2^{-j}\omega_0$, where ω_0 is the central frequency of the chosen mother wavelet. Thanks to the quasi–decorrelation of coefficients $d_{j,\cdot}$, a useful spectral estimator can be designed by performing a time average of the $|d_{j,k}|^2$ at a given scale, that is,

$$\hat{S}\left(2^{-j}\omega_0\right) = \frac{1}{n_j} \sum_k |d_{j,k}|^2$$

where $n_j = 2^{-j}N$ is the available number of wavelet coefficients at octave j. In case of $1/|\omega|^\beta$ processes (as in (3)), it is possible to design an estimator $\hat{H}$ for the parameter H from a simple linear regression of $\log_2\left(\hat{S}\left(2^{-j}\omega_0\right)\right)$ on j, i.e.

$$\log_2\left(\hat{S}\left(2^{-j}\omega_0\right)\right) = \left(2\hat{H} - 1\right) j + c$$

where the constant c is independent of the analyzing scale j (see, for instance, [17,18] for a detailed description).

Although the Hurst parameter is well defined mathematically, it is problematic to measure it properly [7,19] since different methods often produce conflicting results [20]. To highlight this issue, we have applied several of the above-mentioned methods to estimate the Hurst parameter in real traces collected in IITis Pan and synthetic traces generated with MMPP and fractal Gaussian models, discussed later in the article.

Table 1 gives the obtained Hurst parameters for one day IITiS traces: trace 1 (6 002 874 samples), trace 2 (13 874 610 samples), and trace 3 (36 135 490 samples). In the case of Wavelets based method a 95 % confidence interval should be interpreted only as a confidence interval on the fitted line. Our previous work [15] did not confirm the relationship between the degree of LRD and the number of transmitted packet of a given type. One can see a variance of estimators obtained with different methods.

Table 1. Hurst parameter estimates for IITiS data traces

	trace 1	trace 2	trace 3
Estimator	Hurst parameter		
R/S method	0.74	0.655	0.763
Aggregate variance method	0.912	0.817	0.933
Periodogram method	0.781	0.715	0.84
Whittle method	0.714	0.599	0.761
Wavelet-based method	0.681 ± 0.013	0.61 ± 0.027	0.71 ± 0.017

Table 2 shows the obtained Hurst parameters for MMPP data traces: MMPP 1 ($H = 0.75$ and $\rho = 0.6$), MMPP 2 ($H = 0.6$ and $\rho = 0.6$), MMPP 3 ($H = 0.79$ and $\rho = 0.0213$), while Table 3 refers to fGn traces. The fGn and MMPP 3 models were used to simulate the LRD data using a theoretical Hurst parameter and the same theoretical mean as the IITiS data. Different Hurst parameters have been chosen to represent a low and a high level of long-range dependence in data. The models are run to produce 200 000 packets. As can be seen, the MMPP model is more inconsistent than fGn.

Table 2. Hurst parameter estimates for MMPP data traces

	MMPP 1	MMPP 2	MMPP 3
Estimator	Hurst parameter		
R/S method	0.757	0.659	0.798
Aggregate variance method	0.665	0.586	0.715
Periodogram method	0.83	0.549	0.831
Whittle method	0.678	0.57	0.728
Wavelet-based method	0.851 ± 0.036	0.601 ± 0.011	0.841 ± 0.036

Table 3. Hurst parameter estimates for fGn data traces

	fGn 1	fGn 2	fGn 3
Estimator	Hurst parameter		
R/S method	0.605	0.701	0.939
Aggregate variance method	0.507	0.642	0.88
Periodogram method	0.521	0.661	0.991
Whittle method	0.688	0.75	0.882
Wavelet-based method	0.574 ± 0.028	0.698 ± 0.017	0.937 ± 0.009

4   Non-Markov LRD Traffic Models

In this section we recall other traffic models that exhibit LRD, for sake of brevity just mentioning their key features:

- **Fractional Brownian Traffic** The most famous (and the simplest) LRD traffic model is the Fractional Brownian Traffic [21], in which the cumulative arrival process is defined as follows:

$$N(t) \;=\; mt + \sqrt{am}Z_H(t)$$

  where $Z_H(t)$ denotes the normalized (i.e., $\sigma^2 = 1$) fractional Brownian motion (fBm), with Hurst parameter $H \in [1/2, 1)$; $m > 0$ is the mean input rate and $a > 0$ is the variance coefficient.
- **fractional ARIMA(p,d,q) processes**: they are the extension (for fractional values of d) of the traditional ARIMA processes; unlike fGn, there are additional degrees of freedom related to the autoregressive components, whose coefficients can be set to model the short term correlations of actual traffic data [22,23]
- **heavy-tailed On-Off models**: they can be seen as a generalization of MMPP(2), in which the sojourn time in at least one of two states is heavy-tailed, i.e. the right tail of its distribution decays to zero as

$$\mathbb{P}\left[X > x\right] \simeq x^{-\alpha} L(x)$$

where $L(x)$ is a slowly varying function (at infinity), i.e.

$$\lim_{t \to \infty} \frac{L(xt)}{L(t)} = 1 \qquad \text{for any fixed } x > 0$$

These models can be used to describe single connections (indeed, certain traffic features such as file sizes and related transmission times, CPU times, idle times, peak rates and connection times are characterized by heavy-tails) and the actual traffic can be seen as the superposition of many independent on-off sources.

- **α-stable processes**: they permit to combine LRD with heavy-tailed distribution of the arrival process [24].
  It is worth recalling that for α–stable processes (and in general for heavy tailed processes), the autocovariance function is not defined and thus we have to modify the usual techniques (such as Variance Time plot, R/S statistic and so on) in order to estimate the correct value of H [25].
- **α–stable Lévy motion**, a stochastic process characterized by *independent* and stationary increments with α–stable distribution, Note that α–stable Lévy motion is *self-similar* with Hurst parameter $H = 1/\alpha$, but is SRD, since its increments are independent (the case $\alpha = 2$ corresponds to the well-known Brownian motion)
- **Linear Fractional Stable Motion (LFSM)**, the most common extension of fBm to the α-stable case (for $\alpha = 2$, the LFSM process is actually fBm). In this way, we add to the self-similarity and stationary increments property of Fractional Brownian Traffic, the ability to cope with the non Gaussian distribution of actual traffic data (see [24, 26]).
- **multifractal processes**: locally they are similar to a fractal process, but, instead of a single value H of the Hurst parameter as for traditional monofractal processes. The investigations into the multifractal nature of network traffic often use wavelet-based [16] analysis and are related to special classes of multiplicatively generated conservative cascades [27, 28].

5 Markov LRD Models

5.1 MMPP Model

Markov chains and Markov-modulated processes (MMP) are well-known modeling techniques which are successful in wide variety of fields. These models are often motivated by the idea of capturing the long-range dependence which is seen in real internet traffic and replicating the the Hurst parameter which characterizes it [7].

Two-state Markov Modulated Poisson Process (MMPP) is also known as the Switched Poisson Process (SPP). The superposition of MMPP's is also an MMPP which is a special case of Markovian Arrival Process (MAP). Following [29] we use a superposition od d two-state MMPS [15].

A MAP is defined by two square matrices $\mathbf{D_0}$ and $\mathbf{D_1}$ such that $\mathbf{Q} = \mathbf{D_0} + \mathbf{D_1}$ is an irreducible infinitesimal generator for the continuous-time Markov chain (CTMC) underlying the process, and $D_0(i, j)$ (respectively $D_1(i, j)$) is the rate of hidden (respectively observable) transitions from state i to state j [30]. Two-state MAP is a Markovian arrival process with square matrices as follows:

$$\mathbf{D_0} = \begin{bmatrix} -\sigma_1 & \lambda_{1,2} \\ \lambda_{2,1} & -\sigma_2 \end{bmatrix}, \qquad \mathbf{D_1} = \begin{bmatrix} \mu_{1,1} & \mu_{1,2} \\ \mu_{2,1} & \mu_{2,2} \end{bmatrix}$$

where $\lambda_{i,j} \geq 0$, $\mu_{i,j} \geq 0$, for all i, j. The diagonal elements of matrix $\mathbf{D_0}$ are $\sigma_1 = \lambda_{1,2} + \mu_{1,1} + \mu_{1,2} > 0$ and $\sigma_2 = \lambda_{2,1} + \mu_{2,2} + \mu_{2,1} > 0$ such that underlying continuous-time Markov chain Matrix $\mathbf{Q}$ has no absorbing states.

Following the model proposed in [29], a LRD process (used in our study) can be modeled as the superposition of d two-state MMPPs. The i-th MMPP ($1 \leq i \leq d$) can be parameterized by two square matrices:

$$\mathbf{D_0^i} = \begin{bmatrix} -(c_{1i} + \lambda_{1i}) & c_{1i} \\ c_{2i} & -(c_{2i} + \lambda_{2i}) \end{bmatrix}, \qquad \mathbf{D_1^i} = \begin{bmatrix} \lambda_{1i} & 0 \\ 0 & \lambda_{2i} \end{bmatrix}.$$

The element c_{1i} is the transition rate from state 1 to 2 of the i-th MMPP and c_{2i} is the rate out of state 2 to 1. λ_{1i} and λ_{2i} are the traffic rate when the i-th MMPP is in state 1 and 2 respectively. The sum of $\mathbf{D_0^i}$ and $\mathbf{D_1^i}$ is an irreducible infinitesimal generator $\mathbf{Q}^i$ with the stationary probability vector:

$$\overrightarrow{\pi}_i = \left(\frac{c_{2i}}{c_{1i} + c_{2i}}, \frac{c_{1i}}{c_{1i} + c_{2i}} \right)$$

The superposition of these two-state MMPPs is a new MMPP with 2^d states and its parameter matrices, $\mathbf{D_0}$ and $\mathbf{D_1}$, can be computed using the Kronecker sum of those of the d two-state MMPPs [31]:

$$(\mathbf{D_0}, \mathbf{D_1}) = \left(\oplus_{i=1}^d \mathbf{D_0}^i, \oplus_{i=1}^d \mathbf{D_1}^i \right)$$

The article [29] proposed a fitting method for a superposition of two-state MAPs (described in Sect. 5.1) based on Hurst parameter as well as the moments.

For real traffic traces the covariance structure of the counting process is well described by the asymptotic covariance [29]:

$$cov(k) = \psi_{cov} k^{-\beta}$$

where ψ_{cov} jest an absolute measure of the variance, $\beta = 2 - 2H$ and k is the lag. The parameters ψ_{cov} and β should be estimated from the real data traces. The objective of the fitting is to achieve:

$$\gamma(k) = \sum_{i=1}^d \gamma_i(k) \approx \psi_{cov} k^{-\beta}$$

where $1 \leq k \leq 10^n$ and n denotes the number of time scales the model demonstrate self-similar behavior.

Real traces used in our study comprises of Ethernet traffic data of Bellcore Laboratory and data captured on the gateway of IITiS.

Bellcore Laboratory data was already interpreted in multiple studies [9,32]. A dedicated hardware has been built for measuring each packet arrival and the measurements were performed without losses and with high precision. The large part of collected data is available by Internet. In this study we use the file: OctExt.TL. It contains the first million of external arrivals gathered during 35 h. Other data set used in our study has been collected on the Internet gateway of our Institute [33]. The traffic approximately stands for the few dozen of researchers.

The Hurst parameter was estimated by the *Aggregated Variance* method. Figure 1 presents the normalized variance of the aggregated series as a function of time scale in log-log coordinates. The slope of IITiS curve (estimated by the least squares method) is equal to −0.42, which gives the Hurst parameter equal to 0.79. The slope of Bellcore curve is equal to −0.3, which gives the Hurst parameter equal to 0.85. For comparison, the same plot is also drawn for the Poisson process. This line has the slope −1, which gives the Hurst parameter equal to 0.5 (non-self-similar process). Figure 2 shows the autocorrelations of the five IPP's.

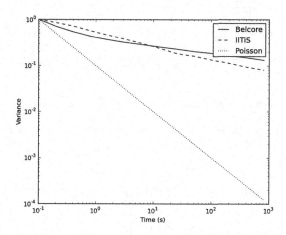

Fig. 1. Variance-time plot in log-log scale (IITiS data, May 2012)

Table 4 presents the parameters obtained from the fitting approach. The fitting procedure was also applied to the traces of IP traffic measured at Bellcore and IITiS. The parameters of the superposition of two-state MMPP's were fitted to those obtained from real data traffic. The superposition of five MMPP's is sufficient to model asymptotic second-order self-similarity of the counting process over five time-scales. Numerous numerical examples we performed prove the efficiency of this approach.

5.2   Hidden Markov Model

Hidden Markov Model(HMM) may be seen as a probabilistic function of a (hidden) Markov chain [34]. This Markov chain is composed of two variables:

- the hidden-state variable, whose temporal evolution follows a Markov-chain behavior ($x_n \in \{s_1, \ldots, s_N\}$ represent the (hidden) state at discrete time n with N being the number of states)
- the observe variable which stochastically depends on the hidden state ($y_n \in \{o_1, \ldots, o_M\}$ and represents the observable at discrete time n with M being the number of observables)

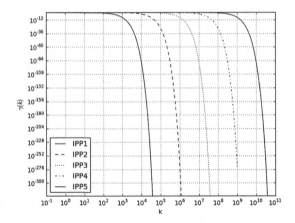

Fig. 2. Autocorrelation of the number of arrivals in a time unit (five IPP's)

Table 4. Obtained parameters of source fitted to second-order self-similarity with input parameters: $H = 0.6$ and $\rho = 0.6$.

	λ_i^{IPP}	c_{1i}	c_{2i}
IPP$_1$	27.646	7.241×10^{-1}	7.590×10^{-2}
IPP$_2$	6.944	2.290×10^{-2}	2.400×10^{-3}
IPP$_3$	1.746	7.241×10^{-4}	7.590×10^{-5}
IPP$_4$	0.434	2.290×10^{-5}	2.400×10^{-6}
IPP$_5$	0.119	7.241×10^{-7}	7.590×10^{-8}
Poisson	$\lambda_p = 0$		

An HMM is characterized by the set of parameters:

$$\lambda = \{\mathbf{u}, \mathbf{A}, \mathbf{B}\}$$

where:

- $\mathbf{u}$ is the initial state distribution, where $u_i = Pr(x_1 = s_i)$
- $\mathbf{A}$ is the $N \times N$ state transition matrix, where $A_{i,j} = Pr(x_n = s_j | x_{n-1} = s_i)$
- $\mathbf{B}$ is the $N \times M$ observable generation matrix, where $B_{i,j} = Pr(y_n = o_j | x_n = s_i)$

Given a sequence of observable variables $y = (y_1, y_2, \ldots, y_L)$ referred to as the *training sequence*, we want to find the set of parameters such that the likelihood of the model $L(\mathbf{y}; \lambda) = Pr(\mathbf{y}|\lambda)$ is maximum. We solved it via the Baum-Welch algorithm, a special case of the Expectation-Maximization algorithm which iteratively updates the parameters in order to find a local maximum point of the parameter set.

We used the well-known Bellcore trace of Internet traffic: OctExt.TL. Each line of this file contains a floating-point time stamp (representing the time in seconds since the start of a trace) and an integer length (representing the Ethernet data length in bytes). We translated the sequence of time stamps into the sequence of inter-arrival times. Then we apply a scheme using *Vector Quantization* (VQ) to translate the obtained sequence of inter-arrival times into a sequence of symbols, and training a HMM for this sequence. The quantization algorithm used is Linde-Buzo-Gray (LBG) algorithm of Vector Quantization is a clustering technique commonly used in compression, image recognition and stream encoding. It is the general approach to map a space of vector valued data to a finite set of distinct symbols, in a way to minimize distortion associated with this mapping.

We consider an HMM in which the state and the observable variables are discrete. A little portion of the sequences was used as the training sequence to learn model parameters. Performance of trained model are tested on the remaining portions of the sequences.

Then we can use the HMM trained with the Bellcore data as the Internet traffic source model. Figure 3 displays exemplary series of inter-arrival times which are obtained from the Bellcore trace and from our HMM traffic source, both have the same Hurst parameter.

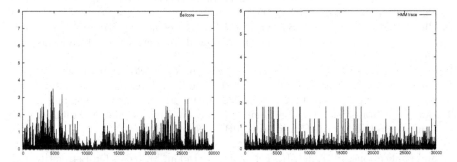

Fig. 3. The sequence of inter-arrival times for Bellcore trace and HMM traffic source trace

6   Conclusions

The article represents a practical approach to the problem of modeling stochastic features of Internet traffic. We summarize the main methods to determine the Hurst parameter and to include it in traffic models. In particular we show that we are able to construct efficiently Markov models of traffic with LRD, also with the use of Hidden Markov Chains. These Markov models may be a part of a computer network models aiming to evaluate its performance. Of course, the complexity of traffic models enlarges the size of the entire state space to be considered and hence the number of equations to be solved numerically. Therefore we are developing a software tool able to cope with models having hundreds of millions states.

References

1. Floyd, S., Kohler, E., Padhye, J.: Profile for datagram congestion control protocol (dccp) congestion control id 3: Tcp-friendly rate control (tfrc). RFC 4342 (Proposed Standard) Updated by RFCs 5348, 6323, March 2006
2. Leland, W., Wilson, D.: High time-resolution measurement and analysis of LAN traffic: implications for LAN interconnection. In: INFOCOM 1991 Proceedings of Tenth Annual Joint Conference of the IEEE Computer and Communications Societies. Networking in the 90s, vol. 3, pp. 1360–1366. IEEE (1991)
3. Leland, W.E., Taqqu, M.S., Willinger, W., Wilson, D.V.: On the self-similar nature of Ethernet traffic (extended version). IEEE/ACM Trans. Netw. 2(1), 1–15 (1994)
4. Erlang, A.K.: The theory of probabilities and telephone conversations. Nyt Tidsskrift for Matematik 20, 33–39 (1909)
5. Hayes, J., Babu, T.: Modeling and Analysis of Telecommunications Networks. Wiley, New York (2004)
6. Gusella, R.: Characterizing the variability of arrival processes with indices of dispersion. IEEE J. Sel. Areas Commun. 9, 203–211 (1990)
7. Clegg, R.G.: A practical guide to measuring the hurst parameter. Int. J. Simul. 7(2), 3–14 (2006)
8. Gong, W.B., Liu, Y., Misra, V., Towsley, D.: Self-similarity and long range dependence on the internet: a second look at the evidence, origins and implications. Comput. Netw. 48, 377–399 (2005)
9. Bhattacharjee, A., Nandi, S.: Statistical analysis of network traffic inter-arrival. In: 12th International Conference on Advanced Communication Technology, pp. 1052–1057 (2010)
10. Nogueira, A., Salvador, P., Valadas, R., Pacheco, A.: Markovian modelling of internet traffic. In: Kouvatsos, D.D. (ed.) Next Generation Internet: Performance Evaluation and Applications. LNCS, vol. 5233, pp. 98–124. Springer, Heidelberg (2011)
11. Beran, J.: Statistics for Long-Memory Processes. Monographs on Statistics and Applied Probability 61. Chapman and Hall, New York (2004)
12. Taqqu, M., Teverovsky, V.: A practical guide to heavy tails: statistical techniques and applications. In: Adler, R., Feldman, R., Taqqu, M. (eds.) On Estimating the Intensity of Long-Range Dependence in Finite anf Infinite Variance Time Series, pp. 177–217. Birkhauser Boston Inc., Boston (1998)
13. Park, C., Hernandez-Campos, F., Long, L., Marron, J., Park, J., Pipiras, V., Smith, F., Smith, R., Trovero, M., Zhu, Z.: Long range dependence analysis of internet traffic. J. Appl. Stat. 38(7), 1407–1433 (2011)
14. Taqqu, M.S., Teverovsky, V., Willinger, W.: Estimators for long-range dependence: an empirical study. Fractals 3, 785–798 (1995)
15. Domanska, J., Domanski, A., Czachorski, T.: A few investigation of long-range dependence in network traffic. Inf. Sci. Syst. 2014, 137–144 (2014)
16. Daubechies, I.: Ten Lectures on Wavelets. CBMS-NSF Regional Conference Series in Applied Mathematics. Society for Industrial and Applied Mathematics, Philadelphia (1992)
17. Abry, P., Veitch, D.: Wavelet analysis of long-range-dependent traffic. IEEE Trans. Inf. Theory 44(1), 2–15 (1998)
18. Roughan, M., Veitch, D., Abry, P.: Real-time estimation of the parameters of long-range dependence. IEEE/ACM Trans. Netw. 8(4), 467–478 (2000)
19. Clegg, R.G.: Markov-modulated on/off processes for long-range dependent internet traffic. Computing Research Repository, CoRR (2006). arXiv:cs/0610135

20. Domanska, J., Domanski, A., Czachorski, T.: Estimating the intensity of long-range dependence in real and synthetic traffic traces. Commun. Comput. Inf. Sci. **522**, 11–22 (2015)

21. Norros, I.: On the use of fractional brownian motion in the theory of connectionless networks. IEEE J. Sel. A. Commun. **13**(6), 953–962 (2006)

22. Granger, C.W.J., Joyeux, R.: An introduction to long-memory time series models and fractional differencing. J. Time Ser. Anal. **1**(1), 15–29 (1980)

23. Samorodnitsky, G., Taqqu, M.S.: Stable non-Gaussian random processes: stochastic models with infinite variance. Chapman and Hall, New York (1994)

24. Gallardo, J.R., Makrakis, D., Orozco-Barbosa, L.: Use of α-stable self-similar stochastic processes for modeling traffic in broadband networks. Perform. Eval. **40**(1–3), 71–98 (2000)

25. Taqqu, M.S., Teverovsky, V.: On estimating long-range dependence in finite and infinite variance case. In: Adler, R.J., Feldman, R.E., Taqqu, M.S. (eds.) A Pratical Guide To Heavy Tails: Statistical Techniques and Applications, pp. 177–217. Birkhauser, Boston (1998)

26. Karasaridis, A., Hatzinakos, D.: A non-gaussian self-similarity processes for broadband heavy-traffic modeling. In: Proceedings of GLOBECOM 1998, Sidney, Australia, pp. 2995–3000 (1998)

27. Riedi, R.H., Crouse, M., Ribeiro, V., Baraniuk, R.: A multifractal wavelet model with application to network traffic. IEEE Trans. Inf. Theory **45**(3), 992–1018 (1999)

28. Gilbert, A., Willinger, W., Feldmann, A.: Scaling analysis of conservative cascades, with applications to network traffic. IEEE Trans. Inf. Theory **45**(3), 971–991 (1999)

29. Andersen, A.T., Nielsen, B.F.: A markovian approach for modeling packet traffic with long-range dependence. IEEE J. Sel. Areas Commun. **16**(5), 719–732 (1998)

30. Casale, G.: Building accurate workload models using markovian arrival processes. In: SIGMETRICS 2011, San Jose, USA (2011)

31. Fischer, W., Meier-Hellstern, K.: The markov-modulated poisson process (mmpp) cookbook. Perform. Eval. **18**(2), 149–171 (1993)

32. Domańska, J., Domański, A.: The influence of traffic self-similarity on qos mechanism. In: International Symposium on Applications and the Internet, SAINT, Trento, Italy (2005)

33. Foremski, P.: Mutrics: Multilevel traffic classification. http://mutrics.iitis.pl

34. Domańska, J., Domański, A., Czachórski, T.: Internet traffic source based on hidden markov model. In: Balandin, S., Koucheryavy, Y., Hu, H. (eds.) NEW2AN 2011 and ruSMART 2011. LNCS, vol. 6869, pp. 395–404. Springer, Heidelberg (2011)

Propabilistic Analysis for European Exotic Option on Stock Market Index Research

Elena Daniliuk[(⊠)]

Department of Applied Mathematics and Cybernetics,
Tomsk State University, Lenina Ave. 36, 634050 Tomsk, Russia
daniluc_elena@sibmail.com

Abstract. European exotic put option with payment limitation for issuer and guaranteed income for holder of the security is researched when base risk active is share index. The equitable option price, the securities portfolio structure and a size of the capital answered the hedging strategy are founded for the option under consideration on diffusion (B, S)-financial market. Comparative price analysis for two option classes is carried out and specific properties of decision and decision under limiting are explored.

Keywords: Stochastic financial mathematics · Financial market · Option price · Hedging strategy · European exotic put option · Put option with guaranteed income for holder of the security · Share index

1 Introduction

Complexity of a market economy and the current state of requirements to it encourage more serious analysis techniques (based on real processes modelling) of it's theoretical and practical problems. One of the grave units of mathematical simulation in economy is queueing theory, presenting theoretical foundations of the effective designing and exploitation of the queuing systems [1].

The set of the investigation subjects of the queueing theory is broad enough [2]. For example, primary securities market and secondary securities market are of interest to the theory. In [3] authors point at applicability of the queueing theory to investigation of the securities price dynamics with agent-based models which estimate characteristics of investor behavior and the complex interactions between market participants and they are typically queueing-type models, that is, models of order flows.

So, to reach exchange more effectively it is needed the following: first, constant monitoring and forecasting the market of financial assets; second, comprehensive planning of the exchange work and operative management by the transactions with financial instruments; third, timely and qualitative client servicing. Since trades are executed in electronic form with special-purpose programs the queueing theory results (assumed as a basis of the programs) and results of the financial markets investigation can considerably increase the volume of deals and quality of deals support.

© Springer International Publishing Switzerland 2015
A. Dudin et al. (Eds.): ITMM 2015, CCIS 564, pp. 304–313, 2015.
DOI: 10.1007/978-3-319-25861-4_26

So, the present paper produces results that can be used as preparatory stage of financial market investigations by queueing theory methods. As of today the financial instruments of trading and risks hedging [4] on the derivatives market are presented by futures, forwards and options, particularly the exotic options [5–9]. The lasts are of interest for investor due to variety of the option's payment liabilities [10] and are the stochastic financial mathematics object [11]. An European put option is a derivative (secondary) security, it is the contract giving option's buyer (the holder) the right to sell stipulated underlying asset by a certain date for a certain price, and option's seller must satisfy an agreement when exercising for an option premium [4].

The research is devoted to European exotic put option with payment limitation for issuer and guaranteed income for holder of the security when base risk active is share index. The payoff function determined the payment size when the option under consideration exercising is

$$f_T\left(i_T\right) = \min\left\{\left(K_1 - i_T\right)^+, K_2\right\},\tag{1}$$

where i_T is risk asset's spot price at expiration date T; K_1 is exercise price or strike price; K_2 is contracted constant restricted payment of the option writer, on the one hand, and guaranteed income for option buyer, on the other hand; $a^+ = \max\left(a; 0\right)$. Stock market index in (1) is a measurement of the value of a section of the stock market. It is a tool used by investors and financial managers to describe the market, and to compare the return on specific investments [12,13]. It is computed from the prices of selected stocks (typically a weighted average) and it can be determined at the current time t as

$$i_t = \frac{1}{a}\sum_{k=1}^{n} V_k S_t^k,\tag{2}$$

where S_t^k and V_k are the current price and quantity of share with a number k entered into a so called index basket respectively; $a = \sum_{k=1}^{n} V_k$.

In accordance to (1) the European standard put option payoff liability assumed as $f_T\left(i_T\right) = \left(K_1 - i_T\right)^+$ [10] is base for the exotic option under study. If at the moment T the market state such as $K_1 > i_T$ then the option holder gets the size $K_1 - i_T$ if $K_1 - i_T > K_2$ or size K_2 if $K_1 - i_T < K_2$; in other cases (if $K_1 < i_T$) the option buyer earns nothing.

We denote the mathematical expectation by $E\left\{\cdot\right\}$, the normal (Gaussian) density with the parameters a and b by $N\left\{a; b\right\}$, $\Phi\left(x\right) = \int_{-\infty}^{x} \varphi\left(y\right) dy$, $\varphi\left(y\right) = \frac{1}{\sqrt{2\pi}}\exp\left\{-\frac{y^2}{2}\right\}$ are Laplace distribution function and probability density function respectively.

The general information from the financial obligations theory is presented in the Sect. 2. In the Sect. 3 statements needed for the further investigation are cited. The decision of the proplem under study is given in the Sect. 4. Analysis of the option price sensitivity to the model and derivative parameters is performed in the Sect. 5. Economic interpretation based on the numerical results (Sect. 6) prove the obtained theoretical results.

2   Statement of the Problem

Let us consider complete, without arbitrage and risk-neutral financial market of two assets, notably: risk (share index) and risk free (bank deposit) active. The stock market index value is termed index price.

According to (2) the index price is detrmined by stocks prices from index basket. By-turn the every stock price evolution is given on stochastic basis $\left(\Omega, F, \mathbf{F} = (F_t)_{t>0}, \mathbf{P}\right)$ [8,10]. The current prices of the securities S_t and B_t, $t \in [0, T]$, are specified by (3) and (4) respectively

$$dS_t^k = S_t^k \left(\mu_k dt + \sigma_k dW_t\right), \quad S_t^k = S_0^k \exp\left\{\left(\mu_k - \frac{\sigma_k^2}{2}\right) t + \sigma_k W_t\right\}, \quad (3)$$

$$dB_t = r B_t dt, \quad B_t = B_0 \exp\{rt\}, \quad (4)$$

where $W = (W_t)_{t\geq0}$ is a standard Wiener process, $S_0^k > 0$ is the stock initial cost numbered k, $\mu_k \in R = (-\infty, +\infty)$ is the percentage drift of the stock initial cost numbered k, $\sigma_k > 0$ is the percentage volatility of the stock initial cost numbered k in a geometric Brownian motion, $B_0 > 0$ is the risk free asset initial price, $r > 0$ is interest rate, $k = 1, 2, ..., n$.

During time interval $t \in [0, T]$ the investor forms self-financing portfolio $\pi_t = (\beta_t, \gamma_t)$, where F_t−measurable processes β_t and γ_t are parts of the risk free and risk assets at investment portfolio respectively, and this portfolio secures investor capital $X_t = \beta_t B_t + \gamma_t i_t$.

The problem involves the fact that to form the portfolio (hedging strategy) $\pi_t = (\beta_t, \gamma_t)$, the evolution of the capital X_t has option price $P_T = X_0$ in accordance to the payoff function (1), as well as, the hedging strategy and corresponding capital, ensuring the fulfillment of payment liability $X_T = f_T(i_T)$.

3   Preliminary Results

All results below are obtained on the assumption of the sole risk-neutral measure $\mathbf{P}^*$ existence. Relative to this measure the process of the risk asset capitalized price $\hat{S}_t^k = S_t^k / B_t$ that forms basic asset (2) is martingale, and that condition guarantees the assigned problem solvability [6–8,10,11]. Theorems are proved with a glance of the base financial relations (5)-(7) [6–8,10,11]

$$P_T = e^{-rT} E^* \{f_T(i_T)\}, \quad (5)$$

$$X_t = E^* \left\{e^{-r(T-t)} f_T(i_T) | i_t\right\}, \quad (6)$$

$$\beta_t = \left.\frac{\partial X_t(s)}{\partial s}\right|_{s=B_t}, \gamma_t = \frac{X_t - \beta_t B_t}{i_t}. \quad (7)$$

where E^* is a risk-neutral measure averaging.

Proposition 1. *Let us that risk-neutral (martingale) measure P^* is associated with source measure P by transformation which looks like*

$$d P_t^* = Z_t dP_t, \tag{8}$$

where

$$Z_t = \exp\left\{ -\frac{\mu - r}{\sigma} W_t - \frac{1}{2}\left(\frac{\mu - r}{\sigma}\right)^2 t \right\} \tag{9}$$

is a Girsanovs exponent.
  Then stochastic properties of the process $S(\mu, r)$ defined by equation

$$dS_t^k (\mu_k, r) = S_t^k (\mu_k, r) (rdt + \sigma_k dW_t^*), \tag{10}$$

with regard to measure P^* are coinciding with properties of the process $S(r)$ defined by equation

$$dS_t^k (r) = S_t^k (r) (rdt + \sigma_k dW_t), \tag{11}$$

relative to measure P where

$$W_t^* = W_t + \frac{\mu - r}{\sigma} t \tag{12}$$

is Wiener process in reference to measure P^*.

4   Main Results

Theorem 1. *For the European exotic put option with payoff function (1) the current value of the minimal investment portfolio X_t is described by Eq. (13)*

$$X_t = K_1 e^{-r(T-t)} [\Phi(y_1(T-t)) - \Phi(y_0(T-t))] + K_2 e^{-r(T-t)} \Phi(y_0(T-t))$$

$$- \sum_{k=1}^{n} B_t^k \left[\Phi\left(y_1(T-t) - \sigma_k\sqrt{T-t}\right) - \Phi\left(y_0(T-t) - \sigma_k\sqrt{T-t}\right) \right], \quad (13)$$

where

$$y_0(T-t) = \min\left[\frac{\ln\left(\frac{K_1 - K_2}{B_t^k}\right) - \left(r - \frac{\sigma_k^2}{2}\right)t}{\sigma_k\sqrt{T-t}} \right], k = 1, 2, ..., n, \tag{14}$$

$$y_1(T-t) = \min\left[\frac{\ln\left(\frac{K_1}{B_t^k}\right) - \left(r - \frac{\sigma_k^2}{2}\right)t}{\sigma_k\sqrt{T-t}} \right], k = 1, 2, ..., n, \tag{15}$$

$$B_t^k = \frac{V_k S_t^k}{a}, a = \sum_{k=1}^{n} V_k. \tag{16}$$

Proof. According to (1), (2), (6) and using change of variables $y = x/\sqrt{T-t}$ and (16) we obtain

$$X_t = e^{-r(T-t)} E^* \left\{ \min \left\{ (K_1 - i_T)^+, K_2 \right\} | i_t \right\}$$

$$= \frac{e^{-r(T-t)}}{\sqrt{2\pi (T-t)}} \int\limits_{-\infty}^{+\infty} \min \left\{ \left(K_1 - \frac{1}{a} \sum_{k=1}^{n} V_k S_t^k \exp \left\{ \left(r - \frac{\sigma_k^2}{2} \right) (T-t) \right. \right. \right.$$

$$\left. \left. \left. + \sigma_k (W_T - W_t) \right\} \right)^+, K_2 \right\} \exp \left\{ -\frac{x^2}{2(T-t)} \right\} dx = \frac{e^{-r(T-t)}}{\sqrt{2\pi}}$$

$$\times \int\limits_{-\infty}^{+\infty} \min \left\{ \left(K_1 - \frac{1}{a} \sum_{k=1}^{n} V_k S_t^k \exp \left\{ \left(r - \frac{\sigma_k^2}{2} \right) (T-t) + y\sigma_k \sqrt{T-t} \right\} \right)^+, K_2 \right\}$$

$$\times \exp \left\{ -\frac{y^2}{2} \right\} dy = \frac{e^{-r(T-t)}}{\sqrt{2\pi}} \int\limits_{-\infty}^{+\infty} \min \left\{ \left(K_1 - \sum_{k=1}^{n} B_t^k \exp \left\{ \left(r - \frac{\sigma_k^2}{2} \right) (T-t) \right. \right. \right.$$

$$\left. \left. \left. + y\sigma_k \sqrt{T-t} \right\} \right)^+, K_2 \right\} \exp \left\{ -\frac{y^2}{2} \right\} dy. \qquad (17)$$

Using (17) write

$$\min \left\{ \left(K_1 - \sum_{k=1}^{n} B_t^k \exp \left\{ \left(r - \frac{\sigma_k^2}{2} \right) (T-t) + y\sigma_k \sqrt{T-t} \right\} \right)^+, K_2 \right\}$$

$$= \begin{cases} K_2, & \text{if} \quad sum < K_1 - K_2, \\ K_1 - sum, & \text{if} \quad K_1 - K_2 < sum < K_1, \\ 0, & \text{if} \quad sum > K_1 \end{cases} \qquad (18)$$

where $sum = \sum\limits_{k=1}^{n} B_t^k \exp \left\{ \left(r - \frac{\sigma_k^2}{2} \right) (T-t) + y\sigma_k \sqrt{T-t} \right\}$.

Under (17), (18) it is obviously that (14), (15) are roots of inequalities below

$$\sum_{k=1}^{n} B_t^k \exp \left\{ \left(r - \frac{\sigma_k^2}{2} \right) (T-t) + y\sigma_k \sqrt{T-t} \right\} < K_1,$$

$$\sum_{k=1}^{n} B_t^k \exp \left\{ \left(r - \frac{\sigma_k^2}{2} \right) (T-t) + y\sigma_k \sqrt{T-t} \right\} < K_1 - K_2,$$

respectively.

So, from (17), (18) we get

$$X_t = \frac{e^{-r(T-t)}}{\sqrt{2\pi}} \int\limits_{-\infty}^{y_0(T-t)} K_2 \exp \left\{ -\frac{y^2}{2} \right\} dy + \frac{e^{-r(T-t)}}{\sqrt{2\pi}} \int\limits_{y_0(T-t)}^{y_1(T-t)} K_1 \exp \left\{ -\frac{y^2}{2} \right\} dy$$

$$-\frac{e^{-r(T-t)}}{\sqrt{2\pi}} \int_{y_0(T-t)}^{y_1(T-t)} \sum_{k=1}^{n} B_t^k \exp\left\{\left(r - \frac{\sigma_k^2}{2}\right)(T-t) + y\sigma_k\sqrt{T-t}\right\} e^{-\frac{y^2}{2}} dy$$

$$= X_t^1 + X_t^2 - X_t^3. \tag{19}$$

Summands X_t^1, X_t^2 and X_t^3 from (19) are defined by the formulas

$$X_t^1 = K_2 e^{-r(T-t)} \Phi\left(y_0\left(T-t\right)\right), \tag{20}$$

$$X_t^2 = K_1 e^{-r(T-t)} \left[\Phi\left(y_1\left(T-t\right)\right) - \Phi\left(y_0\left(T-t\right)\right)\right], \tag{21}$$

$$X_t^3 = \frac{e^{-r(T-t)}}{\sqrt{2\pi}} \int_{y_0(T-t)}^{y_1(T-t)} \sum_{k=1}^{n} B_t^k \exp\left\{\left(r - \frac{\sigma_k^2}{2}\right)(T-t) + y\sigma_k\sqrt{T-t}\right\} e^{-\frac{y^2}{2}} dy$$

$$= \frac{e^{-r(T-t)} e^{r(T-t)}}{\sqrt{2\pi}} \int_{y_0(T-t)}^{y_1(T-t)} \sum_{k=1}^{n} B_t^k \exp\left\{-\frac{1}{2}\left(y - \sigma_k\sqrt{T-t}\right)^2\right\} dy$$

$$= \sum_{k=1}^{n} B_t^k \frac{1}{\sqrt{2\pi}} \int_{y_0(T-t)}^{y_1(T-t)} \exp\left\{-\frac{1}{2}\left(y - \sigma_k\sqrt{T-t}\right)^2\right\} dy$$

$$= \sum_{k=1}^{n} B_t^k \left[\Phi\left(y_1\left(T-t\right) - \sigma_k\sqrt{T-t}\right) - \Phi\left(y_1\left(T-t\right) - \sigma_k\sqrt{T-t}\right)\right]. \tag{22}$$

Then, (13) holds if we substitute (20), (21), (22) into (19).

Theorem 2. *The value of the European exotic put option with payment limitation for issuer and guaranteed income for holder of the security on share index is defined as (23)*

$$P_T = K_1 e^{-r(T)} \left[\Phi\left(y_1\left(T\right)\right) - \Phi\left(y_0\left(T\right)\right)\right] + K_2 e^{-r(T)} \Phi\left(y_0\left(T\right)\right)$$

$$- \sum_{k=1}^{n} B_0^k \left[\Phi\left(y_1\left(T\right) - \sigma_k\sqrt{T}\right) - \Phi\left(y_0\left(T\right) - \sigma_k\sqrt{T}\right)\right], \tag{23}$$

where $y_0\left(T\right)$, $y_1\left(T\right)$, B_0^k follow from (14), (15), (16) when $t = 0$, respectively.

Proof. In accordance with (5) formula (23) arises from $P_T = X_0$ [10] and (13) with $t = 0$.

Theorem 3. *For the European exotic put option with payoff function (1) the current value of the optimal hedging portfolio $\pi_t = (\beta_t, \gamma_t)$ defined by*

$$\beta_t = K_1 e^{-rT} \left[\Phi\left(y_1\left(T-t\right)\right) - \Phi\left(y_0\left(T-t\right)\right)\right] + K_2 e^{-rT} \Phi\left(y_1\left(T-t\right)\right)$$

$$+ e^{-rT} \sum_{k=1}^{n} B_t^k \left[\frac{\varphi\left(y_1\left(T-t\right) - \sigma_k\sqrt{T-t}\right)}{\sigma_m\sqrt{T-t}} - \frac{\varphi\left(y_0\left(T-t\right) - \sigma_k\sqrt{T-t}\right)}{\sigma_l\sqrt{T-t}}\right]$$

$$- K_1 e^{-rT} \left[\frac{\varphi \left(y_1 \left(T - t \right) \right)}{\sigma_m \sqrt{T - t}} - \frac{\varphi \left(y_0 \left(T - t \right) \right)}{\sigma_l \sqrt{T - t}} \right] - \frac{K_2 e^{-rT}}{\sigma_l \sqrt{T - t}} \varphi \left(y_1 \left(T - t \right) \right), \quad (24)$$

$$\gamma_t = \frac{1}{i_t} \left\{ K_1 e^{-r(T-t)} \frac{\varphi \left(y_1 \left(T - t \right) \right)}{\sigma_m \sqrt{T - t}} - \left(K_1 - K_2 \right) e^{-r(T-t)} \frac{\varphi \left(y_0 \left(T - t \right) \right)}{\sigma_l \sqrt{T - t}} \right.$$

$$+ \sum_{k=1}^{n} B_t^k \left[\frac{\varphi \left(y_0 \left(T - t \right) - \sigma_k \sqrt{T - t} \right)}{\sigma_l \sqrt{T - t}} - \frac{\varphi \left(y_1 \left(T - t \right) - \sigma_k \sqrt{T - t} \right)}{\sigma_m \sqrt{T - t}} \right.$$

$$\left. \left. - \Phi \left(y_1 \left(T - t \right) - \sigma_k \sqrt{T - t} \right) + \Phi \left(y_0 \left(T - t \right) - \sigma_k \sqrt{T - t} \right) \right] \right\} \qquad (25)$$

where σ_l and σ_m, $l, m = 1, 2, ..., n$ are volatilities in which expressions

$$\frac{1}{\sigma_k \sqrt{T - t}} \left[\ln \left(\frac{K_1 - K_2}{B_t^k} \right) - \left(r - \frac{\sigma_k^2}{2} \right) t \right], k = 1, 2, ..., n,$$

$$\frac{1}{\sigma_k \sqrt{T - t}} \left[\ln \left(\frac{K_1}{B_t^k} \right) - \left(r - \frac{\sigma_k^2}{2} \right) t \right], k = 1, 2, ..., n,$$

are minimal for every points of time $t \in [0, T]$.

Proof. In consideration of form of functions (14), (15) and using expression of B_t from (4) we have

$$\frac{\partial y_0 \left(T - t \right)}{\partial e^{rt}} = \frac{\partial}{\partial e^{rt}} \left\{ \min \left[\frac{1}{\sigma_k \sqrt{T - t}} \ln \left(\frac{K_1 - K_2}{B_t^k} \right) + \frac{\sigma_k \sqrt{T - t}}{2} t \right. \right.$$

$$\left. \left. - \frac{1}{\sigma_k \sqrt{T - t}} \ln \left(e^{rt} \right) \right] \right\} = - \frac{1}{\sigma_l \sqrt{T - t}} \exp\{-rt\}; k, l = 1, 2, ..., n,$$

$$\frac{\partial y_1 \left(T - t \right)}{\partial e^{rt}} = \frac{\partial}{\partial e^{rt}} \left\{ \min \left[\frac{1}{\sigma_k \sqrt{T - t}} \ln \left(\frac{K_1}{B_t^k} \right) + \frac{\sigma_k}{2 \sqrt{T - t}} t \right. \right.$$

$$\left. \left. - \frac{1}{\sigma_k \sqrt{T - t}} \ln \left(e^{rt} \right) \right] \right\} = - \frac{1}{\sigma_m \sqrt{T - t}} \exp\{-rt\}; k, m = 1, 2, ..., n,$$

and with (7), (13) we find (24), (25).

Expression of a part of the risk asset in optimal hedge follows from (7), (24).

5   Decision Properties

Theorem 4. *Sensivity coefficients that determine the dependences of the European exotic put option value with payoff function (1) on the strike price $P_T^{K_1} = \partial P_T / \partial K_1$; on the contracted constant restricted payment of the option writer, on the one hand, and guaranteed income for option $P_T^{K_1} = \partial P_T / \partial K_1$; on the expiration time T are defined like this*

$$P_T^{K_1} = e^{-rT} \left[\Phi \left(y_1 \left(T \right) \right) - \Phi \left(y_0 \left(T \right) \right) \right] + \frac{K_2 e^{-rT}}{\left(K_1 - K_2 \right)} \frac{\varphi \left(y_0 \left(T \right) \right)}{\sigma_l \sqrt{T - t}}$$

$$+K_1 e^{-rT} \left[\frac{1}{K_1} \frac{\varphi(y_1(T))}{\sigma_m \sqrt{T}} - \frac{1}{(K_1 - K_2)} \frac{\varphi(y_0(T))}{\sigma_l \sqrt{T}} \right]$$

$$-\sum_{k=1}^{n} B_0^k \left[\frac{1}{K_1} \frac{\varphi\left(y_1(T) - \sigma_k \sqrt{T}\right)}{\sigma_m \sqrt{T}} - \frac{1}{(K_1 - K_2)} \frac{\varphi\left(y_0(T) - \sigma_k \sqrt{T}\right)}{\sigma_l \sqrt{T}} \right], \quad (26)$$

$$P_T^{K_2} = \frac{[\varphi(y_0(T)) + \Phi(y_0(T))]}{e^{rT} \sigma_l \sqrt{T}} - \sum_{k=1}^{n} \frac{B_0^k}{(K_1 - K_2)} \frac{\varphi\left(y_0(T) - \sigma_k \sqrt{T}\right)}{\sigma_l \sqrt{T}}, \quad (27)$$

$$P_T^T = (K_1 - K_2) e^{-rT} \left[r\Phi(y_0(T)) + \frac{y_0(T)\varphi(y_0(T))}{2T} \right]$$

$$-K_1 e^{-rT} \left[r\Phi(y_1(T)) + \frac{y_1(T)\varphi(y_1(T))}{2T} \right] + \frac{1}{2T} \sum_{k=1}^{n} B_0^k \left[\left(y_1(T) - \sigma_k \sqrt{T}\right) \right.$$

$$\left. \times \varphi\left(y_1(T) - \sigma_k \sqrt{T}\right) - \left(y_0(T) - \sigma_k \sqrt{T}\right) \varphi\left(y_0(T) - \sigma_k \sqrt{T}\right) \right], \quad (28)$$

$k, l, m = 1, 2, ..., n, \ t = 0$.

Proof. The format of (26), (27), (28) follows from the definition of $P_T^{K_1}$, $P_T^{K_2}$, P_T^T with (23), (14), (15) with $t = 0$ and expressions below

$$\frac{\partial y_0(T)}{\partial K_1} = \frac{\partial \left(y_0(T) - \sigma_k \sqrt{T}\right)}{\partial K_1} = \frac{1}{(K_1 - K_2)} \frac{1}{\sigma_l \sqrt{T}},$$

$$\frac{\partial y_1(T)}{\partial K_1} = \frac{\partial \left(y_1(T) - \sigma_k \sqrt{T}\right)}{\partial K_1} = \frac{1}{K_1} \frac{1}{\sigma_m \sqrt{T}},$$

$$\frac{\partial y_0(T)}{\partial K_2} = \frac{\partial \left(y_0(T) - \sigma_k \sqrt{T}\right)}{\partial K_2} = -\frac{1}{(K_1 - K_2)} \frac{1}{\sigma_l \sqrt{T}},$$

$$\frac{\partial y_1(T)}{\partial K_2} = \frac{\partial \left(y_1(T) - \sigma_k \sqrt{T}\right)}{\partial K_2} = 0,$$

$$\frac{\partial y_j(T)}{\partial T} = -\frac{y_j(T)}{2T}, \frac{\partial \left(y_j(T) - \sigma_k \sqrt{T}\right)}{\partial T} = -\frac{\left(y_j(T) - \sigma_k \sqrt{T}\right)}{2T}, j = 1, 2.$$

Since it was not answered to get option price's P_T (23) explicit dependence on share index initial cost i_0, so sensitivity coefficient of the price P_T to numbered $p, p = 1, 2, ..., n$, share initial cost S_0^p as part of the share index (2) was obtained and presented in Theorem 5.

Theorem 5. *Sensivity coefficient $P_T^{S_0^p} = \partial P_T / \partial S_0^p$ determined the dependence of the European exotic put option value with payoff function (1) on the numbered p, $p = 1, 2, ..., n$ share initial cost S_0^p is defined (29) with (2)*

$$
P_T^{S_0^p} = \begin{cases}
- (V_p/a) \left[\Phi \left(y_1 (T) - \sigma_p \sqrt{T} \right) - \Phi \left(y_0 (T) - \sigma_p \sqrt{T} \right) \right], \text{ if } p \neq l, m, \\
- \left(S_0^p \sigma_p \sqrt{T} \right)^{-1} \sum\limits_{k=1}^{n} B_0^k \varphi \left(y_0 (T) - \sigma_k \sqrt{T} \right) - (V_p/a) \\
\times \left[\Phi \left(y_1 (T) - \sigma_p \sqrt{T} \right) - \Phi \left(y_0 (T) - \sigma_p \sqrt{T} \right) \right], \text{ if } p = l, p \neq m, \\
\left(S_0^p \sigma_p \sqrt{T} \right)^{-1} \sum\limits_{k=1}^{n} B_0^k \varphi \left(y_0 (T) - \sigma_k \sqrt{T} \right) + (V_p/a) \\
\times \left[\Phi \left(y_1 (T) - \sigma_p \sqrt{T} \right) - \Phi \left(y_0 (T) - \sigma_p \sqrt{T} \right) \right], \text{ if } p \neq l, p = m.
\end{cases} \tag{29}
$$

Proof. The format of (29) follows from the definition of $P_T^{S_0^k}$ with (23), (14), (15) with $t = 0$ and expressions below

$$
\frac{\partial y_j (T)}{\partial S_0^p} = \frac{\partial \left(y_j (T) - \sigma_k \sqrt{T} \right)}{\partial S_0^p} = 0, \, j = 1, 2, \text{ if } p \neq l, m,
$$

$$
\frac{\partial y_1 (T)}{\partial S_0^p} = \frac{\partial \left(y_1 (T) - \sigma_k \sqrt{T} \right)}{\partial S_0^p} = 0, \text{ if } p = l, p \neq m,
$$

$$
\frac{\partial y_0 (T)}{\partial S_0^p} = \frac{\partial \left(y_0 (T) - \sigma_k \sqrt{T} \right)}{\partial S_0^p} = - \left(S_0^p \sigma_p \sqrt{T} \right)^{-1}, \text{ if } p = l, p \neq m,
$$

$$
\frac{\partial y_0 (T)}{\partial S_0^p} = \frac{\partial \left(y_0 (T) - \sigma_k \sqrt{T} \right)}{\partial S_0^p} = 0, \text{ if } p \neq l, p = m,
$$

$$
\frac{\partial y_1 (T)}{\partial S_0^p} = \frac{\partial \left(y_1 (T) - \sigma_k \sqrt{T} \right)}{\partial S_0^p} = - \left(S_0^p \sigma_p \sqrt{T} \right)^{-1}, \text{ if } p \neq l, p = m.
$$

6   Conclusions

Numerical calculations of sensitivity coefficients $P_T^{K_1}$, $P_T^{K_2}$, P_T^T and $P_T^{S_0^k}$ showed that European exotic put option price with payoff function (1) is increasing function of strike price $P_T^{K_1} > 0$, decreasing function of limiting the payment option value $P_T^{K_2} < 0$ and expiration time $P_T^T < 0$. Economic interpretation of these properties is the following: strike price K_1 increment leads to probability that ranks over i_T increase. Thus, payment size under exercising increases and derivative cost increases too. The more size of the K_2 guaranteed income for option buyer the more payment size for option emitter respectively. Option buyer risk decreases, and for less risk should pay more. When expiration time T increase it is difficult to pretend share index movement. So, option buyer risk increases, and for more risk should pay less.

It is not succeed to establish analytically derivative value dependence on share index initial cost. But derivative value dependence on numbered k stock initial cost S_0^k from (29) shows that it can be $P_T^{S_0^k} > 0$ (and $P_T^{i_0} > 0$) or $P_T^{S_0^k} < 0$ (and $P_T^{i_0} < 0$). These properties can be explained as follows: on the one hand, at the average spot price S_T^k increment is expected when value S_0^k is more. Probability that $i_T = \left(\sum_{k=1}^{n} V_k S_T^k \right) / \sum_{k=1}^{n} V_k$ ranks over exercise price K_1 increases. In this case, option buyer risk increases, and for this risk should pay less. On the other hand, the case when $P_T^{S_0^k} > 0$ meets the situation $p \neq l, p = m$ that is impossible.

Acknowledgments. All comments and suggestions by critical readers and editorial assistance are acknowledged with gratitude.

References

1. Saakyan, G.R.: Queueing theory : South-Russian State University of economy and service, Shahty (2006) (in Russian)
2. Adan, I., Resing, J.: Queueing Systems. Eindhoven University of Technology, Eindhoven (2015)
3. Bayraktar, E., Horst, U., Sircar, R.: Queueing theoretic approaches to financial price fluctuations. Handbooks Oper. Res. Manag. Sci. Handbook Financ. Eng. **15**, 637–677 (2007)
4. Hull, J.: Options, futures and other derivatives. Williams, Moscow (2013)
5. Rubinstein, M.: Exotic Options. Finance working paper. 220, pp. 5–43 (1991)
6. Burenin, A.N.: Equity market and derivatives market. NTO, Moscow (2011). (in Russian)
7. Burenin, A.N.: Forwards, futures, options, exotic and annual derivatives. NTO, Moscow (2011). (in Russian)
8. Shiryaev, A.N., Kabanov, Y.M., Kramkov, D.O., Melnikov, A.V.: Toward the theory of pricing of options of both european and american types. II. continuous time. Theor. Probab. Appl. **39**(1), 61–102 (2006). (in Russian)
9. Zang, P.G.: An introduction to exotic options. Euro. Finan. Manage. **1**(1), 87–95 (1995)
10. Shiryaev, A.N.: Essentials of Stochastic Finance: Facts, Models Theory. World Scientific Publishing Company, Hackensack (1999)
11. Melnikov, A.V., Volkov, S.N., Nechaev, M.L.: Mathematics of financial obligations. American Mathematical Society, Providence (2002)
12. Amenc, N., Goltz, F., Le Sourd, V.: Assessing the Quality of Stock Market Indices. EDHEC Publication, Roubaix (2006)
13. Wikipedia, the Free Encyclopedia. https://en.wikipedia.org/wiki/Stock_market_index

Mathematical Model of a Type $M/M/1/\infty$ Queuing System with Request Rejection: A Retail Facility Case Study

Natalya Stepanova[1], Mais Farkhadov[2(✉)], and Svetlana Paul[3]

[1] Altai Economics and Law Institute, Barnaul, Russia
natalia0410@rambler.ru
[2] V.A. Trapeznikov Institute of Control Sciences of Russian Academy of Sciences,
Moscow, Russia
mais.farhadov@gmail.com
[3] Tomsk State University, Tomsk, Russia
paulsv82@mail.ru

Abstract. The model of retail outlet in form of queueing system of $M/M/1/\infty$ type with request rejection is proposed. The output flow of the system and the rejected request flow are researched. Average number of events occurred in these flows is determined. In conditions of increasing observation time the asymptotic distributions of probabilities of number of events that occurred in studied flows are found by means of asymptotic analysis.

Keywords: Queueing system · Method of asymptotic analysis · Method of torques · Fourier transformation

1 Introduction

Due to development of economic systems, mathematical models of which could be single-line queueing systems with request rejection, the latter are pretty common in practice. The "request rejection" means customer impatience and his unwillingness to stay in queue which may lead to him refusing to stay in queue.

The subject of research is the output flow of served demands and the flow of demands which refused to stay in queue, because information about output flow properties is very useful. That way, knowing properties of output flow it is possible to draw conclusions about the quality of performance of the system and to analyze its effectiveness.

Output flows research is not getting enough attention due to lack of general approach to their research. Thus the task of modification of existing methods of output flow research and development of new ones is pretty relevant [1].

2 Mathematical Model

Mathematical models of queuing systems (QS) are widely used to investigate various systems with request inflow. The models can be used to describe the

© Springer International Publishing Switzerland 2015
A. Dudin et al. (Eds.): ITMM 2015, CCIS 564, pp. 314–329, 2015.
DOI: 10.1007/978-3-319-25861-4_27

operation of a retail facility. In this work we consider one of these models, taking into account "impatient" customers.

Our mathematical model of this situation is a queuing system, which is fed a simple request flow with the parameter λ. In this paper, we consider the case when the service time has an exponential distribution with the parameter μ.

Our model has the following service discipline: if an incoming request encounters i requests already in the system, then the request is rejected (and leaves the system) with the probability r_i, $0 \leq r_i \leq 1$; on the other hand, the request is accepted for service with the probability $1 - r_i$.

Our notation is:

- $m(t)$ – the number of requests that were refused to be serviced during the time t (the output flow);
- $n(t)$ – the number of requests that have been serviced during the time t;
- $i(t)$ – the number of requests in the system at the time t.

In our QS, as the parameters λ, μ, r_i are specified, the process $i(t)$ is a continuous time Markov chain (birth and death process)[2]; the process is controlled by means of the flows $m(t)$ and $n(t)$. Thus both of these flows are MAP-processes [3].

3   Investigation of the Output Request Flow

Since the two dimensional random process is a Markov chain, the probability distribution of the process

$$P(i, n, t) = P\{i(T) = i, n(t) = n\}.$$

We can write down the following system of Kolmogorovs differential equations:

$$\begin{cases} \dfrac{\partial P(i, n, t)}{\partial t} = -[\lambda(1 - r_i) + \mu]P(i, n, t) + \lambda(1 - r_{i-1})P(i - 1, n, t) \\ +\mu P(i + 1, n - 1, t), \\ \dfrac{\partial P(0, n, t)}{\partial t} = -\lambda(1 - r_0)P(0, n, t) + \mu P(1, n - 1, t), \end{cases} \quad (1)$$

To solve this system we introduce the following function [4]:

$$H(i, u, t) = \sum_{n=0}^{\infty} e^{jun} P(i, n, t),$$

where $j = \sqrt{-1}$. Then we obtain the following system of equations for these functions

$$\begin{cases} \dfrac{\partial H(i, u, t)}{\partial t} = -[\lambda(1 - r_i) + \mu]H(i, u, t) + \lambda(1 - r_{i-1})H(i - 1, u, t) \\ +\mu e^{ju}H(i + 1, u, t), \\ \dfrac{\partial H(0, u, t)}{\partial t} = -\lambda(1 - r_0)H(0, u, t) + \mu e^{ju}H(1, u, t), \end{cases} \quad (2)$$

Lets introduce the following row-vector

$$\mathbf{H}(u,t) = \{H(0,u,t)H(1,u,t),\dots\}$$

and rewrite the system (2) as

$$\frac{\partial \mathbf{H}(u,t)}{\partial t} = \mathbf{H}(u,t)\{\mathbf{Q} + \mu e^{ju}\mathbf{B}\}, \tag{3}$$

where $\mathbf{Q}$ is a three-diagonal matrix for the birth-and-death process $i(t)$; the matrix looks the following way

$$\mathbf{Q} = \begin{bmatrix} -\lambda(1-r_0) & \lambda(1-r_0) & 0 & 0 & \dots \\ \mu & -[\lambda(1-r_1)+\mu] & \lambda(1-r_1) & 0 & \dots \\ 0 & \mu & -[\lambda(1-r_2)+\mu] & \lambda(1-r_2) & \dots \\ 0 & 0 & \mu & -[\lambda(1-r_3)+\mu] & \dots \\ \dots & \dots & \dots & \dots & \dots \end{bmatrix} \tag{4}$$

In the matrix $\mathbf{B}$ the lower sub-diagonal elements equal 1, while the rest equal zero. Finally, lets introduce the column-vector $\mathbf{E}$ that is the all-one column $\mathbf{E} = (1,1,1,\dots)^T$. Then it is easy to see that $\mathbf{QE} = 0$.

We solve the differential-matrix equation under the following initial conditions:

1. $n(0) = 0$ with probability 1.
2. Assume that at $t = 0$, the birth-and-death process $i(t)$ has a stationary probability distribution $P(i(t) = i) = R(i)$, that we will obtain later in this work. If we set $t = 0$, then $P(i,n,0) = R(i)\delta_{n0}$ and thus $H(i,u,0) = R(i)$. Lets introduce the row-vector

$$\mathbf{R} = (R(0), R(1), R(2),\dots),$$

then

$$\mathbf{H}(u,0) = \mathbf{R}$$

Next, if we set $u = 0$, then

$$H(i,0,t) = \sum_{n=0}^{\infty} P(i,n,t) = P(i,t) = R(i)$$

because $R(i)$ is stationary distributed. So the following relation is true: $\mathbf{H}(0,t) = \mathbf{R}$.

Thus, the initial conditions for the system (3) are

$$\mathbf{H}(u,0) = \mathbf{H}(0,t) = \mathbf{R}.$$

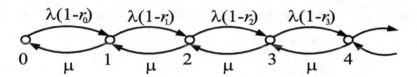

Fig. 1. The transition graph of the process $i(t)$

4 Final Probability Distribution of the Process $i(t)$

In a stationary state, the transition graph of the process $i(t)$ looks like (1).

This graph begets the following finite difference system of equations for the final probability distribution of $R(i)$

$$\begin{cases} \lambda(1 - r_0)R(0) = \mu R(1), \\ \lambda(1 - r_{i-1})R(i - 1) - [\lambda(1 - r_i) + \mu]R(i) + \mu R(i + 1) = 0. \end{cases} \quad (5)$$

Notice, that this system can be written in the matrix form $\mathbf{RQ} = 0$. Lets rewrite (5) as

$$\lambda(1 - r_{i-1})R(i - 1) - \mu R(i) = \lambda(1 - r_i)R(i) - \mu R(i + 1),$$

then it follows that

$$\lambda(1 - r_{i-1})R(i - 1) - \mu R(i) = Const.$$

From the first equation of the system (5) it follows that $Const = 0$, so

$$\lambda(1 - r_{i-1})R(i - 1) = \mu R(i).$$

it comes out that

$$R(i) = \rho(1 - r_{i-1})R(i - 1) = \cdots = R(0)\rho^i \prod_{k=0}^{i-1}(1 - r_k), \quad (6)$$

where $\rho = \lambda/\mu$.

The constant $R(0)$ can be obtained from the normalization condition $\sum_{i=0}^{\infty} R(i) = 1$, that can be written as $\mathbf{RE} = 1$. Its explicit form is $\mathbf{RE} = 1$

$$R(0) = \frac{1}{1 + \sum_{i=0}^{\infty} \rho^i \prod_{k=0}^{i-1}(1 - r_k)}. \quad (7)$$

Specifically, from (7) it follows that a stationary probability distribution in our QS exists if

$$\sum_{i=0}^{\infty} \rho^i \prod_{k=0}^{i-1}(1 - r_k) < +\infty$$

Thus for the row-vector $\mathbf{H}(u,t)$ there exists the following Cauchy problem

$$\begin{cases} \dfrac{\partial \mathbf{H}(u,t)}{\partial t} = \mathbf{H}(u,t)\left\{\mathbf{Q} + \mu\left(e^{ju} - 1\right)\mathbf{B}\right\}, \\ \mathbf{H}(u,0) = \mathbf{R}. \end{cases} \tag{8}$$

The solution of the system $\mathbf{H}(u,t)$ specifies the characteristic function of $n(t)$. Indeed, if we expand

$$H(i,u,t) = \sum_{n=0}^{\infty} e^{jun} P(i,n,t),$$

and sum up over i, we get

$$M\left\{e^{jun(t)}\right\} = \sum_{n=0}^{\infty} e^{jun} \sum_{i=0}^{\infty} P(i,n,t) = \sum_{i=0}^{\infty} H(i,u,t) = \mathbf{H}(u,t)\mathbf{E}, \tag{9}$$

where $\mathbf{E}$ is the all-ones row-vector.

5   Mean Number of Serviced Requests

Utilizing the properties of the characteristic function, we get the following expression for $M\{n(t)\}$

$$M\{n(t)\} = \frac{1}{j}\left.\frac{\partial M\left\{e^{jun(t)}\right\}}{\partial u}\right|_{u=0} = \frac{1}{j}\left.\frac{\partial \mathbf{H}(u,t)}{\partial u}\right|_{u=0}\mathbf{E}.$$

Lets denote

$$n_1(t) = \frac{1}{j}\left.\frac{\partial \mathbf{H}(u,t)}{\partial u}\right|_{u=0}E,$$

so that

$$M\{n(t)\} = n_1(t)\mathbf{E}.$$

Then from (8) we get

$$\frac{dn_1(t)}{dt} = n_1(t)\mathbf{Q} + \mu\mathbf{RB},$$

hence $H(0,t) = \mathbf{R}$. Multiplying both sides on , while keeping in mind that $\mathbf{QE} = 0$, we get

$$\frac{dM\{n(t)\}}{dt} = \mu\mathbf{RBE},$$

alongside the initial condition $M\{n(0)\} = 0$. Thus

$$M\{n(t)\} = \mu\mathbf{RBE}\cdot t,$$

From now on we denote the product $\mu\mathbf{RBE}$ as κ_1.

Since

$$\mathbf{BE} = \begin{bmatrix} 0 & 0 & 0 & 0 & \ldots \\ 1 & 0 & 0 & 0 & \ldots \\ 0 & 1 & 0 & 0 & \ldots \\ 0 & 0 & 1 & 0 & \ldots \\ \cdots\cdots\cdots\cdots \end{bmatrix} \cdot \begin{bmatrix} 1 \\ 1 \\ 1 \\ 1 \\ \cdots \end{bmatrix} = \begin{bmatrix} 0 \\ 1 \\ 1 \\ 1 \\ \cdots \end{bmatrix}$$

then

$$\mathbf{RBE} = [R(0), R(1), R(2), \ldots] \cdot \begin{bmatrix} 0 \\ 1 \\ 1 \\ 1 \\ \cdots \end{bmatrix} = \sum_{i=1}^{\infty} R(i)$$

$$= 1 - R(0) = \frac{\displaystyle\sum_{i=0}^{\infty} \rho^i \prod_{k=0}^{i-1}(1 - r_k)}{1 + \displaystyle\sum_{i=0}^{\infty} \rho^i \prod_{k=0}^{i-1}(1 - r_k)}$$

so finally we obtain

$$M\{n(t)\} = \kappa_1 t = \mu(1 - R(0))t. \tag{10}$$

6   A Solution by Fourier Method

Let $\mathbf{Y}(u, \alpha)$ be the Fourier transform of the vector-function $\mathbf{H}(u, t)$ over t

$$\mathbf{Y}(u, \alpha) = \int_0^{\infty} e^{j\alpha t} \mathbf{H}(u, t) dt. \tag{11}$$

Then, integrating this by parts, we get

$$\int_0^{\infty} e^{j\alpha t} \frac{\partial \mathbf{H}(u, t)}{\partial t} dt = \int_0^{\infty} e^{j\alpha t} d_t \mathbf{H}(u, t) dt = -\mathbf{R} - j\alpha \mathbf{Y}(u, \alpha),$$

and from (8) we get

$$-\mathbf{R} - j\alpha \mathbf{Y}(u, \alpha) = \mathbf{Y}(u, \alpha)\{\mathbf{Q} + \mu(e^{ju} - 1)\mathbf{B}\}. \tag{12}$$

Its solution $\mathbf{Y}(u, \alpha)$ has the following form

$$\mathbf{Y}(u, \alpha) = \mathbf{R} \sum_{n=0}^{\infty} e^{jun} \left[(\mu\mathbf{B} - \mathbf{Q} - j\alpha\mathbf{I})^{-1} \mu\mathbf{B}\right]^n (\mu\mathbf{B} - \mathbf{Q} - j\alpha\mathbf{I})^{-1}.$$

Here $\mathbf{I}$ the identity matrix. This expression, the definition of $\mathbf{H}(i, u, t)$, and the expression (12) give us the following formula for the Fourier transform of $P(n, t)$:

$$\mathbf{P}(n, t) = \frac{1}{2\pi} \int_{-\infty}^{\infty} e^{j\alpha t} \mathbf{R} \left[(\mu \mathbf{B} - \mathbf{Q} - j\alpha \mathbf{I})^{-1} \mu \mathbf{B} \right]^n (\mu \mathbf{B} - \mathbf{Q} - j\alpha \mathbf{I})^{-1} \mathbf{E} d\alpha.$$

7 A Long Time Asymptotic Solution of the Problem (8)

In this part the limiting case $\to \infty$ is investigated. We call the condition $t = \tau T$ the asymptotic condition of increasing time. The problem is analyzed by means of A.A. Nazarov asymptotic method [4].

The first order asymptotics of the characteristic function

$$\mathbf{H}(u, t)\mathbf{E} = M e^{jun(t)},$$

where $n(t)$ is the number of events that took place in the output flow during the time t, is a function $h_1(u, t)$ of the form

$$h_1(u, t) = \exp\{ju\kappa_1 t\},$$

where κ_1 has already been determined by the method of moment;

$$\kappa_1 = \mu \cdot \mathbf{RBE} = \mu(1 - R(0)).$$

7.1 Second Order Asymptotic

To obtain the second order asymptotic $h_2(u, t)$ in Eq. (8) lets do the substitution

$$\mathbf{H}(u, t) = \mathbf{H}_2(u, t) \exp\{ju\kappa_1 t\}. \tag{13}$$

Then for $\mathbf{H}_2(u, t)$ we get the equation

$$\frac{\partial \mathbf{H}_2(u, t)}{\partial t} = \mathbf{H}_2(u, t) \left\{ \mathbf{Q} + \mu \left(e^{ju} - 1 \right) \mathbf{B} - ju\kappa_1 \mathbf{I} \right\}, \tag{14}$$

where $\mathbf{I}$ is the identity matrix.

It follows from (13) that the initial condition for the solution $\mathbf{H}_2(u, t)$ is the same as the initial condition for the function $\mathbf{H}(u, t)$ in the problem (8)

$$\mathbf{H}_2(u, t) = \mathbf{R}.$$

Lets introduce ε such that $\varepsilon^2 = 1/T$; then we plug

$$t\varepsilon^2 = \tau, u = \varepsilon w, \mathbf{H}_2(u, t) = \mathbf{F}_2(w, \tau, \varepsilon). \tag{15}$$

into equation (14). The substitution begets the following differential equation

$$\varepsilon^2 \frac{\partial \mathbf{F}_2(w, \tau, \varepsilon)}{\partial t} = \mathbf{F}_2(w, \tau, \varepsilon) \left\{ \mathbf{Q} + \mu \left(e^{j\varepsilon w} - 1 \right) \mathbf{B} - j\varepsilon w \kappa_1 \mathbf{I} \right\}, \tag{16}$$

We solve this equation in two steps.

Step 1. The solution $\mathbf{F}_2(w, \tau, \varepsilon)$ of Eq. (16) can be written in the following form

$$\mathbf{F}_2(w, \tau, \varepsilon) = \Phi_2(w, \tau)\left\{\mathbf{R} + j\varepsilon w \mathbf{f}\right\} + O(\varepsilon^2). \qquad (17)$$

At first lets find the vector $\mathbf{f}$, while the scalar function $\Phi_2(w, \tau)$ will be obtained on the next step. Equation (16) can be rewritten as

$$O(\varepsilon^2) = j\varepsilon w \left\{\mathbf{f}\mathbf{Q} + \mathbf{R}(\mu\mathbf{B} - \kappa_1\mathbf{I})\right\},$$

where we took into account that $\mathbf{RQ} = 0$.

It follows that the vector $\mathbf{f}$ is a solution of the inhomogeneous system of equations

$$\mathbf{f}\mathbf{Q} + \mathbf{R}(\mu\mathbf{B} - \kappa_1\mathbf{I}) = 0. \qquad (18)$$

As the matrix $\mathbf{Q}$ is degenerate, we have to impose additional restrictions on $\mathbf{f}$ for the vector to be determined uniquely. Let this restriction be

$$\mathbf{f}\mathbf{E} = 0. \qquad (19)$$

Step 2. Multiplying the matrix differential Eq. (16) on $\mathbf{E}$, we get

$$\varepsilon^2 \frac{\partial \mathbf{F}_2(w, \tau, \varepsilon)}{\partial t}\mathbf{E} = \mathbf{F}_2(w, \tau, \varepsilon)\left\{\mathbf{QE} + \mu\left(e^{j\varepsilon w} - 1\right)\mathbf{BE} - j\varepsilon w \kappa_1\mathbf{E}\right\}$$
$$= \mathbf{F}_2(w, \tau, \varepsilon)\left\{j\varepsilon w(\mu\mathbf{B} - \kappa_1\mathbf{I})\mathbf{E} + \mu\frac{(j\varepsilon w)^2}{2}\mathbf{BE}\right\} + O(\varepsilon^3).$$

Substituting the expansion (17) into this equation, it comes out that

$$\varepsilon^2 \frac{\partial \Phi_2(w, \tau)}{\partial t}\mathbf{RE} = \Phi_2(w, \tau)\frac{(j\varepsilon w)^2}{2}\left\{j\varepsilon w(\mu\mathbf{RBE} + 2\mu\mathbf{fBE}\right\} + O(\varepsilon^3).$$

Let the $\varepsilon \to 0$ to zero in the last equation. This expression gives the equation to determine the scalar function $\Phi_2(w, \tau)$

$$\frac{\partial \Phi_2(w, \tau)}{\partial t} = \Phi_2(w, \tau)\frac{(jw)^2}{2}\kappa_2).$$

where

$$\kappa_2 = \mu\mathbf{RBE} + 2\mu\mathbf{fBE} = \kappa_1 + 2\mu\mathbf{fBE}. \qquad (20)$$

Here the vector $\mathbf{f}$ is the solution of (18)-(19).

Obviously, $\Phi_2(w, \tau)$ has the following form

$$\Phi_2(w, \tau) = \exp\left\{\frac{(jw)^2}{2}\kappa_2)\tau\right\}$$

Substituting this expression in (17) and multiplying by $\mathbf{E}$, we obtain

$$\mathbf{F}_2(w, \tau, \varepsilon)\mathbf{E} = \Phi_2(w, \tau)\left\{\mathbf{RE} + j\varepsilon w \mathbf{fE}\right\} + O(\varepsilon^2)$$
$$= \Phi_2(w, \tau) + O(\varepsilon^2) = \exp\left\{\frac{(jw)^2}{2}\kappa_2)\tau\right\} + O(\varepsilon^2).$$

We get

$$\mathbf{H}_2(u,t)\mathbf{E} = \exp\left\{\frac{(ju)^2}{2}\kappa_2)t\right\} + O\left(\frac{1}{T}\right). \tag{21}$$

Plugging this expression into (13), we get the second order asymptotic $h_2(u,t)$ for the characteristic function of $n(t)$

$$h_2(u,t) = \exp\left\{ju\kappa_1 t + \frac{(ju)^2}{2}\kappa_2)t\right\}. \tag{22}$$

Here κ_2 is determined from (20).

It follows that for large enough t we have $Dt = \kappa_2 t$. Obviously that if $\mathbf{fBE} = -f(0) \neq 0$, then the flow $n(t)$ is not Poisson since the necessary condition $M\{n(t)\} \neq D\{n(t)\}$ is violated.

Thus, we find that the asymptotic probability distribution of the number of applications, have completed service in the system during the time t in a growing period of observation is normal with parameters $\kappa_1 t$ and $\kappa_2 t$ [5].

7.2   Determination of the Vector F

The explicit expression of (19) is

$$-\lambda(1-r_0)f(0) + \mu f(1) + \mu R(1) - \kappa_1 R(0) = 0,$$
$$\lambda(1-r_0)f(0) - [\lambda(1-r_1)+\mu]\,f(1) + \mu f(2) + \mu R(2) - \kappa_1 R(1) = 0,$$
$$\cdots$$
$$\lambda(1-r_{i-1})f(i-1) - [\lambda(1-r_i)+\mu]\,f(i) + \mu f(i+1) + \mu R(i+1) - \kappa_1 R(i) = 0,$$
$$\cdots \tag{23}$$

Summing up the first i equations of this system, we get

$$-\lambda(1-r_0)f(0) + \mu f(1) + \mu R(1) - \kappa_1 R(0) = 0,$$
$$\lambda(1-r_1)f(0) + \mu f(2) + \mu\,[R(1)+R(2)] - \kappa_1\,[R(0)+R(1)] = 0,$$
$$\cdots$$
$$\lambda(1-r_{i-1})f(i-1) + \mu f(i) + \sum_{\nu=1}^{i} R(\nu) - \kappa_1 \sum_{\nu=0}^{i-1} R(\nu) = 0, \tag{24}$$
$$\cdots$$

So, there are the following recurrent relations for $f(i)$, where $\rho = \lambda/\mu$:

$$f(i) = \rho(1-r_{i-1})f(i-1) + \frac{\kappa_1}{\mu}\sum_{\nu=0}^{i-1} R(\nu) - \sum_{\nu=1}^{i} R(\nu). \tag{25}$$

Lets introduce the function $b(i)$ such that

$$b(i) = \frac{\kappa_1}{\mu}\sum_{\nu=0}^{i-1} R(\nu) - \sum_{\nu=1}^{i} R(\nu) = (1-R(0))\sum_{\nu=0}^{i-1} R(\nu) - \sum_{\nu=1}^{i} R(\nu)$$
$$= R(0) - R(i) - R(0)\sum_{\nu=0}^{i-1} R(\nu) = R(0)\sum_{\nu=i}^{\infty} R(\nu) - R(i). \tag{26}$$

Hence, (25) becomes

$$f(i) = \rho(1 - r_{i-1})f(i-1) + b(i), \quad \rho = \lambda/\mu, \quad b(i) = R(0)\sum_{\nu=i}^{\infty} R(\nu) - R(i).$$

It follows that

$$f(1) = \rho(1 - r_0)f(0) + b(1),$$
$$f(2) = \rho(1 - r_1)\rho(1 - r_0)f(0) + \rho(1 - r_1)b(1) + b(2),$$
$$f(3) = \rho(1 - r_2)\rho(1 - r_1)\rho(1 - r_0)f(0)+$$
$$+\rho(1 - r_2)\rho(1 - r_1)b(1) + \rho(1 - r_2)b(2) + b(3).$$

The general form of these expressions is

$$f(i) = f(0)\rho^i \prod_{k=0}^{i-1}(1 - r_k) + \sum_{\nu=1}^{i-1} b(\nu) \prod_{k=\nu}^{i-1}(1 - r_k) + b(i). \tag{27}$$

To get $f(0)$, we use the condition $\mathbf{fE} = 0$, that is $\sum_{i=0}^{\infty} f(i) = 0$. Then

$$0 = \sum_{i=0}^{\infty} f(i)$$

$$= f(0) + \sum_{i=0}^{\infty} \left\{ f(0)\rho^i \prod_{k=0}^{i-1}(1 - r_k) + \sum_{\nu=1}^{i-1} b(\nu) \prod_{k=\nu}^{i-1}(1 - r_k) + b(i) \right\} \tag{28}$$

$$= f(0) \left\{ 1 + \sum_{i=0}^{\infty} \rho^i \prod_{k=0}^{i-1}(1 - r_k) \right\} + \sum_{i=0}^{\infty} \left\{ \sum_{\nu=1}^{i-1} b(\nu) \prod_{k=\nu}^{i-1}(1 - r_k) + b(i) \right\}.$$

Finally, we obtain

$$-f(0) = \frac{\displaystyle\sum_{i=0}^{\infty} \left\{ \sum_{\nu=1}^{i-1} b(\nu) \prod_{k=\nu}^{i-1}(1 - r_k) + b(i) \right\}}{\displaystyle 1 + \sum_{i=0}^{\infty} \rho^i \prod_{k=0}^{i-1}(1 - r_k)} = \mathbf{fBE}. \tag{29}$$

This expression determines κ_2 and, accordingly, $\mathrm{D}\{n(t)\}$.

7.3 A Specific Case

Let

$$r_i = \begin{cases} 0, & \text{if } i < N, \\ 1, & \text{if } i = N, \end{cases}$$

In this case

$$R(i) = R(0)\rho^i, \quad R(0) = \frac{1}{\sum_{k=0}^{N} \rho^k} = \frac{1 - \rho}{1 - \rho^{N+1}}.$$

Next

$$f(i) = f(0)\rho^i + \sum_{\nu=1}^{i-1} b(\nu) + b(i) = f(0)\rho^i + \sum_{\nu=1}^{i-1} b(\nu), \quad i \le N,$$

$$0 = \sum_{i=0}^{N} f(i) = f(0) \sum_{i=0}^{N} \rho^i + \sum_{\nu=1}^{i-1} (N+1-\nu)b(\nu).$$

It follows that

$$-f(0) = \frac{\sum\limits_{\nu=1}^{i-1} (N+1-\nu)b(\nu)}{\sum\limits_{i=0}^{N} \rho^i}.$$

Lets compute $b(i)$. So, finally,

$$- f(0) = \frac{\sum\limits_{\nu=1}^{i-1} (N+1-\nu)b(\nu)}{\sum\limits_{i=0}^{N} \rho^i}, \quad b(i) = R(0)(\rho^i - 1)\frac{\rho^{N+1}}{\rho^{N+1}}. \tag{30}$$

If $f(0) \ne 0$, then the output flow is not Poisson.

8   Specific Cases of the Flow $m(t)$

Lets investigate two simplest specific cases of the output flow $m(t)$. Recall that the flow deals with rejected requests.

1. Let $\forall i \quad r_i = r$. Then the flow $m(t)$ is the simplest one with the parameter λr (simplest sifted flow).
2. Let

$$r_i = \begin{cases} 0, & \text{if } i < N, \\ 1, & \text{if } i = N, \end{cases}$$

this means that we deal with a M/M/1/N system. It follows that the flow $m(t)$ is a recurrent phase flow, since the lengths of its intervals match the time it takes the process $i(t)$ to return back to the state N before the first request is lost.

Indeed, lets take a look at the interval between the time t and the time t when the request leaves the system.

We adopt the following notation for $i \leq N$

$$g_i(\alpha, t) = M\{e^{j\alpha(t_n - t)} \mid i(t) = i\}.$$

Then if $i < N$, we have

$$g_0(\alpha, t - \Delta t) = (1 - \lambda \Delta t)e^{j\alpha \Delta t}g_0(\alpha, t) + \lambda \Delta t g_1(\alpha, t) + o(\Delta t),$$
$$g_i(\alpha, t - \Delta t) = [1 - (\lambda + \mu)\Delta t] e^{j\alpha \Delta t}g_i(\alpha, t) + \lambda \Delta t g_{i+1}(\alpha, t)$$
$$+ \mu \Delta t g_{i-1}(\alpha, t) + o(\Delta t),$$
$$g_N(\alpha, t - \Delta t) = [1 - (\lambda + \mu)\Delta t] e^{j\alpha \Delta t}g_N(\alpha, t) + \lambda \Delta t \cdot 1$$
$$+ \mu \Delta t g_{N-1}(\alpha, t) + o(\Delta t).$$

Set $g_i(\alpha, t) \equiv g_i(\alpha)$, then

$$(\lambda + j\alpha)g_0(\alpha) = \lambda g_1(\alpha),$$
$$(\lambda + \mu + j\alpha)g_i(\alpha) = \lambda g_{i+1}(\alpha, t) + \mu g_{i-1}(\alpha), \quad 0 < i < N,$$
$$(\lambda + \mu + j\alpha)g_N(\alpha) = \lambda + \mu g_{N-1}(\alpha).$$

From this system we get the conditional characteristic function

$$g_N(\alpha, t) = M\{e^{j\alpha(t_n - t)} \mid i(t) = N\}.$$

The function is of the length of the time interval between t (when our QS is in the state N) and t_n (when the request leaves the system). Since the exponential distribution has no long-term memory, the distribution of the length of the remaining interval matches that of the length of the full interval.

9 Investigation of the Output Request Flow

The two dimensional random process $\{i(t), m(t)\}$ is a Markov chain. For the process probability distribution function

$$P(i, m, t) = P\{i(t) = i, m(t) = m\}$$

we can write the following expression

$$\begin{cases} \dfrac{\partial P(i, m, t)}{\partial t} = -[\lambda + \mu]P(i, m, t) + \lambda r_i P(i, m - 1, t) \\ + \lambda(1 - r_{i-1})P(i - 1, m, t) + \mu P(i + 1, m, t), \\ \dfrac{\partial P(0, m, t)}{\partial t} = -\lambda P(0, m, t) + \lambda r_0 P(0, m - 1, t) + \mu P(1, m, t). \end{cases} \tag{31}$$

Denote the sum

$$H(i, u, t) = \sum_{n=0}^{\infty} e^{jum}P(i, m, t), \tag{32}$$

then we get the following system of equations

$$
\begin{cases}
\dfrac{\partial H(0,u,t)}{\partial t} = -\lambda(1-r_0)H(0,u,t) + \lambda r_0(e^{ju}-1)H(0,u,t) + \mu H(1,u,t), \\
\dfrac{\partial H(i,u,t)}{\partial t} = \lambda(1-r_{i-1})H(i-1,u,t) - [\lambda(1-r_i)+\mu]H(i,u,t) \\
+\mu H(i+1,u,t) + (e^{ju}-1)\lambda r_i H(i,u,t).
\end{cases}
$$

We can combine $H(i,u,t)$ into the row-vector

$$
\mathbf{H}(u,t) = \{H(0,u,t)H(1,u,t),\dots\}
$$

so that the system becomes

$$
\frac{\partial \mathbf{H}(u,t)}{\partial t} = \mathbf{H}(u,t)\{\mathbf{Q} + \lambda(e^{ju}-1)\mathbf{r}\}, \tag{33}
$$

where $\mathbf{r}$ is the diagonal matrix with elements r_i, the matrix $\mathbf{Q}$ is the three-diagonal infinitesimal matrix of the birth-and-death process $i(t)$; the matrix is shown in (4).

Just as we did for the serviced request flow, we take the following initial condition for the differential-matrix equation (33)

$$
\mathbf{H}(u,0) = \mathbf{R} = \mathbf{H}(0,t),
$$

where $\mathbf{R}$ is the row-vector of the stationary probability distribution of the Markov chain of the process $i(t)$; recall that R was obtained already and has the following properties: $\mathbf{RQ} = 0$, $\mathbf{RE} = 1$.

Thus, for the row-vector $= \mathbf{H}(u,t)$ we have the following Cauchy problem

$$
\begin{cases}
\dfrac{\partial \mathbf{H}(u,t)}{\partial t} = \mathbf{H}(u,t)\{\mathbf{Q} + \lambda(e^{ju}-1)\mathbf{r}\}, \\
\mathbf{H}(u,0) = \mathbf{R}.
\end{cases} \tag{34}
$$

The solution of this problem uniquely determines the characteristic function of $m(t)$ by means of the relation

$$
M\left\{e^{ium(t)}\right\} = \mathbf{H}(u,t)\mathbf{E}. \tag{35}
$$

9.1   Method of Moments

Lets denote

$$
\mathbf{m}_1(t) = \frac{1}{j}\frac{\partial \mathbf{H}(u,t)}{\partial u}\bigg|_{u=0}
$$

then from (34) we get

$$
\frac{d\mathbf{m}_1(t)}{dt} = \mathbf{m}_1(t)\mathbf{Q} + \lambda \mathbf{R}\mathbf{r}.
$$

Since

$$M\{m(t)\} = \mathbf{m}_1(t)\mathbf{E},$$

then

$$M\{m(t)\} = \lambda \mathbf{R}\mathbf{r}\mathbf{E} \cdot t = \kappa_1 t \qquad (36)$$

and

$$\kappa_1 = \lambda \mathbf{R}\mathbf{r}\mathbf{E} = \lambda \sum_{i=0}^{\infty} r_i R(i).$$

This expression can be simplified. As we derived the stationary distribution $R(i)$, we obtained the following relation

$$\kappa_1 = \lambda - \mu(1 - R(0)), \quad M\{m(t)\} = [\lambda - \mu(1 - R(0))] \cdot t. \qquad (37)$$

Notice that the following relation is true

$$M\{n(t)\} + M\{m(t)\} = \lambda t$$

that is quite natural.

9.2 Solution of the Problem (34) by Means of Fourier Transform

Lets do the Fourier transform of $\mathbf{H}(u, t)$ over t

$$\mathbf{Y}(u, \alpha) = \int_0^{\infty} e^{j\alpha t} \mathbf{H}(u, t) dt. \qquad (38)$$

Then, similar to the findings of paragraph Sect. 6 of this article, we get

$$\mathbf{P}(m, t) = \frac{1}{2\pi} \int_{-\infty}^{\infty} e^{j\alpha t} \mathbf{R} \left[(\lambda \mathbf{r} - \mathbf{Q} - j\alpha \mathbf{I})^{-1} \lambda \mathbf{r} \right]^m (\lambda \mathbf{r} - \mathbf{Q} - j\alpha \mathbf{I})^{-1} \mathbf{E} d\alpha.$$

It is problematic to compute this formula numerically since we need to compute the product of the inverse of the infinitely large matrices and to compute the improper integrals. So we seek an approximate asymptotic solution of (34) as $t \to \infty$.

9.3 Asymptotic Solution of (34)

Let be large enough. We investigate the limit $T \to \infty$. We call the condition $t = \tau T$, where $0 \le \tau < \infty$ the asymptotic condition of increasing time. Recall that $m(t)$ is the number of events that appeared in the unserviced request flow during the time t. Also recall that

$$\mathbf{H}(u, t) \cdot \mathbf{E} = M\left\{ e^{ium(t)} \right\}$$

is the characteristic function of $m(t)$. Lets call $h_1(u,t)$ the first order asymptotic of $\mathbf{H}(u,t)$

$$h_1(u,t) = \exp\{ju\kappa_1 t\},$$

where the constant κ_1 was already determined and it value is

$$\kappa_1 = \lambda\mathbf{RrE} = \lambda - \mu(1 - R(0)).$$

9.4   Second Order Asymptotic

To obtain the second order asymptotic in the equation of the problem (34) lets make the substitution

$$\mathbf{H}(u,t) = \mathbf{H}_2(u,t)\exp\{ju\kappa_1 t\}. \tag{39}$$

Then for the function $\mathbf{H}_2(u,t)$ we get the equation

$$\frac{\partial\mathbf{H}_2(u,t)}{\partial t} = \mathbf{H}_2(u,t)\left\{\mathbf{Q} + \lambda\left(e^{ju} - 1\right)\mathbf{r} - ju\kappa_1\mathbf{I}\right\},$$

In this equation we make the substitutions

$$t\varepsilon^2 = \tau, u = \varepsilon w, \mathbf{H}_2(u,t) = \mathbf{F}_2(w,\tau,\varepsilon). \tag{40}$$

where $\varepsilon^2 = 1/T$, to get

$$\varepsilon^2\frac{\partial\mathbf{F}_2(w,\tau,\varepsilon)}{\partial t} = \mathbf{F}_2(w,\tau,\varepsilon)\left\{\mathbf{Q} + \lambda\left(e^{j\varepsilon w} - 1\right)\mathbf{r} - j\varepsilon w\kappa_1\mathbf{I}\right\}, \tag{41}$$

This equation is similar to the solution of Eq. (16) is solved in two steps, then you can write the asymptotic $h_2(u,t)$ of the second order for the characteristic function of $m(t)$

$$h_2(u,t) = \exp\left\{\frac{(ju)^2}{2}\kappa_2)t\right\}.$$

Here, the value κ_2 is defined

$$\kappa_2 = \lambda\mathbf{RrE} + 2\lambda\mathbf{f_2rE} = \kappa_1 + 2\lambda\mathbf{f_2rE},$$

where $\mathbf{f_2}$ is defined by
    The general form of these expressions is

$$f_2(i) = f_2(0)\rho^i\prod_{k=0}^{i-1}(1 - r_k) + \sum_{\nu=1}^{i-1}b(\nu)\prod_{k=\nu}^{i-1}(1 - r_k) + b(i).$$

$$-f(0) = \frac{\displaystyle\sum_{i=0}^{\infty}\left\{\sum_{\nu=1}^{i-1}b(\nu)\prod_{k=\nu}^{i-1}(1 - r_k) + b(i)\right\}}{1 + \displaystyle\sum_{i=0}^{\infty}\rho^i\prod_{k=0}^{i-1}(1 - r_k)},$$

where

$$R(0) \sum_{\nu=i}^{\infty} R(\nu) - R(i) = b(i).$$

If follows that as $t \to \infty$, the variable $m(t)$ is asymptotically normal with the mean $\{m(t)\} = \kappa_1 t$ and variance $D\{m(t)\} = \kappa_2 t$.

Obviously, if $\mathbf{f2rE} \neq 0$, then the flow $m(t)$ is not Poisson, since the necessary condition $M\{m(t)\} = D\{m(t)\}$ is violated.

10   Conclusion

1. So, we presented a new model of a retail facility. The model is a type M/M/1/∞ queuing system with request rejection.
2. The output flow of the system and the flow of flow of rejected requests are researched. Exact formulas for average number of events that occurred in both flows are determined. Prelimit distributions of probabilities of number of events that occurred in these flows in form of integral transformation are found.
3. The asymptotic distributions of probabilities of number of events that occurred in flows $n(t)$ and $m(t)$ are found by means of asymptotic analysis proposed by A.A. Nazarov.

References

1. Lapatin I.L.: Mathematical models of output flows of queueing systems with an unlimited number of devices: the dissertation ... The candidate of physical and mathematical sciences: 05.13.18.Tomsk (2012) (in Russian)
2. Nazarov A.A., Terpugov A.F.: Queueing theory: educational material. Tomsk. NTL, 228 (2004) (in Russian)
3. Lopuhova S.V.: The asymptotic and numerical methods for the study of special flows of events: the thesis ... The candidate of physical and mathematical sciences: 05.13.18. Tomsk (2008) (in Russian)
4. Nazarov, A.A., Moiseeva, S.P.: Method of asymptotic analysis in queueing theory. NTL, Tomsk (2006). (in Russian)
5. Lopuhova, S.V., Nazarov, A.A.: Research MAP flow method of asymptotic analysis of N-ro order. J.Tomsk State Univ. Series Inf. Cybern. Math. **293**, 110–115 (2006). (in Russian)

Performance Analysis and Statistical Modeling of the Single-Server Non-reliable Retrial Queueing System with a Threshold-Based Recovery

Dmitry Efrosinin[1]([✉]) and Janos Sztrik[2]

[1] Johannes Kepler University, Altenbergerstrasse 69, 4040, Linz, Austria
dmitry.efrosinin@jku.at
http://www.jku.at, http://www.unideb.hu
[2] Debrecen University, Egyetem Ter. 1, 4032 Debrecen, Hungary
sztrik.janos@inf.unideb.hu

Abstract. In this paper we study a single-server Markovian retrial queueing system with non-reliable server and threshold-based recovery policy. The arrived customer finding a free server either gets service immediately or joins a retrial queue. The customer at the head of the retrial queue is allowed to retry for service. When the server is busy, it is subject to breakdowns. In a failed state the server can be repaired with respect to the threshold policy: the repair starts when the number of customers in the system reaches a fixed threshold level. Using a matrix-analytic approach we perform a stationary analysis of the system. The optimization problem with respect to the average cost criterion is studied. We derive expressions for the Laplace transforms of the waiting time. The problem of estimation and confidence interval construction for the fully observable system is studied as well.

Keywords: Quasi-birth-and-death process · Retrial queues · Performance analysis · Confidence intervals

1 Introduction

Different types of single server retrial queueing systems have found applications in local area networks and communication protocols. In a retrial queue a customer who finds the server busy is assigned to a queue of retrial customers. It is assumed that the arrived customer finding a free server with probability p gets service immediately or joins a retrial queue with probability $1 - p$. Many

D. Efrosinin—This work was funded by the COMET K2 Center "Austrian Center of Competence in Mechatronics (ACCM)", funded by the Austrian federal government, the federal state Upper Austria, and the scientific partners of the ACCM.

J. Sztrik— The research is supported by the Austro-Hungarian Cooperation Grant No. 90u6, OMAA 2014, Stiftung Aktion Österreich-Ungarn.

© Springer International Publishing Switzerland 2015
A. Dudin et al. (Eds.): ITMM 2015, CCIS 564, pp. 330–343, 2015.
DOI: 10.1007/978-3-319-25861-4_28

papers study the case $p = 1$, where customers have a direct access to the server, or $p = 0$, when a customer upon arrival goes always to the retrial queue. The bibliography for these two particular cases as well as a description of a general model for arbitrary value p can be found in [2].

In our model the customer at the head of the retrial queue is allowed to retry for service, i.e. the system has a retrial queue with a constant retrial policy or FCFS retrial queue. The constant retrial policy was introduced by [8] and it was used in many applications to local area networks and communication protocols, e.g. in [3,4,6,10]. The system with constant retrial policy is simpler to analyze that one with the classical retrial policy assuming the state-dependent retrial intensity, since in the latter case the QBD process with three diagonal block infinitesimal matrix can be constructed. Moreover the constant retrial policy can be used in a truncation model of classical policy exhibiting spatial homogeneity from some orbit level upwards.

The systems with an unreliable server have been studied extensively. But the systems which combines server breakdowns with a retrial effect are still not exhaustively examined. We refer the interested readers to the papers of [1,9] and bibliographies therein. The system under study is assumed to be controllable in the sense that the repair process in a failed state starts according to the threshold-based recovery policy. This policy prescribes to switch on/off the repair facility if the number of customers in the system is higher or lower than a fixed threshold level $q_r \geq 1$. The threshold-based recovery was first introduced by [5] in case of the system with an ordinary queue. Then the obtained results were generalized by [7] to case of the retrial queue with a constant retrial rate.

Whenever a queueing system is fully observable with respect to their random time periods such as inter-arrival time, service time, time to failure, repair time, inter-retrial time and so on, standard parametric estimation methods of mathematical statistics seems to be quite appropriate. But the most papers include only the results about transient and stationary solutions and very few consider the associated statistical problems. [14] have evaluated confidence intervals for the mean waiting time of the single server and tandem queues with blocking. Maximum likelihood estimates of multi-server system with heterogeneous servers were obtained by [13]. In [12] have studied the estimation of arrival and service rates for queues based only on queue length data.

The analysis of the presented retrial queue with constant retrial rate, non-reliable server and threshold-based recovery includes the following contributions:

(a) We model the system as a quasi-birth-and-death (QBD) process with threshold dependent block-tridiagonal infinitesimal matrix and apply a general theory of matrix-analytic solutions to derive the stationary distribution of the system states and stability condition.

(b) We formulate optimization problem to calculate a threshold level which minimizes the long-run average cost per unit of time for the given cost structure.

(c) We derive the main performance characteristics of the system for the given threshold policy.

(d) We obtain the Laplace transforms of the waiting time distribution.
(e) We perform a parameter estimation and construct confidence intervals for the performance measures.

In further sections we will use the notation I for the identity matrix, O and $\mathbf{0}$ – respectively for the square matrix and row vector with zero entries. Furthermore $\mathbf{e}$ will denote a column vector of ones and $\mathbf{e}_j$ – a column vector with 1 in the j-th position and 0 elsewhere. Vectors and matrices are assumed to have an appropriate size. The symbol ∇ will stand for the gradient.

2   Mathematical Model and Stability

We consider a $M/M/1$ queueing system illustrated in Fig. 1. Customers arrive to the system according to a Poisson stream with intensity $\lambda > 0$. The server servers the customers according to an exponentially distributed time with parameter $\mu > 0$. If the server is idle at the time of an external arrival, the customer proceeds to the server with probability p or to the orbit with probability $1 - p$. In particular case $p = 1$ we get a system with a direct access to the server which was already studied in [7]. The servicing customer leaves the system after service completion. If the server is found to be blocked, i.e. busy or failed, the customer has to enter the infinite capacity retrial queue. We assume a constant retrial policy, i.e. FCFS discipline for the retrial queue, when the customer at the head of the queue repeats its requests for service in exponentially distributed retrial times with intensity $\tau > 0$. The server is assumed to be unreliable. During a service process it may fail in exponentially distributed time with intensity $\alpha > 0$. The repair time is again exponentially distributed with intensity $\beta > 0$. The system under study is regulated by a controller who switches the repair facility on only when the number of customers in the system reaches a fixed threshold level $q_r \geq 1$ and switch it off if the number of customers decreases below this level. The inter-arrival times, intervals of successive retrials, service, breakdown and repair times are assumed to be mutually independent.

The system states at time t are described by random vector $\{N(t), D(t)\}_{t \geq 0}$, where $N(t)$ – the number of customers in the queueing system and $D(t)$ – the

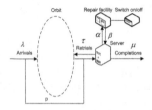

Fig. 1. Scheme of the queueing system

server state, where

$$D(t) = \begin{cases} 0 & \text{the server is idle,} \\ 1 & \text{the server is busy,} \\ 2 & \text{the server is failed.} \end{cases}$$

Note that if $D(t) = 0$, the component $N(t) \in \mathbb{N}_0$, otherwise $N(t) \in \mathbb{N}$. The random process

$$\{X(t)\}_{t \geq 0} = \{N(t), D(t)\}_{t \geq 0} \tag{1}$$

is an irreducible continuous-time Markov chain with a state space

$$E = \{x = (n, d); n \geq 0, d = 0 \vee n \geq 1, d \in \{1, 2\}\} \tag{2}$$

and transition intensities $\lambda_{xy}(q_r)$ from state $x = (n, d) \in E$ to state $y = (n', d') \in E$,

$$\lambda_{xy}(q_r) = \begin{cases} \lambda p, & n' = n+1, \, d' = 1, \, d = 0, \, n \geq 0, \\ \lambda(1-p), & n' = n+1, \, d' = d = 0, \, n \geq 0, \\ \lambda, & n' = n+1, \, d' = d \in \{1, 2\}, \, n \geq 0, \\ \mu, & n' = n-1, \, d' = 0, \, d = 1, \, n > 0, \\ \tau, & n' = n, \, d' = 1, \, d = 0, \, n > 0, \\ \alpha, & n' = n, \, d' = 2, \, d = 1, \, n > 0, \\ \beta, & n' = n, \, d' = 1, \, d = 2, \, n \geq q_r. \end{cases} \tag{3}$$

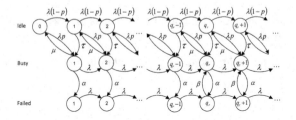

Fig. 2. The state-transition-intensity diagram for the given threshold q_r

Figure 2 illustrates the state transition rates. Now we define a macro-state $\mathbf{n}$ consisting of three states,

$$\mathbf{n} = \{(n, 0), (n+1, 1), (n+1, 2)\}, \, n \geq 0.$$

Define by $\boldsymbol{\pi} = (\boldsymbol{\pi}_0, \boldsymbol{\pi}_1, \boldsymbol{\pi}_2, \dots)$ a row vector of stationary state probabilities with subvectors $\boldsymbol{\pi}_n = (\pi_{(n,0)}, \pi_{(n+1,1)}, \pi_{(n+1,2)})$ for the macro-state $\mathbf{n}$, where

$$\pi_{(n,d)} = \lim_{t \to \infty} \mathbb{P}[N(t) = n, D(t) = d].$$

Theorem 1. *For the fixed threshold q_r the Markov chain (1) belongs to a class of the QBD processes with boundary states and block tri-diagonal infinitesimal matrix $\Lambda = [\lambda_{xy}(q_r)]$,*

$$
\Lambda = \left.\begin{pmatrix}
Q_{1,0} & Q_0 & O & O & O & O & O & \cdots \\
Q_2 & Q_{1,1} & Q_0 & O & O & O & O & \cdots \\
O & Q_2 & Q_{1,1} & Q_0 & O & O & O & \cdots \\
& \ddots & & \ddots & \ddots & \ddots & & \cdots \\
O & O & O & Q_2 & Q_{1,2} & Q_0 & O & \cdots \\
O & O & O & O & Q_2 & Q_{1,2} & Q_0 & \cdots \\
& \ddots & & & & \ddots & \ddots & \ddots
\end{pmatrix}\right\}q_r - 1,
$$

where

$$
Q_{1,j} = \begin{pmatrix}
-(\lambda + \tau 1_{\{j\neq 0\}}) & \lambda p & 0 \\
\mu & -(\alpha + \lambda + \mu) & \alpha \\
0 & \beta 1_{\{j=2 \vee q_r=1\}} & -(\lambda + \beta 1_{\{j=2 \vee q_r=1\}})
\end{pmatrix},
$$

$$
Q_0 = \begin{pmatrix}
\lambda(1-p) & 0 & 0 \\
0 & \lambda & 0 \\
0 & 0 & \lambda
\end{pmatrix}, \quad
Q_2 = \begin{pmatrix}
0 & \tau & 0 \\
0 & 0 & 0 \\
0 & 0 & 0
\end{pmatrix},
$$

and row vector $\boldsymbol{\pi}$ satisfies the matrix system

$$
\boldsymbol{\pi}\Lambda = \mathbf{0}, \quad \boldsymbol{\pi}\mathbf{e} = 1.
$$

Proof. The result follows by arranging the balance equations according to the macro states **n** and collecting them in matrix form.

The next statement reveals the condition that is necessary and sufficient to ensure system stability.

Theorem 2. *The necessary and sufficient stability condition for the process $\{X(t)\}_{t\geq 0}$ is given by*

$$
\rho = \frac{\lambda}{\mu}\left(1 + \frac{\alpha}{\beta}\right) + \frac{\lambda}{\lambda p + \tau} < 1. \tag{4}
$$

Proof. Consider the matrix $A = Q_0 + Q_{1,2} + Q_2$, composed of matrices defined above, and let $\boldsymbol{p}$ be its stationary distribution. Since A is irreducible, the vector $\boldsymbol{p}$ exists such that $\boldsymbol{p}A = \mathbf{0}$ and $\boldsymbol{p}\mathbf{e} = 1$. It is given by

$$
\boldsymbol{p} = \frac{1}{\alpha(p\lambda + \tau) + \beta(p\lambda + \mu + \tau)}(\beta\mu, \beta(p\lambda + \tau), \alpha(p\lambda + \tau)). \tag{5}
$$

According to the mean drift result of [11] the stability condition is given by the inequality $\boldsymbol{p}Q_2\mathbf{e} > \boldsymbol{p}Q_0\mathbf{e}$ which leads to the proposed inequality.

3   Evaluation of Performance Measures

Using the general theory of the QBD processes (see e.g. [11]) we get the following result.

Theorem 3. *Subvectors* $\pi_n, n \geq 0$, *of stationary state probabilities are calculated by*

$$\pi_n = \pi_{q_r} \prod_{i=1}^{q_r-n} M_{q_r-i}, \quad 0 \leq n \leq q_r - 1, \tag{6}$$

$$\pi_n = \pi_{q_r} R^{n-q_r}, \quad n \geq q_r,$$

where the matrices M_i *are given by*

$$M_0 = -Q_2 Q_{1,0}^{-1}, \tag{7}$$

$$M_i = -Q_2 (M_{i-1} Q_0 + Q_{1,1})^{-1}, \quad 1 \leq i \leq q_r - 2,$$

$$M_{q_r-1} = -Q_2 (M_{q_r-2} Q_0 + Q_{1,2})^{-1}.$$

The vector π_{q_r} *is a unique solution of the system of equations*

$$\pi_{q_r} (M_{q_r-1} Q_0 + Q_{1,2} + R Q_2) = \mathbf{0}, \tag{8}$$

$$\pi_{q_r} \left(\sum_{n=0}^{q_r-1} \prod_{i=1}^{q_r-n} M_{q_r-i} + (I - R)^{-1} \right) \mathbf{e} = 1. \tag{9}$$

Matrix R *is the minimal non-negative solution to matrix equation* $R^2 Q_2 + R Q_{1,2} + Q_0 = O$ *and has the following explicit representation,*

$$R = \begin{pmatrix} \frac{\lambda(1-p)}{\tau} & \frac{\lambda^2(1-p)}{\mu\tau} & \frac{\alpha\lambda^2(1-p)}{\mu\tau(\beta+\lambda)} \\ \frac{\lambda}{\tau} & \frac{\lambda(\lambda+\theta)}{\mu\tau} & \frac{\lambda(\lambda+\tau)\alpha}{\mu\tau(\beta+\lambda)} \\ \frac{\lambda}{\tau} & \frac{\lambda(\lambda+\tau)}{\mu\tau} & \frac{\lambda(\lambda+\tau)\alpha+\lambda\mu\tau}{\mu\tau(\beta+\lambda)} \end{pmatrix}. \tag{10}$$

Proof. Due to the general theory of the QBD processes (see [11, Chapter 3, pp. 82–83.]), subvectors π_n which correspond to the macro-states $\mathbf{n}$ with homogeneous blocks in the matrix Λ, have geometric structure,

$$\pi_n = \pi_{q_r} R^{n-q_r}, \quad n > q_r.$$

For the probabilities π_n, $0 \leq n \leq q_r$, of the boundary states the system of balance equations can be transformed in the form

$$\pi_0 Q_{1,0} + \pi_1 Q_2 = \mathbf{0},$$

$$\pi_{n-1} Q_0 + \pi_n Q_{1,1} + \pi_{n+1} Q_2 = \mathbf{0}, \ 1 \leq n \leq q_r - 2,$$

$$\pi_{q_r-2} Q_0 + \pi_{q_r-1} Q_{1,2} + \pi_{q_r} Q_2 = \mathbf{0}.$$

The last system implies the recurrent relation

$$\pi_n = \pi_{n+1} M_n, \quad 0 \le n \le q_r - 1,$$

where matrices M_n can be evaluated recursively using (7). For the boundary state $n = q_r$ we get

$$\pi_{q_r-1} Q_0 + \pi_{q_r} Q_{1,2} + \pi_{q_r+1} Q_2 = 0.$$

Subsequent substitution of $\pi_{q_r-1} = \pi_{q_r} M_{q_r-1}$ and $\pi_{q_r+1} = \pi_{q_r} R$ to the last equality leads to (8). This equation can be solved with respect to the last unknown vector π_{q_r} together with the normalizing condition which follows from the relation

$$\sum_{n=0}^{\infty} \pi_n \mathbf{e} = \sum_{n=0}^{q_r-1} \pi_n \mathbf{e} + \pi_{q_r} \sum_{n=q_r}^{\infty} R^{n-q_r} \mathbf{e} = 1.$$

Finally, the structure of matrices $Q_0, Q_{1,2}, Q_2$ together with a relation $RQ_2\mathbf{e} = Q_0\mathbf{e}$ implies the form (10).

Corollary 1. *Using the probabilities* $\pi_n,\ n \ge 0$, *we can evaluate different performance measures:*

Utilization of the system

$$U = 1 - \pi_0 \mathbf{e}_1. \tag{11}$$

Probability of a server being blocked

$$P_{blocking} = \Big(\sum_{n=0}^{q_r-1} \pi_n + \pi_{q_r} (I - R)^{-1} \Big) (\mathbf{e}_2 + \mathbf{e}_3). \tag{12}$$

Mean number of customers in the queue

$$\bar{Q} = \Big(\sum_{n=0}^{q_r-1} n\pi_n + \pi_{q_r} (q_r(I - R) + R)(I - R)^{-2} \Big) \mathbf{e}. \tag{13}$$

Mean number of customers in the system

$$\bar{N} = \bar{Q} + P_{blocking}. \tag{14}$$

Mean waiting and sojourn times

$$\bar{W} = \frac{\bar{N}}{\lambda}, \quad \bar{T} = \frac{\bar{Q}}{\lambda}. \tag{15}$$

A natural question that may arise in practice is a calculation of an optimal threshold policy which leads to the minimum of the system operating costs per unit of time. To find the optimal threshold q^* the following cost structure is introduced: c_0 – holding cost per unit time for each customer in the system, $c_{0,0}$, $c_{1,0}$ and $c_{2,0}$ – usage costs per unit time if the server is idle, busy or failed for $N(t) < q_r$. For $N(t) \ge q_r$ the costs $c_{0,1}$, $c_{1,1}$ $c_{2,1}$ – usage costs together with the operational costs of the repair facility.

Corollary 2. *The average cost function $g(q_r)$ is of the form*

$$g(q_r) = c_0\bar{N} + \sum_{n=0}^{q_r-2} \boldsymbol{\pi}_n(c_{0,0}, c_{1,0}, c_{2,0})' + \boldsymbol{\pi}_{q_r-1}(c_{0,0}, c_{1,1}, c_{2,1})' \qquad (16)$$
$$+ \boldsymbol{\pi}_{q_r}(I - R)^{-1}(c_{0,1}, c_{1,1}, c_{2,1})',$$

where $(c_{0,i}, c_{1,i}, c_{2,i})'$ is a column-vector of the costs per unit of time, $i = 0, 1$.

In many cases a simple exhaustion method is quite appropriate to calculate the optimal value q_r^*. Setting $\frac{d}{dq_r}g(q_r) = 0$ the optimal threshold level q_r^* can also be numerically evaluated.

4   The Waiting Time Distribution

Here we want to calculate the distribution function of the waiting time W of the customer in the retrial queue. In comparison to the classical queue, where the conditional waiting time of the tagged customer is Erlang distributed, the conditional waiting time in a present system will depend on the future arrivals. It happens due to the presence of the threshold-based policy for the recovering of the server and due to the fact that with probability p a new arrival is served according to the LCFS (last come first served) discipline. Therefore it is required to observe the state of the tagged customer up to the time where its service begins. The further calculation is performed by analyzing of the auxiliary Markov chain just after an arrival of the tagged customer at time t^+,

$$\{\hat{X}(t)\}_{t \geq t^+} = \{N(t), D(t), M(t)\}_{t \geq t^+}.$$

The state space of this process is

$$\hat{E} = \{\hat{x} = (n, d, m) | n \geq 0, d = 0 \vee n \geq 1, d \in \{1, 2\}, 0 \leq m \leq n\}$$

with an absorption states with $m = 0$ when the tagged customer receives the service. The component $M(t)$ of the process denotes the position of the tagged customer in the list of waiting customers at time t. This component can only decrease at retrial time when the server is idle. If the server is busy or failed then we obviously have $M(t^*) = N(t^*) - 1$. The process is absorbed when the component $M(t)$ becomes equal to zero.

The waiting time distribution of the tagged customer is obtained as follows. First we calculate the Laplace transform of the conditional waiting time distribution given the system state and the position of the tagged customer after the arrival. Using the law of total probability and the state distribution just after the arrival of the tagged customer, the conditioning is removed. Numerical inversion of the Laplace transform completes the calculation.

Denote by $w_{(n,d,m)}(t)$ the probability density function of the conditional waiting time given state $\hat{x} = (n, d, m) \in \hat{E}$ and $\tilde{w}_{n,d,m}(s) = \int_0^\infty e^{-st} w_{(n,d,m)}(t)dt$ the corresponding Laplace transform (LT). Then due to the PASTA property and the law of total probability the unconditional LT is of the form

$$\tilde{w}(s) = \sum_{n=0}^{\infty} \pi_{(n,0)} p + \sum_{n=0}^{\infty} \pi_{(n,0)}(1-p)\tilde{w}_{(n+1,0,n+1)}(s) \tag{17}$$

$$+ \sum_{n=1}^{\infty} \pi_{(n,1)}\tilde{w}_{(n+1,1,n)}(s) + \sum_{n=1}^{\infty} \pi_{(n,2)}\tilde{w}_{(n+1,2,n)}(s),$$

where the first summand represents the stationary probability that a tagged customer does not have to wait for service, i.e. $W = 0$; the last three terms represent the LT of the waiting time given $W > 0$.

Now we partition the conditional LT $\tilde{w}_{(n,d,m)}(s)$, $(n, d, m) \in \hat{E}$, according to the number of customers in the system and define the column-vectors

$$\tilde{\mathbf{w}}_{n,m}(s) = (\tilde{w}_{(n,0,m)}(s), \tilde{w}_{(n+1,1,m)}(s), \tilde{w}_{(n+1,2,m)})', \ m \le n \le q_r + m - 1.$$

For the calculation of the conditional waiting time we make use of the Laplace transform of conditional service time for $n \ge q_r$: Let $h_1(t)$ and $h_2(t)$ denote the probability density functions from the start in an busy or failed state to the next departure. Obviously we have

$$h_1(t) = \frac{\mu}{\mu+\alpha}(\mu+\alpha)e^{-(\mu+\alpha)t} + \frac{\alpha}{\mu+\alpha}\int_0^t (\mu+\alpha)e^{-(\mu+\alpha)x}h_f(t-x)dx,$$

$$h_2(t) = \int_0^t \beta e^{-\beta x} h_1(t-x)dx.$$

Denote by $\tilde{h}_1(s)$ and $\tilde{h}_2(s)$ the corresponding LT. For these functions we get

$$\tilde{h}_1(s) = \frac{\mu(\beta+s)}{(\mu+\alpha+s)(\beta+s)-\alpha\beta}, \quad \tilde{h}_2(s) = \frac{\beta}{\beta+s}\tilde{h}_1(s).$$

Theorem 4. *The vector of conditional Laplace transforms $\tilde{\mathbf{w}}_{n,m}(s)$, $m \le n \le q+m-2$ under stability condition satisfy the following recurrent relations*

$$\tilde{\mathbf{w}}_{n,m}(s) = M_1^{q-n-1}(s)M_2^m(s)\tilde{\mathbf{w}}_{q+m-1,m}(s) \tag{18}$$

$$+M_1^{q-n-1}(s)\sum_{r=0}^{m-1} M_2^r(s)L_2(s)\tilde{\mathbf{w}}_{q+r-1,m-1}(s)$$

$$+ \sum_{r=0}^{q-n-2} M_1^r(s)L_1(s)\tilde{\mathbf{w}}_{n+r-1,m-1}(s), \ n \le q-2,$$

$$\tilde{\mathbf{w}}_{n,m}(s) = M_2^{q-n+m-1}(s)\tilde{\mathbf{w}}_{q+m-1,m}(s)$$

$$+ \sum_{r=0}^{q-n+m-2} M_2^r(s)L_2(s)\tilde{\mathbf{w}}_{n+r-1,m-1}(s), \ n \ge q-1,$$

$$\tilde{\mathbf{w}}_{q+m-1,m}(s) = \tilde{\vartheta}_1^{m-1}(s)(\tilde{\vartheta}_0(s), \tilde{\vartheta}_1(s), \tilde{\vartheta}_2(s))', \ \tilde{\mathbf{w}}_{n,0}(s) = (0,1,1)',$$

where

$$M_j(s) = -(Q_{1,j} - sI)^{-1}Q_0, \ L_j(s) = -(Q_{1,j} - sI)^{-1}Q_2, \ i = 1, 2, \qquad (19)$$

$$\tilde{\vartheta}_0(s) = \frac{\tau}{\tau + \lambda p(1 - \tilde{h}_1(s)) + s}, \ \tilde{\vartheta}_1(s) = \tilde{h}_1(s)\tilde{\vartheta}_0(s), \ \tilde{\vartheta}_2(s) = \tilde{h}_2(s)\tilde{\vartheta}_0(s). \quad (20)$$

Proof. The Markov property of the process $\{\hat{X}(t)\}$ implies the following system

$$(\lambda + \theta + s)\tilde{w}_{(n,0,m)}(s) = \lambda p\tilde{w}_{(n+1,1,m)}(s) + \lambda(1 - p)\tilde{w}_{(n+1,0,m)} + \theta\tilde{w}_{(n,1,m-1)}(s),$$
$$(\alpha + \lambda + \mu + s)\tilde{w}_{(n+1,1,m)}(s) = \alpha\tilde{w}_{(n+1,2,m)}(s) + \lambda\tilde{w}_{(n+2,1,m)}(s) + \mu\,\tilde{w}_{(n,0,m)}(s),$$
$$(\lambda + \beta I_{\{n \geq q-1\}} + s)\tilde{w}_{(n+1,2,m)}(s) = \lambda\tilde{w}_{(n+2,2,m)}(s) + \beta\tilde{w}_{(n+1,1,m)}(s)I_{\{n \geq q-1\}},$$

where $\tilde{w}_{(n,1,0)}(s) = \tilde{w}_{(n,2,0)}(s) = 1$ and $\tilde{w}_{(n,0,0)}(s) = 0$. After routing block identification taking into account the difference of the transition rates for the states below and above threshold level, this system can be expressed in matrix form

$$(Q_{1,1}I_{\{n \leq q_r-2\}} + Q_{1,2}I_{\{n \geq q_r-1\}} - sI)\tilde{\mathbf{w}}_{n,m}(s) + Q_0\tilde{\mathbf{w}}_{n+1,m}(s) + Q_2\tilde{\mathbf{w}}_{n-1,m-1}(s) = 0.$$

The recursive forward substitution applied $q_r + m - 1 - n$ times using the notations (19) leads to the expressions (18). Note that the Laplace transforms $\tilde{\mathbf{w}}_{q_r+m-1,m}(s)$ do not depend on future arrivals to the queue, since the number of customers in the retrial queue always exceeds the given threshold level q_r during the waiting time of the tagged customer. To calculate the components of this vector we derive the Laplace transforms $\tilde{\vartheta}_d(s)$ of the waiting time for the customer at the head of the retrial queue given the initial server state $d \in \{0, 1, 2\}$. Obviously these LTs satisfies (20), since the random time to absorption is equal to the sum of the service time given states 1 or 2 of the server plus the time to absorption given state 0.

If we take into account the sequence of epochs at which the queue size decreases in one unit, we easily find the expression (18).

Corollary 3. *For the unconditional LT of the waiting time distribution we have*

$$\tilde{W}(s) = \frac{1}{s}(1 - \boldsymbol{\pi}_W\mathbf{e} + \boldsymbol{\pi}_W\tilde{\mathbf{w}}(s)),$$

where the contribution $1 - \boldsymbol{\pi}_W\mathbf{e}$ *is equal to the first summand of (17) and* $\boldsymbol{\pi}_W\tilde{\mathbf{w}}(s)$ *stands for the last three terms defined in (17).*

5   Confidence Intervals for Performance Measures

Consider a real life system which runs without control, i.e. $q_r = 1$, and system parameters are unknown. Using random samples drawn from observed data we derive simple parameter estimators. An estimator $\hat{q}_r^*$ for the optimal threshold q_r^* is calculated from them next. Given a system which runs under threshold $\hat{q}_r^*$ we provide consistent and asymptotically normal estimators and corresponding confidence intervals for its performance measures. Numerical examples illustrate the performance improvement due to introduced control at the end of the section.

5.1   System Parameter Estimators

Let $(X_1, X_2, \ldots, X_n), (Y_1, Y_2, \ldots, Y_n), (Z_1, Z_2, \ldots, Z_n), (U_1, U_2, \ldots, U_n)$ and $(H_1, H_2, \ldots, H_n)$ each be random samples of size n, which, respectively, are drawn from different exponentially distributed inter-arrival time populations with parameter λ, exponentially distributed service time populations with parameter μ, exponentially distributed time to failure populations with parameter α, exponentially distributed repair time populations with parameter β and exponentially distributed inter-retrial time populations with parameter τ. It follows that $\mathbb{E}[\bar{X}] = \frac{1}{\lambda}$, $\mathbb{E}[\bar{Y}] = \frac{1}{\mu}$, $\mathbb{E}[\bar{Z}] = \frac{1}{\alpha}$, $\mathbb{E}[\bar{U}] = \frac{1}{\beta}$ and $\mathbb{E}[\bar{H}] = \frac{1}{\tau}$, where $\bar{X}, \bar{Y}, \bar{Z}, \bar{U}$ and $\bar{H}$ are the sample means of inter-arrival time, service time, time to failure, repair time and inter-retrial time. It is obvious that $\bar{X}, \bar{Y}, \bar{Z}, \bar{U}$ and $\bar{H}$ are the maximum likelihood estimators of $\frac{1}{\lambda}, \frac{1}{\mu}, \frac{1}{\alpha}, \frac{1}{\beta}$ and $\frac{1}{\tau}$. Let $(J_1, J_2, \ldots, J_n)$ be the random sample of size n with

$$J_i = \begin{cases} 1 & \text{if the i-th arrived customer finding server idle proceeds to the server,} \\ 0 & \text{if the i-th arrived customer finding server idle proceeds to the orbit.} \end{cases}$$

Obviously,

$$\bar{J} \cdot n \xrightarrow{d} \mathcal{B}(n, p)$$

is binomially distributed with parameters n and p. It follows that $\mathbb{E}[\bar{J}] = p$ and $\mathbb{V}[\bar{J}] = \frac{p(1-p)}{n}$. Thus, this relative frequency $\bar{J}$ serves as an unbiased estimator for probability p.

5.2   Optimal Threshold Estimator

We use the average cost function $g(q_r)$ from (16) to derive an estimator $\hat{q}_r^*$ for the optimal threshold q_r^*. For this reason we transform the optimization problem

$$\min_{q_r \in \mathbb{N}} g(q_r) = \min_{q_r \in \mathbb{N}} g(q_r, \lambda, \mu, \alpha, \beta, \tau, p) = g(q_r^*) \tag{21}$$

into

$$\min_{q_r \in \mathbb{N}} g(q_r, \bar{X}^{-1}, \bar{Y}^{-1}, \bar{Z}^{-1}, \bar{U}^{-1}, \bar{H}^{-1}, \bar{J}) = g(\hat{q}_r^*) \tag{22}$$

and numerically evaluate $\hat{q}_r^*$.

5.3   The Consistent and Asymptotically Normal Estimator

Let $\phi(\lambda, \mu, \alpha, \beta, \tau, p)$ denote any function from corollaries 3.1 and 3.2, which characterizes the performance of the system which runs under threshold $\hat{q}_r^*$. For example it can be the cost function $g(\hat{q}_r^*)$ or the mean number of customers in the system $\bar{N}(\hat{q}_r^*)$ with $q_r = \hat{q}_r^*$. In order to derive an estimator for ϕ we linearize

$$\hat{\phi}(\bar{X}, \bar{Y}, \bar{Z}, \bar{U}, \bar{H}, \bar{J}) = \phi(\bar{X}^{-1}, \bar{Y}^{-1}, \bar{Z}^{-1}, \bar{U}^{-1}, \bar{H}^{-1}, \bar{J})$$

around the point $\boldsymbol{\mu} = (\lambda^{-1}, \mu^{-1}, \alpha^{-1}, \beta^{-1}, \tau^{-1}, p)$ and get the approximation

$$\hat{\phi}(\bar{X}, \bar{Y}, \bar{Z}, \bar{U}, \bar{H}, \bar{J}) \approx \hat{\phi}(\boldsymbol{\mu}) - \boldsymbol{\mu} \, \nabla\hat{\phi}(\boldsymbol{\mu}) + (\bar{X}, \bar{Y}, \bar{Z}, \bar{U}, \bar{H}, \bar{J}) \, \nabla\hat{\phi}(\boldsymbol{\mu}). \qquad (23)$$

The random vector

$$(\bar{X}, \bar{Y}, \bar{Z}, \bar{U}, \bar{H}, \bar{J}) \xrightarrow{d} \mathcal{MN}(\boldsymbol{\mu}, \Sigma)$$

is asymptotically multi-normal distributed with mean vector $\boldsymbol{\mu}$ and covariance matrix $\Sigma = diag(\frac{1}{\lambda^2 n}, \frac{1}{\mu^2 n}, \frac{1}{\alpha^2 n}, \frac{1}{\beta^2 n}, \frac{1}{\tau^2 n}, \frac{p(1-p)}{n})$ due to the multivariate central limit theorem. We employ the theorem of the affine transformation on the above approximation and get

$$
\begin{aligned}
\hat{\phi}(\bar{X}, \bar{Y}, \bar{Z}, \bar{U}, \bar{H}, \bar{J}) &\xrightarrow{d} \mathcal{N}\left(\hat{\phi}(\boldsymbol{\mu}), \nabla\hat{\phi}^t(\boldsymbol{\mu}) \, \Sigma \, \nabla\hat{\phi}(\boldsymbol{\mu})\right) \\
&= \mathcal{N}\left(\phi(\lambda, \mu, \alpha, \beta, \tau, p), \nabla\hat{\phi}^t(\boldsymbol{\mu}) \, \Sigma \, \nabla\hat{\phi}(\boldsymbol{\mu})\right).
\end{aligned}
\qquad (24)
$$

Hence, it is a consistent and asymptotically normal estimator of any performance measure $\phi(\lambda, \mu, \alpha, \beta, \tau, p)$.

5.4   Confidence Intervals for Performance Measures

Using Slutsky's theorem we get from (24)

$$\frac{\left(\hat{\phi}(\bar{X}, \bar{Y}, \bar{Z}, \bar{U}, \bar{H}, \bar{J}) - \phi(\lambda, \mu, \alpha, \beta, \tau, p)\right)}{\sqrt{\nabla\hat{\phi}^t(\bar{X}, \bar{Y}, \bar{Z}, \bar{U}, \bar{H}, \bar{J}) \, \bar{\Sigma} \, \nabla\hat{\phi}(\bar{X}, \bar{Y}, \bar{Z}, \bar{U}, \bar{H}, \bar{J})}} \xrightarrow{d} \mathcal{N}(0,1)$$

with $\bar{\Sigma} = \frac{1}{n} \cdot diag(\bar{X}^2, \bar{Y}^2, \bar{Z}^2, \bar{U}^2, \bar{H}^2, \bar{J}(1 - \bar{J}))$. In other words we have

$$\mathbb{P}\left[n_{\frac{\alpha}{2}} < \frac{\left(\hat{\phi}(\bar{X}, \bar{Y}, \bar{Z}, \bar{U}, \bar{H}, \bar{J}) - \phi(\lambda, \mu, \alpha, \beta, \tau, p)\right)}{\sqrt{\nabla\hat{\phi}^t(\bar{X}, \bar{Y}, \bar{Z}, \bar{U}, \bar{H}, \bar{J}) \, \bar{\Sigma} \, \nabla\hat{\phi}(\bar{X}, \bar{Y}, \bar{Z}, \bar{U}, \bar{H}, \bar{J})}} < -n_{\frac{\alpha}{2}}\right] = 1 - \alpha$$

where $n_{\frac{\alpha}{2}}$ is obtained from Normal tables. This implies the following $100(1-\alpha)\%$ asymptotic confidence interval:

$$\phi(\lambda, \mu, \alpha, \beta, \tau, p) \in$$

$$\left[\hat{\phi}(\bar{X}, \bar{Y}, \bar{Z}, \bar{U}, \bar{H}, \bar{J}) + n_{\frac{\alpha}{2}} \sqrt{\nabla\hat{\phi}^t(\bar{X}, \bar{Y}, \bar{Z}, \bar{U}, \bar{H}, \bar{J}) \, \bar{\Sigma} \, \nabla\hat{\phi}(\bar{X}, \bar{Y}, \bar{Z}, \bar{U}, \bar{H}, \bar{J})},\right.$$

$$\left.\hat{\phi}(\bar{X}, \bar{Y}, \bar{Z}, \bar{U}, \bar{H}, \bar{J}) - n_{\frac{\alpha}{2}} \sqrt{\nabla\hat{\phi}^t(\bar{X}, \bar{Y}, \bar{Z}, \bar{U}, \bar{H}, \bar{J}) \, \bar{\Sigma} \, \nabla\hat{\phi}(\bar{X}, \bar{Y}, \bar{Z}, \bar{U}, \bar{H}, \bar{J})}\right].$$

5.5   Numerical Examples

The following numerical examples show the performance improvement due to introduced optimal threshold $\hat{q}_r^*$. We use the general representation of the confidence interval derived above in order to calculate confidence intervals for different performance measures, e.g. the mean number of customers in the system $\bar{N}$ or the average cost function g.

In order to get some numerical results we simulate our system with parameters fixed in the following way:

$$\lambda = 1.0,\ \mu = 2.0,\ \alpha = 0.2,\ \beta = 5.0,\ \tau = 10.0,\ p = 0.5.$$

To perform a simulation we use the probability of an initial state of system which is calculated by (6). The samples $(X_1, X_2, \ldots, X_n)$, $(Y_1, Y_2, \ldots, Y_n)$, $(Z_1, Z_2, \ldots, Z_n)$, $(U_1, U_2, \ldots, U_n)$ and $(H_1, H_2, \ldots, H_n)$ are drawn as described in Subsect. 5.1. The costs are: $c_0 = 0.1$, $c_{0,0} = 0.5$, $c_{1,0} = 0.5$, $c_{2,0} = 0.5$, $c_{0,1} = 2.0$, $c_{1,1} = 2.0$, $c_{2,1} = 2.0$.

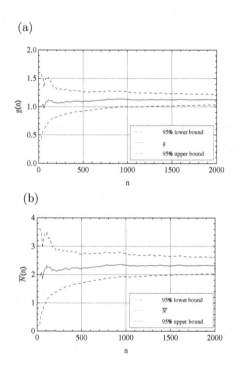

Fig. 3. Confidence intervals for g (a) and $\bar{N}$ (b)

Figure 3(a) illustrates the confidence intervals for the long-run average cost $g(\hat{q}_r^*)$ versus sample size n. Take note that exact value q_r^* as well as the estimation $\hat{q}_r^*$ equal to 3 and the exact value of the cost function $g(3) = 1.107$. For the repair threshold $q_r = 1$ we have $g(1) = 2.14$. Thus, an optimized threshold considerably reduces the system costs. The costs estimation becomes stable for $n \geq 500$.

On the other hand, the confidence intervals for the mean number of customers in the system $\bar{N}(\hat{q}^*)$ versus n are illustrated in Fig. 3(b). The exact value $\bar{N}(q_r^*) = 2.25$ for $q_r^* = 3$. For the threshold level $q_r = 1$ we have $\bar{N}(1) = 1.42$. Thus, the average number of customers in the system increases with q_r. The estimation becomes stable for $n \geq 500$.

References

1. Aissani, A., Artalejo, J.: On the single server retrial queue subject to breakdowns. Queueing Syst. **30**(3–4), 309–321 (1998)
2. Artalejo, J., Gomez-Corral, A., Neuts, M.F.: Analysis of multiserver queues with constant retrial rate. Eur. J. Oper. Res. **135**, 569–581 (2001)
3. Choi, B.D., Rhee, K.H., Park, K.K.: The $M/G/1$ retrial queue with retrial rate control policy. Probab. Eng. Inf. Sci. **7**, 29–46 (1993)
4. Choi, B.D., Shin, Y.W., Ahn, W.C.: Retrial queues with collision arising from unslotted CSMA/CD protocol. Queueing Syst. **11**, 335–356 (1955)
5. Efrosinin, D., Semenova, O.: An M/M/1 system with an unreliable device and threshold recovery policy. J. Commun. Technol. Electron. **55**(12), 1526–1531 (2010)
6. Efrosinin, D., Sztrik, J.: Performance analysis of a two server heterogeneous retrial queue with threshold policy. Qual. Technol. Quant. Manag. **8**, 211–236 (2011)
7. Efrosinin, D., Winkler, A.: Queueing system with a constant retrial rate, non-reliable server and threshold-based recovery. Eur. J. Oper. Res. **210**, 594–605 (2011)
8. Fayolle, G.: A simple telephone exchange with delayed feedbacks. In: Cohen, J.W., Tijms, M.C. (eds.) Teletraffic Analysis and Computer Performance Evaluation OJ Boxma, pp. 245–253. North-Holland, Amsterdam (1986)
9. Li, W., Zhao, Y.Q.: A retrial queue with a constant retrial rate, server downs and impatient customers. Stoch. Models **21**, 531–550 (2005)
10. Martin, M., Artalejo, J.: Analysis of an $M/G/1$ queue with two types of impatient units. Adv. Appl. Probab. **27**, 840–861 (1995)
11. Neuts, M.F.: Matrix-geometric Solutions in Stochastic Models. The John Hopkins University Press, Baltimore (1981)
12. Ross, J.V., Taimre, T., Pollett, P.K.: Estimation for queues from queue length data. Queueing Syst. **55**, 131–138 (2007)
13. Wang, T.Y., Ke, J.C., Wang, K.H., Ho, S.C.: Maximum likelihood estimates and confidence intervals of an $M/M/R$ queue with heterogeneous servers. Math. Methods Oper. Res. **63**, 371–384 (2006)
14. Yadavalli, V.S.S., Adendorff, K., Erasmus, G., Chandrasekhar, P., Deepa, S.P.: Confidence limits for expected waiting time of two queueing models. Oper. Res. Soc. S. Afr. **20**(1), 1–6 (2004)

The Second Order Asymptotic Analysis Under Heavy Load Condition for Retrial Queueing System MMPP/M/1

Ekaterina Fedorova[(⊠)]

Tomsk State University, Tomsk, Russia
moiskate@mail.ru

Abstract. In the paper, the retrial queueing system of $MMPP|M|1$ type is studied by means of the second order asymptotic analysis method under heavy load condition. During the investigation, the theorem about the form of the asymptotic characteristic function of the number of calls in the orbit is formulated and proved. The asymptotic distribution is compared with the exact one obtained by means of numerical algorithm. The conclusion about method application area is made.

Keywords: Retrial queue · Asymptotic analysis · Heavy load

1 Introduction

In queueing theory, there are two classes of queueing systems: systems with queue and loss systems. In real systems, there are situations when queue cannot be explicitly identified, but also call is not lost if it comes when the service device is unavailable. Often primary call does not refuse to be serviced and performs repeated calls to get the service after random time intervals. Examples of these situations are telecommunication systems, cellular networks, call-centres. Thus a new class of queueing systems has been appeared: systems with repeated calls or retrial queueing systems.

The first papers about retrial queues were published in the middle of 20th century. The most of them were devoted to practical problems and influence of repeated attempts on telephone traffic, communication systems etc. [1–4]. The most comprehensive description and detailed comparison of classical queueing systems and retrial queues are contained in books and papers authored by J.R. Artalejo, A. Gomez-Corral, G.I. Falin and J.G.C. Templeton [5–7].

Today there are many papers devoted to these systems. Scientists from different countries study different types of retrial queues, develop methods of their investigation, solve practical and theoretical problems in this area. But the majority of studies of retrial queueing systems are performed numerically or via computer simulation [8–10]. Belarusian researchers A.N. Dudin and V.I. Klimenok [11] mainly use matrix methods in their works. Also matrix methods for retrial queues analysis were used by M.F. Neuts, J.R. Artalejo, A. Gomez-Corral [12],

© Springer International Publishing Switzerland 2015
A. Dudin et al. (Eds.): ITMM 2015, CCIS 564, pp. 344–357, 2015.
DOI: 10.1007/978-3-319-25861-4_29

J.E. Diamond, A.S. Alfa [13], etc. Asymptotic and approximate methods were applied by G.I. Falin [14], V.V. Anisimov [15], T. Yang [16], J.E. Diamond [17], B. Pourbabai [18], etc. But analytical results were obtained only in cases of simple input and service processes (e.g. stationary Poisson input process or the exponential distribution of service law) [6].

In this paper, we study the retrial queueing system $MMPP|M|1$ by means of the second order asymptotic analysis method under heavy load condition. Characteristics of performance of retrial queueing systems under heavy and light loads were studied by G.I. Falin [14], V.V. Anisimov [15] and A. Aissani [19]. Also S.N. Stepanov's work [20] is devoted to investigation under "extreme" load (the intensity of primary calls tends to infinity or zero).

In the paper we use the asymptotic analysis method developed by Tomsk scientific group for investigation of all types of queueing system and networks [21,22]. Principle of the method is derivation of asymptotic equations from the systems of equations determined models states and then getting formulas for asymptotic functions.

In a number of our previous papers (eg. [23]) devoted to the study of various single-server retrial queueing system, we applied the asymptotic analysis method for retrial queueing systems under a heavy load condition. We obtained formulas for asymptotic characteristic functions of the probability distribution of the number of calls in the orbit in systems with different input processes and services laws: $M|M|1$, $M|GI|1$, $MMPP|M|1$, $MMPP|GI|1$. However, we have demonstrated that the proposed method has a fairly narrow range of applicability: for the load rate $\rho > 0.95$, Kolmogorov distance between exact and asymptotic distributions has values $\Delta \leq 0.05$. In this regard, we propose to increase the accuracy of the approximation by getting the second order asymptotic formula.

The rest of the paper is organized as follows. In the Sect. 2, the description of the mathematical model of retrial queue $MMPP|M|1$ is presented and the process of the system states is analysed. In the Sect. 3, we introduce asymptotic functions and determine the limit condition of heavy load, then the theorem about the formula for the asymptotic characteristic function is formulated and proved. The last Sect. 4 is devoted to the numerical comparison of the asymptotic distribution with exact one.

2   Mathematical Model and the Process Under Study

In the paper, retrial queueing system of $MMPP|M|1$ type is analyzed. The input process is Markov Modulated Poisson Process which is a particular case of Markovian Arrival Process (MAP) and it is defined by matrix $\mathbf{D_0}$ and $\mathbf{D_1}$ [24,25]. The underlying process $n(t)$ is Markov chain with continuous time and finite set of states $n = 1, 2, \ldots, W$.

We denote the generator of the underlying process $n(t)$ by matrix $\mathbf{Q} = \mathbf{D_0} + \mathbf{D_1}$. And the matrix $\mathbf{Q}$ has elements q_{mv} where $m, v = 1, 2, \ldots, W$.

$\mathbf{D_1}$ is a diagonal matrix with elements $\rho\lambda_n$ where $n = 1, 2, \ldots, W$ and ρ is some parameter defined below. We introduce a matrix $\mathbf{\Lambda} = \text{diag}\{\lambda_n\}$. Then the following equality holds: $\mathbf{D_1} = \rho\mathbf{\Lambda}$.

The vector-row $\boldsymbol{\theta}$ is the stationary probability distribution of the underlying process $n(t)$. $\boldsymbol{\theta}$ is defined as the unique solution of the system:

$$\begin{cases} \boldsymbol{\theta}\mathbf{Q} = \mathbf{0}, \\ \boldsymbol{\theta}\mathbf{e} = 1 \end{cases} \tag{1}$$

where $\mathbf{e}$ is unit column-vector, $\mathbf{0}$ is zero row-vector.

The service time of each call is distributed by exponential law with parameter μ. If a call arrives when a service device (server) is free, the call occupies the device for the service. If the server is busy, the call goes to the orbit (source of repeated calls) where it is staying during a random time distributed exponentially with parameter σ. After this random time, the call from the orbit makes an attempt to reach the device. If the device is free, the call occupies it, otherwise the call immediately returns to the orbit. Structure of the system is presented in Fig. 1.

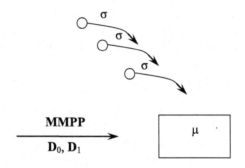

Fig. 1. Retrial queueing system $MMPP|M|1$

The rate of MMPP is defined as $\lambda = \boldsymbol{\theta} \cdot \rho\boldsymbol{\Lambda} \cdot \mathbf{e}$.

Let the system parameters be such that the following equation holds:

$$\boldsymbol{\theta} \cdot \boldsymbol{\Lambda} \cdot \mathbf{e} = \mu. \tag{2}$$

So, the parameter ρ is calculated as $\rho = \dfrac{\lambda}{\boldsymbol{\theta} \cdot \boldsymbol{\Lambda} \cdot \mathbf{e}} = \dfrac{\lambda}{\mu}$ and it is called the load of the system. Thus the stationary state of the system exists when $\rho < 1$. And the heavy load condition is determined by limit condition $\rho \uparrow 1$.

Let $i(t)$ be the random process described the number of calls in the orbit and by $k(t)$ be the random process defined the server state as follows:

$$k(t) = \begin{cases} 0, \text{if device is free}, \\ 1, \text{if device is busy at the moment } t. \end{cases}$$

The problem is to find the probability distribution of the number of calls in the orbit in this system.

However, the process $i(t)$ is not Markovian. So firstly we will consider the multidimensional process $\{k(t), n(t), i(t)\}$ which is a continuous time Markov chain.

We denote the probability that the device is in the state k, there are i calls in the orbit and the underlying process in the state n at the time moment t by $P(k, n, i, t) = P\{k(t) = k, n(t) = n, i(t) = i\}$. So the following direct system of Kolmogorov differential equations for the system states probability distribution $P(k, n, i, t)$ can be written:

$$\begin{cases} \dfrac{\partial P(0, n, i, t)}{\partial t} = -(\rho\lambda_n + i\sigma - q_{nn})P(0, n, i, t) + \mu P(1, n, i, t) \\ \quad + \sum_{v \neq n} P(0, v, i, t)q_{vn}, \\ \dfrac{\partial P(1, n, i, t)}{\partial t} = -(\rho\lambda_n + \mu - q_{nn})P(1, n, i, t) \\ \quad + \rho\lambda_n P(1, n, i-1, t)(1 - \delta_{i,0}) + \rho\lambda_n P(0, n, i, t) \\ \quad + (i+1)\sigma P(0, n, i+1, t) + \sum_{v \neq n} P(1, v, i, t)q_{vn}, \text{ for } i \geq 0, n = \overline{1, N} \end{cases} \quad (3)$$

where $\delta_{i,0}$ is Kronecker symbol which is defined as $\delta_{i,j} = \begin{cases} 0, \text{ if } i \neq j, \\ 1, \text{ if } i = j. \end{cases}$

We denote row-vectors $\mathbf{P}(k, i) = \{P(k, 1, i), P(k, 2, i), \ldots, P(k, N, i)\}$ where $P(k, n, i) = \lim_{t \to \infty} P(k, n, i, t)$. Then in stationary state, the system (3) has the following matrix form:

$$\begin{cases} \mathbf{P}(0, i)(\mathbf{Q} - \rho\mathbf{\Lambda} - i\sigma\mathbf{I}) + \mu\mathbf{P}(1, i) = \mathbf{0}, \\ \mathbf{P}(1, i)(\mathbf{Q} - \rho\mathbf{\Lambda} - \mu\mathbf{I}) + \mathbf{P}(0, i)\rho\mathbf{\Lambda} + (1 - \delta_{i,0})\mathbf{P}(1, i-1)\rho\mathbf{\Lambda} \\ \quad + \sigma(i+1)\mathbf{P}(0, i+1) = \mathbf{0}, \text{ for } i \geq 0 \end{cases} \quad (4)$$

where $\mathbf{I}$ is the identity matrix.

So we have the system of matrix difference equations.

3   Asymptotic Analysis Method Under Heavy Load Condition

We introduce the partial characteristic functions:

$$\mathbf{H}(k, u) = \sum_i e^{jui}\mathbf{P}(k, i), \text{ for } k = 0, 1$$

where $j = \sqrt{-1}$ is the imaginary unit.

Then the system (4) is rewritten as the following system:

$$\begin{cases} \mathbf{H}(0, u)(\mathbf{Q} - \rho\mathbf{\Lambda}) + j\sigma\dfrac{\partial\mathbf{H}(0, u)}{\partial u} + \mu\mathbf{H}(1, u) = \mathbf{0}, \\ \mathbf{H}(1, u)(\mathbf{Q} - \rho\mathbf{\Lambda} - \mu\mathbf{I}) + \mathbf{H}(0, u)\rho\mathbf{\Lambda} + e^{ju}\mathbf{H}(1, u)\rho\mathbf{\Lambda} \\ \quad - j\sigma e^{-ju}\dfrac{\partial\mathbf{H}(0, u)}{\partial u} = \mathbf{0}. \end{cases} \quad (5)$$

We will solve the system (5) by the method of asymptotic analysis under heavy load condition. The heavy load condition is defined by the assumption that $\rho \uparrow 1$ or $\varepsilon \downarrow 0$ where ε is an infinitesimal variable $\varepsilon = 1 - \rho > 0$.

First of all, we introduce notations:

$$u = \varepsilon w, \quad \mathbf{H}(0, u) = \varepsilon \mathbf{G}(w, \varepsilon), \quad \mathbf{H}(1, u) = \mathbf{F}(w, \varepsilon).$$

Then the system (5) can be rewritten as:

$$\begin{cases} \varepsilon \mathbf{G}(w, \varepsilon)(\mathbf{Q} - (1 - \varepsilon)\mathbf{\Lambda}) + j\sigma \dfrac{\partial \mathbf{G}(w, \varepsilon)}{\partial w} + \mu \mathbf{F}(w, \varepsilon) = 0, \\ \mathbf{F}(w, \varepsilon)(\mathbf{Q} + (1 - \varepsilon)(e^{j\varepsilon w} - 1)\mathbf{\Lambda} - \mu \mathbf{I}) \\ \quad + (1 - \varepsilon)\varepsilon \mathbf{G}(w, \varepsilon)\mathbf{\Lambda} - j\sigma e^{-j\varepsilon w} \dfrac{\partial \mathbf{G}(w, \varepsilon)}{\partial w} = 0. \end{cases} \tag{6}$$

For obtaining the second order asymptotic formula, it is necessary to consider following expansions of functions:

$$\mathbf{G}(w, \varepsilon) = \mathbf{G}(w) + \varepsilon \mathbf{g}(w) + \varepsilon^2 \mathbf{g}_2(w) + O(\varepsilon^3), \tag{7}$$

$$\mathbf{F}(w, \varepsilon) = \mathbf{F}(w) + \varepsilon \mathbf{f}(w) + \varepsilon^2 \mathbf{f}_2(w) + O(\varepsilon^3) \tag{8}$$

where $O(\varepsilon^3)$ is an infinitesimal variable of order ε^3.

The characteristic function of the number of calls in the orbit $h(u) = Me^{ju \cdot i(t)}$ can be presented by introduced notations in the following form:

$$h(u) = [\mathbf{H}(0, u) + \mathbf{H}(1, u)]\,\mathbf{e} = \left[\varepsilon \mathbf{G}\left(\frac{u}{\varepsilon}, \varepsilon\right) + \mathbf{F}\left(\frac{u}{\varepsilon}, \varepsilon\right)\right]\mathbf{e}.$$

Using expansions (7) and (8), the characteristic function of the number of calls in the orbit is presented as

$$h(u) = \mathbf{F}\left(\frac{u}{\varepsilon}\right)\mathbf{e} + \varepsilon \left[\mathbf{G}\left(\frac{u}{\varepsilon}\right) + \varepsilon \mathbf{f}\left(\frac{u}{\varepsilon}\right)\right]\mathbf{e} + O(\varepsilon^2)$$

where functions $\mathbf{F}(w), \mathbf{G}(w)$ and $\mathbf{f}(w)$ are defined in expansions (7) and (8), and the parameter $\varepsilon = 1 - \rho$.

Then we will call the function $h_1(u) = \mathbf{F}\left(\dfrac{u}{\varepsilon}\right)\mathbf{e}$ as the first order asymptotic characteristic function and the function

$$h_2(u) = \mathbf{F}\left(\frac{u}{\varepsilon}\right)\mathbf{e} + \left[\varepsilon \mathbf{G}\left(\frac{u}{\varepsilon}\right) + \varepsilon \mathbf{f}\left(\frac{u}{\varepsilon}\right)\right]\mathbf{e} \tag{9}$$

as the second order asymptotic characteristic function.

In the paper [23], we found that the first order asymptotic characteristic function $h_1(u)$ has the form of the characteristic function of gamma distribution:

$$h_1(u) = \mathbf{F}\left(\frac{u}{1 - \rho}\right)\mathbf{e} = \left(1 - \frac{ju}{(1 - \rho)\beta}\right)^{-\alpha}$$

where

$$\alpha = 1 + \frac{\mu}{\sigma}\beta, \quad \beta = \frac{\mu}{v\mathbf{\Lambda}e - \mu v e + \mu}, \tag{10}$$

and the vector $\mathbf{v}$ is a solution of the inhomogeneous system $\mathbf{v}\mathbf{Q} = \boldsymbol{\theta}(\mu\mathbf{I} - \boldsymbol{\Lambda})$.

The second order asymptotic characteristic function $h_2(u)$ is defined by the following theorem.

Theorem 1. *The second-order asymptotic characteristic function has the following form*

$$h_2(u) = \left(1 - \frac{ju}{(1-\rho)\beta}\right)^{-\alpha}\left\{1 + (1-\rho)\left[\frac{ju}{(1-\rho)}\mathbf{v}\mathbf{e} - j\int_0^{\frac{ju}{(1-\rho)}}\frac{a(y)}{(jy-\beta)}dy\right]\right\}$$

where function $a(w)$ is presented as follows:

$$a(w) = \frac{\alpha}{\beta}\left(1 - \frac{jw}{\beta}\right)^{-1}\left[-jw\frac{2\mathbf{v}\boldsymbol{\Lambda}\mathbf{e} - \mu\mathbf{v}\mathbf{e}}{\mu} + (jw)^2\frac{\delta}{\mu}\right]$$

$$-\frac{2\mathbf{v}\boldsymbol{\Lambda}\mathbf{e} - \mu\mathbf{v}\mathbf{e}}{\mu} + 2jw\left(\frac{\delta}{\mu} - \frac{\mu}{\sigma}\right) - 2\left(1 + \frac{\mu}{\sigma}\right)\left(1 - \frac{jw}{\beta}\right) + jw\mathbf{v}\mathbf{e}\frac{\mu}{\sigma},$$

α and β are described by formula (10), the constant δ is defined as

$$\delta = \mu\mathbf{v}\mathbf{e} + \mathbf{v}_1(\boldsymbol{\Lambda}\mathbf{e} - \mu\mathbf{e}) - \frac{\mu}{2}$$

and $\mathbf{v}_1$ is a solution of the inhomogeneous system

$$\mathbf{v}_1\mathbf{Q} = \frac{\mu}{\beta}\boldsymbol{\theta} - \frac{1}{2}(\boldsymbol{\theta}\boldsymbol{\Lambda} - \mu\boldsymbol{\theta}) - (\mathbf{v}\boldsymbol{\Lambda} - \mu\mathbf{v}).$$

Proof. The proof will be carried out in several steps.

Step 1: Derivation of asymptotic equations.

Substituting expansions (7) and (8) into the system (6), performing some transformations, and equating the coefficients under the same powers of ε, we obtain the following system of equations for unknown functions $\mathbf{F}(w)$, $\mathbf{G}(w)$, $\mathbf{f}(w)$, $\mathbf{g}(w)$, $\mathbf{f_2}(w)$ and $\mathbf{g_2}(w)$:

$$\begin{cases} j\sigma\mathbf{G}'(w) + \mu\mathbf{F}(w) = 0, \\ \mathbf{F}(w)(\mathbf{Q} - \mu\mathbf{I}) - j\sigma\mathbf{G}'(w) = 0, \\ \mathbf{G}(w)(\mathbf{Q} - \boldsymbol{\Lambda}) + j\sigma\mathbf{g}'(w) + \mu\mathbf{f}(w) = 0, \\ jw\mathbf{F}(w)\boldsymbol{\Lambda} + \mathbf{f}(w)(\mathbf{Q} - \boldsymbol{\Lambda}) + \mathbf{G}(w)\boldsymbol{\Lambda} + j\sigma jw \cdot \mathbf{G}'(w) - \mathbf{g}'(w) = 0, \\ \mathbf{G}(w)\boldsymbol{\Lambda} + \mathbf{g}(w)(\mathbf{Q} - \boldsymbol{\Lambda}) + \mathbf{f_2}(w)\boldsymbol{\Lambda} + \mu\mathbf{f_2}(w) = 0, \qquad (11) \\ \left(-jw + \frac{(jw)^2}{2}\right)\mathbf{F}(w)\boldsymbol{\Lambda} + jw\mathbf{f}(w)\boldsymbol{\Lambda} + \mathbf{f_2}(w)(\mathbf{Q} - \mu\mathbf{I}) - \mathbf{G}(w)\boldsymbol{\Lambda} \\ +\mathbf{g}(w)\boldsymbol{\Lambda} - j\sigma\frac{(jw)^2}{2} \cdot \mathbf{G}'(w) + j\sigma jw\mathbf{g}'(w) - j\sigma\mathbf{g}_2'(w) = 0. \end{cases}$$

To get one more scalar equation, we sum equations of the system (6) and multiply the result equation by the unit column-vector $\mathbf{e}$. Taken into account that $\mathbf{Q}\mathbf{e} = 0$, we obtain equation:

$$\mathbf{F}(w,\varepsilon)(1-\varepsilon)\boldsymbol{\Lambda}\mathbf{e} + j\sigma e^{-j\varepsilon w}\frac{\partial\mathbf{G}(w,\varepsilon)}{\partial w}\mathbf{e} = 0.$$

We substitute expansions (7) and (8) into obtained equation and again equate coefficients under the same powers of ε. As the result, we write the following system:

$$
\begin{cases}
\mathbf{F}(w)\mathbf{\Lambda e} + j\sigma\mathbf{G}'(w)\mathbf{e} = 0, \\
-\mathbf{F}(w)\mathbf{\Lambda e} + \mathbf{f}(w)\mathbf{\Lambda e} - j\sigma jw\mathbf{G}'(w)\mathbf{e} + j\sigma\mathbf{g}'(w)\mathbf{e} = 0, \\
-\mathbf{f}(w)\mathbf{\Lambda e} + \mathbf{f}_2(w)\mathbf{\Lambda e} + j\sigma\dfrac{(jw)^2}{2}\mathbf{G}'(w)\mathbf{e} - j\sigma jw\mathbf{g}'(w)\mathbf{e} + j\sigma\mathbf{g}'_2(w)\mathbf{e} = 0.
\end{cases}
$$

The first two equations are linearly dependent on the first four equations of the system (11), so we will use for further derivations only the last equation:

$$
-\mathbf{f}(w)\mathbf{\Lambda e} + \mathbf{f}_2(w)\mathbf{\Lambda e} + j\sigma\frac{(jw)^2}{2}\mathbf{G}'(w)\mathbf{e} - j\sigma jw\mathbf{g}'(w)\mathbf{e} + j\sigma\mathbf{g}'_2(w)\mathbf{e} = 0. \quad (12)
$$

Six matrix equations in the system (11) and one scalar equation (12) are enough to find functions $\mathbf{F}(w)$, $\mathbf{G}(w)$ and $\mathbf{f}(w)$ which are necessary for obtaining the second order asymptotic characteristic function $h_2(u)$.

Step 2: Multiplicative form of functions $\mathbf{F}(w)$, $\mathbf{G}(w)$.

Obviously, summing the first and second equations of the system (11), we can write:

$$
\mathbf{F}(w) = \boldsymbol{\theta}\Phi(w) \quad (13)
$$

where the unknown scalar function $\Phi(w)$ is defined as $\Phi(w) = \mathbf{F}(w)\mathbf{e}$.

Then the first equation of the system (11) has the form:

$$
\mathbf{G}'(w) = j\frac{\mu}{\sigma}\mathbf{F}(w) = j\frac{\mu}{\sigma}\boldsymbol{\theta}\Phi(w). \quad (14)
$$

Step 3: Determination of functions $\mathbf{G}(w)$ and $\mathbf{f}(w)$.

Summing up the third and the fourth equations of the system (11), we obtain

$$
\{\mathbf{G}(w) + \mathbf{f}(w)\}\mathbf{Q} + jw\mathbf{F}(w)\mathbf{\Lambda} + j\sigma jw\mathbf{G}'(w) = \mathbf{0}.
$$

Given the formula (14), it is easy to show that

$$
\{\mathbf{G}(w) + \mathbf{f}(w)\}\mathbf{Q} = -jw\Phi(w)\boldsymbol{\theta}\{\mathbf{\Lambda} - \mu\mathbf{I}\}. \quad (15)
$$

Let the solution of the Eq. (14) with respect to the vector $\mathbf{G}(w) + \mathbf{f}(w)$ has the form:

$$
\mathbf{G}(w) + \mathbf{f}(w) = jw\Phi(w)\mathbf{v} + \varphi(w)\boldsymbol{\theta} \quad (16)
$$

where $\varphi(w)$ is an arbitrary scalar function and vector $\mathbf{v}$ is a solution of the following system:

$$
\mathbf{v}\mathbf{Q} = \boldsymbol{\theta}(\mu\mathbf{I} - \mathbf{\Lambda}). \quad (17)
$$

For existence of the solution of the system (15), it is necessary that the rank of the augmented matrix be equal to the rank of the matrix $\mathbf{Q}$. Because the determinant $\det(\mathbf{Q}) = 0$ the rank of the augmented matrix must be less than the dimension of the system. Then it is sufficient that the following condition should hold:

$$(\mu\boldsymbol{\theta} - \boldsymbol{\theta}\boldsymbol{\Lambda})\mathbf{e} = 0,$$

what is true due to the condition (2).

So from the Eq. (15), it follows that

$$\mathbf{f}(w) = jw\Phi(w)\mathbf{v} + \varphi(w)\boldsymbol{\theta} - \mathbf{G}(w). \tag{18}$$

Step 4: Obtaining of expression for the function $\mathbf{g}'(w)$.

From the third equation of the system (11), it follows that:

$$j\sigma\mathbf{g}'(w) = \mathbf{G}(w)(\boldsymbol{\Lambda} - \mathbf{Q}) - \mu\mathbf{f}(w).$$

By substituting the expression (17) into this formula, we get:

$$j\sigma\mathbf{g}'(w) = \mathbf{G}(w)(\boldsymbol{\Lambda} - \mathbf{Q} + \mu\mathbf{I}) - \mu jw\Phi(w)\mathbf{v} - \mu\varphi(w)\boldsymbol{\theta}. \tag{19}$$

Step 5: Derivation of the explicit expression for the scalar function $\Phi(w)$ *and calculation of functions* $\mathbf{F}(w)$ *and* $\mathbf{G}(w)$.

Summing up the fifth and the sixth equations of the system (11) and multiplying the result by the vector $\mathbf{e}$, we can write:

$$\mathbf{f}(w)\boldsymbol{\Lambda}\mathbf{e} + j\sigma\mathbf{g}'(w)\mathbf{e} + j\sigma(1 - jw)\mathbf{G}'(w)\mathbf{e} = 0.$$

We substitute formulas (14) and (19) into the last expression and take into account the expression (2). So, the following equation is derived:

$$jw\Phi(w)(\mathbf{v}\boldsymbol{\Lambda}\mathbf{e} - \mu) + j\sigma(1 - jw)\mathbf{G}'(w)\mathbf{e} + \mu\mathbf{G}(w)\mathbf{e} = 0.$$

We differentiate this equation:

$$j\Phi(w)(\mathbf{v}\boldsymbol{\Lambda}\mathbf{e} - \mu) + jw\Phi'(w)(\mathbf{v}\boldsymbol{\Lambda}\mathbf{e} - \mu) + \sigma\mathbf{G}'(w)\mathbf{e} + j\sigma(1 - jw)\mathbf{G}''(w)\mathbf{e} + \mu\mathbf{G}'(w)\mathbf{e} = 0.$$

So the following equation can be obtained by performing some transformations:

$$\Phi(w)\left[jv\boldsymbol{\Lambda}\mathbf{e} - j\mu\mathbf{v}\mathbf{e} + j\mu + j\frac{j\mu^2}{\sigma}\right] \tag{20}$$
$$= \Phi'(w)\left[-jwv\boldsymbol{\Lambda}\mathbf{e} + jw\mu\mathbf{v}\mathbf{e} + \mu - jw\mu\right].$$

Denote $\beta = \dfrac{\mu}{v\boldsymbol{\Lambda}\mathbf{e} - \mu\mathbf{v}\mathbf{e} + \mu}$, $\alpha = 1 + \dfrac{\mu^2}{\sigma(v\boldsymbol{\Lambda}\mathbf{e} - \mu\mathbf{v}\mathbf{e} + \mu)}$. Then the formula (20) can be rewritten as:

$$\Phi(w)j\alpha = \Phi'(w)(\beta - jw).$$

The solution of this equation has the form:

$$\Phi(w) = c(w + j\beta)^{-\alpha} \tag{21}$$

where c is an arbitrary constant and it is equal to $(j\beta)^\alpha$ from the initial condition $\Phi(0) = 1$.

So, the formula (21) is rewritten as

$$\Phi(w) = \left(1 + \frac{jw}{\beta}\right)^{-\alpha}. \tag{22}$$

Turning to expressions (13) and (14), we can obtain functions $\mathbf{F}(w)$, $\mathbf{G}(w)$:

$$\begin{cases} \mathbf{F}(w) = \boldsymbol{\theta}\left(1 + \dfrac{jw}{\beta}\right)^{-\alpha}, \\ \mathbf{G}(w) = \boldsymbol{\theta}\left(1 + \dfrac{jw}{\beta}\right)^{-\alpha+1}. \end{cases} \tag{23}$$

Step 6: Getting of the expression for the function $\mathbf{f}_2(w)$.

From the fifth equation of system (11), we obtain the following expression:

$$j\sigma \mathbf{g}_2'(w) = \mathbf{g}(w)(\boldsymbol{\Lambda} + \mathbf{Q}) - \mathbf{G}(w)\boldsymbol{\Lambda} - \mu \mathbf{f}_2(w). \tag{24}$$

Substituting the expressions (18), (23) and (24) in the sixth equation of the system (11), the following equation is obtained:

$$[\mathbf{g}(w) + \mathbf{f}_2(w)]\mathbf{Q} = jw\Phi(w)(\boldsymbol{\theta}\boldsymbol{\Lambda} - \mu\boldsymbol{\theta})$$

$$+ (jw)^2 \Phi(w)\left[\frac{\mu}{\beta}\boldsymbol{\theta} - \frac{1}{2}(\boldsymbol{\theta}\boldsymbol{\Lambda} + \mu\boldsymbol{\theta}) - (\mathbf{v}\boldsymbol{\Lambda} - \mu\mathbf{v})\right] - jw\varphi(w)(\boldsymbol{\theta}\boldsymbol{\Lambda} - \mu\boldsymbol{\theta}).$$

Let the solution of this equation with respect to the vector $\mathbf{g}(w) + \mathbf{f}_2(w)$ has the form:

$$\mathbf{g}(w) + \mathbf{f}_2(w) = (jw)^2 \Phi(w)\mathbf{v}_1 - jw\Phi(w)\mathbf{v} + jw\varphi(w)\mathbf{v} + \varphi_2(w)\boldsymbol{\theta} \tag{25}$$

where $\varphi_2(w)$ is an arbitrary scalar function, $\mathbf{v}$ is a solution of the system (16) and vector $\mathbf{v}_1$ is a solution of the following system:

$$\mathbf{v}_1 \mathbf{Q} = \frac{\mu}{\beta}\boldsymbol{\theta} - \frac{1}{2}(\boldsymbol{\theta}\boldsymbol{\Lambda} + \mu\boldsymbol{\theta}) - (\mathbf{v}\boldsymbol{\Lambda} - \mu\mathbf{v}).$$

For existence of a solution of the system (17), it is necessary that the rank of the augmented matrix be equal to the rank of the matrix $\mathbf{Q}$. Because the determinant $\det(\mathbf{Q}) = 0$, the rank of the augmented matrix must be less than the dimension of the system. Then it is sufficient that the following condition should hold:

$$\left[\frac{\mu}{\beta}\boldsymbol{\theta} - \frac{1}{2}(\boldsymbol{\theta}\boldsymbol{\Lambda} + \mu\boldsymbol{\theta}) - (\mathbf{v}\boldsymbol{\Lambda} - \mu\mathbf{v})\right]\mathbf{e} = 0.$$

It is easy to show that this condition is satisfied.

Then from the Eq. (25), we have

$$\mathbf{f}_2(w) = (jw)^2 \Phi(w) \cdot \mathbf{v}_1 - jw\Phi(w)\mathbf{v} + jw\varphi(w)\mathbf{v} + \varphi_2(w)\boldsymbol{\theta} - \mathbf{g}(w). \tag{26}$$

Step 7: Derivation of the explicit expression for the scalar function $\varphi(w)$.

Substituting all found expressions in the Eq. (12), the following equation can be obtain:

$$\Phi(w)\left[-jw(2\mathbf{v\Lambda e} - \mu\mathbf{ve}) + (jw)^2\left(\mu\mathbf{ve} + \mathbf{v}_1(\mathbf{\Lambda e} - \mu\mathbf{e}) - \frac{\mu}{2}\right)\right]$$

$$+\varphi(w)\left[-\boldsymbol{\theta}\mathbf{\Lambda e} + jw(\mu + \mathbf{v\Lambda e} - \mu\mathbf{ve})\right] - jw\mathbf{G}(w)(\mathbf{\Lambda e} + \mu\mathbf{e}) + \mu\mathbf{g}(w)\mathbf{e} = 0.$$

We denote $\delta = \mu\mathbf{ve} + \mathbf{v}_1(\mathbf{\Lambda e} - \mu\mathbf{e}) - \frac{\mu}{2}$.

Differentiating the equation, we obtain:

$$\Phi'(w)\left[-jw(2\mathbf{v\Lambda e} - \mu\mathbf{ve}) + (jw)^2\delta\right] + \Phi(w)\left[-j(2\mathbf{v\Lambda e} - \mu\mathbf{ve}) + 2j^2w\delta\right]$$

$$+\varphi'(w)\left[-\boldsymbol{\theta}\mathbf{\Lambda e} + jw\frac{\mu}{\beta}\right] + \varphi(w)j\frac{\mu}{\beta} - jw\mathbf{G}'(w)(\mathbf{\Lambda e} + \mu\mathbf{e})$$

$$-j\mathbf{G}(w)(\mathbf{\Lambda e} + \mu\mathbf{e}) + \mu\mathbf{g}'(w)\mathbf{e} = 0.$$

Taking into account formulas (2), (10), (18) and (19), the following differential equation is obtained:

$$\varphi'(w)\left(1 - \frac{jw}{\beta}\right) - j\varphi(w)\frac{\alpha}{\beta} = j\Phi(w)a(w) \tag{27}$$

where

$$a(w) = \frac{\alpha}{\beta}\left(1 - \frac{jw}{\beta}\right)^{-1}\left[-jw\frac{2\mathbf{v\Lambda e} - \mu\mathbf{ve}}{\mu} + (jw)^2\frac{\delta}{\mu}\right] - \frac{2\mathbf{v\Lambda e} - \mu\mathbf{ve}}{\mu}$$

$$+2jw\left(\frac{\delta}{\mu} - \frac{\mu}{\sigma}\right) - 2\left(1 + \frac{\mu}{\sigma}\right)\left(1 - \frac{jw}{\beta}\right) + jw\mathbf{ve}\frac{\mu}{\sigma}.$$

The solution of the inhomogeneous differential equation (27) has form:

$$\varphi(w) = e^{j\int_0^w \frac{\alpha/\beta}{1 - \frac{jx}{\beta}}dx}\left\{\varphi(0) + j\int_0^w e^{-j\int_0^y \frac{\alpha/\beta}{1 - \frac{jx}{\beta}}dx}\frac{\Phi(y)a(y)}{1 - jy/\beta}dy\right\}. \tag{28}$$

Given normalization condition for the function $\mathbf{F}(w)$: $\mathbf{F}(0)\mathbf{e} = 1$, from the expression(16) we have $\varphi(0) = 0$.

It is easy to show that

$$\int_0^w \frac{\alpha/\beta}{1 - \frac{jx}{\beta}}dx = j\alpha\ln\left(1 - \frac{jw}{\beta}\right)$$

and

$$\int_0^w \left(1 - \frac{jy}{\beta}\right)^\alpha \frac{\Phi(y)a(y)}{1 - \frac{jy}{\beta}}dy = \int_0^w \frac{a(y)}{1 - \frac{jy}{\beta}}dy.$$

So, the solution (28) has the following form:

$$\varphi(w) = j\left(1 - \frac{jw}{\beta}\right)^{-\alpha} \int_0^w \frac{a(y)}{1 - \frac{jy}{\beta}}dy = -j\Phi(w)\int_0^w \frac{a(y)}{jy - \beta}dy.$$

Step 8: Obtaining of the final formula for the function $h_2(u)$.

Turning to the formula (16), we have

$$\{\mathbf{G}(w) + \mathbf{f}(w)\}\mathbf{e} = jw\Phi(w)\mathbf{v}\mathbf{e} + \varphi(w)$$
$$= \left(1 - \frac{jw}{\beta}\right)^{-\alpha}\left\{jw\mathbf{v}\mathbf{e} - j\int_0^w \frac{a(y)}{jy-\beta}dy\right\}. \tag{29}$$

From the formula (9), the second order asymptotic characteristic function for retrial queueing system $MMPP|M|1$ is represented as

$$h_2(u) = \mathbf{F}\left(\frac{u}{1-\rho}\right)\mathbf{e} + (1-\rho)\left[\mathbf{G}\left(\frac{u}{1-\rho}\right)\mathbf{e} + \mathbf{f}\left(\frac{u}{1-\rho}\right)\mathbf{e}\right].$$

Taking into account expressions (23) and (29), we obtain that the function $h_2(u)$ has the following form:

$$h_2(u) = \left(1 - \frac{ju}{(1-\rho)\beta}\right)^{-\alpha}\left\{1 + (1-\rho)\left[\frac{ju}{1-\rho}\mathbf{v}\mathbf{e} - j\int_0^{\frac{u}{1-\rho}} \frac{a(y)}{jy-\beta}dy\right]\right\}.$$

So the theorem is proved.

Having the second order asymptotic characteristic function $h_2(u)$, we can construct the asymptotic probability distribution $P_2(i)$ of the number of calls in the orbit by means of the formula of inverse Fourier transform.

4   Numerical Analysis of the Results

To determine the applicability range of the proposed method, we compare the obtained asymptotic distribution and the distribution obtained by numerical solution of the system of linear algebraic equations (4) and calculate Kolmogorov distance between distributions.

Consider an example. Let the system parameters be the following:

$$\mu = 1, \sigma = 1,$$

$$\Lambda = \begin{pmatrix} 0.588 & 0 & 0 \\ 0 & 0.980 & 0 \\ 0 & 0 & 1.373 \end{pmatrix}, \mathbf{Q} = \begin{pmatrix} -0.5 & 0.2 & 0.3 \\ 0.1 & -0.3 & 0.2 \\ 0.3 & 0.2 & -0.5 \end{pmatrix}.$$

In Figs. 2 and 3, comparison of distributions are shown for different value of the load ρ (D_n is exact distribution which is obtained numerically, $P1_n$ and $P2_n$ are the first order and the second order asymptotic distributions, respectively).

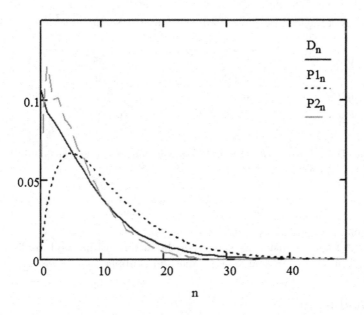

Fig. 2. Comparison of asymptotic and exact distributions for $\rho = 0.8$

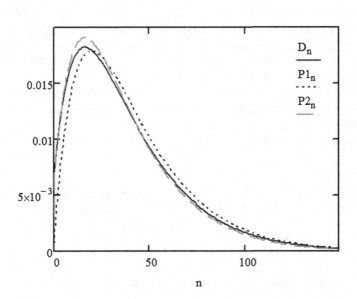

Fig. 3. Comparison of asymptotic and exact distributions for $\rho = 0.95$

Table 1. Kolmogorov distance between asymptotic and exact distributions

Values of the load rate	First-order asymptotic distribution	Second-order asymptotic distribution
$\rho = 0.7$	0.350	0.118
$\rho = 0.8$	0.235	0.050
$\rho = 0.9$	0.114	0.026
$\rho = 0.95$	0.050	0.018

In the Table 1 we show the Kolmogorov distance between asymptotic and exact distributions:

$$\Delta = \max_{0 \leq i \leq N} \left| \sum_{n=0}^{i} D_n - \sum_{n=0}^{i} P_n \right|$$

for different values of the parameter of load ρ.

We chose the condition $\Delta \leq 0.05$ as the criteria of method application. So the second order asymptotic analysis method is applied for $\rho \geq 0.8$.

5   Conclusion

In the paper, we study the retrial queueing system $MMPP|M|1$ by means of the second order asymptotic analysis method under heavy load condition. During the investigation, the asymptotic characteristic function of the number of calls in the orbit is obtained. Numerical comparison of asymptotic distributions (of the 1st and the 2nd orders) with the exact one is performed. The comparison shows that the application area of the second order asymptotic method increases by 4 times than first order asymptotic results: for load rate $\rho > 0.8$ Kolmogorov distance has values $\Delta \leq 0.05$.

In this regard, there is the question about the increasing the accuracy of the method by means of obtaining the third order asymptotic formula. However, studies have shown that this approximation does not increase the range of the method applicability. So for $\rho \leq 0.8$ it need to develope other methods of system studying.

Acknowledgments. This work is performed under the state order of the Ministry of Education and Science of the Russian Federation No. 1.511.2014/K.

References

1. Wilkinson, R.I.: Theories for toll traffic engineering in the USA. Bell Sys. Tech. J. **35**(2), 421–507 (1956)
2. Cohen, J.W.: Basic problems of telephone trafic and the influence of repeated calls. Philips Telecommun. Rev. **18**(2), 49–100 (1957)
3. Elldin, A., Lind, G.: Elementary Telephone Trafic Theory. Ericsson Public Telecommunications, Stockholm (1971)

4. Gosztony, G.: Repeated call attempts and their efect on trafic engineering. Budavox Telecommun. Rev. **2**, 16–26 (1976)
5. Artalejo, J.R., Gomez-Corral, A.: Retrial Queueing Systems. A Computational Approach. Springer, Heidelberg (2008)
6. Falin, G.I., Templeton, J.G.C.: Retrial queues. Chapman & Hall, London (1997)
7. Artalejo, J.R., Falin, G.I.: Standard and retrial queueing systems: a comparative analysis. Revista Matematica Complutense **15**, 101–129 (2002)
8. Stepanov, S.N.: Numerical methods of calculation of retrial queues. Nauka, Moscow (1983). (In Russian)
9. Neuts, M.F., Rao, B.M.: Numerical investigation of a multiserver retrial model. Queueing Sys. **7**(2), 169–189 (2002)
10. Ridder, A.: Fast simulation of retrial queues. In: Third Workshop on Rare Event Simulation and Related Combinatorial Optimization Problems, pp. 1–5, Pisa (2000)
11. Dudin, A.N., Klimenok, V.I.: Queueing system BMAP/G/1 with repeated calls. Math. Comput. Model. **30**(3–4), 115–128 (1999)
12. Artalejo, J.R., Gomez-Corra, A., Neuts, M.F.: Analysis of multiserver queues with constant retrial rate. Euro. J. Oper. Res. **135**, 569–581 (2001)
13. Diamond, J.E., Alfa, A.S.: Matrix analytical methods for M/PH/1 retrial queues. Stochast. Models **11**, 447–470 (1995)
14. Falin, G.I.: M/G/1 queue with repeated calls in heavy trafic. Moscow Univ. Math. Bull. **35**(6), 48–50 (1980)
15. Anisimov, V.V.: Asymptotic Analysis of Reliability for Switching Systems in Light and Heavy Trafic Conditions. Statistics for Industry and Technology. Birkhauser Boston, Boston (2000)
16. Yang, T., Posner, M.J.M., Templeton, J.G.C., Li, H.: An approximation method for the M/G/1 retrial queue with general retrial times. Euro. J. Oper. Res. **76**, 552–562 (1994)
17. Diamond, J.E., Alfa, A.S.: Approximation method for M/PH/1 retrial queues with phase type inter-retrial times. Euro. J. Oper. Res. **113**, 620–631 (1999)
18. Pourbabai, B.: Asymptotic analysis of G/G/K queueing-loss system with retrials and heterogeneous servers. Int. J. Sys. Sci. **19**, 1047–1052 (1988)
19. Aissani, A.: Heavy loading approximation of the unreliable queue with repeated orders. In: Actes du Colloque Methodes et Outils d'Aide 'a la Decision (MOAD 1992), pp. 97–102, Bejaa (1992)
20. Stepanov, S.N.: Asymptotic analysis of models with repeated calls in case of extreme load. Prob. Inf. Transm. **29**(3), 248–267 (1993)
21. Pankratova, E., Moiseeva, S.: Queueing System $MAP/M/\infty$ with n Types of Customers. In: Dudin, A., et al. (eds.) Information Technologies and Mathematical Modelling. CCIS, vol. 487, pp. 356–366. Springer International Publishing, Switzerland (2014)
22. Nazarov, A., Moiseev, A.: Analysis of an open non-Markovian $GI - (GI|\infty)^K$ queueing network with high-rate renewal arrival process. Prob. Inf. Trans. **49**(2), 167–178 (2013)
23. Nazarov, A.A., Moiseeva, E.A.: The research of retrial queueing system $MMPP|M|1$ by the method of asymptotic analysis under heavy load. Bull. Tomsk Polytechnic Univ. **322**(2), 19–23 (2013). (In Russian)
24. Neuts, M.F.: Versatile Markovian point process. J. Appl. Probab. **16**(4), 764–779 (1979)
25. Lucantoni, D.M.: New results on the single server queue with a batch Markovian arrival process. Stoch. Models **7**, 1–46 (1991)

Comparison of Polling Disciplines When Analyzing Waiting Time for Signaling Message Processing at SIP-Server

Yuliya Gaidamaka and Elvira Zaripova[✉]

Department of Applied Probability and Informatics,
Peoples' Friendship University of Russia, Miklukho-Maklaya Street 6,
117198 Moscow, Russia
{ygaidamaka,ezarip}@sci.pfu.edu.ru

Abstract. The main signaling protocol for IP Multimedia Subsystem is Session Initiation Protocol. A typical procedure for a session establishment involves two types of signaling messages with different priority: Invite message, which initiate a session, and non-Invite messages, which continue session establishment. In the paper for the analysis of two types signaling messages service process the single server asymmetric polling system with two infinite queues and a non-zero switching time is proposed. We estimate the waiting time for the gated and the exhaustive service disciplines using input data applicable to signaling traffic analysis. For the exhaustive discipline the formulas for the second moments of the waiting time were obtained and calculation of the first and the second moments of the waiting time was carried out.

Keywords: SIP · Polling system · Gated · Exhaustive · Symmetric · Switching time · Second moment · Queue

1 Introduction

The service of several types customer can be implemented via polling system with multiple queues of finite or infinite capacity, non-zero switching time and one or more servers which are common for all queues. Polling systems were studied starting from 1950's and the number of works in this area is quite large. One can find a comprehensive review of results on the polling systems in [12,14] indicating work with the theoretical results and the application of polling systems to the analysis of technical systems, including public health systems, air and railway transportation, and telecommunication systems.

In our paper we study a polling model applicable to the analysis of signaling traffic in telecommunication networks. One of the elements of modern telecommunication network is a server of Session Initiation Protocol [10], which today

The reported study was partially supported by RFBR, research projects No. 15-07-03051, 15-07-03608.

A. Dudin et al. (Eds.): ITMM 2015, CCIS 564, pp. 358–372, 2015.
DOI: 10.1007/978-3-319-25861-4_30

is one of the major signaling protocols in the Next Generation Networks. The waiting time of SIP signaling messages at SIP-server is an important performance measure because of limited lifetime of the SIP messages. Constructing a model for the analysis of SIP message waiting time we consider two types of SIP-messages: low priority Invite messages for the session initiation and so-called non-Invite messages for the session establishment which have a high priority.

To describe SIP messages service by a SIP-server and analyze the random service time of a SIP message we used the classical queueing theory for the study of complex systems [7,8]. We develop the mathematical model of a SIP-server as a polling system with two infinite queues, non-zero switching time and a single server. The purpose of our investigation is to estimate numerical characteristics of the random waiting time of a customer in the queueing system for two service disciplines - gated and exhaustive.

The polling system with two queues and the both service disciplines was investigated in [12] where the formulas for the mean waiting time in case of a symmetric model with non-zero switching time and in case of an asymmetric model with zero switching time were obtained. For the exhaustive service discipline in [12] they obtain the mean waiting time both for symmetric and asymmetric models with non-zero switching time. Part of these results is also published in earlier studies [3,4,13]. In this paper, explicit expressions for the mean waiting time in an asymmetric polling system with two queues, non-zero switching time and the gated service discipline are presented. In [6] the formulas for the second waiting time moments were obtained for a symmetric polling systems with two, three and four queues with zero and non-zero switching time for the both service disciplines. They notice in [6] that most research has focused only on the mean waiting times due to prohibitively growing complexity in computing higher-order moments of the waiting time. With regard to the analysis of SIP signaling traffic in [5,11] it was shown that the exhaustive service discipline corresponds to lower blocking probability and lower queue length and therefore lower waiting time for the priority customers in comparison with the gated service discipline. So the contribution of this paper is developing explicit expressions for the second moments of the waiting time distribution in an asymmetric polling system with two queues, non-zero switching time and the exhaustive service discipline which is more effective applying to the SIP signaling traffic analysis.

The paper is organized as follows. In Sect. 2 we introduce all the necessary concepts and denotation and describe the polling system with non-zero switching time and two queues. In Sects. 3 and 4 we carry out the mathematical analysis of the model for the gated and the exhaustive service disciplines, respectively. Section 5 provides examples of numerical analysis, were we analyze the effectiveness of the gated and the exhaustive service disciplines in terms of waiting time. Finally, we summarize our key results.

2   The Polling System with Non-zero Switching Time and Two Queues

We consider the asymmetric polling system with two queues and non-zero switching time classified according to the Basharin-Kendall's notation as $M_2|G_2|1|\infty$ (Fig. 1).

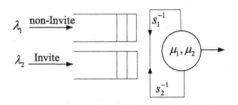

Fig. 1. Polling system with two queues and non-zero switching time.

According to the gated service discipline only customers standing in front of the gate are served during this server's visit. According to the exhaustive service discipline the server serves customers until the queue is emptied.

We assume the Poisson incoming flow of customers with parameter $\lambda = \lambda_1 + \lambda_2$. The mean service time of a customers at Q_i queue is $b_i = \int_0^\infty t\, dB_i(t)$ with distribution function (DF) $B_i(t)$ [7]. We denote kth moment of service time by $b_i^{(k)} = \int_0^\infty t^k\, dB_i(t)$, $k \geqslant 2$, the Laplace-Stieltjes transform (LST) of DF by $\widetilde{B}_i(x) = \int_0^\infty e^{-xt} dB_i(t)$, $x \geqslant 0$, $i = 1, 2$. The value $\rho_i = \lambda_i b_i^{(1)}$ is the load from Q_i customers, $i = 1, 2$, the value $\rho = \sum_{i=1}^2 \lambda_i b_i^{(1)}$ forms the system total load. The necessary and sufficient stability conditions obey the inequality $\rho < 1$. The mean, kth moment and LST of switching time to queue Q_i, $i = 1, 2$ are denoted by s_i, $s_i^{(k)}$ and $\widetilde{S}_i(x)$, $i = 1, 2$, respectively. The value $s = s_1 + s_2$ is total switching time, and s_i - switching time to queue Q_i, $i = 1, 2$. Hamiltonian cycle is equal $C = \rho C + s$, and $C = \frac{s}{1-\rho}$. The notation corresponds to [14].

3   Gated Service Discipline

In this section an asymmetric gated polling system with two queues and non-zero switching time is considered. We obtain the expressions for the first moments of the waiting time distribution for both queues. For special cases of a symmetric system and for zero switching time the presented expressions coincide with the formulas obtained in [12].

Let X_i^j denote the number of customers in the queue Q_j when the server serves customers at the queue Q_i, $i, j = 1, 2$. Let $A_i(T)$ denote the number of customers arriving to the queue Q_i during a time interval of length T, B_{ik} — k-th customers service time at the queue Q_i, $i = 1, 2$.

The values X_{i+1}^j and X_i^j are related as follows:

$$X_{i+1}^j = \begin{cases} X_i^j + A_j \left(\sum\limits_{k=1}^{X_i^i} B_{ik} + S_{i+1} \right), & j \neq i, \\[4mm] A_i \left(\sum\limits_{k=1}^{X_i^i} B_{ik} + S_{i+1} \right), & j = i. \end{cases} \tag{1}$$

Let $p_i(n_1, n_2)$ be the stationary probability, that n_1 customers are in the queue Q_1 and n_2 customers are in the queue Q_2 at polling instant of the queue Q_i, $n_1, n_2 \geqslant 0$, $i = 1, 2$. The generating function (GF) of the number of customers in the system at polling instant of the queue Q_i is denoted by $G_i(z_1, z_2)$:

$$\begin{cases} G_1(z_1, z_2) = G_2 \left(z_1, \widetilde{B}_2 \left(\lambda_1(1 - z_1) + \lambda_2(1 - z_2) \right) \right) \\[2mm] \qquad\qquad \times \widetilde{S}_1 \left(\lambda_1(1 - z_1) + \lambda_2(1 - z_2) \right), \\[2mm] G_2(z_1, z_2) = G_1 \left(\widetilde{B}_1 \left(\lambda_1(1 - z_1) + \lambda_2(1 - z_2) \right), z_2 \right) \\[2mm] \qquad\qquad \times \widetilde{S}_2 \left(\lambda_1(1 - z_1) + \lambda_2(1 - z_2) \right). \end{cases} \tag{2}$$

We denote the mean number of customers in the queue Q_j at polling instant of the queue Q_i as $f_i(j)$:

$$f_i(j) := \mathrm{M}\left[X_i^j \right] = \left. \frac{\partial G_i(z_1, z_2)}{\partial z_j} \right|_{(z_1, z_2) = (1,1)}, \qquad i, j = 1, 2 \tag{3}$$

We use the following notation: $v = \lambda_1(1 - z_1) + \lambda_2(1 - z_2)$, $\widetilde{B}_i(0) = 1$,

$$\left. \frac{d\widetilde{B}_i(v)}{dv} \right|_{v=0} = -b_i, \qquad \left. \frac{d\widetilde{S}_i(v)}{dv} \right|_{v=0} = -s_i, \qquad \frac{\partial v}{\partial z_i} = -\lambda_i,$$

$$\left. \frac{d^2\widetilde{S}_i(v)}{(dv)^2} \right|_{v=0} = s_i^{(2)}, \qquad \left. \frac{d^2\widetilde{B}_i(v)}{(dv)^2} \right|_{v=0} = b_i^{(2)},$$

if $(z_1, z_2) = (1, 1)$ then $v(1, 1) = 0$, $G_i(1, 1) = 1$, $\widetilde{S}_i(0) = 1$, $i = 1, 2$.

From (4) we obtain all values $f_i(j)$ $i, j = 1, 2$:

$$\begin{cases} f_1(1) = \lambda_1 C, \\ f_1(2) = \lambda_2 \left(\rho_2 C + s_1 \right), \\ f_2(2) = \lambda_2 C, \\ f_2(1) = \lambda_1 \left(\rho_1 C + s_2 \right). \end{cases} \tag{4}$$

The second moments (5) of the X_i^j are also obtained from GF $G_i(z_1, z_2)$, $i, j = 1, 2$:

$$f_i(j, k) = \left. \frac{\partial^2 G_i(z_1, z_2)}{\partial z_j \partial z_k} \right|_{(z_1, z_2) = (1,1)}, \qquad f_i(i, i) = \left. \frac{\partial^2 G_i(z_1, z_2)}{(\partial z_i)^2} \right|_{(z_1, z_2) = (1,1)}. \tag{5}$$

For the gated service disciplines we get the following system:

$$
\begin{cases}
f_1(1,1) = f_2(1,1) + 2f_2(1,2)b_2\lambda_1 + f_2(2,2)b_2^2\lambda_1^2 + f_2(2)b_2^{(2)}\lambda_1^2 \\
\qquad + 2s_1\lambda_1\left(f_2(1) + f_2(2)b_2\lambda_1\right) + s_1^{(2)}\lambda_1^2, \\
f_1(1,2) = f_2(1,2)\rho_2 + f_2(2,2)b_2\lambda_1\rho_2 + f_2(2)b_2^{(2)}\lambda_1\lambda_2 \\
\qquad + s_1\lambda_2 f_2(1) + 2s_1\lambda_1\rho_2 f_2(2) + s_1^{(2)}\lambda_1\lambda_2, \\
f_1(2,1) = f_2(1,2)\rho_2 + f_2(2,2)b_2\lambda_1\rho_2 + f_2(2)b_2^{(2)}\lambda_1\lambda_2 \\
\qquad + s_1\lambda_2 f_2(1) + 2s_1\lambda_1\rho_2 f_2(2) + s_1^{(2)}\lambda_1\lambda_2, \\
f_1(2,2) = f_2(2,2)b_2^2\lambda_2^2 + f_2(2)b_2^{(2)}\lambda_2^2 + 2f_2(2)b_2\lambda_2^2 s_1 + s_1^{(2)}\lambda_2^2, \\
f_2(1,1) = f_1(1,1)b_1^2\lambda_1^2 + f_1(1)b_1^{(2)}\lambda_1^2 + 2f_1(1)b_1\lambda_1^2 s_2 + s_2^{(2)}\lambda_1^2, \\
f_2(1,2) = f_1(1,2)\rho_1 + f_1(1,1)b_1\lambda_2\rho_1 + f_1(1)b_1^{(2)}\lambda_1\lambda_2 \\
\qquad + s_2\lambda_1 f_1(2) + 2s_2\lambda_2\rho_1 f_1(1) + s_2^{(2)}\lambda_1\lambda_2, \\
f_2(2,1) = f_1(1,2)\rho_1 + f_1(1,1)b_1\lambda_2\rho_1 + f_1(1)b_1^{(2)}\lambda_1\lambda_2 \\
\qquad + s_2\lambda_1 f_1(2) + 2s_2\lambda_2\rho_1 f_1(1) + s_2^{(2)}\lambda_1\lambda_2, \\
f_2(2,2) = f_1(2,2) + 2f_1(1,2)b_1\lambda_2 + f_1(1,1)b_1^2\lambda_2^2 \\
\qquad + f_1(1)b_1^{(2)}\lambda_2^2 + 2s_2\lambda_2\left(f_1(2) + f_1(1)b_1\lambda_2\right) + s_2^{(2)}\lambda_2^2.
\end{cases}
\tag{6}
$$

The value $f_1(1,1)$ satisfies the following equation:

$$
\frac{f_1(1,1)}{\lambda_1^2} = \frac{\left(Cb_1^{(2)}\lambda_1 + s_1^{(2)}\right)\left(1 + 2\rho_2 - 2\rho_2^3 - \rho_1\rho_2 - 2\rho_1\rho_2^2\right)}{(1-\rho)(1-\rho_1\rho_2)(2\rho_1\rho_2 + \rho + 1)}
$$

$$
+ \frac{\left(Cb_2^{(2)}\lambda_2 + s_1^{(2)}\right)(1+\rho_1\rho_2)}{(1-\rho)(1-\rho_1\rho_2)(2\rho_1\rho_2 + \rho + 1)} + \frac{2s_1^2\left(\rho - \rho_1^2\rho_2^2\right)}{(1-\rho)^2(1-\rho_1\rho_2)(2\rho_1\rho_2 + \rho + 1)} \tag{7}
$$

$$
+ \frac{2s_1 s_2\left(1 + \rho + \rho_1\rho_2\left(1 - \rho - 2\rho_1\rho_2\right)\right)}{(1-\rho)^2(1-\rho_1\rho_2)(2\rho_1\rho_2 + \rho + 1)}
$$

$$
+ \frac{2s_2^2\left(\rho_2^2 + \rho_2^3 - \rho_2^4 + \rho_1 + 2\rho_1\rho_2 - 2\rho\rho_1\rho_2^2 - \rho_1^2\rho_2\right)}{(1-\rho)^2(1-\rho_1\rho_2)(2\rho_1\rho_2 + \rho + 1)}.
$$

The equations for $f_2(2,2)$ can be obtained in the same way.

The LST of the DF for the waiting time with the gated service discipline is given by

$$
\widetilde{W}_i(x) = \frac{1-\rho}{s}\frac{G_i\left(\widetilde{B}_i(x)\right) - G_i(1 - x/\lambda_i)}{x - \lambda_i + \lambda_i\widetilde{B}_i(x)}. \tag{8}
$$

From (9) we get the mean waiting time for an asymmetric polling system with the gated service discipline in the form of

$$
w_i = -\frac{d}{dx}\widetilde{W}_i(x)\bigg|_{x=0} = \frac{f_i(i,i)}{2\lambda_i^2 C}(1+\rho_i), \quad i=1,2. \tag{9}
$$

In case of an asymmetric gated polling system with two queues and non-zero switching time we obtain the following equations:

$$\begin{cases} \omega_1 = \dfrac{(1+\rho_1)f_1(1,1)}{2\lambda_1^2 C}, \\ \omega_2 = \dfrac{(1+\rho_2)f_2(2,2)}{2\lambda_2^2 C}. \end{cases} \tag{10}$$

From (10) it follows that

$$\begin{aligned} \omega_1 = &\frac{(1+\rho_1)}{2s(1-\rho_1\rho_2)(2\rho_1\rho_2+\rho+1)} \times \left\{ \left(Cb_2^{(2)}\lambda_2 + s_1^{(2)} \right)(1+\rho_1\rho_2) \right. \\ &+ \left(Cb_1^{(2)}\lambda_1 + s_2^{(2)} \right)(1+2\rho_2-2\rho\rho_2^2-\rho_1\rho_2) \\ &+ \frac{2s_1^2\left(\rho-\rho_1^2\rho_2^2\right)}{(1-\rho)} + \frac{2s_1s_2\left(1+\rho+\rho_1\rho_2\left(1-\rho-2\rho_1\rho_2\right)\right)}{(1-\rho)} \\ &\left. + \frac{2s_2^2\left(\rho_2^2+\rho_2^3-\rho_2^4+\rho_1+2\rho_1\rho_2-2\rho\rho_1\rho_2^2-\rho_1^2\rho_2\right)}{(1-\rho)} \right\}, \end{aligned} \tag{11}$$

$$\begin{aligned} \omega_2 = &\frac{(1+\rho_2)}{2s(1-\rho_1\rho_2)(2\rho_1\rho_2+\rho+1)} \times \left\{ \left(Cb_2^{(2)}\lambda_1 + s_2^{(2)} \right)(1+\rho_1\rho_2) \right. \\ &+ \left(Cb_2^{(2)}\lambda_2 + s_1^{(2)} \right)(1+2\rho_1-2\rho\rho_1^2-\rho_1\rho_2) \\ &+ \frac{2s_1^2\left(\rho_1^2+\rho_1^3-\rho_1^4+\rho_2+2\rho_1\rho_2-2\rho\rho_1^2\rho_2-\rho_1\rho_2^2\right)}{(1-\rho)} \\ &\left. + \frac{2s_1s_2\left(1+\rho+\rho_1\rho_2\left(1-\rho-2\rho_1\rho_2\right)\right)}{(1-\rho)} + \frac{2s_2^2\left(\rho-\rho_1^2\rho_2^2\right)}{(1-\rho)} \right\}. \end{aligned} \tag{12}$$

In case of a symmetric gated polling system with two queues and non-zero switching time using (11) with $\rho_1 = \rho_2 = \rho_i$, $s_1 = s_2 = s_i$, $b_1^{(2)} = b_2^{(2)} = b_i^{(2)}$, $\lambda_1 = \lambda_2 = \lambda_i$ we obtain the mean waiting time as follows:

$$\omega_{gated\_symm} = \frac{b_i^{(2)}\lambda_i}{(1-2\rho_i)} + \frac{s_i^{(2)}}{2s_i} + \frac{s_i(4\rho_i+1)}{2(1-2\rho_i)}. \tag{13}$$

In case of zero switching time the mean waiting time for the first queue from (11) with $s_1 = s_2 = 0$ is equal to

$$\begin{aligned} &\omega_1(s_i=0) \\ &= \frac{(1+\rho_1)\left(b_1^{(2)}\lambda_1\left(1+2\rho_2-2\rho_2^3-\rho_1\rho_2-2\rho_1\rho_2^2\right)+b_2^{(2)}\lambda_2\left(1+\rho_1\rho_2\right)\right)}{2(1-\rho_1-\rho_2)(1-\rho_1\rho_2)(2\rho_1\rho_2+\rho_1+\rho_2+1)}. \end{aligned} \tag{14}$$

The special cases (13) and (14) are also obtained in [12].

4   Exhaustive Service Discipline

In this section an asymmetric exhaustive polling system with two queues and non-zero switching time is considered. The expressions for the first moments of the waiting time distribution are known from [12]. Using the approach of [6] we obtain the expressions for the second moments of the waiting time distribution for an asymmetric system. For the special case of a symmetric system the presented expressions for the second moments of the waiting time distribution coincide with the formulas obtained in [6].

The values X_i^j and X_{i+1}^j for the exhaustive service discipline are linked by the following system:

$$
\begin{cases}
X_1^1 = X_2^1 + A_1 \left(\sum_{k=1}^{X_2^2} \Theta_{2k} + S_1 \right), \\
X_2^1 = A_1(S_2), \\
X_2^2 = X_1^2 + A_2 \left(\sum_{k=1}^{X_1^1} \Theta_{1k} + S_2 \right), \\
X_1^2 = A_2(S_1),
\end{cases}
\tag{15}
$$

where $A_i(T)$ is the same as for the gated discipline, Θ_{ik} is the k-th customer service time at queue Q_i, $i = 1, 2$.

Values Θ_{ik} are independent and distributed identically with LST $\widetilde{\theta}_i(x)$, which is the solution of expression (16) and correspond to the LST of the DF for the length of a busy period at the queue Q_i in polling system with the exhaustive service discipline [2]:

$$
\widetilde{\theta}_i(x) = \widetilde{B}_i \left(x + \lambda_i - \lambda_i \widetilde{\theta}_i(x) \right).
\tag{16}
$$

The first three moments of the busy period in Eqs. (17)–(19) are necessary for the moments of waiting time:

$$
\theta_i = -\widetilde{\theta}'(0) = \frac{b_i}{1 - \rho_i},
\tag{17}
$$

$$
\theta_i^{(2)} = \widetilde{\theta}''(0) = \frac{b_i^{(2)}}{(1 - \rho_i)^3},
\tag{18}
$$

$$
\theta_i^{(3)} = -\widetilde{\theta}'''(0) = \frac{b_i^{(3)}}{(1 - \rho_i)^4} + \frac{3\lambda_i \left(b_i^{(2)} \right)^2}{(1 - \rho_i)^5}.
\tag{19}
$$

The GF $G_i(z_1, z_2)$ of the number of customers in the system at polling instant of the queue Q_i is given by

$$
\begin{cases}
G_1(z_1, z_2) = G_2 \left(z_1, \widetilde{\theta}_2(\lambda_1(1 - z_1)) \right) \cdot \widetilde{S}_1(\lambda_1(1 - z_1) + \lambda_2(1 - z_2)), \\
G_2(z_1, z_2) = G_1 \left(\widetilde{\theta}_1(\lambda_2(1 - z_2)), z_2 \right) \cdot \widetilde{S}_2(\lambda_1(1 - z_1) + \lambda_2(1 - z_2)).
\end{cases}
\tag{20}
$$

We denote $\widetilde{S}_i(\lambda_1(1 - z_1) + \lambda_2(1 - z_2))$ is the LST of the DF $S_i(t)$. The $S_i(t)$ is the DF of switching time to queue Q_i. The values $f_i(j)$ of the mean number of customers we estimate by derivation of GF $G_1(z_1, z_2)$ and $G_2(z_1, z_2)$, $i, j = 1, 2$:

$$
f_i(j) = M \left[X_i^j \right] = \left. \frac{\partial G_i(z_1, z_2)}{\partial z_j} \right|_{(z_1, z_2) = (1, 1)}.
\tag{21}
$$

We use the following notation: $\rho_i = \lambda_i b_i$, $\eta_i = \lambda_i(1 - z_i)$, $v = \eta_1 + \eta_2$,

$$\left.\frac{d\widetilde{S}_i(v)}{dv}\right|_{v=0} = -s_i, \quad \frac{\partial v}{\partial z_i} = -\lambda_i, \quad \left.\frac{d^2\widetilde{S}_i(v)}{(dv)^2}\right|_{v=0} = s_i^{(2)},$$

$$\left.\frac{d^2\widetilde{\theta}_i(\eta_{i+1})}{(d(\eta_{i+1}))^2}\right|_{\eta_{i+1}=0} = \frac{b_i^{(2)}}{(1 - \rho_i)^3}.$$

If $(z_1, z_2) = (1, 1)$, then $v(1, 1) = 0$, $G_i(1, 1) = 1$, $\widetilde{S}_i(0) = 1$, $\widetilde{\theta}_i(0) = 1$, $i = 1, 2$.

Note that for the exhaustive service discipline $\frac{\partial G_i(z_1, z_2)}{\partial z_i} = 0$, so the exhaustive service discipline is easier in calculations that the gated service discipline. The values $f_i(j)$ expressed in the following explicit form [14]:

$$\begin{cases} f_1(1) = \lambda_1 C(1 - \rho_1), \\ f_1(2) = \lambda_2 s_1, \\ f_2(1) = \lambda_1 s_2, \\ f_2(2) = \lambda_2 C(1 - \rho_2). \end{cases} \tag{22}$$

The second moments (23) of the X_i^j are also defined in the same way as for the gated service discipline.

$$\begin{cases} f_1(2, 2) = \lambda_2^2 s_1^{(2)} \\ f_2(1, 1) = \lambda_1^2 s_2^{(2)} \\ f_2(1, 2) = \lambda_1 \lambda_2 s_1 s_2 + \rho_1 \lambda_1 \lambda_2 s_2 C + s_2^{(2)} \lambda_1 \lambda_2 \\ f_2(2, 1) = \lambda_1 \lambda_2 s_1 s_2 + \rho_1 \lambda_1 \lambda_2 s_2 C + s_2^{(2)} \lambda_1 \lambda_2 \\ f_1(1, 2) = \lambda_1 \lambda_2 s_1 s_2 + \rho_2 \lambda_1 \lambda_2 s_1 C + s_1^{(2)} \lambda_1 \lambda_2 \\ f_1(2, 1) = \lambda_1 \lambda_2 s_1 s_2 + \rho_2 \lambda_1 \lambda_2 s_1 C + s_1^{(2)} \lambda_1 \lambda_2 \\ f_2(2, 2) = \lambda_2^2 s_2^{(2)} + f_1(1, 1)\left(\frac{b_1 \lambda_2}{1-\rho_1}\right)^2 + \lambda_1 C \lambda_2^2 \left[\frac{b_1^{(2)}}{(1-\rho_1)^2} + 2s_2 b_1\right] \\ \quad + 2\left(\lambda_1 \lambda_2 s_1 s_2 + \rho_2 \lambda_1 \lambda_2 s_1 C + s_1^{(2)} \lambda_1 \lambda_2\right)\frac{b_1 \lambda_2}{1-\rho_1} + \lambda_2^2 s_1^{(2)} + 2\lambda_2^2 s_1 s_2 \\ f_1(1, 1) = \lambda_1^2 s_1^{(2)} + f_2(2, 2)\left(\frac{b_2 \lambda_1}{1-\rho_2}\right)^2 + \lambda_2 C \lambda_1^2 \left[\frac{b_2^{(2)}}{(1-\rho_2)^2} + 2s_1 b_2\right] \\ \quad + 2\left(\lambda_1 \lambda_2 s_1 s_2 + \rho_1 \lambda_1 \lambda_2 s_2 C + s_2^{(2)} \lambda_1 \lambda_2\right)\frac{b_2 \lambda_1}{1-\rho_2} + \lambda_1^2 s_2^{(2)} + 2\lambda_1^2 s_1 s_2 \end{cases} \tag{23}$$

Using the system (23), we get $f_1(1, 1)$ and $f_2(2, 2)$ in the explicit form, by way of example

$$\begin{aligned} \frac{f_1(1, 1)}{\lambda_1^2} =\ & \frac{s_1^{(2)}(1 - \rho_1)(1 - \rho - \rho_2 + 2\rho\rho_2)}{(1 - \rho + 2\rho_1\rho_2)(1 - \rho)} + \frac{s_2^{(2)}(1 - \rho_1)^2}{(1 - \rho + 2\rho_1\rho_2)(1 - \rho)} \\ & + \frac{b_1^{(2)}\lambda_1\rho_2^2 C}{(1 - \rho + 2\rho_1\rho_2)(1 - \rho)} + \frac{b_2^{(2)}\lambda_2 C(1 - \rho_1)^2}{(1 - \rho + 2\rho_1\rho_2)(1 - \rho)} \\ & + \frac{2s_1^2(1 - \rho_1)\rho_2\left(1 - \rho - \rho_2 + 2\rho_1\rho_2 + \rho_2^2\right)}{(1 - \rho + 2\rho_1\rho_2)(1 - \rho)^2} + \frac{2s_2^2\rho_1\rho_2(1 - \rho_1)^2}{(1 - \rho + 2\rho_1\rho_2)(1 - \rho)^2} \\ & + \frac{2s_1 s_2(1 - \rho_1)\left(-2\rho_1^2\rho_2 + \rho_1^2 + 3\rho_1\rho_2 - \rho_1 - \rho + 1\right)}{(1 - \rho)^2(1 - \rho + 2\rho_1\rho_2)}. \end{aligned} \tag{24}$$

The LST of the DF for the waiting time with the exhaustive service discipline is given by

$$\tilde{W}_i\left(x\right) = \frac{1-\rho}{s}\frac{1-G_i\left(1-x/\lambda_i\right)}{x-\lambda_i+\lambda_i\tilde{B}_i(x)}. \tag{25}$$

From (25) we get for exhaustive polling system the first ω_i and the second $W_i^{(2)}$ moments of the waiting time distribution at the queue Q_i, $i = 1, 2$, in the form of

$$\omega_i = -\frac{d}{dx}\tilde{W}_i(x)\Big|_{x=0} = \frac{f_i(i,i)}{2\lambda_i^2(1-\rho_i)C} + \frac{\lambda_i b_i^{(2)}}{2\left(1-\rho_i\right)}, \tag{26}$$

$$
\begin{aligned}
W_i^{(2)} &= \frac{d^2}{(dx)^2}\tilde{W}_i(x)\Big|_{x=0} \\
&= \frac{\lambda_i b_i^{(3)}}{3(1-\rho_i)} + \frac{\left(\lambda_i b_i^{(2)}\right)^2}{2(1-\rho_i)^2} + \frac{f_i(i,i,i)}{3(1-\rho_i)\lambda_i^3 C} + \frac{f_i(i,i)b_i^{(2)}}{2(1-\rho_i)^2\lambda_i C}.
\end{aligned} \tag{27}
$$

From (26) it follows that the mean waiting time for exhaustive polling system and non-zero switching time at the queue Q_1 have the following form:

$$
\begin{aligned}
\omega_1 = {} & \frac{s_1^{(2)}\left(1-\rho-\rho_2+2\rho\rho_2\right)}{2s\left(1-\rho+2\rho_1\rho_2\right)} + \frac{s_2^{(2)}\left(1-\rho_1\right)}{2s\left(1-\rho+2\rho_1\rho_2\right)} \\
& + \frac{\lambda_1 b_1^{(2)}\left(1-\rho-\rho_2+2\rho_2\rho\right)+b_2^{(2)}\lambda_2\left(1-\rho_1\right)}{2\left(1-\rho+2\rho_1\rho_2\right)\left(1-\rho\right)} \\
& + \frac{s_2^2\rho_1\rho_2\left(1-\rho_1\right)}{s\left(1-\rho+2\rho_1\rho_2\right)\left(1-\rho\right)} + \frac{s_1 s_2\left(1-\rho_1\right)}{s\left(1-\rho\right)} \\
& + \frac{s_1^2\rho_2\left(1-\rho-\rho_2+2\rho_1\rho_2+\rho_2^2\right)}{s\left(1-\rho+2\rho_1\rho_2\right)\left(1-\rho\right)}.
\end{aligned} \tag{28}
$$

The expression for ω_2 can be obtained in the same way.

In case of a symmetric exhaustive polling system with two queues and non-zero switching time using (28) with $\rho_1 = \rho_2 = \rho_i$, $\lambda_1 = \lambda_2 = \lambda_i$, $s_1 = s_2 = s_i$, $b_1^{(2)} = b_2^{(2)} = b_i^{(2)}$, $s_1^{(2)} = s_2^{(2)} = s_i^{(2)}$ we obtain the mean waiting time as follows:

$$\omega_{exhaustive\_symm} = \frac{b_i^{(2)}\lambda_i}{(1-2\rho_i)} + \frac{s_i^{(2)}}{2s_i} + \frac{s_i}{2\left(1-2\rho_i\right)}. \tag{29}$$

Special case (29) is also obtained in [12].

In case of zero switching time the mean waiting time for the first queue from (28) with $s_1 = s_2 = 0$ is equal to

$$\omega_1\left(s_i = 0\right) = \frac{\lambda_1 b_1^{(2)}\left(1-\rho-\rho_2+2\rho_2\rho\right)+b_2^{(2)}\lambda_2\left(1-\rho_1\right)}{2\left(1-\rho+2\rho_1\rho_2\right)\left(1-\rho\right)}. \tag{30}$$

Comparing the expressions (13) for the gated service discipline and (29) for the exhaustive service discipline we found that

$$\omega_{gated\_symm} - \omega_{exhaustive\_symm} = \frac{4\rho_i s_i}{2(1-2\rho_i)}. \tag{31}$$

In case of non-zero switching time and $\rho_i > 0$, $\rho = 2\rho_i$, $\rho < 1$, the value of (31) is positive. Hence for a symmetric polling system the exhaustive service discipline is more effective in terms of the mean waiting time. Therefore, we will obtain formulas for the second moments of the waiting time distribution for an asymmetric polling system with two queues and equal non-zero switching time for the exhaustive service discipline. We use the following notation: $s_1 = s_2 = s_0$, $s_1^{(k)} = s_2^{(k)} = s_0^{(k)}$, $k = \overline{2,3}$.

To estimate the second moment of the waiting time it is necessary to know the values of the third moments of the customer's number at the queues:

$$f_i(j,k,l) = \left. \frac{\partial^3 G_i(z_1, z_2)}{\partial z_j \partial z_k \partial z_l} \right|_{(z_1, z_2)=(1,1)}. \tag{32}$$

By way of example the value $f_1(1,1,1)$ satisfy the following equation:

$$\frac{f_1(1,1,1)}{\lambda_1^3} = \frac{s_0^{(3)}(1-\rho_1)}{(1-\rho)A}$$
$$\times \frac{(3\rho_1^2\rho_2^2 + 3\rho_1\rho_2^3 - 3\rho_1^2\rho_2 - 6\rho_1\rho_2^2 + 2\rho_1^2 + 6\rho_1\rho_2 + 3\rho_2^2 - 4\rho_1 - 3\rho_2 + 2)}{(1-\rho)A}$$
$$+ \frac{2s_0 b_1^{(3)}\lambda_1\rho_2^3}{(1-\rho)^2 A} + \frac{2s_0 b_2^{(3)}\lambda_2(1-\rho_1)^3}{(1-\rho)^2 A}$$
$$+ \frac{6s_0\left(b_1^{(2)}\right)^2 \lambda_1^2\rho_2^3(2\rho_1\rho_2 + \rho_2^2 - \rho_1 - 2\rho_2 + 1)}{(1-\rho+2\rho_1\rho_2)(1-\rho)^3 A}$$
$$+ \frac{6s_0 b_1^{(2)}\lambda_1 b_2^{(2)}\lambda_2\rho_2(1-\rho_1)(-\rho_1^2\rho_2 + \rho_1\rho_2^2 + \rho_1^2 + 2\rho_1\rho_2 - 2\rho_1 - \rho_2 + 1)}{(1-\rho+2\rho_1\rho_2)(1-\rho)^3 A}$$
$$+ \frac{6s_0\left(b_2^{(2)}\right)^2 \lambda_2^2(\rho_1^2 + 2\rho_1\rho_2 - 2\rho_1 - \rho_2 + 1)(1-\rho_1)^3}{(1-\rho+2\rho_1\rho_2)(1-\rho)^3 A}$$
$$+ \frac{3s_0^{(2)}b_1^{(2)}\lambda_1\rho_2^3(-\rho_1^2 + 2\rho_1\rho_2 + \rho_2^2 - 2\rho_2 + 1)}{(1-\rho+2\rho_1\rho_2)(1-\rho)^2 A}$$
$$+ \frac{3s_0^{(2)}b_2^{(2)}(1-\rho_1)^3\lambda_2(-\rho_2^2 + 2\rho_1\rho_2 + \rho_1^2 - 2\rho_1 + 1)}{(1-\rho+2\rho_1\rho_2)(1-\rho)^2 A}$$
$$+ \frac{3s_0^2\lambda_1 b_1^{(2)}\rho_2^2}{(1-\rho)^3 A(2\rho_1\rho_2 - \rho_1 - \rho_2 + 1)}(-12\rho_1^3\rho_2^2 + 13\rho_1^3\rho_2 + 27\rho_1^2\rho_2^2 + \rho_1\rho_2^3$$
$$- \rho_2^4 - 4\rho_1^3 - 35\rho_1^2\rho_2 - 20\rho_1\rho_2^2 + \rho_2^3 + 12\rho_1^2 + 31\rho_1\rho_2 + 5\rho_2^2 - 12\rho_1 - 9\rho_2 + 4)$$
$$+ \frac{3s_0^2\lambda_2 b_2^{(2)}(1-\rho_1)^3}{(1-\rho)^3 A(2\rho_1\rho_2 - \rho_1 - \rho_2 + 1)}(12\rho_1^2\rho_2^2 + \rho_2^3 - 9\rho_2^2\rho_1 - 11\rho_2\rho_1^2 - \rho_1^3 + \rho_2^2$$
$$+ 16\rho_1\rho_2 + 5\rho_1^2 - 5\rho_2 - 7\rho_1 + 3) + \frac{3s_0 s_0^{(2)}(1-\rho_1)}{(1-\rho)^2 A(1-\rho+2\rho_1\rho_2)}$$

$$\times(-6\rho_1^4\rho_2^3 + 6\rho_1^2\rho_2^5 + 15\rho_1^4\rho_2^2 + 15\rho_1^3\rho_2^3 - 3\rho_1^2\rho_2^4 - 3\rho_1\rho_2^5 - 9\rho_1^4\rho_2$$
$$-44\rho_1^3\rho_2^2 - 15\rho_1^2\rho_2^3 + 2\rho_1\rho_2^4 + 2\rho_1^4 + 32\rho_1^3\rho_2 + 48\rho_1^2\rho_2^2 + 9\rho_1\rho_2^3+$$
$$+\rho_2^4 - 8\rho_1^3 - 42\rho_1^2\rho_2 - 24\rho_1\rho_2^2 - 3\rho_2^3 + 12\rho_1^2 + 24\rho_1\rho_2 + 5\rho_2^2 - 8\rho_1 - 5\rho_2 + 2)$$
$$+\frac{6s_0^3\rho_2(1-\rho_1)}{(1-\rho+2\rho_1\rho_2)A(1-\rho)^3} \times (4\rho_1^5\rho_2^2 - 6\rho_1^4\rho_2^3 + 2\rho_1^2\rho_2^5 - 3\rho_1^5\rho_2 - 2\rho_1^4\rho_2^2$$
$$+18\rho_1^3\rho_2^3 - \rho_1\rho_2^5 + \rho_2^5 + 6\rho_1^4\rho_2 - 20\rho_1^3\rho_2^2 - 22\rho_1^2\rho_2^3 - \rho_1\rho_2^4 - 3\rho_1^4 + 4\rho_1^3\rho_2$$
$$+36\rho_1^2\rho_2^2 + 14\rho_1\rho_2^3 + \rho_2^4 + 2\rho_1^3 - 18\rho_1^2\rho_2 - 24\rho_1\rho_2^2 - 4\rho_2^3 + 2\rho_1^2 + 15\rho_1\rho_2$$
$$+6\rho_2^2 - 3\rho_1 - 4\rho_2 + 1),$$

where $A = 3\rho_1^2\rho_2^2 - 3\rho_1^2\rho_2 - 3\rho_1\rho_2^2 + \rho_1^2 + 5\rho_1\rho_2 + \rho_2^2 - 2\rho_1 - 2\rho_2 + 1$ is the load parameter.

The second moment $W_1^{(2)}$ of the waiting time distribution of customers at the first queue in an asymmetric polling system with two queues and identical switching times and the exhaustive service discipline is given by

$$
\begin{aligned}
W_1^{(2)} =\ & \frac{s_0^{(3)}A_1}{6s_0 A} + \frac{b_1^{(3)}\lambda_1 A_2}{3(1-\rho)A} \\
&+\frac{b_2^{(3)}\lambda_2(1-\rho_1)^2}{3(1-\rho)A} + \frac{\left(\lambda_1 b_1^{(2)}\right)^2 A_3}{2(1-\rho)^2(2\rho_1\rho_2 - \rho_1 - \rho_2 + 1)} \\
&+\frac{b_1^{(2)}\lambda_1 b_2^{(2)}\lambda_2 A_4}{2(1-\rho+2\rho_1\rho_2)(1-\rho)^2 A} + \frac{\left(b_2^{(2)}\right)^2\lambda_2^2 A_5}{(1-\rho+2\rho_1\rho_2)(1-\rho)^2 A} \\
&+\frac{s_0^{(2)}b_2^{(2)}\lambda_2 A_6}{2s_0(1-\rho+2\rho_1\rho_2)(1-\rho)A} + \frac{s_0^{(2)}b_1^{(2)}\lambda_1 A_7}{2s_0(1-\rho+2\rho_1\rho_2)(1-\rho)A} \\
&+\frac{s_0\lambda_1 b_1^{(2)}A_8}{2(1-\rho)^2 A(1-\rho+2\rho_1\rho_2)} + \frac{s_0\lambda_2 b_2^{(2)}(1-\rho_1)^2 A_9}{2(1-\rho)^2 A(1-\rho+2\rho_1\rho_2)} \\
&+\frac{s_0^{(2)}A_{10}}{2(1-\rho)A(1-\rho+2\rho_1\rho_2)} + \frac{s_0^2\rho_2 A_{11}}{(1-\rho+2\rho_1\rho_2)A(1-\rho)^2},
\end{aligned}
\tag{33}
$$

where the loading parameters $A, A_1, ..., A_{11}$ are listed below:

$$A = \rho_1^2\rho_2^2 - 3\rho_1^2\rho_2 - 3\rho_1\rho_2^2 + \rho_1^2 + 5\rho_1\rho_2 + \rho_2^2 - 2\rho_1 - 2\rho_2 + 1,$$
$$A_1 = 3\rho_1^2\rho_2^2 + 3\rho_1\rho_2^3 - 3\rho_1^2\rho_2 - 6\rho_1\rho_2^2 + 2\rho_1^2 + 6\rho_1\rho_2 + 3\rho_2^2 - 4\rho_1 - 3\rho_2 + 2,$$
$$A_2 = 3\rho_1^2\rho_2^2 + 3\rho_1\rho_2^3 - 3\rho_1^2\rho_2 - 6\rho_1\rho_2^2 + \rho_1^2 + 6\rho_1\rho_2 + 3\rho_2^2 - 2\rho_1 - 3\rho_2 + 1,$$
$$
\begin{aligned}
A_3 =\ & 6\rho_1^3\rho_2^3 + 12\rho_1^2\rho_2^4 + 6\rho_1\rho_2^5 - 9\rho_1^3\rho_2^2 - 27\rho_1^2\rho_2^3 - 18\rho_1\rho_2^4 + 5\rho_1^3\rho_2 \\
&+29\rho_1^2\rho_2^2 + 33\rho_1\rho_2^3 + 4\rho_2^4 - \rho_1^3 - 15\rho_1^2\rho_2 - 31\rho_1\rho_2^2 - 11\rho_2^3 + 3\rho_1^2 \\
&+15\rho_1\rho_2 + 11\rho_2^2 - 3\rho_1 - 5\rho_2 + 1,
\end{aligned}
$$
$$
\begin{aligned}
A_4 =\ & -3\rho_1^3\rho_2^2 - 3\rho_1^2\rho_2^3 + 3\rho_1^3\rho_2 + 7\rho_1^2\rho_2^2 + 5\rho_1\rho_2^3 - \rho_1^3 - 7\rho_1^2\rho_2 - 5\rho_1\rho_2^2 \\
&-\rho_2^3 + 3\rho_1^2 + 5\rho_1\rho_2 + \rho_2^2 - 3\rho_1 - \rho_2 + 1,
\end{aligned}
$$
$$A_5 = \left(\rho_1^2 + 2\rho_1\rho_2 - 2\rho_1 - \rho_2 + 1\right)(1-\rho_1)^2,$$
$$A_6 = \left(-\rho_2^2 + 2\rho_1\rho_2 + \rho_1^2 - 2\rho_1 + 1\right)(1-\rho_1)^2,$$
$$
\begin{aligned}
A_7 =\ & 3\rho_1^3\rho_2^3 + 6\rho_1^2\rho_2^4 + 3\rho_1\rho_2^5 - 6\rho_1^3\rho_2^2 - 15\rho_1^2\rho_2^3 - 9\rho_1\rho_2^4 + 4\rho_1^3\rho_2 \\
&+19\rho_1^2\rho_2^2 + 19\rho_1\rho_2^3 + 2\rho_2^4 - \rho_1^3 - 12\rho_1^2\rho_2 - 20\rho_1\rho_2^2 - 6\rho_2^3 + 3\rho_1^2 \\
&+12\rho_1\rho_2 + 7\rho_2^2 - 3\rho_1 - 4\rho_2 + 1,
\end{aligned}
$$

$$A_8 = -3\rho_1^4\rho_2^3 + 3\rho_1^3\rho_2^4 + 3\rho_1^2\rho_2^5 - 3\rho_1\rho_2^6 + 6\rho_1^4\rho_2^2 + 6\rho_1^3\rho_2^3 + 6\rho_1^2\rho_2^4$$
$$+6\rho_1\rho_2^5 - 4\rho_1^4\rho_2 - 19\rho_1^3\rho_2^2 - 17\rho_1^2\rho_2^3 - 16\rho_1\rho_2^4 - 2\rho_2^5 + \rho_1^4 + 15\rho_1^3\rho_2$$
$$+27\rho_1^2\rho_2^2 + 25\rho_1\rho_2^3 + 8\rho_2^4 - 4\rho_1^3 - 21\rho_1^2\rho_2 - 21\rho_1\rho_2^2 - 11\rho_2^3 + 6\rho_1^2$$
$$+13\rho_1\rho_2 + 7\rho_2^2 - 4\rho_1 - 3\rho_2 + 1,$$
$$A_9 = 12\rho_1^2\rho_2^2 + \rho_2^3 - 9\rho_1\rho_2^2 - 11\rho_1^2\rho_2 - \rho_1^3 + \rho_2^2 + 16\rho_1\rho_2 + 5\rho_1^2$$
$$-5\rho_2 - 7\rho_1 + 3,$$
$$A_{10} = -6\rho_1^4\rho_2^3 + 6\rho_1^2\rho_2^5 + 15\rho_1^4\rho_2^2 + 15\rho_1^3\rho_2^3 - 3\rho_1^2\rho_2^4 - 3\rho_1\rho_2^5 - 9\rho_1^4\rho_2$$
$$-44\rho_1^3\rho_2^2 - 15\rho_1^2\rho_2^3 + 2\rho_1\rho_2^4 + 2\rho_1^4 + 32\rho_1^3\rho_2 + 48\rho_1^2\rho_2^2 + 9\rho_1\rho_2^3$$
$$+\rho_2^4 - 8\rho_1^3 - 42\rho_1^2\rho_2 - 24\rho_1\rho_2^2 - 3\rho_2^3 + 12\rho_1^2 + 24\rho_1\rho_2 + 5\rho_2^2$$
$$-8\rho_1 - 5\rho_2 + 2,$$
$$A_{11} = 4\rho_1^5\rho_2^2 - 6\rho_1^4\rho_2^3 + 2\rho_1^2\rho_2^5 - 3\rho_1^5\rho_2 - 2\rho_1^4\rho_2^2 + 18\rho_1^3\rho_2^3 - \rho_1\rho_2^5$$
$$+\rho_1^5 + 6\rho_1^4\rho_2 - 20\rho_1^3\rho_2^2 - 22\rho_1^2\rho_2^3 - \rho_1\rho_2^4 - 3\rho_1^4 + 4\rho_1^3\rho_2 + 36\rho_1^2\rho_2^2$$
$$+14\rho_1\rho_2^3 + \rho_2^4 + 2\rho_1^3 - 18\rho_1^2\rho_2 - 24\rho_1\rho_2^2 - 4\rho_2^3 + 2\rho_1^2 + 15\rho_1\rho_2$$
$$+6\rho_2^2 - 3\rho_1 - 4\rho_2 + 1.$$

In case of a symmetric exhaustive polling system with two queues and non-zero switching time the second moment $W_i^{(2)}$, $i = 1, 2$, of the waiting time distribution of customers from (33) with $\rho_1 = \rho_2 = \rho_i$, $b_1^{(k)} = b_2^{(k)} = b_i^{(k)}$, $s_0^{(k)} = s_i^{(k)}$, $\lambda_1 = \lambda_2 = \lambda_i$ is equal to

$$W_i^{(2)} =$$
$$= \frac{\left(2\rho_i^2 - \rho_i + 2\right)}{\left(\rho_i^2 - \rho_i + 1\right)} \left(\frac{b_i^{(3)}\lambda_i}{3\left(1 - 2\rho_i\right)} + \frac{s_i^{(3)}}{6s_i} + \frac{\left(b_i^{(2)}\right)^2 \lambda_i^2}{\left(1 - 2\rho_i\right)^2} + \frac{s_i^{(2)}b_1^{(2)}\lambda_i}{2s_i\left(1 - 2\rho_i\right)} \right)$$
$$+ \frac{s_ib_i^{(2)}\lambda_i\left(6\rho_i^2 - 5\rho_i + 4\right)}{2\left(1 - 2\rho_i\right)^2\left(\rho_i^2 - \rho_i + 1\right)} + \frac{s_i^{(2)}\left(4\rho_i^2 - 3\rho_i + 2\right)}{2\left(1 - 2\rho_i\right)\left(\rho_i^2 - \rho_i + 1\right)}$$
$$+ \frac{s_i^2\rho_i\left(2\rho_i^2 - 2\rho_i + 1\right)}{\left(1 - 2\rho_i\right)^2\left(\rho_i^2 - \rho_i + 1\right)}. \tag{34}$$

The special case (34) is also obtained in [6].

5 Numerical Experiment

In this section we present results of computation of the first and the second moments of the waiting time distribution for the gated and the exhaustive service disciplines using input data applicable to signaling traffic analysis [5,9,11]. Let $s_i = 2$ ms be the mean switching time to the queue Q_i, $i = 1, 2$, $b_1 = 4$ ms be the mean service time of the first type customers (non-Invite), $b_2 = 10$ ms be the mean service time of the second type customers (Invite). For typical session initiation procedure [9] the values of input flow rates λ_1 and λ_2 are related as follows: $\lambda_1 = 6\lambda_2$.

Figure 2 shows the dependence of the mean waiting times ω_1 and ω_2 at the queues Q_1 and Q_2 on the total input load ρ. Two service disciplines, gated and

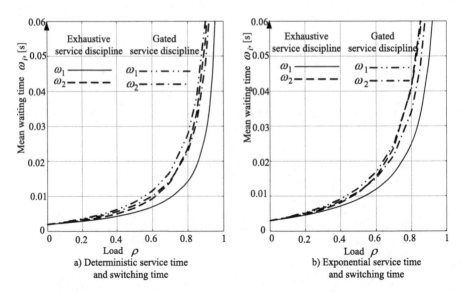

Fig. 2. Mean waiting time for gated and exhaustive service disciplines.

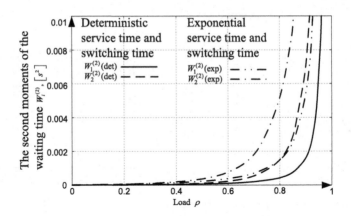

Fig. 3. The second moments of the waiting time distribution for exhaustive service discipline.

exhaustive, were compared under the exponential service time and the deterministic service time with the same mean values of b_1 and b_2.

The first thing to notice is that the mean waiting time for an exponential service time is higher than those for a deterministic service time for the both service disciplines. As might be expected, the mean waiting time for the exhaustive service discipline is always less than those for the gated service discipline. It follows from the fact that the value of (31) is positive with the above input data.

Figure 3 illustrates the calculation of the second moments $W_1^{(2)}$ and $W_2^{(2)}$ of the waiting time distribution for the exhaustive service discipline using formula (33). Figure 3 shows the dependence of $W_1^{(2)}$ and $W_2^{(2)}$ at the queues Q_1 and Q_2 on the total input load ρ. The calculations were made for the exhaustive service discipline because it is more effective than the gated service discipline applying to the SIP signaling traffic analysis.

6   Conclusion

In this paper we study a single server asymmetric polling system with non-zero switching time for two service disciplines - gated and exhaustive. The waiting time is one of the key performance characteristics of the system related to SIP signaling traffic analysis. First we analyzed the mean waiting time for both the gated and the exhaustive service disciplines. Numerical analysis shows that the exhaustive service discipline is preferable to gated one in terms of the mean waiting time. So after that we analyzed the second moments of the waiting time distribution only for the exhaustive service discipline. The paper provides the explicit expressions for the second moments of the waiting time distribution for a single server asymmetric polling system with non-zero switching time and the exhaustive service discipline.

We thank Professor Konstantin Samouylov from Peoples' Friendship University of Russia for comments that greatly improved the paper.

References

1. Boxma, O.J.: Workloads and waiting times in single-server systems with multiple customer classes. Queueing Syst. **5**(1–3), 185–214 (1989)
2. Dorsman, J.L., Boxma, O.J., Van der Mei, R.D.: On two-queue Markovian polling systems with exhaustive service. Queueing Syst. **78**(4), 287–311 (2014)
3. Eisenberg, M.: Queues with periodic service and changeover times. Oper. Res. **20**(2), 440–451 (1972)
4. Eisenberg, M.: Two queues with changeover times. Oper. Res. **19**(2), 386–401 (1971)
5. Gaidamaka, Y.V.: Model with threshold control for analyzing a server with an SIP protocol in the overload mode. Autom. Control Comput. Sci. **47**(4), 211–218 (2013)
6. Kudoh, S., Takagi, H., Hashida, O.: Second moments of the waiting time in symmetric polling systems. J. Oper. Res. Soc. Jpn. **43**(2), 306–316 (2000)
7. Moiseeva, S.P., Zakhorolnaya, I.A.: Mathematical model of parallel retrial queueing of multiple requests. Optoelectron. Instrum. Data Process. **47**(6), 567–572 (2011)
8. Nazarov, A.A., Moiseev, A.N.: Distributed system of processing of data of physical experiments. Russ. Phys. J. **57**(7), 984–990 (2014)
9. Ohta, M.: Overload control in a SIP signalling network. Int. J. Electr. Electron. Eng. **3**(2), 87–92 (2009)
10. Rosenberg J., Schulzrinne H., Camarillo G., et al.: SIP: Session initiation protocol, in RFC (Request for Comments) 3261 (2002)

11. Shorgin, S., Samouylov, K., Gaidamaka, Y., Etezov, S.: Polling system with threshold control for modeling of sip server under overload. In: Swiątek, J., Grzech, A., Swiątek, P., Tomczak, J.M. (eds.) Advances in Systems Science. AISC, vol. 240, pp. 97–107. Springer, Heidelberg (2014)
12. Takagi, H., Kleinrock, L.: A tutorial on the analysis of polling systems. Computer Science Department, University of California, Los Angeles. - Report No. CSD-850005, pp. 1–172 (1985)
13. Takagi, H.: Mean message waiting time in symmetric polling system. In: Gelenbe, E. (ed.) Performance'84, pp. 293–302. Elsevier Science Publishers, Amsterdam (1984)
14. Vishnevskii, V.M., Semenova, O.V.: Mathematical methods to study the polling systems. Autom. Remote Control 67(2), 173–220 (2006)

Stationary Distribution Insensitivity of a Closed Multi-regime Queueing Network with Non-active Customers

Julia Kruk[1,2] and Yuliya Dudovskaya[1,2]($\boxtimes$)

[1] Belarusian National Technical University, Minsk, Belarus
juls1982@list.ru
[2] Francisk Skorina Gomel State University, Gomel, Belarus
dudovskaya@gmail.com

Abstract. Stationary functioning of a closed queueing network with temporarily non-active customers and multi-regime service strategies is analyzed. Non-active customers are located in the network nodes in queues, being not serviced. For a customer, the opportunity of passing from its ordinary state to the temporarily non-active state (and backwards) is provided. Quantity of work for customer service is a random distributed value. Stationary distribution insensitivity with respect to functional form of distribution of work quantity for customer service is established.

Keywords: Closed queueing network · Non-active customers · Multi-regime service · Stationary distribution insensitivity

1 Introduction

Currently, attention to queueing theory is mainly stimulated by the need to apply results of this theory to important practical problems. During the past years, an important research effort has been devoted to the problem of queueing systems reliability. In practical terms, it is important to consider several different approaches: the queueing system can break down totally or partially. Yu.V. Malinkovsky have introduced into consideration queuing networks with multi-regime service strategies: systems at such networks can operate at several regimes. Each regime corresponds to a certain degree of service efficiency.

Herewith, the problem of customer reliability becomes relevant too. Indeed not only queueing system can break down. Customers may also lose their quality indicators. Queueing network with temporarily non-active customers is a model with customers, which are partly unreliable. The necessity of their study was caused by practical considerations, because such networks allow us to consider models with partially unreliable customers. Non-active customers are located in the network systems in queues, being not serviced. For a customer, the opportunity of passing from its ordinary state to the temporarily non-active state (and backwards) is provided. Non-active customers can be interpreted as customers

A. Dudin et al. (Eds.): ITMM 2015, CCIS 564, pp. 373–383, 2015.
DOI: 10.1007/978-3-319-25861-4_31

with defect that makes them unfit for service. G. Tsitsiashvili and M. Osipova [1,2] have observed an open exponential queueing network with non-active customers and have established the form of stationary distribution.

The standard assumption in analysis of classical queueing networks [3,4] is that service time is exponentially distributed random value. But real numerous statistical data prove the opposite. Therefore there is an actual problem to develop an analytical apparatus for the study of queueing networks with arbitrary functions of service time distribution. Currently, this problem attracts increasing attention of researchers. The first result about stationary distribution insensitivity belongs to B.A. Sevastyanov, who has observed queueing system M/G/m/0 and has proved stationary distribution insensitivity [5]. BCMP-theorem (Baskett, Chandy, Muntz, Palacios) [6] is the first result about stationary distribution insensitivity for queueing networks. We have generalized the result [1,2] in the case of random distributed service times [7–9]. We have established stationary distribution insensitivity with respect to functional form of service time distribution.

V.A. Ivnitsky [10] has considered quite interesting class of queueing networks: customer service has not "temporal" but so-called "energetical" interpretation. Every service operation is characterized by the random variable of work to be performed. Stationary distribution insensitivity with respect to functional form of distribution of work quantity for customer service has been obtained for different classes of open and closed queueing networks [10] and for closed queueing networks with non-active customers [11].

This paper provides stationary functioning of a closed queueing network with temporarily non-active customers and multi-regime service strategies. Quantity of work for customer service is a random distributed value. Stationary distribution insensitivity with respect to functional form of distribution of work quantity for customer service is established.

2   Queueing Network Description

A closed queueing network with the set of systems $J = \{1, 2, \ldots, N\}$ is considered. M customers are circulating in the network. Non-active customers are located in the network systems in queues, being not serviced. There are input Poisson flows of signals with rates ν_i and φ_i, $i \in J$. When arriving at the system $i \in J$ the signal with rate ν_i induces an ordinary customer, if any, to become a non-active. When arriving at the system $i \in J$ the signal with rate φ_i induces an non-active customer, if any, to become an ordinary. Signals do not need service.

Let $n_i(t), n_i'(t)$ are numbers of ordinary and non-active customers at the system $i \in J$ at time t accordingly and $n_i''(t)$ – the number of service regime. Stochastic process $z(t) = ((n_i(t), n_i'(t), n_i''(t)), \ i \in J)$ is considered. Space of states for process $z(t)$ is $Z = \{(z = (n_1, n_1', n_1''), \ldots, (n_N, n_N', n_N''))| \ n_i, n_i' \geq 0, \sum_{i \in J}(n_i + n_i') = M, n_i'' = 0, \ldots, r_i, i \in J\}$.

Numbering of ordinary customers in the system queue is made from the "tail" of the queue to the device. Non-active customers in the queue of the system $i \in J$ are numbered as follows: a customer, which has become non-active in the last turn, has number n'_i. When arriving at the system $i \in J$ the signal with rate ν_i induces an ordinary customer with number 1 to become a non-active customer with number $n'_i + 1$. When arriving at the system $i \in J$ the signal with rate φ_i induces a non-active customer with number n'_i to become an ordinary customer with number 1. So, the set of customers numbers in the system $i \in J$ is $(1, \ldots, n'_i, 1, \ldots, n_i)$.

The discipline of service is LCFS-PR. When arriving at the system $i \in J$ a customer receives immediate service and gets number $n_i + 1$. Displaced customer keeps number n_i and becomes the first in the queue to finish its service. Customer service has not "temporal" but so-called "energetical" interpretation. Every service operation is characterized by the random variable of work to be performed. Quantities of work for customer service are independent random distributed values $\eta_i(n_i + n'_i)$ with functions of distribution $B_i(n_i + n'_i, z)$ ($B_i(n_i + n'_i, 0) = 0, i \in J$) and expected values $\tau_i(n_i + n'_i) < \infty$. The speed of customer service is $\alpha_i(n_i + n'_i, n''_i)$, $i \in J$. Here n_i, n'_i are numbers of ordinary and non-active customers in the system $i \in J$ accordingly, n''_i – the number of regime. After the service in the system $i \in J$ the customer passes to the system $j \in J$ with the probability $p_{i,j}$ ($\sum_{j=1}^{N} p_{i,j} = 1$). Let $p_{i,i} = 0$, $i \in J$.

Each system can operate at several regimes corresponding to different degrees of its efficiency. Time of regime switching is an exponentially distributed random value. Switching is possible only to neighboring regimes. System i has a single device, which can operate at $r_i + 1$ regimes. Denote 0 as basic service regime. The work time at the basic regime is an exponentially distributed random value with the rate $\sigma_i(n_i + n'_i, 0)$, then the device is switched to regime 1. For states (n_i, n'_i, n''_i), where $0 \leq n''_i \leq r_i - 1$, the work time at the regime n''_i is also an exponentially distributed random value, the device is switched to regime $n''_i + 1$ with the rate $\sigma_i(n_i + n'_i, n''_i)$ or to regime $n''_i - 1$ with the rate $\rho_i(n_i + n'_i, n''_i)$. The work time at the regime r_i is an exponentially distributed random value with the rate $\rho_i(n_i + n'_i, r_i)$, then the device is switched to regime $r_i - 1$. During switching the number of customers does not change.

The switching from the regime 0 to 1 can be interpreted as partial working capacity decline, so the speed of customer service decreases from the value $\alpha_i(n_i + n'_i, 0)$ to $\alpha_i(n_i + n'_i, 1)$. Analogically the transition from the regime n''_i to the regime $n''_i + 1$ means reduction of service speed. Transition from the regime n''_i to the regime $n''_i - 1$ means the recovery of working capacity, which was lost after switching from the regime $n''_i - 1$ to n''_i.

A traffic equations system is:

$$\varepsilon_i = \sum_{j=1}^{N} \varepsilon_j p_{j,i}, \quad i \in J. \tag{1}$$

It has been proved [4], that traffic equations system has the unique non-trivial solution up to constant.

3　Stationary Distribution Insensitivity

We consider a closed queueing network with multi-regime service strategies, quantities of work for customer service are independent random distributed values. In this instance $z(t)$ is not a Markov process. Denote by $\psi_{i,k}(t)$ – the remaining quantity of work for service of the customer, which has position k in the system i at time t, $\psi_i(t) = (\psi_{i,1}(t), \ldots, \psi_{i,n_i+n_i'}(t))$, $i \in J$.

$$\frac{d\psi_{i,n_i+n_i'}(t)}{dt} = -\alpha_i(n_i + n_i'), \ i \in J.$$

So we introduce into consideration Markov process $\zeta(t) = (z(t), \psi(t))$, where $\psi(t) = (\psi_1(t), \ldots, \psi_N(t))$.
Denote by

$$F(z, x) = F(z, x_{1,1}, \ldots, x_{1,n_1+n_1'}; x_{2,1}, \ldots, x_{2,n_2+n_2'}; \ldots; x_{N,1}, \ldots, x_{N,n_N+n_N'})$$

$$= \lim_{t \to \infty} P\{z(t) = z, \psi_{i,1}(t) < x_{i,1}, \ldots, \psi_{i,n_i+n_i'}(t) < x_{i,n_i+n_i'}, i \in J\}, \ z \in Z,$$

$$x_{k,l} \in \mathbb{R} \ \forall \ k = \overline{1, N}, \ l = \overline{1, n_k + n_k'}.$$

Functions $F(z, x)$ are called stationary functions of probabilities states distribution of the process $\zeta(t)$.

The model of closed queueing network with temporarily non-active customers has been considered in [11]. Quantity of work for customer service was a random distributed value. The following theorem has been proved.

Theorem 1. *Markov process $\zeta(t)$ is ergodic. Stationary functions of probabilities states distribution of the process $\zeta(t)$ are:*

$$F(z, x) = G^{-1}(M, N)p_1(n_1, n_1')p_2(n_2, n_2') \ldots p_N(n_N, n_N') \times$$

$$\times \prod_{i=1}^{N} \prod_{s=1}^{n_i+n_i'} \frac{1}{\tau_i(s)} \int_0^{x_{i,s}} (1 - B_i(s, u))du, \ z \in Z,$$

where

$$p_i(n_i, n_i') = \varepsilon_i^{n_i} \left(\frac{\varepsilon_i \nu_i}{\varphi_i}\right)^{n_i'} \prod_{s=1}^{n_i+n_i'} \frac{\tau_i(s)}{\alpha_i(s)},$$

ε_i is the traffic equations system solution. $G(M, N)$ is a normalizing constant.

For a closed queueing network with multi-regime service strategies the following theorem is true.

Theorem 2. *Markov process $\zeta(t)$ is ergodic. Under conditions*

$$\sigma_i(n_i + n_i', n_i'' - 1)\alpha_i(n_i + n_i', n_i'')\rho_i(n_i + n_i' - 1, n_i'') \tag{2}$$

$$= \sigma_i(n_i + n_i' - 1, n_i'' - 1)\alpha_i(n_i + n_i', n_i'' - 1)\rho_i(n_i + n_i', n_i''),$$

$$1 \le n_i + n_i' \le M, \quad 1 \le n_i'' \le r_i, \quad 1 \le i \le N,$$

stationary functions of probabilities states distribution of the process $\zeta(t)$ are:

$$F(z, x) = G^{-1}(M, N) p_1(n_1, n_1', n_1'') p_2(n_2, n_2', n_2'') \dots p_N(n_N, n_N', n_N'') \quad (3)$$

$$\times \prod_{i=1}^{N} \prod_{s=1}^{n_i+n_i'} \frac{1}{\tau_i(s)} \int_0^{x_{i,s}} (1 - B_i(s, u)) du, \ z \in Z,$$

where

$$p_i(n_i, n_i', n_i'') = \varepsilon_i^{n_i} \left(\frac{\varepsilon_i \nu_i}{\varphi_i} \right)^{n_i'} \prod_{s=1}^{n_i+n_i'} \frac{\tau_i(s)}{\alpha_i(s, n_i'')} \prod_{k=1}^{n_i''} \frac{\sigma_i(0, k-1)}{\rho_i(0, k)}, \quad (4)$$

ε_i *is the traffic equations system solution (1). $G(M, N)$ is a normalizing constant, which can be found from the following condition*

$$\sum_{((n_1, n_1', n_1''), \dots, (n_N, n_N', n_N'')) \in Z} p((n_1, n_1', n_1'), \dots, (n_N, n_N', n_N'')) = 1. \quad (5)$$

Proof. Denote by $e_i \in Z$ – the vector, which coordinates equal 0 with the exception of $(n_i, n_i', n_i'') = (1, 0, 0)$, denote by $e_i' \in Z$ – the vector, which coordinates equal 0 with the exception of $(n_i, n_i', n_i'') = (0, 1, 0)$, analogically denote by $e_i'' \in Z$ – the vector, which coordinates equal 0 with the exception of $(n_i, n_i', n_i'') = (0, 0, 1)$, $i \in J$.

We consider the process $\zeta(t)$. In the case of exponentially distributed service times the process $z(t)$ is ergodic by ergodic Markov theorem. The process $\zeta(t)$ is also ergodic, because $\zeta(t)$ is obtained from $z(t)$ by adding of continuous components.

The process $\zeta(t)$ can change its states due to incoming signals or regime switching. Such changes we call spontaneous changes.

Suppose that h is a small time interval and consider the probability

$$P\{z(t + h) = z, \psi_{i,1}(t + h) < x_{i,1}, \dots, \psi_{i,n_i+n_i'}(t + h) < x_{i,n_i+n_i'}, \ i \in J\}.$$

This event may occur in the following ways:

1. From the moment t during time h there were no spontaneous changes and service in any system was not over. The probability of this event is

$$P\{z(t) = z, \psi_{i,1}(t) < x_{i,1}, \dots, \alpha_i(n_i + n_i', n_i'') h I_{n_i > 0} \le \psi_{i,n_i+n_i'}(t)$$

$$< x_{i,n_i+n_i'} + \alpha_i(n_i + n_i', n_i'') h I_{n_i > 0}, \ i \in J\}$$

$$\times (1 - \sum_{i=1}^{N} (\nu_i I_{n_i > 0} + \varphi_i I_{n_i' > 0} + \sigma_i(n_i + n_i', n_i'')$$

$$+ \rho_i(n_i + n_i', n_i'')) h + o(h)).$$

2. During time h a customer has been serviced in the system $j \in J$ and has been routed to the system $i \in J$. There were no spontaneous changes.

$$P\{z(t) = z - e_i + e_j, \psi_{k,1}(t) < x_{k,1}, \ldots, \alpha_k(n_k + n'_k, n''_k)hI_{n_k > 0}$$

$$\leq \psi_{k,n_k+n'_k}(t) < x_{k,n_k+n'_k} + \alpha_k(n_k + n'_k, n''_k)hI_{n_k>0}, \; k \in J, \; k \neq i, \; k \neq j,$$

$$\psi_{j,1}(t) < x_{j,1}, \ldots, \psi_{j,n_j+n'_j}(t) < x_{j,n_j+n'_j}, \psi_{j,n_j+n'_j+1}(t)$$

$$< \alpha_j(n_j + n'_j + 1, n''_j)(h - \theta),$$

$$\psi_{i,1}(t) < x_{i,1}, \ldots, \alpha_i(n_i + n'_i - 1, n''_i)(h - \theta)I_{n_i>1} \leq \psi_{i,n_i+n'_i-1}(t)$$

$$< x_{i,n_i+n'_i-1} + \alpha_i(n_i + n'_i - 1, n''_i)(h - \theta)I_{n_i>1}\}$$

$$\times B_i(n_i + n'_i, x_{i,n_i+n'_i} + \alpha_i(n_i + n'_i, n''_i)\theta)p_{j,i}I_{n_i>0}.$$

3. During time h an informational signal with rate ν_i has arrived at the system $i \in J$. There were no other spontaneous changes. No customer was serviced.

$$P\{z(t) = z + e_i - e'_i, \psi_{k,1}(t) < x_{k,1}, \ldots, \alpha_k(n_k + n'_k, n''_k)hI_{n_k>0}$$

$$\leq \psi_{k,n_k+n'_k}(t) < x_{k,n_k+n'_k} + \alpha_k(n_k + n'_k, n''_k)hI_{n_k>0}, \; k \in J, \; k \neq i,$$

$$\psi_{i,1}(t) < x_{i,1}, \ldots, \alpha_i(n_i + n'_i, n''_i)h \leq \psi_{i,n_i+n'_i}(t)$$

$$< x_{i,n_i+n'_i} + \alpha_i(n_i + n'_i, n''_i)h\}(\nu_i h + o(h))I_{n'_i>0}.$$

4. During time h an informational signal with rate φ_i has arrived at the system $i \in J$. There were no other spontaneous changes. No customer was serviced.

$$P\{z(t) = z - e_i + e'_i, \psi_{k,1}(t) < x_{k,1}, \ldots, \alpha_k(n_k + n'_k, n''_k)hI_{n_k>0}$$

$$\leq \psi_{k,n_k+n'_k}(t) < x_{k,n_k+n'_k} + \alpha_k(n_k + n'_k, n''_k)hI_{n_k>0}, \; k \in J, \; k \neq i,$$

$$\psi_{i,1}(t) < x_{i,1}, \ldots, \alpha_i(n_i + n'_i, n''_i)hI_{n_i>1} \leq \psi_{i,n_i+n'_i}(t)$$

$$< x_{i,n_i+n'_i} + \alpha_i(n_i + n'_i, n''_i)hI_{n_i>1}\}(\varphi_i h + o(h))I_{n_i>0}.$$

5. During time h service regime of system i was increased by 1. There were no other spontaneous changes. No customer was serviced.

$$P\{z(t) = z - e''_i, \psi_{k,1}(t) < x_{k,1}, \ldots, \alpha_k(n_k + n'_k, n''_k)hI_{n_k>0} \leq \psi_{k,n_k+n'_k}(t)$$

$$< x_{k,n_k+n'_k} + \alpha_k(n_k + n'_k, n''_k)hI_{n_k>0}, \; k \in J, \; k \neq i,$$

$$\psi_{i,1}(t) < x_{i,1}, \ldots, \alpha_i(n_i + n'_i, n''_i - 1)(h - \theta) + \alpha_i(n_i + n'_i, n''_i)\theta \leq \psi_{i,n_i+n'_i}(t)$$

$$< x_{i,n_i+n'_i} + \alpha_i(n_i + n'_i, n''_i - 1)(h - \theta) + \alpha_i(n_i + n'_i, n''_i)\theta\}$$

$$\times (\sigma_i(n_i + n'_i, n''_i - 1)h + o(h)).$$

6. During time h service regime of system i was decreased by 1. There were no other spontaneous changes. No customer was serviced.

$$P\{z(t) = z + e_i'', \psi_{k,1}(t) < x_{k,1}, \ldots, \alpha_k(n_k + n_k', n_k'')hI_{n_k>0} \leq \psi_{k,n_k+n_k'}(t)$$

$$< x_{k,n_k+n_k'} + \alpha_k(n_k + n_k', n_k'')hI_{n_k>0}, \; k \in J, \; k \neq i,$$

$$\psi_{i,1}(t) < x_{i,1}, \ldots, \alpha_i(n_i + n_i', n_i'' + 1)(h - \theta) + \alpha_i(n_i + n_i', n_i'')\theta \leq \psi_{i,n_i+n_i'}(t)$$

$$< x_{i,n_i+n_i'} + \alpha_i(n_i + n_i', n_i'' + 1)(h - \theta) + \alpha_i(n_i + n_i', n_i'')\theta\}$$

$$\times (\rho_i(n_i + n_i', n_i'' + 1)h + o(h)).$$

Hereinbefore $0 < \theta < h$.

7. During time h there were more than two changes of queueing network condition. This probability is $o(h)$.

Therefore

$$P\{z(t + h) = z, \psi_{i,1}(t + h) < x_{i,1}, \ldots, \psi_{i,n_i+n_i'}(t + h) < x_{i,n_i+n_i'}, \; i \in J\}$$

$$= P\{z(t) = z, \psi_{i,1}(t) < x_{i,1}, \ldots, \alpha_i(n_i + n_i', n_i'')hI_{n_i>0} \leq \psi_{i,n_i+n_i'}(t)$$

$$< x_{i,n_i+n_i'} + \alpha_i(n_i + n_i', n_i'')hI_{n_i>0}, \; i \in J\}$$

$$\times (1 - \sum_{i=1}^{N}(\nu_i I_{n_i>0} + \varphi_i I_{n_i'>0} + \sigma_i(n_i + n_i', n_i''))$$

$$+ \rho_i(n_i + n_i', n_i''))h + o(h)) + \sum_{i=1}^{N}\sum_{j=1, j \neq i}^{N} P\{z(t) = z - e_i + e_j, \psi_{k,1}(t)$$

$$< x_{k,1}, \ldots, \alpha_k(n_k + n_k', n_k'')hI_{n_k>0}$$

$$\leq \psi_{k,n_k+n_k'}(t) < x_{k,n_k+n_k'} + \alpha_k(n_k + n_k', n_k'')hI_{n_k>0}, \; k \in J, \; k \neq i, \; k \neq j,$$

$$\psi_{j,1}(t) < x_{j,1}, \ldots, \psi_{j,n_j+n_j'}(t) < x_{j,n_j+n_j'}, \psi_{j,n_j+n_j'+1}(t)$$

$$< \alpha_j(n_j + n_j' + 1, n_j'')(h - \theta),$$

$$\psi_{i,1}(t) < x_{i,1}, \ldots, \alpha_i(n_i + n_i' - 1, n_i')(h - \theta)I_{n_i>1} \leq \psi_{i,n_i+n_i'-1}(t) \qquad (6)$$

$$< x_{i,n_i+n_i'-1} + \alpha_i(n_i + n_i' - 1, n_i'')(h - \theta)I_{n_i>1}\}$$

$$\times B_i(n_i + n_i', x_{i,n_i+n_i'} + \alpha_i(n_i + n_i', n_i'')\theta)p_{j,i}I_{n_i>0}$$

$$+ \sum_{i=1}^{N} P\{z(t) = z + e_i - e_i', \psi_{k,1}(t) < x_{k,1}, \ldots, \alpha_k(n_k + n_k', n_k'')$$

$$\times hI_{n_k>0} \leq \psi_{k,n_k+n_k'}(t) < x_{k,n_k+n_k'} + \alpha_k(n_k + n_k', n_k'')hI_{n_k>0},$$

$$k \in J, \; k \neq i, \; \psi_{i,1}(t) < x_{i,1}, \ldots, \alpha_i(n_i + n_i', n_i'')h \leq \psi_{i,n_i+n_i'}(t)$$

$$< x_{i,n_i+n_i'} + \alpha_i(n_i + n_i', n_i'')h\}(\nu_i h + o(h))I_{n_i'>0}$$

$$+ \sum_{i=1}^{N} P\{z(t) = z - e_i + e_i', \psi_{k,1}(t) < x_{k,1}, \ldots, \alpha_k(n_k + n_k', n_k'')hI_{n_k>0}$$

$$\leq \psi_{k,n_k+n'_k}(t) < x_{k,n_k+n'_k} + \alpha_k(n_k + n'_k, n''_k)hI_{n_k>0}, \ k \in J, \ k \neq i,$$

$$\psi_{i,1}(t) < x_{i,1}, \ldots, \alpha_i(n_i + n'_i, n''_i)hI_{n_i>1} \leq \psi_{i,n_i+n'_i}(t)$$

$$< x_{i,n_i+n'_i} + \alpha_i(n_i + n'_i, n''_i)hI_{n_i>1}\}(\varphi_i h + o(h))I_{n_i>0}$$

$$+\sum_{i=1}^{N} P\{z(t) = z - e''_i, \psi_{k,1}(t) < x_{k,1}, \ldots, \alpha_k(n_k + n'_k, n''_k)hI_{n_k>0}$$

$$\leq \psi_{k,n_k+n'_k}(t) < x_{k,n_k+n'_k} + \alpha_k(n_k + n'_k, n''_k)hI_{n_k>0}, \ k \in J, \ k \neq i,$$

$$\psi_{i,1}(t) < x_{i,1}, \ldots, \alpha_i(n_i + n'_i, n''_i - 1)(h - \theta) + \alpha_i(n_i + n'_i, n''_i)\theta \leq \psi_{i,n_i+n'_i}(t)$$

$$< x_{i,n_i+n'_i} + \alpha_i(n_i + n'_i, n''_i - 1)(h - \theta) + \alpha_i(n_i + n'_i, n''_i)\theta\}$$

$$\times(\sigma_i(n_i + n'_i, n''_i - 1)h + o(h)) +$$

$$+\sum_{i=1}^{N} P\{z(t) = z + e''_i, \psi_{k,1}(t) < x_{k,1}, \ldots, \alpha_k(n_k + n'_k, n''_k)hI_{n_k>0}$$

$$\leq \psi_{k,n_k+n'_k}(t) < x_{k,n_k+n'_k} + \alpha_k(n_k + n'_k, n''_k)hI_{n_k>0}, \ k \in J, \ k \neq i,$$

$$\psi_{i,1}(t) < x_{i,1}, \ldots, \alpha_i(n_i + n'_i, n''_i + 1)(h - \theta) + \alpha_i(n_i + n'_i, n''_i)\theta \leq \psi_{i,n_i+n'_i}(t)$$

$$< x_{i,n_i+n'_i} + \alpha_i(n_i + n'_i, n''_i + 1)(h - \theta) + \alpha_i(n_i + n'_i, n''_i)\theta\}$$

$$\times(\rho_i(n_i + n'_i, n''_i + 1)h + o(h))I_{n_i>0} + o(h).$$

Every probability from (6) may be expressed in terms of functions

$$F_t(z,x) = P\{z(t) = z, \psi_{i,1}(t) < x_{i,1}, \ldots, \psi_{i,n_i+n'_i}(t) < x_{i,n_i+n'_i}, \ i \in J\}.$$

Consider the decomposition of $F_t(z,x)$ in a Taylor series, taking into consideration that

$$P\{z(t) = z, \psi_{i,1}(t) < x_{i,1}, \ldots, \alpha_i(n_i + n'_i, n''_i)h \leq \psi_{i,n_i+n'_i}(t) < x_{i,n_i+n'_i}$$

$$+\alpha_i(n_i + n'_i, n''_i)h, \ i \in J\} = F_t(z, x_{i,1}, \ldots, x_{i,n_i+n'_i} + \alpha_i(n_i + n'_i, n''_i)h, \ i \in J)$$

$$-\sum_{k=1}^{N} F_t(z, x_{i,1}, \ldots, x_{i,n_i+n'_i} + \alpha_i(n_i + n'_i, n''_i)h, \ i \in J, i \neq k; x_{k,1}, \ldots, x_{k,n_k+n'_k-1},$$

$$+\alpha_k(n_k + n'_k, n''_k)h) + \ldots + F_t(z, x_{i,1}, \ldots, x_{i,n_i+n'_i-1}, \alpha_i(n_i + n'_i, n''_i)h, \ i \in J).$$

Therefore

$$P\{z(t) = z, \psi_{i,1}(t) < x_{i,1}, \ldots, \alpha_i(n_i + n'_i, n''_i)h \leq \psi_{i,n_i+n'_i}(t) < x_{i,n_i+n'_i}$$

$$+\alpha_i(n_i + n'_i, n''_i)h, i \in J\} = F_t(z, x_{i,1}, \ldots, x_{i,n_i+n'_i}, \ i \in J)$$

$$+\sum_{i=1}^{N} \frac{\partial F_t(z, x_{i,1}, \ldots, x_{i,n_i+n'_i}, \ i \in J)}{\partial x_{i,n_i+n'_i}} \alpha_i(n_i + n'_i, n''_i)h$$

$$-\sum_{i=1}^{N} \frac{\partial F_t(z, x_{l,1}, \ldots, x_{l,n_l+n'_l}, \ l \in J, l \neq i; x_{i,1}, \ldots, x_{i,n_i+n'_i-1}, 0)}{\partial x_{i,n_i+n'_i}}$$

$$\times \alpha_i(n_i + n_i', n_i'')h + o(h).$$

We consider $B_i(n_i + n_i', x_{i,n_i+n_i'} + \theta)$ as a function of the variable θ, use its decomposition in a Taylor series and let t tend to infinity. So we obtain the following differential equations system:

$$F(z,x) = F(z,x) + h \sum_{i=1}^{N} \alpha_i(n_i + n_i', n_i'') \left(\frac{\partial F(z,x)}{\partial x_{i,n_i+n_i'}} \right.$$

$$- \left(\frac{\partial F(z,x)}{\partial x_{i,n_i+n_i'}} \right)_{x_{i,n_i+n_i'}=0} \right) I_{n_i>0} - \left(\sum_{i=1}^{N} \left(\nu_i I_{n_i>0} + \varphi_i I_{n_i'>0} \right. \right.$$

$$\left. + \sigma_i(n_i + n_i', n_i'') + \rho_i(n_i + n_i', n_i'') \right) h + o(h) \bigg) F(z,x)$$

$$+ h \sum_{j=1}^{N} \sum_{i=1,i\neq j}^{N} \alpha_j(n_j + n_j' + 1, n_j'')p_{j,i}B_i(n_i + n_i', x_{i,n_i+n_i'}) \tag{7}$$

$$\times \left(\frac{\partial F(z + e_j - e_i, x)}{\partial x_{j,n_j+n_j'+1}} \right)_{x_{j,n_j+n_j'+1}=0} I_{n_i>0} + \sum_{i=1}^{N} F(z + e_i - e_i', x)(\nu_i h + o(h))I_{n_i'>0}$$

$$+ \sum_{i=1}^{N} F(z - e_i + e_i', x)(\varphi_i h + o(h))I_{n_i>0} + \sum_{i=1}^{N} F(z - e_i'', x)(\sigma_i(n_i + n_i', n_i'' - 1)h + o(h))$$

$$+ \sum_{i=1}^{N} F(z + e_i'', x)(\rho_i(n_i + n_i', n_i'' + 1)h + o(h)) + o(h).$$

Subtracting $F(z,x)$ from both sides of (7), dividing both sides of (7) by h and letting h tend to zero, we obtain the following differential equations system:

$$F(z,x) \sum_{i=1}^{N} \left(\nu_i I_{n_i>0} + \varphi_i I_{n_i'>0} + \sigma_i(n_i + n_i', n_i'') + \rho_i(n_i + n_i', n_i'') \right)$$

$$= \sum_{i=1}^{N} \alpha_i(n_i + n_i') \left(\frac{\partial F(z,x)}{\partial x_{i,n_i+n_i'}} \left(\frac{\partial F(z,x)}{\partial x_{i,n_i+n_i'}} \right)_{x_{i,n_i+n_i'}=0} \right) I_{n_i>0}$$

$$+ \sum_{j=1}^{N} \sum_{i=1,i\neq j}^{N} \alpha_j(n_j + n_j' + 1)p_{j,i}B_i(n_i + n_i', x_{i,n_i+n_i'})$$

$$\times \left(\frac{\partial F(z + e_j - e_i, x)}{\partial x_{j,n_j+n_j'+1}} \right)_{x_{j,n_j+n_j'+1}=0} I_{n_i>0} \tag{8}$$

$$+ \sum_{i=1}^{N} F(z + e_i - e_i', x)\nu_i I_{n_i'>0} + \sum_{i=1}^{N} F(z - e_i + e_i', x)\varphi_i I_{n_i>0}$$

$$+ \sum_{i=1}^{N} F(z - e_i'', x)\sigma_i(n_i + n_i', n_i'' - 1) + \sum_{i=1}^{N} F(z + e_i'', x)\rho_i(n_i + n_i', n_i'' + 1).$$

Divide (8) into the next local balance equations:

$$F(z,x)\big(\nu_i I_{n_i>0} + \varphi_i I_{n'_i>0}\big) = F(z+e_i-e'_i,x)\nu_i I_{n'_i>0} + F(z-e_i+e'_i,x)\varphi_i I_{n_i>0},$$
(9)

$$\alpha_i(n_i + n'_i)\left(\left(\frac{\partial F(z,x)}{\partial x_{i,n_i+n'_i}}\right)_{x_{i,n_i+n'_i}=0} - \frac{\partial F(z,x)}{\partial x_{i,n_i+n'_i}}\right)I_{n_i>0}$$

$$= \sum_{j=1,j\neq i}^{N} \alpha_j(n_j + n'_j + 1)p_{j,i}B_i(n_i + n'_i, x_{i,n_i+n'_i})$$
(10)

$$\times\left(\frac{\partial F(z+e_j-e_i,x)}{\partial x_{j,n_j+n'_j+1}}\right)_{x_{j,n_j+n'_j+1}=0} I_{n_i>0},\ i \in J.$$

$$F(z,x)\big(\sigma_i(n_i+n'_i,n''_i)I_{n_i>0} + \rho_i(n_i+n'_i,n''_i)\big) = F(z-e''_i,x)\sigma_i(n_i+n'_i,n''_i-1)I_{n_i>0} +$$
$$+ F(z+e''_i,x)\rho_i(n_i+n'_i,n''_i+1)I_{n_i>0}.$$
(11)

Substituting $F(z,x)$, determined by means of (3), (4), into local balance Eqs. (9), (10) and (11), considering (2) and traffic equation system (1), we obtain identity. □

Denote by $\{p(z),\ z \in Z\}$ – stationary distribution of the process $z(t)$. From the foregoing theorem, considering equality $p(z) = F(z,+\infty)$, we obtain

Corollary 1. *Process $z(t)$ is ergodic. Under conditions (2) $z(t)$ has stationary distribution*

$$p(z) = G^{-1}(M,N)p_1(n_1,n'_1,n''_1)p_2(n_2,n'_2,n''_2)\ldots p_N(n_N,n'_N,n''_N),\ z \in Z,$$

which does not depend on functional form of $B_i(s,x)$, $i \in J$. Probabilities $p_i(n_i,n'_i,n''_i)$, $i \in J$, may be found by means of (4).

4   Conclusion

We have considered stationary functioning of a closed queueing network with temporarily non-active customers and multi-regime service strategies. Expression for stationary distribution has been derived. Finally, stationary distribution insensitivity with respect to functional form of distribution of work quantity for customer service is established. Research results have practical importance and may be used for real networks investigation.

References

1. Tsitsiashvili, G.S., Osipova, M.: Distributions in Stochastic Network Models. Nova Publishers, Inc(US) (2008)
2. Tsitsiashvili, G.S., Osipova, M.: Queueing models with different schemes of customers transformations. In: 19th International Conference "Mathematical Methods for Increasing Efficiency of Information Telecommunication Networks", pp. 128–133. BSU, Minsk (2007)

3. Jackson, J.R.: Network of waiting lines. Oper. Res. **4**, 518–521 (1957)
4. Gordon, W.J., Newell, G.F.: Closed queueing networks with exponential servers. Oper. Res. **15**, 252–267 (1967)
5. Sevastyanov, B.A.: An ergodic theorem for Markov processes and its application to telephone systems with refusals. Theo. Probab. Appl. **2**, 104–112 (1957)
6. Baskett, F.: Open, closed and mixed networks of queues with different classes of customers. J. Assoc. Comput. Mach. **22**, 248–260 (1975)
7. Boyarovich, Y.S.: The stationary distribution invariance of states in a closed queueing network with temporarily non-active customers. Autom. Remote Control **73**, 1616–1623 (2012)
8. Bojarovich, J., Malinkovsky, Y.V.: Stationary distribution invariance of an open queueing network with temporarily non-active customers. J. Control Comput. Sci. **20**, 62–70 (2012)
9. Bojarovich, J., Malinkovsky, Y.: Stationary distribution invariance of an open queueing network with temporarily non-active customers. In: Dudin, A., Klimenok, V., Tsarenkov, G., Dudin, S. (eds.) BWWQT 2013. CCIS, vol. 356, pp. 26–32. Springer, Heidelberg (2013)
10. Ivnitsky, V.A.: Theory of Queueing Networks. Fizmatlit, Moscow (2004)
11. Bojarovich, J., Dudovskaya, Y.: Stationary distribution insensitivity of a closed queueing network with non-active customers. In: Dudin, A., Nazarov, A., Yakupov, R., Gortsev, A. (eds.) ITMM 2014. CCIS, vol. 487, pp. 50–58. Springer, Heidelberg (2014)

Analysis of Dynamic and Adaptive Retrial Queue System with the Incoming MAP-Flow of Requests

Tatyana Lyubina$^{(\boxtimes)}$ and Yana Bublik

Branch of Kemerovo State University in Anzhero-Sudzhensk,
Anzhero-Sudzhensk, Russia
lyubina_tv@mail.ru

Abstract. In the paper, the dynamic and adaptive RQ-systems with incoming MAP-flow of requests are investigated with the method of asymptotic analysis. It is shown that the dynamic and the adaptive RQ-systems are asymptotically equivalent. The results obtained by investigating the dynamic RQ-systems can be used to determine the probability distribution of customers in the orbit of the adaptive RQ-systems.

Keywords: Retrial queue system · Highly-loaded RQ-system · Throughput capacity

1 Introduction

Retrial Queue Systems (RQ-systems) are relevant to describe telephone networks, local area networks with random multiple access protocols, broadcast and cellular radio networks, technological and transport systems and others. RQ-systems have been investigated by J.R. Artalejo [1], B.D. Choi, G.I. Falin [2], I.I. Khomichkov, A.N. Dudin [3], A.A. Nazarov [4–6], V.I. Klimenok [7]. There is a difference between RQ-systems and classical queueing systems. In RQ-systems, the requests entering the system and finding the service device busy, do not leave the system but join the orbit to retry to occupy the service device later [8–10].

Static random access protocols have been analyzed by A.N. Tuenbayeva, N.M. Yurevich, A.A. Nazarov and others. These investigations show that static RQ-systems with conflicts and announcing are not steady even under the infinitely small load.

To solve the problems of information loss and stabilization of RQ-systems the modifications of random multiple access protocols are given:

- dynamic access protocols for the stationary Poisson flow are considered by I.I. Khomichkov, Y.D. Odyshev, S.L. Shokhor and others;
- adaptive access protocols for the systems with the stationary Poisson flow are studied by R.L. Rivest, V.A. Mikhaylov, D.Y. Kuznetsov [11].

© Springer International Publishing Switzerland 2015
A. Dudin et al. (Eds.): ITMM 2015, CCIS 564, pp. 384–392, 2015.
DOI: 10.1007/978-3-319-25861-4_32

Models of the stationary Poisson [12] in RQ-systems are not relevant for real telecommunication streams and so, it causes the task to investigate RQ-systems with correlated incoming flow (for example, MMPP, MAP) [6].

That is why in this paper we study dynamic and adaptive RQ-systems with incoming MAP-flow (Markov Arrival Process).

2 Mathematical Model of Dynamic RQ-System

In this paper under the term RQ-system we mean the queue system with the orbit and incoming MAP-flow of requests controlled by the dynamic access protocol [13].

MAP-flow of requests comes to the system input from the external source. This inflow is defined by matrix $\mathbf{Q}$ of infinitesimal characteristics q_{vn} of Markov chain $n(t)$. The chain controls MAP-flow. Also we give the set of non-negative numbers $\rho\lambda_n$ and probabilities $d_{nn} = 0$ defined by matrix $\mathbf{D} = [d_{vn}]$ and scalar matrix $\rho\mathbf{\Lambda}$ of conditional densities $\rho\lambda_n$ on the main diagonal.

If the service device is free at the time of a request arrival the request occupies it to be served for some random amount of time arranged according to the exponential law with parameter μ. Having been successfully served the request leaves the service device. If at the time of a request being served one more request arrives this new request joins the orbit.

From the orbit after some random delay the request with the dynamic intensity γ/i that depends on the state of the orbit retries for service; i – the number of arrivals in the orbit. If the service device is free the request is served, if the service device is occupied the request comes back to the orbit [14].

The system state at the time t is defined by a three-dimensional Markov chain $\{k(t), n(t), i(t)\}$, where $i(t)$ – the number of requests in the orbit, $n(t)$ – the values of a Markov chain, managing MMPP flow, and $k(t)$ defines the service device state as follows: $k(t) = 0$ if the device is free, and $k(t) = 1$ if the device is servicing the request.

Let us denote $P\{k(t) = k, n(t) = n, i(t) = i\} = P(k, n, i, t)$ as the probability of the device state k at the time t, the state of a Markov chain n and the number of requests in the orbit – i. So, the probabilities distribution $P(k, , i, t)$ satisfy the following Kolmogorov differential equation system for probabilities distribution $P(k, n, i, t)$

$$
\begin{cases}
\frac{\partial P(0,n,i,t)}{\partial t} = -(\rho\lambda_n + \gamma)P(0, n, i, t) + \mu P(1, n, i, t) \\
+ \sum_{v} \{P(0, v, i, t)(1 - d_{vn})\} q_{vn}, \\
\frac{\partial P(1,n,i,t)}{\partial t} = -(\rho\lambda_n + \mu)P(1, n, i, t) + \gamma P(0, n, i + 1, t) + \rho\lambda_n P(0, n, i, t) \\
+ \rho\lambda_n P(0, n, i - 1, t) \\
+ \sum_{v} \{P(1, v, i, t)(1 - d_{vn}) + P(0, v, i, t)d_{vn} + P(1, v, i - 1, t)d_{vn}\} q_{vn}.
\end{cases}
\tag{1}
$$

The solution of Kolmogorov simultaneous Eq. (1) determines completely the functioning of dynamic RQ-system with incoming MAP-flow. Prelimit investigation will be performed with the method of generating functions.

3   Research of Dynamic RQ-System with the Method of Generating Functions

We consider the system is functioning in steady regime, i.e. $P(k,n,i,t) \equiv P(k,n,i)$ Let us present system (1) for stationary distribution matrix form. Denoting row-vectors

$$\mathbf{P}(k,i) = \{P(k,1,i), P(k,2,i), ..., P(k,N,i)\},$$

we get $\mathbf{P}(0,0)(\mathbf{Q} - \rho\mathbf{\Lambda}) + \mathbf{P}(1,0)\mu - \mathbf{P}(0,0)\mathbf{A} = 0, \; i = 0,$

$$\mathbf{P}(1,0)(\mathbf{Q} - \rho\mathbf{\Lambda} - \mu\mathbf{I}) + \mathbf{P}(0,0)\rho\mathbf{\Lambda} - \mathbf{P}(1,0)\mathbf{A} + \mathbf{P}(0,0)\mathbf{A} + \mathbf{P}(0,1)\gamma = 0, \quad i = 0,$$

$$\mathbf{P}(0,i)(\mathbf{Q} - \rho\mathbf{\Lambda} - \gamma\mathbf{I}) + \mathbf{P}(1,i)\mu - \mathbf{P}(0,i)\mathbf{A} = 0, \quad i \geq 1, \qquad (2)$$

$$\begin{matrix}\mathbf{P}(1,i)(\mathbf{Q} - \rho\mathbf{\Lambda} - \mu\mathbf{I}) + \mathbf{P}(0,i)\rho\mathbf{\Lambda} + \mathbf{P}(1,i-1)\rho\mathbf{\Lambda} \\ +\mathbf{P}(0,i+1)\gamma - \mathbf{P}(1,i)\mathbf{A} + \mathbf{P}(0,i)\mathbf{A} + \mathbf{P}(1,i-1)\mathbf{A} = 0\end{matrix}, \quad i \geq 1,$$

where matrix $\mathbf{A} = \mathbf{D} * \mathbf{Q}$, is Hadamard product of two matrixes $\mathbf{D}$ and $\mathbf{Q}$.

To solve system (2) we define vector generating functions

$$\mathbf{G}(k,x) = \sum_{i=0}^{\infty} x^i \mathbf{P}(k,i), \quad k = 0,\, 1\,. \qquad (3)$$

Taking into account Eq. (3) we get from system (2) the following system for functions $\mathbf{G}(k,x)$:

$$\begin{cases} \mathbf{G}(0,x)(\mathbf{Q} - \rho\mathbf{\Lambda} - \gamma\mathbf{I} - \mathbf{A}) + \mathbf{G}(1,x)\mu = -\gamma\mathbf{P}(0,0), \\ \mathbf{G}(0,x)((\rho\mathbf{\Lambda} + \mathbf{A})x + \gamma\mathbf{I}) + \mathbf{G}(1,x)(\mathbf{Q} + (x-1)(\rho\mathbf{\Lambda} + \mathbf{A}) - \mu\mathbf{I})x = \gamma\mathbf{P}(0,0)\,. \end{cases} \qquad (4)$$

From system (4) we get expressions for $\mathbf{G}(0,x)$ $\mathbf{G}(1,x)$:

$$\mathbf{G}(0,x) = \mathbf{P}(0,0)\left\{\gamma\mathbf{I} + \frac{\gamma}{\mu}x\left(\mathbf{Q} + (x-1)(\rho\mathbf{\Lambda} + \mathbf{A}) - \mu\mathbf{I}\right)\right\}\left\{(1-x)\gamma\mathbf{I} + x\mathbf{Q}\right.$$

$$\left. -\frac{1}{\mu}(\mathbf{Q} - \rho\mathbf{\Lambda} - \mathbf{A} - \gamma\mathbf{I})(\mathbf{Q} + (x-1)(\rho\mathbf{\Lambda} + \mathbf{A}))x\right\}^{-1}, \qquad (5)$$

$$\mathbf{G}(1,x) = -\frac{1}{\mu}\left[\gamma\mathbf{P}(0,0) + \mathbf{G}(0,x)(\mathbf{Q} - \rho\mathbf{\Lambda} - \mathbf{A} - \gamma\mathbf{I})\right]\,.$$

Let us denote matrixes

$$(x) = \gamma\mathbf{I} + \frac{\gamma}{\mu}x\left(\mathbf{Q} + (x-1)(\rho\mathbf{\Lambda} + \mathbf{A}) - \mu\mathbf{I}\right),$$

$$\mathbf{B}(x) = (1-x)\gamma\mathbf{I} + x\mathbf{Q} - \frac{x}{\mu}(\mathbf{Q} - \rho\mathbf{\Lambda} - \mathbf{A} - \gamma\mathbf{I})(x-1)(\mathbf{Q} + \rho\mathbf{\Lambda} + \mathbf{A}),$$

then Eq. (5) we rewrite as $\mathbf{G}(0,x) = \mathbf{P}(0,0)\mathbf{A}(x)\mathbf{B}^{-1}(x)$.

Generating function $\mathbf{G}(0,x)$ is defined for all the function values $x \in [0,1]$, but matrix $\mathbf{B}(x)$ is confluent with $x = x_\nu$, where x_ν are roots of the equation $|\mathbf{B}(x)| = 0$ in the investigated interval $[0, 1]$.

We write the reciprocal matrix $\mathbf{B}^{-1}(x)$ as $\mathbf{B}^{-1}(x) = \frac{1}{|\mathbf{B}(x)|}\mathbf{D}^T(x)$, where elements $D(x)_{n_1 n_2}$ of matrix $\mathbf{D}(x)$ are cofactors for elements $B(x)_{n_1 n_2}$ of matrix $\mathbf{B}(x)$.

It follows from zero equation of determinant $|\mathbf{B}(x_\nu)| = 0$ that components of vector $\mathbf{P}(0,0)$ satisfy homogeneous linear equation system

$$\mathbf{P}(0,0)\mathbf{A}(x_\nu)\mathbf{B}^T(x_\nu) = 0.$$

This system defines components values of vector $\mathbf{P}(0,0)$ within the accuracy of multiplicative invariable with the values determined by the normalization requirement. So, we managed to find expressions (5) for generating functions $\mathbf{G}(k,x)$.

4   Research of Dynamic RQ-System with the Method of Asymptotic Analysis

Defining explicit expression (5) for generating function in mathematical models of RQ-systems is an unordinary situation. That is why it is necessary to devise other methods of analysis for such models. The method of asymptotic analysis is the most productive [15]. We shall devise it for our model. It will let us display the efficiency of this method by comparing asymptotic results with prelimit ones. Also it will let us compare them with asymptotic results we got for adaptive RQ-system.

We modify system (4) as follows:

$$\begin{cases} \mathbf{H}(0,u)(\mathbf{Q} - \rho\mathbf{\Lambda} - \mathbf{A} - \gamma\mathbf{I}) + \mathbf{H}(1,u)\mu = -\gamma\mathbf{P}(0,0), \\ \mathbf{H}(0,u)(\rho\mathbf{\Lambda} + \mathbf{A} + e^{-ju}\gamma\mathbf{I}) \\ + \mathbf{H}(1,u)(\mathbf{Q} + (e^{ju} - 1)(\rho\mathbf{\Lambda} + \mathbf{A}) - \mu\mathbf{I}) = e^{-ju}\gamma\mathbf{P}(0,0)\,, \end{cases} \tag{6}$$

where ρ – parameter for defining the limit heavy load condition for RQ-system, whereas function $\mathbf{H}(k,u) = \mathbf{G}(k, e^{ju}) = \mathbf{G}(k,x)$.

System (6) will be solved with the method of asymptotic analysis under a heavy load $\rho \uparrow S_1$ [16], where S_1 is throughput capacity of RQ-system. Denoting $\varepsilon = S_1 - \rho$ we consider $\varepsilon \to 0$. And we solve system (6) under this condition. In system (6) we substitute

$$\rho = S_1 - \varepsilon, \quad u = \varepsilon w, \quad \mathbf{H}_k(u) = \mathbf{F}_k(w,\varepsilon), \quad \mathbf{P}(0,0) = \varepsilon\mathbf{\Pi}(\varepsilon).$$

And rewrite

$$\begin{cases} \mathbf{F}_0(w,\varepsilon)(\mathbf{Q} - (S_1 - \varepsilon)\mathbf{\Lambda} - \mathbf{A} - \gamma\mathbf{I}) + \mathbf{F}_1(w,\varepsilon)\mu = -\gamma\varepsilon\mathbf{\Pi}(\varepsilon), \\ \mathbf{F}_0(w,\varepsilon)\left((S_1 - \varepsilon)\mathbf{\Lambda} + e^{-j\varepsilon w}\gamma\mathbf{I} + \mathbf{A}\right) \\ + \mathbf{F}_1(w,\varepsilon)\left(\mathbf{Q} + (e^{j\varepsilon w} - 1)(S_1 - \varepsilon)\mathbf{\Lambda} - \mu\mathbf{I} - \mathbf{A}(1 - e^{j\varepsilon w})\right) = e^{-j\varepsilon w}\gamma\varepsilon\mathbf{\Pi}(\varepsilon)\,. \end{cases} \tag{7}$$

Theorem 1. *The S_1-value of the throughput capacity of the dynamic RQ-system with incoming MAP-flow equals the value of the equation root*

$$\gamma R_0\mathbf{E} - R_1(S_1\mathbf{\Lambda} + \mathbf{A})\mathbf{E} = 0\,, \tag{8}$$

where row-vector $\mathbf{R}_k$ is joint probability distribution of the service device state and the values of Markov chain managing the incoming MAP-flow. It is determined by the equations

$$\begin{aligned}
\mathbf{R}_0\,(S_1) &= \mu\mathbf{R}\left\{(\mu+\gamma)\,\mathbf{I}+S_1\mathbf{\Lambda}+\mathbf{A}-\mathbf{Q}\right\}^{-1}\,, \\
\mathbf{R}_1\,(S_1) &= \mathbf{R}\left\{\mathbf{I}-\mu\left[(\mu+\gamma)\,\mathbf{I}+S_1\mathbf{\Lambda}+\mathbf{A}-\mathbf{Q}\right]^{-1}\right\}\,.
\end{aligned} \tag{9}$$

Proof. There are two stages of proving.

Stage 1. We denote $\lim\limits_{\varepsilon\to 0}\mathbf{F}_k(w,\varepsilon)=\mathbf{F}_k(w)$. Fulfilling this limiting transition we obtain the system

$$\begin{cases}
\mathbf{F}_0\,(w)\,(\mathbf{Q}-S_1\mathbf{\Lambda}-\mathbf{A}-\gamma\mathbf{I})+\mathbf{F}_1\,(w)\,\mu=0\,, \\
\mathbf{F}_0\,(w)\,(S_1\mathbf{\Lambda}+\mathbf{A}+\gamma\mathbf{I})+\mathbf{F}_1\,(w)\,(\mathbf{Q}-\mu\mathbf{I})=0\,.
\end{cases} \tag{10}$$

We solve $\mathbf{F}_k(w)$ system as follows

$$\mathbf{F}_k(w)=\mathbf{R}_k\Phi(w)\,, \tag{11}$$

where function $\Phi(w)$ at infinity is equal to zero, and $\mathbf{R}_k$ is probability distribution of the device state determined by the system

$$\begin{cases}
\mathbf{R}_0\,(\mathbf{Q}-S_1\mathbf{\Lambda}-\mathbf{A}-\gamma\mathbf{I})+\mathbf{R}_1\mu=0\,, \\
\mathbf{R}_0\,(S_1\mathbf{\Lambda}+\mathbf{A}+\gamma\mathbf{I})+\mathbf{R}_1\,(\mathbf{Q}-\mu\mathbf{I})=0\,.
\end{cases} \tag{12}$$

It is easy to show that $(\mathbf{R}_0+\mathbf{R}_1)\,\mathbf{Q}=0$, i.e. $\mathbf{RQ}=0$, where $\mathbf{R}=\mathbf{R}_0+\mathbf{R}_1$ and it satisfies normalization requirement $\mathbf{RE}=1$. Then $\mathbf{R}_0$ and $\mathbf{R}_1$ are dependent on S_1 and are determined by equations

$$\begin{aligned}
\mathbf{R}_0\,(S_1) &= \mu\mathbf{R}\left\{(\mu+\gamma)\,\mathbf{I}+S_1\mathbf{\Lambda}+\mathbf{A}-\mathbf{Q}\right\}^{-1}\,, \\
\mathbf{R}_1\,(S_1) &= \mathbf{R}\left\{\mathbf{I}-\mu\left[(\mu+\gamma)\,\mathbf{I}+S_1\mathbf{\Lambda}+\mathbf{A}-\mathbf{Q}\right]^{-1}\right\}\,,
\end{aligned}$$

coinciding with (9).

Stage 2. Having rewritten (7) as follows

$$\begin{cases}
\mathbf{F}_0\,(w,\varepsilon)\,(\mathbf{Q}-S_1\mathbf{\Lambda}-\mathbf{A}-\gamma\mathbf{I}+\varepsilon\mathbf{\Lambda})+\mathbf{F}_1\,(w,\varepsilon)\,\mu=-\gamma\varepsilon\mathbf{\Pi}(\varepsilon)+O\left(\varepsilon^2\right)\,, \\
\mathbf{F}_0\,(w,\varepsilon)\,(S_1\mathbf{\Lambda}+\gamma\mathbf{I}-\varepsilon(\mathbf{\Lambda}+jw\gamma\mathbf{I})+\mathbf{A}) \\
\quad+\mathbf{F}_1\,(w,\varepsilon)\,(\mathbf{Q}-\mu\mathbf{I}+j\varepsilon w(S_1\mathbf{\Lambda}+\mathbf{A}))=\gamma\varepsilon\mathbf{\Pi}(\varepsilon)+O\left(\varepsilon^2\right)\,,
\end{cases}$$

we sum up all the equations of the system according to k and n and obtain the equation

$$\mathbf{F}_0\,(w,\varepsilon)\,\varepsilon jw\gamma\mathbf{E}-\mathbf{F}_1\,(w,\varepsilon)\,j\varepsilon w(S_1\mathbf{\Lambda}+\mathbf{A})\mathbf{E}=0\,,$$

that lets us get

$$\mathbf{F}_0\,(w,\varepsilon)\,\gamma\mathbf{E}-\mathbf{F}_1\,(w,\varepsilon)\,(S_1\mathbf{\Lambda}+\mathbf{A})\mathbf{E}=0\,.$$

Using (11) we get the equation

$$\mathbf{R}_0\Phi(w)\gamma\mathbf{E}-\mathbf{R}_1\Phi(w)(S_1\mathbf{\Lambda}+\mathbf{A})\mathbf{E}=0\,,$$

that causes the expression

$$\gamma \mathbf{R}_0 \mathbf{E} - \mathbf{R}_1 (S_1 \mathbf{\Lambda} + \mathbf{A}) \mathbf{E} = 0 \,,$$

which coincides with (8) and determines the value of the throughput capacity S of the dynamic RQ-system.

5   Mathematical Model of Adaptive RQ-System

Let us consider unilinear queue system with orbit and incoming MAP-flow managed by the adaptive access protocol. We name this system adaptive RQ-system with incoming MAP-flow.

MAP-flow comes to the system input. This inflow is defined by matrix $\mathbf{Q}$ of infinitesimal characteristics q_{vn} of Markov chain $n(t)$. The chain controls MAP-flow. Also we give the set of non-negative numbers $\rho \lambda_n$ and probabilities $d_{nn} = 0$ defined by matrix $\mathbf{D} = [d_{vn}]$ and scalar matrix $\rho \mathbf{\Lambda}$ of conditional densities $\rho \lambda_n$ on the main diagonal.

If the service device is free at the time of a request arrival the request occupies it to be served for some random amount of time arranged according to the exponential law with parameter μ. Having been successfully served the request leaves the service device. If at the time of a request being served one more request arrives this new request joins the orbit. From the orbit after some random delay the request with the dynamic intensity $1/T$, where T – the adapter condition at the current time we shall determine later, retries for service. If the service device is free the request is served, if the service device is occupied the request comes back to the orbit.

The system state at the time t is defined by a four-dimensional Markov process $\{k(t),\ n(t),\ i(t),\ T(t)\}$, where $k(t)$ determines the device state as follows: $k(t) = 0$ if the device is free, and $k(t) = 1$ if the device is servicing the request. The values of a Markov chain, managing MMPP flow equal $n(t)$, whereas $i(t)$ – the number of requests in the orbit. The adapter during the time t changes its states $T(t)$ as follows: $T(t + \Delta t) = T(t) - \alpha \Delta t$, if $k(t) = 0$, and $T(t + \Delta t) = T(t) + \beta \Delta t$, if $k(t) = 1$, where $\alpha > 0$, $\beta > 0$ – the adapter parameters which values are given.

This adaptive system was investigated with the method of asymptotic analysis under a heavy load condition. We defined the throughput capacity S of an adaptive RQ-system as a supremum of those ρ-values for which there exist stationary function regimes of an adaptive RQ-system model, and considering the asymptotic condition $\rho \uparrow S$ fulfilled. The investigations caused the following theorem.

Theorem 2. *The S_2-value of the throughput capacity of the adaptive RQ-system with incoming MAP-flow of arrivals is determined by the system of equations*

$$\begin{cases} \gamma \mathbf{R}_0 \mathbf{E} - \mathbf{R}_1 (S_2 \mathbf{\Lambda} + \mathbf{A}) \mathbf{E} = 0 \,, \\ \alpha \mathbf{R}_0 \mathbf{E} - \beta \mathbf{R}_1 \mathbf{E} = 0 \,, \end{cases} \tag{13}$$

where $\alpha, \beta-$ adapter parameters, which values are given, $\gamma-$ some positive constant determined also by the system, $\mathbf{R}_k-$ probability distribution of the device states defined by the equations

$$\mathbf{R}_0\left(S_2, \gamma\right) = \mu \mathbf{R}\left\{(\mu+\gamma)\mathbf{I} + S_2\mathbf{\Lambda} + \mathbf{A} - \mathbf{Q}\right\}^{-1},$$
$$\mathbf{R}_1\left(S_2, \gamma\right) = \mathbf{R}\left\{\mathbf{I} - \mu\left[(\mu+\gamma)\mathbf{I} + S_2\mathbf{\Lambda} - \mathbf{Q}\right]^{-1}\right\}. \tag{14}$$

The first equation of system (13) for the defining the throughput capacity of the adaptive RQ-system and equation (8) for defining the throughput capacity of the dynamic RQ-system make us conclude that the throughput capacity of the adaptive RQ-system S_2 is equal to the throughput capacity of the dynamic RQ-system S_1, i.e. $S_1 = S_2$.

Equation (14) for the defining probability distribution of the device states of the adaptive RQ-system coincide with Eq. (9) for the defining probability distribution of the device states of the dynamic RQ-system.

6   Numerical Analysis of the Dynamic RQ-System

We write the vector characteristic function $\mathbf{H}(u)$ for probability distribution $\mathbf{P}(i) = \mathbf{P}(0, i) + \mathbf{P}(1, i)$ of the arrivals' number in orbit as

$$\mathbf{H}(u) = \mathbf{H}(0, u) + \mathbf{H}(1, u) = \mathbf{H}(0, u)\left(\mathbf{I} - \frac{1}{\mu}(\mathbf{Q} - \rho\mathbf{\Lambda} - \mathbf{A} - \gamma\mathbf{I})\right) - \frac{\gamma}{\mu}\mathbf{P}(0, 0).$$

Then the probability distribution $p(i) = \mathbf{P}(i)\mathbf{E}$ of the arrivals' number in the orbit is determined by the inverse Fourier transform from the scalar characteristic function $h(u) = Me^{jui(t)} = \sum_i e^{jui}p(i) = \mathbf{H}(u)\mathbf{E}$:

$$p(i) = \frac{1}{2\pi}\int\limits_{-\pi}^{\pi} e^{-jui}h(u)du. \tag{15}$$

For the given parameters values $\mu = 1$, $\gamma = 3$ and matrixes

$$\mathbf{Q} = \begin{pmatrix} -0.7 & 0.4 & 0.3 \\ 0.1 & -0.4 & 0.3 \\ 0.4 & 0.5 & -0.9 \end{pmatrix}, \quad \mathbf{\Lambda} = \begin{pmatrix} 1 & 0 & 0 \\ 0 & 2 & 0 \\ 0 & 0 & 4 \end{pmatrix},$$

$$\mathbf{E} = \begin{pmatrix} 1 \\ 1 \\ 1 \end{pmatrix}, \quad \mathbf{I} = \begin{pmatrix} 1 & 0 & 0 \\ 0 & 1 & 0 \\ 0 & 0 & 1 \end{pmatrix}, \quad \mathbf{D} = \begin{pmatrix} 0 & 0.5 & 0.3 \\ 0.4 & 0 & 0.6 \\ 0.2 & 0.4 & 0 \end{pmatrix}, \tag{16}$$

which define $h(u)$, by means of numerical integration (15) we obtain probability distribution of the arrivals' number in orbit $p(i)$ (Table 1).

This distribution is characterized by the sequence of $\delta_i = p(i+1)/p(i)$ which rapidly gets steady and with $i \geq 2$ takes the value within the accuracy of three signs after the comma.

There exist the similar results for other values of parameters μ, γ and matrixes $\mathbf{\Lambda}$ and $\mathbf{Q}$. The dynamic RQ-system with the given values of parameters and matrixes has the value of the throughput capacity equal to $S_1 = 0.540$.

Table 1. Probability distribution $p(i)$ of the arrivals' number in orbit, $i = 0,1,2,\ldots$

i	0	1	2	3	4	5	6	7	$\ldots$
$p(i)$	0.052	0.026	0.026	0.025	0.063	0.024	0.023	0.022	$\ldots$
δ_i	0.506	0.970	0.972	0.972	0.881	0.972	0.972	0.972	$\ldots$

7   Numerical Analysis of the Adaptive RQ-System

Having the given parameters of the adaptive RQ-system we define the values of the throughput capacity S_2 and variable γ. Let the values of parameters be defined as $\mu = 1$, $\beta = 1$, and the values of matrixes in form (16).

Solving system (13) according to S_2 and γ we get the following numerical results (Table 2).

Table 2. The values S_2 and γ with different α

α	0.4	0.6	0.8	1	2	3.774	5	10	100	$10 \cdot 10^5$
S_2	0.035	0.125	0.194	0.25	0.416	0.540	0.583	0.659	0.740	0.75
γ	0.116	0.230	0.363	0.510	1.349	3		4.182	9.102	99.011	9.999

Using the data in Table 2 we can conclude that if α/β increases the S_2-value increases too as well as the γ-value.

In Table 2 if $\alpha = 3.774$ the throughput capacity of the adaptive RQ-system $S_2 = 0.540$ and $\gamma = 3$. This goes with the throughput capacity of the dynamic RQ-system $S_1 = 0.540$ when $\gamma = 3$. This supports the asymptotic equivalence of the adaptive and dynamic RQ-systems with incoming MAP-flow of arrivals.

8   Conclusion

In the paper, the dynamic and adaptive RQ-systems with incoming MAP-flow of arrivals with the method of asymptotic analysis under the high load condition are investigated. As a result of the researching the dynamic RQ-system we got probability distribution of arrivals in orbit $p(i)$, Eq. (8) for defining the throughput capacity S_1. Having investigated the adaptive RQ- system we got simultaneous equations (13) to define the throughput capacity S_2 and γ-value. The coincidence of S_1 and S_2 was showed.

Then all the analytical results were presented numerically. Also the equality of the throughput capacities S_1 and S_2 under the certain parameters was given.

References

1. Artalejo, J.R., Gomez-Corral, A.: Retrial Queueing Systems: A Computational Approach. Springer, Heidelberg (2008)
2. Falin, G.I.: A Survey of retrial queues. Queuing Syst. **7**, 127–167 (1990)
3. Dudin, A.N., Klimenok, V.I., Kim, C.S., Lee, M.H.: The SM/PH/N queueing system with broadcasting service. In: Proceedings of the 13th International Conference on Analytical and Stochastic Modeling Techniques and Applications, pp. 8–13, Bonn, Germany (2006)
4. Nazarov, A.A., Sudyko, E.A.: The method of asymptotic semi-invariants in investigating mathematical model of the random access networks. Probl. Inf. Transf. **46**(1), 94–111 (2010)
5. Nazarov, A.A., Semenova, I.A.: Research of RQ-systems with the method of asymptotic semi-invariants. Newsl. Tomsk State Univ. Managing Comput. Inf. Sci. **3**(12), 85–96 (2010)
6. Garayshina, I.R., Moiseeva, S.P., Nazarov, A.A.: Research Methods of Correlated Streams and Special Queue Systems. Publishing house NTL, Tomsk (2010)
7. Klimenok, V.I.: Optimization of dynamic management of the operating mode of data systems with repeat calls. Autom. Control Comput. Sci. **24**(1), 23–28 (1993)
8. Nazarov, A.A., Terpugov, A.F.: The Queuing Theory: Learning Guide - 2 issue. Publishing house NTL, Tomsk (2010)
9. Gnedenko, B.V., Kovalenko, I.I.: Intoduction in the Queuing Theory - 3 issue. KomKniga, Moscow (2005)
10. Saati, T.L.: Elements of the Queuing Theory And Its Supplements. Sovereign Radio, Moscow (1971)
11. Nazarov, A.A., Kusnetsov, D.Y.: Adaptive Random Access Networks. TPU, Tomsk (2002)
12. Lyubina, T.V., Nazarov, A.A.: Research of the Markov dynamic RQ-system with conflicts arrivals. Newsl. Tomsk State Univ. Managing Comput. Inf. Sci. **3**(12), 73–84 (2010)
13. Lyubina, T.V., Nazarov, A.A.: Research of the dynamic and adaptive RQ-systems with incoming MMPP-flow of requests. Newsl. Tomsk State Univ. **3**(24), 104–112 (2013)
14. Lyubina, T.V., Nazarov, A.A.: Research of the non-Markov model of the computer communication networks managing by the dynamic access protocol. Newsl. Tomsk State Univ. Managing Comput. Inf. Sci. **1**(18), 16–27 (2012)
15. Nazarov, A.A., Moiseeva, S.P.: The method of asymptotic analysis in the queuing theory. Publishing house NTL, Tomsk (2006)
16. Lyubina, T.V., Nazarov, A.A.: Research of the non-Markov dynamic RQ-system with conflicts arrivals. Newsl. Kemerovo State Univ. **49**, 38–44 (2012)

Analyzing Blocking Probability in LTE Wireless Network via Queuing System with Finite Amount of Resources

Konstantin Samouylov$^{(\boxtimes)}$, Eduard Sopin, and Olga Vikhrova

Department of Applied Probability and Informatics,
Peoples' Friendship University of Russia, Miklukho-Maklaya Str. 6,
117198 Moscow, Russia
{ksam,esopin}@sci.pfu.edu.ru, o.vikhrova@gmail.com

Abstract. According to analytics, the global mobile date traffic will grow three times faster than fixed traffic by 2019. The number of user's mobile devices is supposed to increase from 4.1 billion to 4.9 billion while the number of mobile device connections can reach even 10 billion. Broadband speeds in wireless networks are expected to double from 1.7 Mbps to 4.0 Mbps to the end of 2019. As it is noted the mobile video traffic will be up to 72 percent of the global mobile traffic. As the number of wireless connections tends to increase significantly, it results in dramatic mobile traffic growth. Mobile service providers face the challenge to utilize limited radio resources efficiently. In this paper, we propose a mathematical model of radio resources allocation in broadband wireless networks such as LTE-Advanced in terms of queuing systems and evaluate blocking probability and average amount of occupied resources.

Keywords: LTE-advanced · Resource allocation policy · Queuing system with limited resources

1 Introduction

High popularity of various multimedia mobile services is conditioned by huge amount of mobile devices and attractiveness of mobile services. Users prefer to use their smart phones, tablets, etc. for business needs and entertaining via wireless networks. To deliver a service with expected quality providers have focused on methods and techniques to utilize resources more efficiently and more flexible. In modern high-speed wireless networks such as LTE and LTE-Advanced, every mobile session will be allocated with radio resources based on type of service, distance between mobile device and base station and multiple access scheme.

We consider that each mobile device can transfer data to the LTE base station with specific data rate. Let's denote maximum bitrate for customer

K. Samouylov—This work was partially supported by RFBR, projects No. 14-07-0090, 15-07-03608, 15-07-03051.

© Springer International Publishing Switzerland 2015
A. Dudin et al. (Eds.): ITMM 2015, CCIS 564, pp. 393–403, 2015.
DOI: 10.1007/978-3-319-25861-4_33

i as follows: $c_i^{\max} = w \log_2 (1 + \gamma_i p_{\max})$, where w is a spectral bandwidth, $p_{\max}$ is maximum transmit power of the base station and γ_i is signal-to-noise ratio for the session, that depends on distance between user device and base station, possible obstacles between them, etc. New session will be accepted only if there is enough resources to serve it, i.e. $\sum_i \frac{c_i}{c_i^{\max}} \leqslant 1$, where A_i is required data rate for session i. In order to evaluate network performance measures we design a mathematical model in terms of queuing system with finite amount of resources to analyze blocking probability and average amount of occupied radio resources.

2   Model Description

In paper [2] multiple servers queuing system is considered where each customer occupies on arrival some random amount of finite resources. At the end of service time the amount of allocated resources is fully released, see Fig. 1. If there is not enough resources to meet customer requirements, the customer is denied service. Random variables (RV) of required resources assumed to be independent of arrival and service processes, mutually independent and identically distributed. In this type of queue we have to remember a vector of allocated resources for each customer. Thus, it significantly complicates state space of the corresponding random process and its analysis.

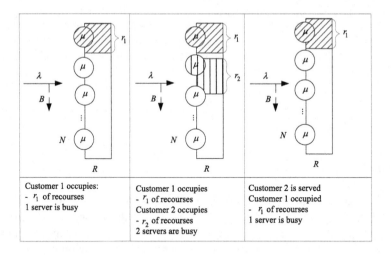

Fig. 1. Diagram of a general model

To simplify the model we offer to track only total amount of occupied resources. As soon as we don't know how many resources have been allocated to each customer, we assume that amount of released resources on a departure of a customer is also random and may differ from allocated one.

Given the total amount of allocated resources and the number of customers in system, RV of released resources are independent from past behaviour of the system and its cumulative distribution function (CDF) can be obtained using the Bayes theorem.

In [2] it was shown that average amount of occupied resources of initial and simplified models are very close to each other in case of Poisson arrival process and exponential service distribution time. Later in [4], simulations showed that not only average values but also steady state distribution of allocated resources for both models are very close. Finally, in [3], it was analytically proved that steady-state distributions of total occupied resources and number of customers are equal. Note, that some generalizations discussed in [6] include a system in which the service time and the amount of resources allocated to the customer are dependent random variables, and each customer has three random characteristics: the number of devices required for the service, resources and service time.

In this paper we analyze simplified model with N servers and limited amount R of a discrete resource (Fig. 2). Customers arrive according to the Poisson process with rate λ. Service times are mutually independent, independent of arriving process and are exponentially distributed with rate μ. Let us denote $\xi(t)$ – number of customers in the system at time t, and $\delta(t) < R$ – amount of total occupied resources. Customer i requires $r_i \geqslant 0$ units of discrete resources, RVs r_i are mutually independent and identically distributed with CDF $F(x)$, mean m and variance σ^2.

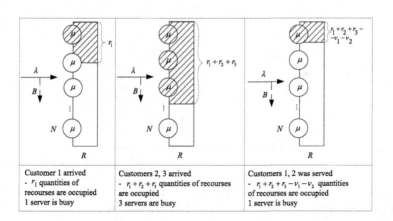

Customer 1 arrived	Customers 2, 3 arrived	Customers 1, 2 was served
- r_1 quantities of recourses are occupied	- $r_1 + r_2 + r_3$ quantities of recourses are occupied	- $r_1 + r_2 + r_3 - \nu_1 - \nu_2$ quantities of recourses are occupied
1 server is busy	3 servers are busy	1 server is busy

Fig. 2. Diagram of simplified model

Customer i will be lost in case the system doesn't have enough resources ($R - \delta(t) < r_i$) or if there is no free servers, i.e. $\xi(t) = N$. Total amount of occupied resources $\delta(t)$ increases by $r_i \geqslant 0$ immediately after arrival of a customer.

Total amount of occupied resources $\delta(\tau_i)$ will decrease by random value ν_i at time τ_i as soon as customer i is served. Given the number of total occupied resources $\delta(\tau_i) = y$ and number of customers in the system $\xi(\tau_i) = k$, RVs ν_i are independent from previous system behavior and have CDF $F_k(x|y) = P(\nu_i \leqslant x|\xi(\tau_i) = k;\ \delta(\tau_i) = y),\ 0 \leqslant x \leqslant y$.

RVs r_i attain values $j = \overline{0, R}$, and $p_j = P(r_i = j)$ is a probability that arriving customer requires j units of resource. Let's denote $p_j^{(k)}$, $j = \overline{0, R}$ a k-fold convolution of probabilities p_j, which allows to calculate the total amount of occupied resources $y = \sum_{i=0}^{k} r_i$. CDF $F_k(x|j)$ of units of resources released by a customer on departure is piecewise-constant function and it has a saltus at $x = i,\ i = \overline{0, R}$ on a value

$$p_{ij}^k = \frac{p_i p_{j-i}^{(k-1)}}{p_j^{(k)}}, \quad 0 < i \leqslant j, \quad j = \overline{0, R}. \tag{1}$$

In general case convolutions $p_j^{(k)}$, $k \geqslant 2$ can be calculated using probabilities p_i, $0 \leqslant i \leqslant R$ according to following recurrent formula:

$$p_j^{(k)} = \sum_{i=0}^{j} p_i p_{j-i}^{(k-1)}. \tag{2}$$

Random process $X(t) = (\xi(t), \delta(t))$ over the set $X = \bigcup_{k=0}^{N} X_k$, $X_k = \{(k, i)\ |\ 0 \leqslant i \leqslant R,\ p_i^{(k)} > 0\}$ is a Markov chain. Figure 3 shows the state transitions diagram, where $1 \leqslant i < N,\ 0 \leqslant s \leqslant j \leqslant q \leqslant R$.

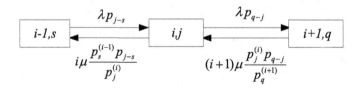

Fig. 3. State transition diagram

Infinitesimal matrix $A = [a((i,j),(k,r))]$ has a block three-diagonal structure with main diagonal blocks $\Psi_0, \Psi_1, ..., \Psi_N$, upper blocks $\Lambda_1, ..., \Lambda_N$ and lower blocks $M_0, ..., M_{N-1}$.

$$\Psi_0 = -\lambda P_R, \tag{3}$$

$$\Lambda_1 = (\lambda_{0,j})_{(j|(1,j)\in X_1)} = \lambda p_j, \tag{4}$$

$$M_0 = (\mu_{i,0})_{(i|(1,i)\in X_1)} = \mu, \tag{5}$$

$$\Psi_n = (\psi_{i,j})_{(i,j|(n,i),(n,j)\in X_k)} \begin{cases} -(\lambda P_{R-i} + n\mu), & i = j \\ 0, & i \neq j \end{cases}, \quad n = \overline{1, N-1}, \tag{6}$$

$$\Psi_N = (\psi_{i,j})_{(i,j|(N,i),(N,j)\in X_N)} \begin{cases} -N\mu, & i = j \\ 0, & i \neq j \end{cases}, \tag{7}$$

$$\Lambda_n = (\lambda_{i,j})_{(i,j|(n,i)\in X_{k-1},(n,j)\in X_k)} \begin{cases} \lambda p_{j-i}, & i \leqslant j \leqslant R \\ 0 \end{cases}, \quad n = \overline{2, N}, \quad (8)$$

$$M_n = (\mu_{i,j})_{(i,j|(n,i)\in X_{k+1},(n,j)\in X_k)} \begin{cases} (n+1)\mu \dfrac{p_j^{(n)} p_{i-j}}{p_i^{(n+1)}}, & j \leqslant i \leqslant R \\ 0 \end{cases}, \quad (9)$$

$n = \overline{1, N-1}$.

Having infinitesimal matrix $\boldsymbol{A}$ we obtain stationary distribution for $X(t) = (\xi(t), \delta(t))$, where

$$q_{0,0} = \lim_{t\to\infty} P\{\xi(t) = 0, \delta(t) = 0\}, \quad (10)$$

$$q_{k,i} = \lim_{t\to\infty} P\{\xi(t) = k, \delta(t) = i\}, \quad 1 \leqslant k \leqslant N, \quad 0 \leqslant i \leqslant R. \quad (11)$$

Based on formulas (3)–(9) the set of equilibrium equations can be written as follows:

$$\lambda P_R q_{0,0} - \mu \sum_{j=0}^{R} q_{1,j} = 0, \quad (12)$$

$$(\lambda P_{R-j} + i\mu) q_{i,j} - \lambda \sum_{s=0}^{j} p_{j-s} q_{i-1,s} - (i+1)\mu \sum_{s=j}^{R} \frac{p_j^{(i)} p_{s-j}}{p_s^{(i+1)}} q_{i+1,s} = 0, \quad (13)$$

$$0 \leqslant j \leqslant R, \quad 1 \leqslant i \leqslant N-1.$$

$$N\mu q_{N,j} - \lambda \sum_{s=0}^{j} p_{j-s} q_{N-1,s} = 0, \quad 0 \leqslant j \leqslant R. \quad (14)$$

Denote vectors of state probabilities $q_0 = q_{0,0}$, $q_i = (q_{i,j})_{j=\overline{0,R}}$, $i = 1, \ldots, N$. Then, the set of equilibrium equations can be written in vector form:

$$q_0 \Psi_0 - q_1^T M_0 = 0, \quad (15)$$

$$q_i^T \Psi_i - q_{i+1}^T M_i - q_{i-1}^T \Lambda_i = 0, \ i = 1, \ldots, N-1, \quad (16)$$

$$q_N^T \Psi_N - q_{N-1}^T \Lambda_N = 0. \quad (17)$$

Matrix $\boldsymbol{A}$ is indecomposable and conservative with three diagonal blocks thus we apply UL matrix decomposition techniques [1]. Given steady-state distribution of vector q we calculate average number of occupied resources in the system:

$$b = \sum_{k=0}^{N} b_k \sum_{i=k}^{N} q_{k,i}, \quad (18)$$

where b_k means number of occupied resources in case of k customers in the system:

$$b_k = \frac{\sum_{i=0}^{N} i p_i^{(k)}}{\sum_{i=0}^{N} p_i^{(k)}}. \quad (19)$$

Blocking probability can be found as:

$$B = 1 - \sum_{k=0}^{N} \sum_{i=k}^{N} q_{k,i} \sum_{j=1}^{N-i} p_j. \tag{20}$$

Average number of customers in the system can be calculated using the following formula:

$$\bar{N} = \sum_{i=0}^{N} i q_i,$$

where

$$q_i = \sum_{j=0}^{R} q_{i,j}. \tag{21}$$

In [3] analytical solution of (3)–(9) was obtained for initial model. In the case of discrete numbers of allocated resources we derive more simple equations to calculate steady-state probabilities distribution:

$$q_{k,\bullet} = \lim_{t \to \infty} P\{\xi(t) = k\} = p_0 \frac{\rho^k}{k!} \sum_{i=0}^{R} p_i^{(k)}, \quad 0 < k \leqslant N, \tag{22}$$

$$q_{k,j} = \lim_{t \to \infty} P\{\xi(t) = k\,;\delta(t) = j\,\} = p_0 \frac{\rho^k}{k!} p_j^{(k)}, \tag{23}$$

$$0 \leqslant j \leqslant R, \quad 0 < k \leqslant N,$$

$$p_0 = \left(1 + \sum_{k=1}^{N} \frac{\rho^k}{k!} \sum_{i=0}^{R} p_i^{(k)}\right)^{-1}. \tag{24}$$

Thus the average amount of occupied resources b is easier to calculate:

$$b = \sum_{k=0}^{N} b_k q_{k,\bullet} = p_0 \sum_{k=0}^{N} \frac{\rho^k}{k!} \sum_{i=0}^{R} i p_i^{(k)}. \tag{25}$$

System blocking probability can also be obtained as follows:

$$B = 1 - \sum_{k=0}^{N-1} \sum_{i=0}^{R} q_{k,i} \sum_{j=0}^{R-i} p_j = 1 - p_0 \sum_{k=0}^{N-1} \frac{\rho^k}{k!} \sum_{i=0}^{R} p_i^{(k+1)}. \tag{26}$$

3   Numerical Results

We considered different distributions for resource allocation such as binomial, shifted binomial and geometric distributions. We also assume that total amount of system resources R is equal to the number of servers N.

To evaluate systems characteristics we calculated k-fold convolutions $p_j^{(k)}$ for each suggested distributions using formula (2).

If RVs $r_j \geqslant 0$ have binomial distribution

$$p_i = \binom{r}{i} p^i (1-p)^{r-i}, \quad 0 \leqslant i \leqslant r, \tag{27}$$

where $p = \frac{m}{r}$, then $p_j^{(k)} = \binom{kr}{j} p^j (1-p)^{kr-j}$. Probability that i units of resource are released on departure of k-th customer in case when these k customers occupy j units of resources is

$$p_{ij}^k = \frac{p_i p_{j-i}^{(k-1)}}{p_j^{(k)}} = \frac{\binom{r}{i} \binom{kr-r}{j-i}}{\binom{kr}{j}}, \tag{28}$$

$0 \leqslant i \leqslant r, \quad i \leqslant j, \quad 0 \leqslant j \leqslant kr$, it depends only from the maximum quantity of resources r which can be allocated to the customer j.

However shifted binomial distribution is more interesting for application, because in real networks resource requirements of any session is always greater than zero. Thus, number of allocated resources to the customer j is only positive number, $r_j > 0$ and it is distributed with

$$p_i = \binom{r-1}{i-1} p^{i-1} (1-p)^{r-i}, \quad 1 \leqslant i \leqslant r, \tag{29}$$

where $p = \frac{m-1}{r-1}$ and $p_j^{(k)} = \binom{k(r-1)}{j-k} p^{j-k} (1-p)^{kr-j}$. Therefore the

$$p_{ij}^k = \frac{\binom{r-1}{i-1} \binom{(k(r-1))-(r-1)}{(j-k)-(i-1)}}{\binom{k(r-1)}{j-k}} \tag{30}$$

is for number of released resources on the departure, where $1 < i \leqslant r, i \leqslant j$, $k \leqslant j \leqslant k(r-1), 1 < k \leqslant j$.

To a better comparative research we explore the key system performance characteristics in case of using geometrical distribution for the quantity of allocated resources. i.e.

$$p_i = p^i (1-p), 1 \leqslant i \leqslant r, \tag{31}$$

where $p_j^{(k)} = \binom{k+j-1}{k} p^j (1-p)^k$ and probabilities for number of released resources

$$p_{ij}^k = \frac{\binom{k+j-i-2}{k-1}}{\binom{k+j-1}{k}} \tag{32}$$

where $i \leqslant j$, $k \leqslant j$, $1 < k \leqslant j$.

We consider the total amount of system resources and the number of servers is equal and $N = R = 100$, while the arriving rate is $\lambda = \{12, 13, \ldots, 20\}$ and average service time $\mu^{-1} = 1$.

Figures 4, 5 and 6 show the relation between average number of customers in the system, average amount of occupied resources, blocking probability and system load ρ correspondingly for three different distributions of r_j. We selected mean $m = 5, 4$, maximum quantity of resources allocated to each customer $r = 18$ for each suggested distributions, but the variances are different.

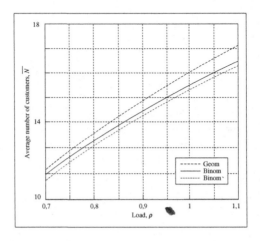

Fig. 4. Average number of customers in the system

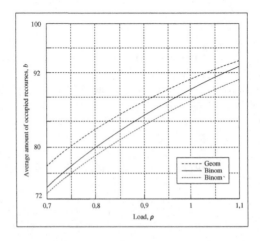

Fig. 5. Average amount of occupied resources

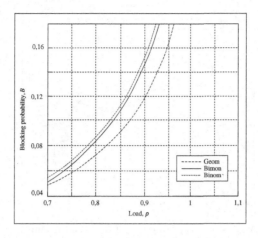

Fig. 6. System blocking probability

As we can see on Fig. 4, average number of customers in the system when $\rho > 1$ grows faster in case of geometrically distributed quantities of allocated resources. Contrary to the average number of customer the graph of average amount of resources for geometrical distribution leans to the graph corresponding to the binominal distribution. As soon as variance of geometrical distribution higher than the variance of binomial distributions we studied the relation between the same system characteristics and variance of distributions under different loads.

Figures 7, 8 and 9 depict the relation between average number of customers in the system, average amount of occupied resources, blocking probability and variance σ.

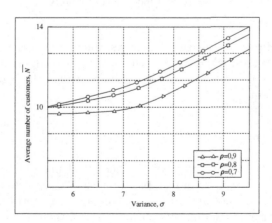

Fig. 7. Average number of customers in the system

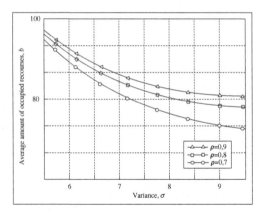

Fig. 8. Average amount of occupied resources

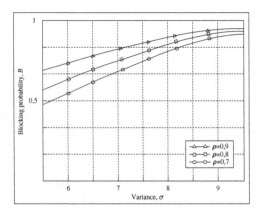

Fig. 9. System blocking probability

As variance of resource distribution grows, the average number of customers in the system increases, while the average amount of occupied resources drops. The more resource is needed to service customer session the higher probability that customer is lost.

4 Conclusion

In this paper, we analyzed simplified mathematical model in case of discrete distribution of radio resources in LTE base station. We suggested analytical and numerical methods to calculate the most interesting characteristics of the model. Besides, we investigated relation between system performance characteristics and distribution of allocated resources.

In result we state that high variance of distribution of allocated resources leads to lower average number of customers in the system and average amount of occupied resources. In heavy load states system will accept customer's sessions with minimal requirements. Our further study will include computing algorithm complexity analysis and design of more effective algorithms.

The authors wish to express their appreciation to Prof. V. Naumov for suggesting the problem and his help during the research.

References

1. Naumov, V.A.: Numerical Methods for Markov Queues Analysis. PFUR, Moscow (1985)
2. Naumov, V.A., Samouylov, K.E.: On the modeling of queuing systems with multiple resources. PFUR Bull. Ser. Inf. Math. Phys. **3**, 58–62 (2014)
3. Naumov, V.A., Samuoylov, K.E., Samuylov, A.K.: On total amount of resources occupied by customers. Automation and Remote Control (in Print)
4. Naumov, V., Samuoylov, K., Sopin, E., Andreev, S.: Two approaches to analysis of queuing systems with limited resources. In: Ultra Modern Telecommunications and Control Systems and Workshops (ICUMT), pp. 485–488. IEEE (2014)
5. Pyattaev, A., Johnsson, K., Surak, A., Florea, R., Andreev, S., Koucheryavy, Y.: Network-assisted D2D communications: implementing a technology prototype for cellular traffic offloading. In: Wireless Communications and Networking Conference (WCNC), pp. 3266–3271. IEEE (2014)
6. Tikhonenko, O.M.: Generalized Erlang problem for service systems with finite total capacity. Probl. Inf. Transm. **41**(3), 243–253 (2005)

Synergetic Effects for Number of Busy Servers in Multiserver Queuing Systems

Gurami Tsitsiashvili[1,2](✉) and Marina Osipova[1,2]

[1] Institute for Applied Mathematics Far Eastern Branch of RAS,
Radio Str. 7, 690041 Vladivostok, Russia
guram@iam.dvo.ru
[2] Far Eastern Federal University, Sukhanova Str. 8, 690091 Vladivostok, Russia
mao1975@list.ru

Abstract. In this paper we concretize a condition when an aggregation of n oneserver queuing systems into multiserver system for $n \to \infty$ leads to an disappearance of a queue (in some probabilistic sense) and to a transformation of multiserver system into a system with infinite number of servers. An initial oneserver system is a system with Poisson input flow or with some modifications of this flow like a regular flow without an aftereffect or with Poisson flow in a random environment. Such formulation of a problem is connected with a large number of articles devoted to a modeling of computer networks by queuing systems with infinite number of servers and to a justification of these models application for real networks with finite number of servers.

Keywords: Poisson flow · Regular flow without aftereffect · Kolmogorov-Chencov condition

1 Introduction

In this paper we analyze a condition when an aggregation of n oneserver queuing systems into multiserver system for $n \to \infty$ leads to an disappearance of a queue (in some probabilistic sense) and to a transformation of multiserver system into a system with infinite number of servers.

Such formulation of a problem is connected with a large number of articles devoted to a modeling of computer networks by queuing systems with infinite number of servers (see for an example [1,2]) and to a justification of these models application for real networks with finite number of servers. Indeed in a queuing system with infinite number of servers each customer passes through the system without any delay. So it is much simpler to calculate different distributions in the system. Another cause of this consideration is a convergence to zero of a probability that a virtual waiting time is positive in described limiting transition and so service quality increases significantly.

An initial oneserver system is a system with Poisson input flow or with some modifications of this flow like a regular flow without an aftereffect or with Poisson

© Springer International Publishing Switzerland 2015
A. Dudin et al. (Eds.): ITMM 2015, CCIS 564, pp. 404–414, 2015.
DOI: 10.1007/978-3-319-25861-4_34

flow in a random environment. All these considerations are closely connected with a concept of Poisson flow which is a foundation of regular random process without aftereffect and of its different modifications.

Considered problem is solved using (1) a limit theorem for a convergence of multiserver queuing system to infinite server system, (2) a concept of C - convergence for random processes, (3) sufficient Kolmogorov-Chencov condition of C - convergence, (4) a concept of ε - entropy for a calculation of an asymptotic of Gaussian random process maximum distribution. Emphasize that regular random process without aftereffect is very convenient for an application of Kolmogorov-Chencov condition.

2   Preliminaries

Denote $\mathcal{F}_1$ metric space of deterministic functions defined on the segment $[0, T]$ with uniform metric . Put $\mathcal{F}$ the set of continuous and bounded by the unit functions defined on $\mathcal{F}_1$. Say that the sequence of random processes $z_n = z_n(t)$, $n \geq 1$, C - converges to the random process $z = z(t)$, $0 \leq t \leq T$, if for any functional $f \in \mathcal{F}$ the convergence of mean values

$$M f(z_n) \to M f(z), \ n \to \infty$$

takes place.

Denote $\mathcal{D}$ the space of random functions on the segment $[0, T]$ which almost surely (a.s.) have not breaks of the second kind and designate $\mathcal{C}$ the space of random functions on the segment $[0, T]$ which a.s. are continuous on the segment $[0, T]$. Represent the following sufficient conditions of C - convergence [3, Chapter 3, Theorems 6, 7, Corollary 1].

Theorem 1. *Assume that random functions $z_n \in \mathcal{D}$, $n \geq 1$, and random function $z \in \mathcal{C}$. For C - convergence $z_n \to z$, $n \to \infty$, it is sufficient that finite dimensional distributions of random functions $z_n(t)$ converge to finite dimensional distributions of random function $z(t)$ and the "moment" condition of Kolmogorov-Chencov [4] is true: for some $C > 0$, $\alpha > 1$, $\beta > 1$, $\lambda > 0$ and for any $0 \leq t_1 < t < t_2 < T$*

$$\sup_n M(|z_n(t_2) - z_n(t)|^\alpha \, |z_n(t) - z_n(t_1)|^\beta) \leq C(t_2 - t_1)^{1+\lambda}. \tag{1}$$

Assume that $z(t)$, $0 \leq t \leq T$, is Gaussian process $Mz(t) = 0$, $Mz^2(t) < a < \infty$ then this process is a.s. continuous. Formulate conditions of the following limit relation for the process $z(t)$.

Theorem 2. *If there is positive number C satisfying the inequality*

$$\varepsilon^2(t, t+u) = M(z(t) - z(t+u))^2 \leq Cu, \ 0 \leq t \leq t+u \leq T,$$

then

$$P\left(\sup_{0 \leq t \leq T} z(t) > L\right) \to 0, \ L \to \infty. \tag{2}$$

Proof. In the condition of Theorem 2 the minimal number $N(r)$ of balls with the radios r in the metric space $([0,T], \varepsilon)$ (here $\varepsilon(t, t+u)$ is a half metric on $[0,T]$) covering the segment $[0,T]$ satisfies the inequality $N(r) \leq TCr^{-2}$ and so Dadly integral $\Psi(z) = \int_0^z (\ln N(r))^{1/2} dr$ constructed by the relative entropy $\ln N(r)$ satisfies the condition: $\Psi(T) < \infty$. Consequently we have the relation (2) from [5], [6, Theorem 1].

Theorem 3. *Assume that the sequence of random processes $z_n(t)$, $n \geq 1$, $0 \leq t \leq T$, C - converges to Gaussian and continuous random process $z(t)$, such that $P\left(\sup_{0 \leq t \leq T} z(t) > L \right) \to 0$, $L \to \infty$ and the sequence of positive numbers $L_n \to \infty$, $n \to \infty$. Then for $n \to \infty$ we have the limit relation*

$$P\left(\sup_{0 \leq t \leq T} z_n(t) \geq L_n \right) \to 0. \tag{3}$$

Proof. Take arbitrary positive number $\varepsilon > 0$ and choose such $L(\varepsilon)$ that

$$P(\sup_{0 \leq t \leq T} z(t) > L(\varepsilon)) < \varepsilon.$$

From C - convergence of $z_n(t)$ to $z(t)$ for $n \to \infty$ it is possible to take such $n(\varepsilon)$, that for $n > n(\varepsilon)$ the inequality

$$\left| P\left(\sup_{0 \leq t \leq T} z(t) > L(\varepsilon) \right) - P\left(\sup_{0 \leq t \leq T} z_n(t) > L(\varepsilon) \right) \right| < \varepsilon$$

is true and so

$$P\left(\sup_{0 \leq t \leq T} z_n(t) > L(\varepsilon) \right) < 2\varepsilon, \ n > n(\varepsilon).$$

From the convergence $L_n \to \infty$, $n \to \infty$, it is possible to find such $n_1(\varepsilon) > n(\varepsilon)$ that for $n > n_1(\varepsilon)$ the inequality $L_n > L(\varepsilon)$ is true, consequently

$$P\left(\sup_{0 \leq t \leq T} z_n(t) \geq L_n \right) \leq P\left(\sup_{0 \leq t \leq T} z_n(t) > L(\varepsilon) \right) \leq 2\varepsilon, \ n > n_1(\varepsilon).$$

The relation (3) is proved.

3   Main Theorem

Consider the series scheme in which the characteristics of n - server queuing system are defined by the parameter $n \to \infty$ which characterizes an intensity of input flow tending to infinity. Denote $e_n(t)$ a number of input flow customers arriving before the moment t, $e_n(0) = 0$. Assume that $q_n(t)$ is a number of busy servers in this system at the moment t, $q_n(0) = 0$, τ_j is the service time of j input flow customer and τ_j, $j \geq 1$, is a sequence of independent and identically distributed random variables (s.i.i.d.r.v.'s) with the distribution function (d.f.) $F(t)$ ($\overline{F} = 1 - F$), which has continuous and bounded by $\bar{f} > 0$ density $f(t)$.

Theorem 4. *Assume that for some $T > 0$ the following conditions are true.*

1. *$Me_n(t) = nm(t)$ where $m(t)$ is differentiable and non decreasing function with continuous and bounded by some positive number $\overline{M}$ derivative.*

2. *For $n \to \infty$ the sequence of random processes $x_n(t) = \dfrac{e_n(t) - nm(t)}{\sqrt{n}}$, $n \geq 1$, C - converges on $[0, T]$ to the random processes $a\xi(g(t)) + b\eta t$. Here $g(t)$, $0 \leq t \leq T$, is differentiable and non decreasing function with bounded by some positive number $\overline{G}$ derivative, $\xi(t)$ is standard Wiener process, $a \neq 0$, b is nonnegative number, η is standard normal random variable independent from $\xi(t)$.*

3. *The inequality $Q(T) = \displaystyle\int_0^T \overline{F}(t)dm(t) < 1$ is true.*

Then

$$P\left(\sup_{0 \leq t \leq T} q_n(t) = n\right) \to 0, \ n \to \infty. \tag{4}$$

Proof. Denote $q_n^\infty(t)$ a number of busy servers at the moment t in the system with input flow described by the process $e_n(t)$ provided that there is infinite number of servers. From the condition 2 and [7, chapter II, Sect. 1, Theorem 1] we have that for $n \to \infty$ the sequence of random processes

$$y_n(t) = \frac{q_n^\infty(t) - nm(t)}{\sqrt{n}}, \ n \geq 1,$$

C - converges on the segment $[0, T]$ to the random process $\zeta(t)$, defined by the equality

$$\zeta(t) = a \int_0^t \overline{F}(t - u)d\xi(g(u)) + b\eta \int_0^t \overline{F}(t - u)du + \Theta(t), \ 0 \leq t \leq T,$$

where $\Theta(t)$ is centered Gaussian process independent from $\xi(t)$ with the covariation function $R(t, t + u) = \displaystyle\int_0^t \overline{F}(v + u)F(v)dm(v)$.

Lemma 1. *The random process $\zeta(t)$ for some $C > 0$ satisfies the inequality*

$$\varepsilon^2(t, t + u) = M(\zeta(t) - \zeta(t + u))^2 \leq Cu, \ 0 \leq t \leq t + u \leq T.$$

Proof. Indeed the equality

$$\varepsilon^2(t, t + u) = a^2 M \left(\int_0^t \overline{F}(t - v)d\xi(g(v)) - \int_0^{t+u} \overline{F}(t + u - v)d\xi(g(u))\right)^2$$

$$+ b^2 M \left(\int_0^t \eta\overline{F}(t - v)dv - \int_0^{t+u} \eta\overline{F}(t + u - v)dv\right)^2 + M(\Theta(t) - \Theta(t + u))^2$$

is true and

$$M(\Theta(t) - \Theta(t + u))^2 = R(t, t) + R(t + u, t + u) - 2R(t, t + u)$$

$$= \int_0^t \overline{F}(v)F(v)dm(v) + \int_0^{t+u} \overline{F}(v)F(v)dm(v) - 2\int_0^t \overline{F}(v+u)F(v)dm(v)$$

$$= 2\int_0^t (\overline{F}(v) - \overline{F}(v+u))dm(v) + \int_t^{t+u} \overline{F}(v)F(v)dm(v) \le \overline{M}(2T\bar{f}+1)u,$$

$$M\left(\int_0^t \eta\overline{F}(t-v)dv - \int_0^{t+u} \eta\overline{F}(t+u-v)dv\right)^2$$

$$= \left(\int_0^t \overline{F}(t-v)dv - \int_0^{t+u} \overline{F}(t+u-v)dv\right)^2 = \left(\int_0^t \overline{F}(t-v)dv\right)^2$$

$$+ \left(\int_0^{t+u} \overline{F}(t+u-v)dv\right)^2 - 2\int_0^t \overline{F}(t-v)dv \int_0^{t+u} \overline{F}(t+u-w)dw$$

$$\le \int_0^t \overline{F}(t-v)dv \int_0^t (\overline{F}(t-w) - \overline{F}(t+u-w))dw \le T^2\bar{f}u.$$

Calculate now

$$M\left(\int_0^t \overline{F}(t-v)d\xi(g(v)) - \int_0^{t+u} \overline{F}(t+u-v)d\xi(g(u))\right)^2$$

$$= M\left[\left(\int_0^t \overline{F}(t-v)d\xi(g(v))\right)^2 + \left(\int_0^{t+u} \overline{F}(t+u-v)d\xi(g(v))\right)^2\right.$$

$$\left. -2\int_0^t \overline{F}(t-v)d\xi(g(v)) \int_0^{t+u} \overline{F}(t+u-w)d\xi(g(w))\right]$$

$$= \int_0^t \overline{F}^2(t-v)dg(v) + \int_0^{t+u} \overline{F}^2(t+u-v)dg(v) - 2\int_0^t \overline{F}(t-v)\overline{F}(t+u-v)dg(v)$$

$$\le \overline{G}\bar{f}Tu + \overline{G}u = \overline{G}(\bar{f}T+1)u,$$

consequently

$$\varepsilon^2(t, t+u) \le Cu, \quad C = a^2\overline{G}(\bar{f}T+1) + \overline{M}(2T\bar{f}+1) + b^2T^2\bar{f}.$$

The process $\zeta(t)$ obviously satisties the conditions

$$M\zeta(t) = 0, \ M\zeta^2(t) \le ag(T) + m(T), \ 0 \le t \le T.$$

Then from Theorem 2 and Lemma 1 we obtain the limit relation

$$P\left(\sup_{0 \le t \le T} \zeta(t) \ge L\right) \to 0, \ L \to \infty.$$

Consequently for $n \to \infty$

$$P\left(\sup_{0 \le t \le T} \zeta(t) \ge \frac{(1 - Q(T))n}{\sqrt{n}}\right) \to 0.$$

And from C - convergence for $n \to \infty$ of random processes sequence $y_n(t)$, $n \ge 1$, to the random process $\zeta(t)$ on the segment $[0, T]$ and from Theorem 3 we obtain the relation $P\left(\sup_{0 \le t \le T} q_n^\infty(t) \ge n\right) \to 0, \ n \to \infty.$ Using the end of the proof in [7, chapter II, Sect. 1, Theorem 1] remark that the random events

$$\{q_n(t) < n, \ 0 \le t \le T\} = \{q_n^\infty(t) < n, \ 0 \le t \le T\}.$$

Consequently from the inequality $q_n(t) \le n$ we obtain the limit relation

$$P\left(\sup_{0 \le t \le T} q_n(t) \ge n\right) = P\left(\sup_{0 \le t \le T} q_n(t) = n\right) \to 0, \ n \to \infty.$$

Remark 1. Theorem 4 statement means a convergence of a virtual waiting time in n - server queuing system on the segment $[0, T]$ to zero and so characterizes a disappearance of a queue after an aggregation of n oneserver queuing systems for $n \to \infty$.

Remark 2. For a validity of the formula (4) it is sufficient a fulfillment of the conditions 1, 2 for arbitrary $T > 0$ and an equity of the condition 3 on T.

4 Examples

In this section examples of input flows which satisfy the conditions 1, 2 of Theorem 4 are considered. All these examples are based on a fulfillment of the "moment" Kolmogov-Chencov condition (1).

4.1 Aggregation of Regular Input Flows Without Aftereffect

Assume that $x(t)$, $0 \le t \le T$, is a number of random flow points on a half interval $[0, t)$, denote $m(t) = Mx(t)$. Call random flow described by the process $x(t)$ regular and without aftereffect if the function $m(t)$ is continuous and for any $0 \le t_1 < t_2 \le t_3 < t_4 \le T$ r.v.'s $x(t_2) - x(t_1)$, $x(t_4) - x(t_3)$ are independent. Following [8] give general representation of the regular flow without aftereffect.

Assume that there are nonnegative continuous and nondecreasing functions $m_k(t)$, $k \geq 1$, and the series $\sum_{k=1}^{\infty} km_k(t) = m(t)$ converges. Define random function $x^k(t)$ as a number of points of Poisson flow with the mean $Mx_k(t) = m_k(t)$ on a half-interval $[0, t)$, $0 \leq t \leq T$, $k \geq 1$. Suppose that random functions $x^k(t)$, $k \geq 1$, are independent and consider the sum $\sum_{k=1}^{\infty} kx^k(t) = x(t)$. In [8, Chapter "Flows of random events without aftereffect", Sect. 4] it is shown that this random sum gives general representation of the regular flow without aftereffect.

Assume that for some nonnegative functions $\lambda_k(t) \leq \lambda_k < \infty$, $k \geq 1$, the functions $m_k(t) = \int_0^t \lambda_k(u)du$, $0 \leq t \leq T$, and the series $\sum_{k=1}^{\infty} k^2 \lambda_k = \overline{G} < \infty$. Then the series

$$\sum_{k=1}^{\infty} km_k(t) = m(t) = Mx(t), \quad \sum_{k=1}^{\infty} k^2 m_k(t) = g(t) = Dx(t)$$

converge and the following inequalities take place

$$\frac{dm(t)}{dt} \leq \sum_{k=1}^{\infty} k\lambda_k = \overline{M} < \infty, \quad \frac{dg(t)}{dt} \leq \sum_{k=1}^{\infty} k^2 \lambda_k = \overline{G} < \infty, \ 0 \leq t \leq T. \quad (5)$$

Put $x_1(t), x_2(t), \ldots$ the sequence of independent copies of the random process $x(t)$. Define the random process $e_n(t)$ which characterizes input flow into aggregated n - server queuing system by the equality $e_n(t) = \sum_{k=1}^{n} x_k(t)$. From this definition and first inequality in (5) we have that the condition 1 of Theorem 4 is true. Consider now the normed sum

$$y_n(t) = \frac{e_n(t) - nm(t)}{\sqrt{n}}, \ n \geq 1.$$

Prove that the random process $y_n(t)$, $n \to \infty$, C - converges to random process $\xi(g(t))$. Indeed from multidimensional central limit theorem finite dimensional distributions of the process $y_n(t)$ tend to multidimensional distributions of the process $\xi(g(t))$. From the inequality (5) we obtain

$$\sup_n M(y_n(t) - y_n(t_1))^2 (y_n(t_2) - y_n(t))^2$$

$$= \sup_n M(y_n(t) - y_n(t_1))^2 M(y_n(t_2) - y_n(t))^2$$

$$= D(x(t) - x(t_1))D(x(t_2) - x(t)) = (g(t) - g(t_1))(g(t_2) - g(t))$$

$$\leq \overline{G}^2 (t_2 - t_1)^2, \ 0 \leq t_1 < t < t_2 \leq T. \quad (6)$$

Consequently for $\alpha = 2$, $\beta = 2$, $\lambda = 1$, $C = \overline{G}^2$ the random processes $y_n(t)$, $n \geq 1$, satisfy the "moment" condition of Kolmogorov-Chencov (1). So from Theorem 1 we have C - convergence of the random process $y_n(t)$, $n \to \infty$, to the random process $\xi(g(t))$. Consequently the condition 2 of Theorem 4 also is fulfilled.

4.2 Poisson Input Flows with Independent Random Intensities

Consider following multiserver queuing system in random environment. Denote $x(t)$, $0 \leq t \leq T$, Poisson flow with the random intensity λ which has the distribution $P(d\lambda)$, $\lambda > 0$, and satisfies the inequality $M\lambda^4 < \infty$. Assume that $x_1(t), x_2(t), \ldots$ are independent copies of the process $x(t)$ with independent random intensities $\lambda_1, \lambda_2, \ldots$ which have the common distriribution $P(d\lambda)$. Take an aggregation of n oneserver queuing systems in which input flow is characterized by the random process $e_n(t) = \sum_{k=1}^{n} x_k(t)$. From the definition of $e_n(t)$ we have

$$Me_n(t) = nm(t), \ m(t) = Mx(t) = tM\lambda, \ \frac{dm(t)}{dt} = M\lambda \qquad (7)$$

and so the random process $e_n(t)$ satisfies the condition 1 of Theorem 4.

Consider the sequence of random processes

$$y_n(t) = \frac{e_n(t) - Me_n(t)}{\sqrt{n}}, \ n \geq 1,$$

and prove that this sequence for $n \to \infty$ C - converges to the random process $\sqrt{M\lambda}\xi(t) + t\eta\sqrt{D\lambda}$. Here $\xi(t)$ is the standard Wiener process, η is the standard Gaussian r.v. and $\xi(t)$, η are independent.

It is obvious that the covariation function $r(t, s)$ of the random process $x(t)$ coincides with the covariation function of the random process $y_n(t)$. Calculate $r(t, s)$, $0 \leq t \leq s = t + \tau$:

$$r(t, s) = Mx(t)x(s) - Mx(t)Mx(s) = Mx^2(t) + Mx(t)(x(s) - x(t))$$

$$-Mx(t)Mx(s) = tM\lambda + t^2M\lambda^2 + t\tau M\lambda^2 - ts(M\lambda)^2 = tM\lambda + tsD\lambda.$$

From the multidimensional central limit theorem we have that multidimensional distributions of the random process $y_n(t)$ for $n \to \infty$ converge to multidimensional distributions of the Gaussian random process $\sqrt{M\lambda}\xi(t) + t\eta\sqrt{D\lambda}$.

Denote $\delta_k(t', t'') = x_k(t') - Mx_k(t') - x_k(t'') + Mx_k(t'')$, $M\delta_k(t', t'') = 0$. Calculate for $0 \leq t_1 \leq t \leq t_2 \leq T$

$$\sup_n M((y_n(t_2) - y_n(t))^2 (y_n(t) - y_n(t_1))^2)$$

$$= \sup_n M \left(\frac{\sum_{k=1}^{n} \delta_k(t, t_2)}{\sqrt{n}} \right)^2 \left(\frac{\sum_{k=1}^{n} \delta_k(t_1, t)}{\sqrt{n}} \right)^2$$

$$= \sup_n \frac{1}{n^2} \sum_{1 \leq k,l,p,q \leq n} M\delta_k(t,t_2)\delta_l(t,t_2)\delta_p(t_1,t)\delta_q(t_1,t)$$

$$= \sup_n \frac{1}{n^2} \left[\sum_{1 \leq k \leq n} M\delta_k^2(t_1,t)\delta_k^2(t,t_2) + \sum_{1 \leq k \neq l \leq n} M\delta_l^2(t_1,t)\delta_k^2(t,t_2) \right.$$

$$\left. +2 \sum_{1 \leq k \neq l \leq n} M\delta_k(t_1,t)\delta_k(t,t_2)\delta_l(t_1,t)\delta_l(t,t_2) \right] = \sup_n \left[\frac{M\delta_1^2(t_1,t)\delta_1^2(t,t_2)}{n} \right.$$

$$\left. +\frac{n(n-1)M\delta_1^2(t_1,t)\delta_2^2(t,t_2)}{n^2} + \frac{2n(n-1)M\delta_1(t_1,t)\delta_1(t,t_2)\delta_2(t_1,t)\delta_2(t,t_2)}{n^2} \right]$$

$$\leq M\delta_1^2(t_1,t)\delta_1^2(t,t_2) + M\delta_1^2(t_1,t)M\delta_2^2(t,t_2) + 2(M\delta_1(t_1,t)\delta_1(t,t_2))^2.$$

Further for a convenience introduce the designations $M\lambda = \overline{\lambda}$, $M\lambda^i = \overline{\lambda^i}$, $i = 2,3,4$. It is clear that

$$(M\delta_1(t_1,t)\delta_1(t,t_2))^2 \leq ((t-t_1)(t_2-t)\overline{\lambda^2})^2 \leq (t_2-t_1)^2 T^2 \overline{\lambda^2}^2,$$

$$M\delta_1^2(t_1,t)M\delta_2^2(t,t_2) \leq ((t-t_1)\overline{\lambda} + (t-t_1)^2\overline{\lambda^2})((t_2-t)\overline{\lambda} + (t_2-t)^2\overline{\lambda^2})$$

$$\leq ((t_2-t_1)\overline{\lambda} + (t_2-t_1)T\overline{\lambda^2})^2 = (t_2-t_1)^2(\overline{\lambda} + T\overline{\lambda^2})^2$$

and

$$M(x_1(t) - x_1(t_1))^2 = \overline{\lambda}(t-t_1) + \overline{\lambda^2}(t-t_1)^2 \leq \overline{\lambda}(t_2-t_1) + \overline{\lambda^2}(t_2-t_1)^2,$$

$$M(x_1(t_2) - x_1(t))^2 = \overline{\lambda}(t_2-t) + \overline{\lambda^2}(t_2-t)^2 \leq \overline{\lambda}(t_2-t_1) + \overline{\lambda^2}(t_2-t_1)^2,$$

$$M(x_1(t) - x_1(t_1))^2(x_1(t_2) - x_1(t))^2$$

$$= \int_0^\infty P(d\lambda)(\lambda(t-t_1) + \lambda^2(t-t_1)^2)(\lambda(t_2-t) + \lambda^2(t_2-t)^2)$$

$$\leq \int_0^\infty P(d\lambda)(\lambda(t_2-t_1) + \lambda^2(t_2-t_1)^2)^2 \leq (t_2-t_1)^2(\overline{\lambda^2} + 2\overline{\lambda^3}T + \overline{\lambda^4}T^2),$$

consequently

$$M\delta_1^2(t_1, t)\delta_1^2(t, t_2)$$

$$\leq M\left((x_1(t) - x_1(t_1))^2 + (t - t_1)^2\overline{\lambda}^2\right)\left((x_1(t_2) - x_1(t))^2 + (t_2 - t)^2\overline{\lambda}^2\right)$$

$$\leq 2(t_2 - t_1)^2\overline{\lambda}^2(\overline{\lambda}(t_2 - t_1) + \overline{\lambda^2}(t_2 - t_1)^2) + (t_2 - t_1)^2(\overline{\lambda^2} + 2\overline{\lambda^3}T + \overline{\lambda^4}T^2)$$

$$+((t_2 - t_1)^2\overline{\lambda}^2)^2 \leq (t_2 - t_1)^2[T^2\overline{\lambda}^4 + 2\overline{\lambda}^2(\overline{\lambda}T + \overline{\lambda^2}T^2) + \overline{\lambda^2} + 2\overline{\lambda^3}T + \overline{\lambda^4}T^2].$$

So for $0 \leq t_1 \leq t \leq t_2 \leq T$ the following inequality is true

$$\sup_n M((y_n(t_2) - y_n(t))^2 (y_n(t) - y_n(t_1))^2) \leq (t_2 - t_1)^2 C,$$

where

$$C = [T^2\overline{\lambda}^4 + 2\overline{\lambda}^2(\overline{\lambda}T + \overline{\lambda^2}T^2) + \overline{\lambda^2} + 2\overline{\lambda^3}T + \overline{\lambda^4}T^2] + 2T^2\overline{\lambda^2}^2 + (\overline{\lambda} + T\overline{\lambda^2})^2$$

Consequently for $\alpha = 2$, $\beta = 2$, $\lambda = 1$ and for calculated here constant C the "moment" condition of Kolmogorov-Chencov (1) is fulfilled. So from Theorem 1 we obtain C - convergence of the random process $y_n(t)$ to the random process $\sqrt{M\lambda}\xi(t) + t\eta\sqrt{D\lambda}$. Then the condition 2 of Theorem 4 is true also.

4.3   Aggregation of Poisson Input Flows with Common Random Intensity

Assume that the intensity λ of Poisson flow is random and has numerable number of meanings $\lambda_1, \lambda_2, \ldots$ with probabilities $p_1, p_2, \ldots$ Consider n Poisson flows which have the same random intensity λ and for any $\lambda = \lambda_k$ are independent (that is input flows are conditionally independent). Denote $x_1(t), x_2(t), \ldots$ random processes characterizing Poisson input flows with the random intensity λ common for all flows. Define input flow to aggregated n - server queuing system by the equality $e_n(t) = \sum_{k=1}^{n} x_k(t)$ and designate conditional random process $(e_n(t)/\lambda = \lambda_k)$.

It is clear that the conditional random process $(e_n(t)/\lambda = \lambda_k)$ for any k satisfies the condition 1 of Theorem 4. Define random process $y_n(t) = \dfrac{e_n(t) - Me_n(t)}{\sqrt{n}}$ and consider for $k = 1, 2, \ldots$ the sequence of conditional random processes

$$\{(y_n(t)/\lambda = \lambda_k), \ n \geq 1\}.$$

Using results of previous subsection it is not difficult to prove that the random process $(y_n(t)/\lambda = \lambda_k)$ for $n \to \infty$ C - converges to the random process

$\sqrt{\lambda_k}\xi(t)$, $0 \le t \le T$. Here $\xi(t)$ is the standard Wiener process. Consequently the conditional random process $(e_n(t)/\lambda = \lambda_k)$ for any k satisfies the conditions 1, 2 of Theorem 4.

If for any $k \ge 1$ the following condition

$$\int_0^T \overline{F}(t)\lambda_k dt < 1 \tag{8}$$

is true then from Theorem 4 we have for $k \ge 1$ the relation

$$P\left(\sup_{0 \le t \le T} q_n(t) = n/\lambda = \lambda_k\right) \to 0, \; n \to \infty. \tag{9}$$

Put arbitrary $\varepsilon > 0$ and define such $K(\varepsilon) > 0$ that $\sum_{k>K(\varepsilon)} p_k < \varepsilon$. Then by chosen $\varepsilon, K(\varepsilon)$ it is possible to find such $N(\varepsilon)$ that for $n > N(\varepsilon)$

$$P\left(\sup_{0 \le t \le T} q_n(t)/\lambda = \lambda_k\right) < \varepsilon, \; 1 \le k \le K(\varepsilon).$$

Consequently we have

$$P\left(\sup_{0 \le t \le T} q_n(t) = n\right) = \sum_{k=1}^{\infty} p_k P\left(\sup_{0 \le t \le T} q_n(t) = n/\lambda = \lambda_k\right) \le 2\varepsilon, \; n > N(\varepsilon).$$

So from the condition (8) random input flow described by the process $e_n(t)$ satisfies to the limit relation $P\left(\sup_{0 \le t \le T} q_n(t) = n\right) \to 0, \; n \to \infty.$

References

1. Zhidkova, L.S., Moiseeva, S.P.: Investigation of fold customers parallel service of simplest flow. Bull. Tomsk Univ. Control Comput. Sci. Inf. **17**(4), 49–54 (2011) (In Russian)
2. Nazarov, A.A., Moiseeva, S.P., Morozova, A.S.: Investigation of queuing systems with reconversion and infinite number of servers by method of limiting decomposition. Calculating Technol. **35**, 88–92 (2008) (In Russian)
3. Borovkov, A.A., Mogulskiy, A.A., Sakhanenko, A.I.: Limit theorems for random processes. Totals of Science and Technique. Series of Modern Problems in Mathematics and Physics. Fundamental Directions, 82, 5–194 (1995) (In Russian)
4. Chencov, N.N.: Weak convergence of random processes without breaks of second kind. Probab. Theor. Appl. 1(1), 154–161 (1956) (In Russian)
5. Dmitrovskiy, V.A.: Condition of boundedness and estimates of maximum distribution for random fields on arbitrary sets. Lectures Academy Sciences of USSR. 253(2), 271–274 (1980) (In Russian)
6. Lifshits, M.A.: On distribution of Gaussian process maximum. Probab. Theor. Appl. 31(1), 134–142 (1986) (In Russian)
7. Borovkov, A.A.: Asymptotic Methods in Queuing Theory. Science, Moscow (1980) (In Russian)
8. Khinchin, A.Y.: Researchs on Mathematical Queuing Theory. Phyzmatlit, Moscow (1963) (In Russian)

Fractal Queues Simulation Peculiarities

Vladimir Zadorozhnyi[(✉)]

Omsk State Technical University, Omsk, Russia
Zwn2015@yandex.ru

Abstract. Relevant to the modern theory of computer networks design
questions of developing adequate service models of fractal traffic are considered in the article. The fidelity criteria of heavy-tailed distributions
(HTD), which take into account the HTD distortion effect on the results
of fractal queues simulation, are offered. The problem of HTD significant
distortions in their realization during simulation is revealed and examined. To solve this problem we also developed the method, which does
not require the use of "long arithmetic".

Keywords: Fractal traffic · Queueing theory · Simulation modeling ·
Random number generators

1 Introduction

The discovery of fractal properties of information networks traffic [1] was the
key to understanding the reason of a number of major failures in projects of network devices aimed at ensuring the quality of information exchange and based
on the classical theory of queues. This discovery led to a radical correcting of
mathematical models of traffic and methods of traffic service. In the description and analysis of fractal traffic such mathematical concepts as self-similar
stochastic process, long-range dependence (LRD) and heavy-tailed distribution
(HTD) have been widely used [1–3]. In these studies much attention has been
given to the methods of fractal traffic identification based on various types of
mathematical models [4].

The queueing systems are the most appropriate mathematical models [3,5–8]
for the purposes of designing the network devices on the system level. This article
discusses and solves some problems caused by peculiarities of simulation of fractal
queueing systems [8]. Fractal systems (FS) are called the systems of $GI|GI|n|m$
class, which have the following properties. The independent interval τ_i between
the time of receipt of the i-th and $(i+1)$-th requests in FS has the same for all
i distribution function (DF) $A(t)$ with mathematical expectation (m.e.) $E(\tau_i) =
\bar{\tau} < \infty$. Independent service interval x_i (processing time) of any i-th request has
DF $B(t)$ with m.e. $E(x_i) = \bar{x} < \infty$. At least one of the DF $A(t), B(t)$ describes
a fractal random variable (r.v.) (i.e. it is asymptotically power [9]) and has an
infinite variance. The load rate ρ of considered FS does not exceed one:

$$\rho = \frac{\bar{x}}{n\bar{\tau}} \leq 1.$$

© Springer International Publishing Switzerland 2015
A. Dudin et al. (Eds.): ITMM 2015, CCIS 564, pp. 415–432, 2015.
DOI: 10.1007/978-3-319-25861-4_35

As FS study with a help of analytical methods is complicated, so the simulation modeling (SM) [6] is widely used for their calculation, and one of the most common tools is a GPSS simulation system [10]. The traditional way to improve the reliability and accuracy of simulation estimates is the development and application of appropriate powerful methods of variance reduction, such as the essential sampling method and the method of stratification; the methods are used to estimate the probability of rare events [11]. Unfortunately, the available literature describing the methods of variance reduction in FS modeling refers mainly to estimated ruin probability in models of risk insurance, or, equivalently, the waiting time in line systems with a single channel. Asmussen et al. [12] described several algorithms for the $M|G|1$ queue, which require an explicit representation of random sums timeout. Boots and Shahabuddin [13] developed a very efficient algorithm for SM of $GI|GI|1$ system with Weibull distribution of service time. The development of efficient parameter estimates for HTD based on the limited samples relates to the same range of issues. The solution of this problem is given in a large number of publications; we point out for example the papers [14,15]. However, we found no studies in which the accuracy of implementation of HTD on the basis of pseudo random number generators has been studied by accurate methods, rather than using sample estimates. The simple accurate methods applied in the article can identify the problem of serious distortions realized in SM of HTD. The problem of HTD distortion is analyzed on the basis of the need to ensure proper FS simulation. There are developed effective ways to solve this problem.

2   FS Simulation Tasks

Typical representatives of FS systems are $Pa|M|n|m$, $M|Pa|n|m$ and $Pa|Pa|n|m$. Here the symbol Pa corresponds to the Pareto distribution (PD). If the buffer size m is finite, it is required for a given m to define the probability P of request loss (direct problem) or to find the smallest size m, guaranteeing that the probability of loss is not greater than P (inverse problem). When $m = \infty$ is then the subject of special interest is the average queue length L or the average queueing time W.

Power and asymptotically power distributions, the typical representative of which is PD, are referred to HTD. PD with the parameters K, α will be denoted in short as $Pa(K, \alpha)$. Its distribution function $F(t)$ has the form

$$F(t) = 1 - \left(\frac{K}{t}\right)^{\alpha}, \quad \alpha > 0, \quad K > 0, \quad t \geq K, \tag{1}$$

where α is the shape parameter, K is the smallest value of r.v. (and the scale parameter).

The initial moment of k-th order $\xi^{(k)}$ for r.v. $\xi \in Pa(K, \alpha)$ is determined taking into account the DF (1) as follows:

$$\xi^{(k)} = \mathrm{E}(\xi^k) = \int\limits_{K}^{\infty} t^k dF(t) = \begin{cases} \frac{\alpha K^k}{\alpha - k}, & \alpha > k, \\ \infty, & \alpha \leq k, \end{cases} \quad (k = 1, 2, ...). \tag{2}$$

Hence, when $\alpha > 1$ we obtain a finite mathematical expectation

$$\xi^{(1)} = \bar{\xi} = \frac{\alpha K}{(\alpha - 1)}, \tag{3}$$

when $\alpha > 2$ – the finite variance

$$\mathrm{Var}(\xi) = \sigma_\xi^2 = \xi^{(2)} - \bar{\xi}^2 = \frac{\alpha K^2}{(\alpha - 1)^2 (\alpha - 2)}. \tag{4}$$

If we consider PD with infinite variance and finite m.e., the shape parameter α is situated in the range $1 < \alpha \leq 2$. This range of values of α is the most relevant in simulation of information networks with fractal traffic. The smaller α, the heavier the tail of PD is. By changing scale parameter K at given α, one can obtain any required m.e.

If both DF $A(t)$ and $B(t)$ are PD, then FS adequately takes into consideration all the main features of the serviced fractal traffic [1] – namely its statistical self-similarity, the presence of specific load fluctuations and LRD on the incoming values.

Adequate reflection of the main peculiarities of the serviced fractal traffic by network devices leads to a wide application of FS to address issues of quality assurance of data exchange. This queueing systems study, in which DF $A(t)$ and/or $B(t)$ are HTD, is also a promising direction of the mathematical queueing theory development [3]. This confirms the relevance of the questions which are considered in the article.

3 The Problem of HTD Distortion Realized in SM

3.1 Displacement of PD Moments

Using the method of the inverse transformation (inversion) DF (1) we obtain the following formula to generate r.v. $\xi \in Pa(K, \alpha)$:

$$\xi = K(1 - u)^{-1/\alpha}. \tag{5}$$

where u is basic r.v. (BRV) uniformly distributed in the area from 0 to 1. As a result of the equivalent replacement distribution $(1 - u) \to u$, formula (5) takes the form

$$\xi = K u^{-1/\alpha}. \tag{6}$$

Formula (6) can be obtained directly by inverse transformation of the tail $\bar{F}(t) = 1 - F(t) = (K/t)^\alpha$:

$$\xi = \bar{F}^{-1}(u) = K u^{-1/\alpha},$$

where $\bar{F}^{-1}$ is inverse to $\bar{F}$ function. We will show in Sect. 5 that the formulas derived from tail inverse transformation are more preferable then the formulas derived from DF inverse transformation not only in terms of high speed but also in terms of the overcoming HTD distortions.

The reason of distortions realized in the SM of HTD is a discreteness of BRV software generators. Software random number generators realizes *discrete* BRV (DBRV) u' with a set of equally probable values $\{0, \varepsilon, 2\varepsilon, 3\varepsilon, ..., 1 - \varepsilon, 1\}$ forming a uniform lattice with step ε (usually zero or one of this set are excluded). For example, in GPSS citeten generator Uniform(1,0,1) realizes DBRV, taking values $\{0.000000, 0.000001, ..., 0.999999\}$ with a step between them $\varepsilon = 10^{-6}$. The value of ε will be further called the sample spacing. Using the expression Uniform(1,0,1)+1E-6#Uniform(1,0,1) in GPSS the DBRV is realized with sample spacing $\varepsilon = 10^{-12}$ (the sign # denotes here multiplication, constant 1E-6 – value 10^{-6}). In other languages a step $\epsilon \approx 10^{-15}$ is usually provided by.

Thus, in the inverse transformation (6) there instead of BRV u there is DBRV u' and, consequently, instead of a continuous r.v. $\xi \in Pa(K, \alpha)$ the discrete r.v. (d.r.v.) ξ' is implemented:

$$\xi' = K(u')^{-1/\alpha}. \tag{7}$$

In SM of FS it turns out that in the case of the PD realization and other HTD (as opposed to the realization of DF with light tails) the step $\varepsilon = 10^{-15}$ and, especially, the step $\varepsilon = 10^{-6}$ is too large and leads to a significant difference the properties of generated d.r.v. ξ' from the properties of r.v. $\xi \in Pa(K, \alpha)$. Let us determine, for example, the exact values of the initial moments of $\xi'^{(k)}$ d.r.v. ξ' (7), taking into account the finite number $N = 1/\varepsilon$ of its possible values, corresponding to the values DBRV u'_i:

$$\xi'^{(k)} = \mathrm{E}(\xi'^k) = \sum_{i=1}^{N} \mathrm{P}(\xi'_i)\xi'^k_i = \sum_{i=1}^{N} \frac{1}{N} K^k u'^{-\frac{k}{\alpha}}_i = K^k N^{-1} \sum_{i=1}^{N} (iN^{-1})^{-\frac{k}{\alpha}}$$

$$= K^k N^{-1} \sum_{i=1}^{N} i^{-\frac{k}{\alpha}} N^{\frac{k}{\alpha}} = K^k N^{\frac{k}{\alpha}-1} \sum_{i=1}^{N} i^{-\frac{k}{\alpha}} \quad (\alpha > k, \ N = 1/\varepsilon); \tag{8}$$

then calculate for different ε m.e. $\mathrm{E}(\xi')$, a variation $\sigma^2_{\xi'}$ and coefficient of variation (c.v.) C'_ξ d.r.v. ξ' on the derived from (8) formulas

$$\mathrm{E}(\xi') = KN^{\frac{1}{\alpha}-1} \sum_{i=1}^{N} i^{-\frac{1}{\alpha}}, \tag{9}$$

$$\sigma^2_{\xi'} = \xi'^{(2)} - \mathrm{E}^2(\xi') = K^2 N^{\frac{2}{\alpha}-1} \left(\sum_{i=1}^{N} i^{-\frac{2}{\alpha}} \right) - \mathrm{E}^2(\xi'), \quad C_{\xi'} = \frac{\sigma_{\xi'}}{\mathrm{E}(\xi')} \tag{10}$$

and compare it with the corresponding numerical characteristics of r.v. $\xi \in Pa(K, \alpha)$ (Table 1). The characteristics $E(\xi)$ and C_ξ of r.v. $\xi \in Pa(K, \alpha)$ are calculated using the formulas (2). All values in the Table are precise or approximated to five significant digits.

Presented in Table 1 the results of calculations using the exact formulas (9) and (10) demonstrate the presence of substantial deviations of characteristics

Table 1. Numerical characteristics d.r.v. ξ', implementing the r.v. $\xi \in Pa(K, \alpha)$ at $K = 1$

α	$E(\xi')$			$E(\xi)$	$C_{\xi'}$			C_ξ
	$\varepsilon = 10^{-6}$	$\varepsilon = 10^{-12}$	$\varepsilon = 10^{-15}$		$\varepsilon = 10^{-6}$	$\varepsilon = 10^{-12}$	$\varepsilon = 10^{-15}$	
1.01	13.415	24.612	29.662	101.00	3.856	$3.9866 \cdot 10^4$	$9.7690 \cdot 10^5$	∞
1.1	8.0297	10.154	10.549	11.000	48.305	$1.0882 \cdot 10^4$	$1.7677 \cdot 10^5$	
1.2	5.4565	5.9457	5.9828	6.0000	26.687	2450.9	$2.4357 \cdot 10^4$	
1.3	4.1777	4.3269	4.3320	4.3333	15.488	618.20	3965.6	
1.4	3.4432	3.4989	3.4999	3.5000	9.5534	182.69	802.54	
1.5	2.9755	2.9998	3.0000	3.0000	6.2716	63.247	200.02	
1.6	2.6548	2.6666	2.6666	2.6666	4.3646	25.390	60.269	
1.7	2.4222	2.4286	2.4286	2.4286	3.2008	11.710	21.649	
1.8	2.2463	2.2500	2.2500	2.2500	2.4563	6.1655	9.2253	
1.9	2.1089	2.1111	2.1111	2.1111	1.9590	3.6806	4.6685	
2	1.9985	2.0000	2.0000	2.0000	1.6136	2.4601	2.7891	

generated by d.r.v. ξ' on the characteristics to be realized by r.v. $\xi \in Pa(K, \alpha)$. Therefore, we will further talk about this d.r.v. ξ' as the r.v. belonging to the discrete Pareto distribution $DPa(K, \alpha, \varepsilon)$ where ε is a sample spacing of used DBRV. In Table 1 the values of $E(\xi')$, $C_{\xi'}$ are the values to be converged to by corresponding sample estimates with the increase of the amount of realized samples to infinity.

Table 1 also shows that the heavier the tail of PD is, the more m.e. of actually realized r.v. ξ' differs from m.e. subject to the realization of r.v. ξ. The ends of the interval $1 < \alpha \leq 2$ are the centers of critical areas: $\alpha = 1$ point is a center of critical area for the realized m.e., $\alpha = 2$ point is for the realized c.v.

The reason for HTD distortions is that for values DBRV u' similar while using a formula of transforming DF to one and using the inversion formula of the tail to zero, with heavy tails of the distribution and the usual step ε there are too many large r.v. values that are essential for the formation of moments, they are not realized simply.

3.2 The Elementary Universal Method of Reducing Distortion

Estimating the general case (for all $\alpha > 0$) the last in (8) the sum at the top and bottom with suitable integrals by means of elementary algebraic transformations there can be derived a simple estimate of the moments displacement $\Delta\xi^{(k)}$:

$$\frac{k}{\alpha}\xi^{(k)}\varepsilon^{1-\frac{k}{\alpha}} < \Delta\xi^{(k)} < \xi^{(k)}\varepsilon^{1-\frac{k}{\alpha}}, \quad (\alpha > K), \tag{11}$$

where $\Delta\xi^{(k)} = \left|\xi'^{(k)} - \xi^{(k)}\right|$. In this case actually realized k-th interval $\xi'^{(k)}$ is always less than the interval to be realized $\xi^{(k)}$. Coefficient of reducing $\gamma = \xi^{(k)}/\xi'^{(k)} \to \infty$ is at $\alpha \downarrow k$. If $\alpha \leq k$, then $\xi^k = \infty$ and $\gamma = \infty$.

The performed calculations and the analysis of the obtained general equation (11) leads to the conclusion that the moment displacement of PD using any fixed

ε (i.e. any length digit grid of generator BRV) can be arbitrarily large. Obviously, there are similar features in other HTD realization.

When realizing light-tailed distributions with discreteness of machine arithmetic does not lead to significant errors. For example, the transformation $(-\lambda^{-1} \times \ln u')$ of DBRV u' in the step $\varepsilon = 10^{-6}$ implements discrete version of exponential r.v. with m.e. $0.999992 \cdot \lambda^{-1}$ and c.v. 0.99995. The relative differences between these values from the corresponding exact characteristics λ^{-1} and 1 of continuous exponential r.v. are comparable to ε and they are sufficiently small to allow to ignore them in the practice of SM systems with queues.

The differences between PD and the realized discrete PD can be estimated by the distance R between the DF $F(t)$ and $F_d(t)$:

$$R = \max_t |F_d(t) - F(t)|, \qquad (12)$$

where $F_d(t)$ is DF r.v. ξ' realized for the given ε transformation (7). The transformation (7) displays a uniform lattice N of equally probable values of d.r.v. u' on irregular lattice N of equally probable values of d.r.v. ξ'. DF $F_d(t)$ is stepped piecewise constant function with altitude jumps $R = \Delta = 1/N = \varepsilon$ in points $t = \xi_i \pm 0$ of r.v. possible values ξ' $(i = 1, ..., N)$. Thus, the distance R (12) between $F(t)$ and $F_d(t)$ is equal to a step ε of BRV.

Hence the universal elementary method of reducing realized HTD distortion consists in the reduction of step ε, i.e. in the digit increase of used generators of standard random numbers. When $\varepsilon \to 0$ the distance $R = \varepsilon \to 0$, and the moments displacement (11) $\Delta \xi^{(k)} \to 0$. By increasing the number of decimal places to one, step ε is reduced by an order.

Unfortunately, this simple method for any fixed number of digits does not guarantee the appropriate accuracy of the PD realization and many other HTD. According to Table 1, if $\alpha \le 1.3$, then 15-digits of generators are not enough number to realize m.e. $E(\xi')$ with acceptable accuracy, providing at least 4 decimal digits of precision. And it is not enough for the realization of the infinite "with the precision", i.e. achieved the value of 10,000, the variation coefficient is near the point $\alpha = 2$. Using formulas (9) and (10), the sum in which can be easily calculated by means of WolframAlpha, it is easy to determine that for 4 exact numbers of realized m.e. when $\alpha = 1.1$ it is necessary to generate DBRV u' with more than 30 decimal digits, and for the realization of acceptable accuracy c.v. when $\alpha = 2$, and 1000 decimal digits of DBRV is not enough. But the main disadvantage of this elementary method is that the increase in the number of digits DBRV u' does nothing if the realization of the inverse transform (5) or (6) does not perform calculations with the same number of significant digits. It requires the use in the generation of r.v. $\xi \in Pa(K, \alpha)$ "long arithmetic" (realized, for example, by presentation of numbers of character strings) that drastically reduces the rate of r.v. generation. In practice, it leads to a reduction by orders of magnitude in the received samples and the effect of verification distributions realized in these samples is not shown.

4   Effect of HTD Distortion on SM of FS Results

4.1   Singularities of Empirical Estimates of PD Parameters

If $\alpha \in (1, 2]$ for a sample $\xi'_1, ..., \xi'_n$ of d.r.v. values $\xi' \in DPa(K, \alpha, \varepsilon))$ the estimation $\hat{E}_{\xi'} = (\xi'_1 + ... + \xi'_n)/n$ is calculated, then $\hat{E}_{\xi'} \to \mathrm{E}(\xi')$ when $n \to \infty$. The convergence estimating $\hat{E}_{\xi'}$ to $E(\xi')$ is very slow due to the large c.v. $C_{\xi'}$. Thus if $\alpha = 1.1$, then even if a relatively large sample spacing $\varepsilon = 10^{-6}$ to obtain an acceptable approximation 8.028 to the exact value of $E(\xi') = 8.0297$ (see Table 1) required volume sample was at $n = 10^9$.

At the same time different ways of the distribution parameters estimating in the sample $\xi'_1, ..., \xi'_n$ lead to different results. Therefore, we must carefully estimate and justify the applied data and cautiously interpret them. For example, if the sample $\xi'_1, ..., \xi'_n$ parameters K and α are estimated by maximum likelihood method, leading to the formulas

$$\hat{\alpha} = n/(\sum\nolimits_{i=1}^{n} \ln \xi'_i - n \ln \hat{K}), \quad \hat{K} = \min\{\xi'_i\}, \tag{13}$$

then such estimates converge to the exact values of K and α sufficiently fast and even for $\varepsilon = 10^{-6}$ are substantially unbiased. With the obtained estimates $\hat{K}, \hat{\alpha}$ it is possible to get an indirect estimate of m.e. in the form $\hat{E}'_{\xi} = \hat{K}\hat{\alpha}/(\hat{\alpha} - 1)$. This estimate, respectively, is sufficiently precise. But it does not converge to the m.e. $E(\xi')$ of sampling units but to m.e. $E(\xi) = \alpha K/(\alpha - 1)$ of r.v. ξ subjected to realization. The reason for this lies in the fact that the method of maximum likelihood uses the probability density of the r.v. $\xi \in Pa(K, \alpha)$.

4.2   Singularities of SM FS $M|Pa|1|\infty$

The system $M|Pa|1|\infty$ service time x appertains to the distribution $Pa(K, \alpha)$ and, if $\alpha \in (1, 2]$, then the stationary average length $L(\rho)$ of requests line is infinite for any $\rho \in (0, 1]$. In fact, according to the Pollaczek-Khinchine formula [5] here for any $\rho \in (0, 1]$ there is

$$L(\rho) = \frac{\rho^2(1 + C_x^2)}{2(1 - \rho)} = \infty, \tag{14}$$

as the variation coefficient of C_x at $\alpha \in (1, 2]$ is infinite. But if FS to be investigated by SM method, then instead of service time $x \in Pa(K, \alpha)$ realized during $x' \in DPa(K, \alpha, \varepsilon)$ is realized, and for example, when $K = 1, \alpha = 2, \varepsilon = 10^{-15}$ in accordance with (14), at a sufficiently high run-length model it is obtain:

$$L(\rho) \approx \frac{\rho^2(1 + 2.79^2)}{2(1 - \rho)} \approx \frac{4.39\rho^2}{1 - \rho} \tag{15}$$

(see Table 1). This conclusion is also confirmed by the results of simulation experiments. The peculiarity of the case is considered a fundamental difference of simulation solutions (15) of the exact (14).

4.3   Singularities of SM FS $Pa|M|1|\infty$

In the system $Pa|M|1|\infty$ the average waiting time can be found accurately [5] by the formula

$$W = \frac{\sigma}{\mu(1 - \sigma)}, \tag{16}$$

where $\mu = 1/\bar{x}$ is the service intensity, σ is the only root of the equation

$$\sigma = A^*(\mu - \mu\sigma) \tag{17}$$

in the area $0 \leq \sigma < 1$, $A^*(s) = \int\limits_0^\infty e^{-st} dA(t)$ is Laplace transform (LT) distribution function $A(t)$.

In this case $A(t)$ is Pareto DF (1). Therefore, LT $A^*(s)$ is as follows:

$$A^*(s) = F^*(s) = E(e^{-s\tau}) = \int\limits_0^\infty e^{-st} dF(t) = \int\limits_K^\infty e^{-st} \frac{\alpha K^\alpha}{t^{\alpha+1}} dt = \alpha K^\alpha \int\limits_K^\infty e^{-st} t^{-\alpha-1} dt,$$

where $\tau \in F(\tau)$. Assuming it the Eq. (17) is reduced to relatively easily solved by numerical method equation:

$$\sigma = \alpha K^\alpha \mu^\alpha (1 - \sigma)^\alpha \Gamma(-\alpha, K\mu(1 - \sigma)), \tag{18}$$

where $\Gamma(c, x)$ is incomplete gamma function: $\Gamma(c, x) = \int_x^\infty t^{c-1} e^{-t} dt$.

Solving the Eq. (18) by the numerical method at $\alpha = 1.1, K = 1$ (i.e. $\bar{\tau} = 11$) for any ρ (i.e. $\mu = 1/(\rho\bar{\tau}) = 1/11/\rho$), we determine the corresponding σ and L, presented in Table 2 (the exact values are approximated). It also presents the results of the FS SM at $\varepsilon = 10^{-12}$ and run-length $n = 10$ million requests – they are estimates L_{SM} of average queue length L and estimates ρ_{SM} of load rates ρ.

Table 2. Results of exact calculation and simulation of FS $Pa|M|1|\infty$ at $\tau' \in DPa(1,\ 1.1, \varepsilon)$, $\varepsilon = 10^{-12}$, $n = 10^7$

ρ	σ	L	L_{SM}	ρ_{SM}	ρ'
0.1	0.259078	0.03497	0.047	0.136	0.108
0.2	0.648676	0.36928	0.505	0.275	0.217
0.3	0.871632	2.03703	2.733	0.402	0.325
0.4	0.964199	10.7730	14.41	0.532	0.433
0.5	0.992816	69.1015	98.14	0.673	0.542
0.6	0.999076	648.429	859.3	0.831	0.650
0.7	0.999939	11526.7	–	–	0.758

Note that all estimates ρ_{SM} in Table 2, despite the high run-length n, are significantly higher than its final values when $n \to \infty$ values $\rho' = \frac{\bar{x}}{E(\tau')} = \frac{\bar{x}}{\bar{\tau}} \frac{\bar{\tau}}{E(\tau')} =$

$\rho\frac{11}{10.154} = 1.0833\, \rho$ (see also Table 1), which illustrates the problem of slow convergence estimates for SM of FS. The reasons for this phenomenon, which quite complicates SM of FS are the infinite variance of distribution $A(T), B(T)$, and the high correlation between the number of requests coming in distant from each other with equal length intervals of time.

To calculate the average queue length realized in SM, you can calculate the LT DF $F_d^*(s)$ r.v. $\tau' \in DPa(K, \alpha, \varepsilon)$ according to the formula

$$F_d^*(s) = \mathrm{E}(e^{-s\tau'}) = \mathrm{E}(e^{-sK(u')^{-1/\alpha}}) = \frac{1}{N}\sum_{i=1}^{N} e^{-sK(\varepsilon i)^{-1/\alpha}} \quad (\varepsilon = 1/N). \quad (19)$$

As the exact calculations show, the LT $F^*(S)$ and LT $F_d^*(s)$ for various α and $\varepsilon \leq 10^{-6}$ coincide over wide ranges of values s with precision up to 6 and more significant digits. However, the impact on the SM results of such a small distortion of LT caused by the discreteness of BRV generators can be significant.

Let us consider a specific example of the SM of system $Pa|M|1|\infty$. Let us suppose that $\alpha = 1.1, K = 1$ (while the average time $\bar{\tau}$ between the requests receipt is 11) is required to find $W = W(\rho)$ when the load rate $\rho = \bar{x}/\bar{\tau} = 0.4$. It occurs when loading $\bar{x} = 0.4\bar{\tau} = 4.4$, i.e. when $\mu = 1/\bar{x} = 0.22727$. Numerically solving equation (17) for this value of μ and $A^*(s) = F_d^*(s)$ (19), we determine $\sigma = 0.964199$, and, according to (16), $W = W(0.4) = \mathbf{118.5}$.

When SM of this FS with generator sample spacing of DBRV $\varepsilon = 10^{-6}$ average time $\bar{\tau}'$ between requests entering will be 8.0297 (see Table 1). The load rate $\rho = 0.4$ is reached at $\bar{x} = 0.4 \cdot 8.0297 = 3.21188$, i.e., when $\mu = 1/\bar{x} = 0.311344$. SM of this FS at such μ (when running about 125 million requests) provides for the load rate ρ the estimation $\rho_{SM} = 0.401 \approx 0.4$, and for W – the estimation $W_{SM} = \mathbf{19.657}$. Thus, the simulation solution of the task of the magnitude of $W_{SM}(\rho)$ at $\rho = 0.4$ obtained by SM is different from the exact solution more than five times.

The obtained by SM method results uniquely are determined by the discreteness of the used BRV generator. Indeed, taking into account the LT (19), the Eq. (17) for the considered FS takes the form

$$\sigma = 10^{-6}\sum_{i=1}^{10^6} 1/\exp[(0.3113441(1-\sigma))(10^{-6}i)^{-1/1.1}], \quad (20)$$

and its exact numerical solution $\sigma = 0.85953$, in accordance with (16), at $\mu = 0.311344$ giving $W' = \mathbf{19.653}$. The obtained above simulation estimation $W_{SM} = \mathbf{19.657}$ is consistent with the exact solution. For the transition from calculated with help of Eq. (20) W', implemented in the SM when $n \to \infty$ the corresponding L', you can use the Little's formula in the form of $L' = W'/\bar{\tau}'$.

No matter how small HTD sampling is it is very important in terms of moments realization, minor LT distortions of realized HTD are important for solution σ to Eq. (17). Finally, we discover the existence of significant, sometimes fundamental deterministic errors in the results of SM FS $M|Pa|1|\infty$ and $Pa|M|1|\infty$.

4.4   Singularities of Other Systems with HTD

The performed for systems $Pa|M|1|\infty$ and $M|Pa|1|\infty$ analysis of deterministic errors of average waiting time W and the average queue length L, realized in the SM shows that these errors are significant and are caused by the final length of digits grid of BRV generators. Obviously, that SM of other FS with power and asymptotically power distributions determined at the input patterns generally also burdens with significant deterministic errors of obtained results in general case. This conclusion can be extended to SM of all other systems as $GI|GI|m|n$ class with asymptotically power distribution and other HTD.

5   Effective Methods to Eliminate HTD Distortions

The control of results accuracy of SM with a help of classical methods such as "three sigma rule" allows evaluating only a random deviation of simulated evaluation of the exact desired values, which is caused by the statistical nature of these estimates. The analysis made in Sect. 4 of the article shows that in FS simulation along with classical control of estimates' statistical error it is necessary to provide also a control of deterministic errors arising due to the distortion of realized HTD or to eliminate the deterministic error through the development of more accurate methods of HTD generation. We offer effective methods for HTD generating, which can eliminate distortions in the sense that the amount of distortion is reduced to the level of the minor computational errors that typical for realization of light-tailed distributions.

5.1   Summation of Scaled Pseudo Random Numbers

A simple and, as the in-depth analysis shows, agreed with machine arithmetic way out is to realize BRV in the form of "long sum" of several appropriately scaled realizations of the standard random numbers. A simple example of such sum in the GPSS language (with use of the Uniform generator twice) is given in Sect. 3.1. Similarly BRV can be realized as the sum of several scaled random numbers obtained with the help of a standard generator.

However, in the above example, the GPSS language the correct application of the sum of two 6 digits uniformly distributed r.v., where the first term is realizing the first 6 digits after the decimal point, and the second – the next 6 digits has no doubt. Indeed, in general, the GPSS uses arithmetic calculates with an accuracy of at least 15 decimal digits. Therefore, this sum is realized exactly as 12 digits BRV. But if you add three independent standard random numbers using the expression Uniform(1,0,1)+1E-6#Uniform(1,0,1)+1E-12#Uniform(1,0,1), then instead of 18-digits BRV it actually formed approximated result with a smaller number of significant digits. Even longer sums will also be approximated to the number of digits, which is realized in the simulation language by the machine of floating-point arithmetic.

However, the use of such "long sum" is realized by formulas, obtained with a help of the tail inverse transformation, like formula (6), and allows you to solve the problem of HTD distortion.

Similarly, in a program with the usual arithmetic (with 15 significant decimal digits) using the standard 15 digits random number generator – let it be a generator $RAND()$ to realize uniform in the interval $(0, 1]$ DBRV u' with sample spacing $\varepsilon = 10^{-15}$, the u'' sum of three terms

$$u'' = RAND() + 10^{-15}RAND() + 10^{-30}RAND() \tag{21}$$

will be a random number with 15 exact significant figures instead of 45. However, if in order to implement HTD the formula obtained by the inverse transformation of the tail is used, then the problem of HTD distortion with a help of long sums as (21) is actually achieved. This occurs because the step of values DBRV u'' lattice decreases as they are closer to zero (see Sect. 3.1). If for values close to one the lattice spacing is equal to 10^{-15} , then it near zero (where the first two terms in (21)vanish), the lattice spacing reaches 10^{-45}, since in mashine arithmetic the first zeros after the decimal point at the mantissa are converted to the value of the number's order. Formula (21) can be used, for example in Excel, wherein step ε is exactly 10^{-15}.

Depending on the method of realization of the standard generator its sample spacing may be different. For example, if the RAND() generator is implemented by multiplicative congruent method with module $m = 2^{31}$, the lattice spacing value will be $\varepsilon = 2^{-31}$. In this case, as the scale factor in terms of additional components of weighted sum of type (21) one should use $2^{-31}, 2^{-62}$, etc.

Let us calculate m.e. of Pareto d.r.v. ξ', realized by formula (6) with $K = 1, \alpha = 1.1$ with use of DBRV u'', which is formed by the "long-sum" (21) .To do this we first assume that all u''_i are 45-bit values with sample spacing $\varepsilon = 10^{-45}$. Then, in accordance with formula (9)

$$E(\xi') = 10^{-45} \sum_{i=1}^{10^{45}} (u''_i)^{-\frac{1}{1.1}} = 10^{45\left(\frac{1}{1.1}-1\right)} \left(\sum_{i=1}^{10^{45}} i^{-\frac{1}{1.1}} \right) = 10.9991540... , \tag{22}$$

that is sufficiently close to the exact value of m.e. subject to the realization of Pareto r.v. $E(\xi) = 11$ (a normal 15-digit generator realizes r.v. ξ' with m.e. 10.549, see Table 1). Now let us take into account that only 10^{15} values of u'', realized by the sum of (21) (when the first two terms are equal to zero), coincide with the corresponding first at the bottom 10^{15} values of 45-digit DBRV. Others 10^{30} values realized by the sum (21) coincide with the corresponding values of 45-digits DBRV only up to 15 significant digits (not counting the first zeros after the decimal point). Applying the sensitivity analysis, it is easy to show that the result of the calculation (22) with such precision representation of values u'' representation will change only after the 15-th decimal figure. However, the general rules of approximation in the calculation prove the same.

However, it means that all the nine shown in (22) digit values $E(\xi')$ are accurate, and therefore, $E(\xi')$ coincides with $E(\xi) = 11$ up to five significant figures. The calculations confirmed that the method of "long sums" really allows us to solve the problem of distortion of HTD in SM.

The advantage of the "long sums" method lies in its simplicity. When using it to solve the problem of HTD distortion we do not need to use the "long arithmetic" and slowing of calculations only occurs during the generation of r.v. with HTD because of use of pseudo random number generator several times. Calculations by analogy with (22) made for 6-digits random number generators indicate that an acceptable deterministic error of m.e. realization of Pareto r.v. ξ if $\alpha \geq 1.1$ is achieved by a "long-sum" six or eight scaled random numbers. The m.e. of realized r.v. ξ' in this case coincides at $\alpha \geq 1.1$ with the m.e. of r.v. ξ wit the accuracy up to four or five significant digits.

Disadvantages of the method consists in the fact that it does not guarantee the absence of moments displacement over the entire range of realized distribution parameters, and in case of "long enough sum" of scaled random numbers the generation of random variables slows down in proportion to the number of terms in this sum.

We emphasize once again that the method of long sums does not work when HTD realizing formulas are received with inverse transformation of DF. For example, when using the formula (5) the method will not work, because without a transition to "long arithmetic" it is not possible to approach to one with a small step. The usual double-precision arithmetic, which provides 15 accurate decimal digits, by subtracting the number less than 10^{-15} from one results in one exactly.

5.2   Basic Cascade Method

Suppose we have a random number generator $RAND1()$ with a number of decimal digits $r \geq 12$ and $RAND2()$, which has at least 6 digits. We divide the area $t \geq K$ of r.v. values $x \in Pa(K, \alpha)$ at intervals with a help of points $K_1 = K, K_2 = K_1 \cdot 10^{6/\alpha}, K_3 = K_2 \cdot 10^{6/\alpha}$, and so on. With regard to (1) the probability of intervals (K_i, K_{i+1}) will amount $p_i = P(K_i \leq x < K_{i+1}) = 0.999999 \cdot 10^{-6(i-1)}, i = 1, 2, \ldots$. Let us imagine the tail $\bar{F}(t) = (K/t)^\alpha$ as a linear combination of the corresponding conventional tails:

$$\bar{F}(t) = \sum_{i=1}^{\infty} p_i \bar{F}_i(t), \tag{23}$$

where

$$\bar{F}_i(t) = \begin{cases} 1, & if \ t < K_i, \\ \frac{\bar{F}(t) - \bar{F}(K_{i+1})}{p_i}, & if \ K_i \leq t < K_{i+1}, \\ 0, & if \ K_{i+1} \leq t. \end{cases} \tag{24}$$

The random variable x with the tail of the distribution $\bar{F}(t) = (K/t)^\alpha$ we will generate as a mixture of r.v. x_i, with tails (24) shown on Fig. 1. Components x_i of the mixture should be selected with the corresponding probabilities p_i.

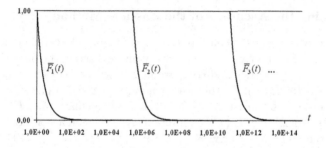

Fig. 1. Tails (24) for $K = 1, \alpha = 1.1$. The scale on the axis t is logarithmic

In accordance with the said procedure of r.v. generation $x \in Pa(K, \alpha)$ can be represented as shown in Fig. 2. The cycle starting at step Step2 (excluding condition test) is performed once approximately one time per every 10^6 procedure calls (twice one time per 10^{12} calls, etc.) to enable on the average a sufficiently rapid generation of r.v. x. Step Step3 realizes a method inverse transformation for the tail $\bar{F}_i(t)$.

```
Input: K, α
output: x
Step1: p ← (1 – 10⁻⁶)
       i ← 1
Step2: If Rand2() < 10⁻⁶
       then p ← p·10⁻⁶
            i ← i + 1
            goto Step2
Step3: u ← Rand1()
       x ← K(up + 10⁻⁶ⁱ)⁻¹/ᵃ
return: x
```

Fig. 2. Basic cascade generation algorithm r.v. with PD

The method is not difficult to modify and for the generator $RAND()$ with a different length of digits grid, with other partitions of axis t at intervals of conditional distributions, and with the other, not decimal system of calculation.

By suitable linear transformations of r.v. x on the output of basic cascade generator versions of PD, other than (1) can be realized. For example, to realize r.v. y that has Pareto DF in version

$$F_y(t) = 1 - \left(\frac{b}{b+t}\right)^\alpha, \quad b > 0, \ t \geq 0,$$

it is enough to put $K = b$, refer to the basic cascade generator and convert the resulting value of x to y by the formula $y = x - b$.

5.3 Checking the Accuracy of the Cascade Method

To estimate the accuracy of the proposed cascade method of HTD realization let us calculate m.e. $E(x')$ d.r.v. x', which is actually realized in the procedure shown in Fig. 2 for $K = 1, \alpha = 1.1$, and compare it m.e. with the exact value of the 11 m.e. of r.v. $x \in Pa(1, 1.1)$. As the used generator RAND1() has $r \geq 12$ decimal digits, we consider the worst case $r = 12$. Let us calculate m.e. $E(x')$ realized through conventional m.e. $E_i(x')$. In interval (K_i, K_{i+1}) in accordance with algorithm step $\underline{Step3}$ is realized d.r.v. $x' = (u' \cdot 10^{-6(i-1)}(1 - 10^{-6}) + 10^{-6i})^{-1/1.1} = (0.999999 \cdot 10^{-6(i-1)}u' + 10^{-6i})^{-1/1.1}$. Hence, running all equally probable values u', we obtain the formula

$$
E_i(x') = 10^{-12} \sum_{n=1}^{10^{12}} \left(0.999999 \cdot 10^{-6(i-1)}10^{-12}n + 10^{-6i}\right)^{-1/1.1}
$$

$$
= 10^{-12} \sum_{n=1}^{10^{12}} \left(0.999999 \cdot 10^{-6(i+1)}n + 10^{-6i}\right)^{-1/1.1}. \tag{25}
$$

The results of calculation of conditional m.e. (25), the corresponding probabilities $p_i = 0.999999 \cdot 10^{-6 \cdot (i-1)}$ of intervals and unconditional m.e. are presented in Table 3. These calculations are made with a large margin of accuracy and in Table 3 are approximated. We see that unconditional m.e. $E(x')$ of actually realized d.r.v. x' up to $6 - 7$ significant figures coincides with the m.e. $E(x) = 11$ of r.v. $x \in Pa(1, 1.1)$.

Table 3. Calculation of unconditional m.e realized d.r.v. x'

i	$E_i = E_i(x')$	p_i	$p_i E_i$	Partial amount of $p_i E_i$
1	7.86717E+00	9.99999E-01	7.867160	7.867161
2	2.24060E+06	9.99999E-07	2.240596	10.10776
3	6.38130E+11	9.99999E-13	0.638130	10.74589
4	1.81742E+17	9.99999E-19	0.181742	10.92763
5	5.17607E+22	9.99999E-25	0.051761	10.97939
6	1.47416E+28	9.99999E-31	0.014742	10.99413
7	4.19847E+33	9.99999E-37	0.004198	10.99833
8	1.19574E+39	9.99999E-43	0.001196	10.99952
9	3.40551E+44	9.99999E-49	0.0003406	10.99986
10	9.69901E+49	9.99999E-55	9.699E-05	10.99996
11	2.76231E+55	9.99999E-61	2.762E-05	10.999997

The calculations show that basic cascaded generator solves the problem of displacement moments realized by PD with high accuracy. There are only minor computational errors, comparable with errors in the realization of light-tailed distributions. These computational errors are negligible when using 12-digits and more conventional 15-digits generators of BRV.

Table 3 also allows to see the cascade method possesses virtually inexhaustible even with modern supercomputers "reserve" of precision: the performance of 10^{18} floating point operations per second, and the continuous generation of samples within a day, getting into the range of conditional distributions with the number $i = 5$ happen only with probability of order of 0.01. Intervals with numbers $i > 5$ will not be represented, and therefore realized m.e. will be somewhat understated. But it will be a consequence of the statistical properties of the realized distribution and not developed into the generation of r.v. deterministic error. Thus, the problem of development of methods for the reduction of variation in SM of FS is reliably separated from the problem of deterministic displacement of realized sample.

Compared with the method of "long sums" the cascade method has potentially higher performance, as it takes on average virtually only two calls to standard random number generator. To achieve a comparable displacement using 15-digits random number generators the method of may require three or more calls, depending on the parameter α. Of course, in practice, it is now easier to use the method of "long sums", because there are no cascade generators in libraries of simulation systems. But it is fixable.

5.4   Realization of Subexponential HTD

Characteristic representatives HTD related to subexponential distributions are Weibull distribution and lognormal distribution. Weibull distribution is the distribution with the tail $\bar{F}(t) = e^{-(t/\lambda)^\beta}$ which is heavy at $0 < \beta < 1$. The smaller β, the heavier the tail is. The method of the inverse transformation of the tail is obtained for generation of r.v. ξ having a Weibull distribution with $\lambda = 1$, the formula $\xi = (-ln(u))^{1/\beta}$. Let us calculate the displacement of m.e. $E(\xi')$ realized r.v. $\xi' = (-ln(u'))^{1/\beta}$:

$$E(\xi') = \varepsilon \sum_{i=1}^{1/\varepsilon} (-\ln(i\varepsilon))^{1/\beta}. \qquad (26)$$

When $\beta = 1/7$ and $\varepsilon = 10^{-12}$ using the formula (26) we calculate $E(\xi') = 5039.990168\ldots$. Such precision of the m.e. realization can be considered as acceptable, since the exact value of m.e. $E(\xi) = \lambda\Gamma(1 + \beta^{-1}) = \Gamma(1 + 7) = 7! = 5040$. When $\beta = 0.01$ we have $E(\xi) = 100! \approx 9.33 \cdot 10^{157}$. An attempt to realize such a distribution by means of a conventional machine arithmetic with double precision leads to a banal overflow of digit gripd. When $\beta = 0.05$ m.e. $E(\xi) = 20! \approx 2.4329 \cdot 10^{18}$, and realized at $\varepsilon = 10^{-15}$ m.e. $E(\xi') \approx 2.4232 \cdot 10^{18}$ has only two precise figures.

Since the basic cascade algorithm, shown in Fig. 2, is not intended to the present case, you can apply the method of "long sums", or develop a cascade method to realize the Weibull distribution.

The lognormal distribution is a the distribution of r.v. $\xi = e^{\sigma x + \mu}$ where x is standard normal r.v. Normal and lognormal distributions possesses DF, which can not be expressed in closed form in terms of elementary functions, so to use the method of the inverse transformation for the realization of these distributions is difficult. Let us consider an embodiment of the lognormal r.v. ξ with the standard normal r.v. x realized by Box-Muller method. Initially, two independent values BRV u_1 and u_2 are converted into two independent realization of x_1 and x_2 standard normal r.v. x by the formulas:

$$x_1 = \sqrt{-2\ln u_1} \cdot \sin(2\pi u_2), \quad x_2 = \sqrt{-2\ln u_1} \cdot \cos(2\pi u_2), \qquad (27)$$

then two corresponding realization of lognormal r.v. ξ are calculated:

$$\xi_1 = e^{\sigma x_1 + \mu}, \quad \xi_2 = e^{\sigma x_2 + \mu}. \qquad (28)$$

The m.e. $E(\xi) = e^{\mu + \frac{\sigma^2}{2}}$. Calculation of double sums which determine the realized m.e. $E(\xi')$ is difficult, but it is a clear indication that the realized displacement of m.e. can be large at any finite number of digits DBRV u_1'. Since large values of ξ' are generated at near zero values of u_1', to reduce the displacement u_1 can be realized by "long-term sums".

With the development of the base cascade algorithm to generate lognormal r.v. we should keep in mind that DF is not expressed in terms of elementary functions.

The considered examples of subexponential distributions are aiming to develop a generic version of the cascade algorithm, which would allow customizing easily the realization of a sufficiently broad class of HTD, including subexponential distributions.

5.5   Universal Version of Cascade Generation Algorithm HTD

The main condition determining the choice of intervals for conditional distributions of mixture (23) is a decrease in the probability of these intervals in geometric sequence. It allows choosing the intervals of conditional distributions that have a low probability by several calls to the DBRV generator with limited digit. Figure 3 shows a universal version of the cascade algorithm, when the main condition for the selection intervals is kept, and at step Step3 we use either inverse transformations of unconditional DF or any other method of generating of r.v. x, where large x corresponds to small u.

With a help of direct verification of expressions received during the realization of the shown in Fig. 3 algorithm (with symbolic given K, α), we can easily see that it is equivalent to the basic cascade algorithm realization of the PD. Setting universal cascade algorithm for the realization of other PD is carried out the recording in step Step3 formula for generating unconditional r.v. with the required distribution when large x correspond to small u. In realizing the lognormal distribution in step Step3 there is need to record consistently four transformations (27) and (28) (replacing K, α in the line **Input** by σ and μ), using the realization of BRV u_1 generator $Rand1()$ (which has at least 12 decimal digits), and to realize BRV u_2 $Rand2()$ can be used.

```
Input: K, α
output: x
Step1:   i ← 1
Step2:   If Rand2( ) < 10⁻⁶
         then i ← i + 1
         goto Step2
         a ← 10⁻⁶ⁱ
         b ← 10⁻⁶⁽ⁱ⁻¹⁾
         u ← (b − a)Rand1( )+a
Step3:   x ← Ku⁻¹/α
return: x
```

Fig. 3. Universal cascade algorithm configured to generate r.v. with PD

5.6 Notes on the Realization of Asymptotic Power HTD

Asymptotic power HTD has a tail $\bar{F}(t) = cL(t)t^{-\alpha}$, where c is constant, $L(t)$ is a slowly varying function, $\alpha > 0$.

If $L(t)$ has the tail that can be reversed, then for the realization of the asymptotic power HTD there can be used universal cascade algorithm. As an example of inverse transformed asymptotic power tail we can cite the tail $\bar{F}(t) = 0.5(1 + t^{-\alpha})t^{-\alpha}$, $t \geq 1$. By the tail inverse transformation to realize the corresponding r.v. we get a formula $x = [(\sqrt{1 + 8u} − 1)/2]^{-1/\alpha}$ used in step Step3 of universal cascade algorithm. The problem of calculating $(\sqrt{1 + 8u} − 1)$ of the formula for small u is solved with its transformation for such u in a Taylor series with a small number of terms.

If the inverse transformation of the tail is difficult, as for example in case of $L(t) = ln(t)$, then it is necessary to use either the appropriate inverse transformation numerical methods, or methods for generating r.v. x by its probability density (for example, the method of rejection sampling).

6 Conclusion

The r.v. generators study used in SM indicates that HTD are generally realized with significant distortions, which leads to significant errors in the results of fractal queueing systems SM.

The reason of HTD distortions is a discreteness of used standard random numbers generators. Smoothing of their discreteness due to the transition to the "long arithmetic" is ineffective because there are the additional hardware costs or loss of productivity.

The article proposed and investigated effective and relatively simple methods of HTD realization, namely the method of summation of scaled realization of the standard random numbers and universal cascade method of HTD realization. These methods do not require the use of "long arithmetic" and result in minimal performance loss.

In practice the SM does not often require samples of such volume at which the merits of the proposed cascade method are manifested regularly and in obvious way. However, this method is useful as it frees a researcher from the need to estimate in simulation of fractal systems the deterministic errors caused by discreteness of random number generators and may occur with finite probability for any length of sample.

References

1. Leland, W.E., Taqqu, M.S., Willinger, W., Wilson, D.V.: On the self-similar nature of ethernet traffic. IEEE/ACM Trans. Netw. 2(1), 1–15 (1994)
2. Crovella, M.E., Taqqu, M., Bestavros, A.: Heavy tailed-probability distributions in the world wide web. IEEE/ACM Trans. Netw. 5(6), 835–846 (1997)
3. Vishnevskiy, V.M.: Theoretical Bases of Designing Computer Networks. Technosphere, Moscow (2003)
4. Resaul, K.M., Grout, V.: A comparison of methods for estimating the tail index of heavy-tailed internet traffic. In: Sobh, T., Elleithy, K., Mahmood, A., Karim, M. (eds.) Innovative Algorithms and Techniques in Automation, Industrial Electronics and Telecommunications, pp. 219–222. Springer, Netherlands (2007)
5. Kleinrock, L.: Queueing Systems: V. II - Computer Applications. Wiley, New York (1976)
6. Zwart, A.P.: Queueing Systems with Heavy Tails. Eindhoven University of Technology, p. 227 (2001)
7. Cheng, C.S., Thomas, J.A., Cheng, C.S., Thomas, J.A.: Effective bandwidth in high-speed digital networks. IEEE J. Selected Areas Commun. 13, 1091–1100 (1995)
8. Zadorozhnyi, V.N.: Simulation modeling of fractal queues, in Dynamics of Systems, Mechanisms and Machines (Dynamics), pp 1–4, December 2014, doi:10.1109/Dynamics.2014.7005703
9. Mandelbrot, B.: The Fractal geometry of nature. WH Freeman and Co, New York (1982)
10. GPSS World reference manual: Minuteman Software, Fifth Edition, Holly Springs. U.S.A, NC (2009)
11. Kleijnen, J.P.C.: Statistical Techniques in Simulation, Part 1. Marcel Dekker, New York (1974)
12. Asmussen, S., Binswanger, K., Hojgaard, B.: Rare events simulation for heavy-tailed distributions. Bernoulli 6(2), 303–322 (2000)
13. Boots, N.K., Shahabuddin, P.: Simulating GI/GI/1 queues and insurance risk processes with subexponential distributions. Unpublished manuscript, Shortened version. In: Proceedings of the 2000 Winter Simulation Conference, pp. 656–665, Free University, Amsterdam (2000)
14. Blanchet, J, Li, C.: Efficient Rare Event Simulation for Heavy-tailed Compound Sums (2008). http://www.columbia.edu/~jb2814/papers/RGSFinalJan08B.pdf (Accessed 30.05.2015)
15. Chan, H.P., Deng, S., Lai, T-L.: Rare-event Simulation of Heavy-tailed random walks by sequential importance sampling and resampling. http://statweb.stanford.edu/~ckirby/lai/pubs/2012_Rare-EventSimulation.pdf (Accessed 30.05.2015)

Author Index

Printed in the United States
By Bookmasters